.

Crop Science: Productivity and Management

Crop Science: Productivity and Management

Editor: Corey Aiken

CALLISTO REFERENCE
www.callistoreference.com

Callisto Reference,
118-35 Queens Blvd., Suite 400,
Forest Hills, NY 11375, USA

Visit us on the World Wide Web at:
www.callistoreference.com

ISBN: 978-1-64116-066-7 (Hardback)

Cataloging-in-Publication Data

Crop science : productivity and management / edited by Corey Aiken.
     p. cm.
Includes bibliographical references and index.
ISBN 978-1-64116-066-7
1. Crop science. 2. Crop yields. 3. Agricultural productivity. 4. Soil management.
I. Aiken, Corey.
SB91 .C76 2019
631--dc21

Table of Contents

Preface

The objective of this book is to give a general view of the different areas of crop science and its applications in the development of modern methods that help in improving crop productivity and management. Crop science is the study of crops and the diverse aspects related to crops such as breeding, physiology, genetics, ecology, etc. Crops are primarily harvested for food. Some popular techniques of crop science include intercropping, crop rotation, sharecropping and multiple cropping. Fundamental theories and concepts of crop science play a central role in the sustenance of a population. This book also brings forth some of the most innovative concepts and elucidates the unexplored aspects of this field. The various advancements in crop science are glanced at and their applications as well as ramifications are looked at in detail. Those in search of information to further their knowledge will be greatly assisted by this book.

This book is a result of research of several months to collate the most relevant data in the field.

When I was approached with the idea of this book and the proposal to edit it, I was overwhelmed. It gave me an opportunity to reach out to all those who share a common interest with me in this field. I had 3 main parameters for editing this text:

1. Accuracy – The data and information provided in this book should be up-to-date and valuable to the readers.

2. Structure – The data must be presented in a structured format for easy understanding and better grasping of the readers.

3. Universal Approach – This book not only targets students but also experts and innovators in the field, thus my aim was to present topics which are of use to all.

Thus, it took me a couple of months to finish the editing of this book.

I would like to make a special mention of my publisher who considered me worthy of this opportunity and also supported me throughout the editing process. I would also like to thank the editing team at the back-end who extended their help whenever required.

Editor

Evaluation of Mutant Lines of *Rosa* Species

Atif sarwar* and Shahid Javed Butt

Department of Horticulture, PMAS-Arid Agriculture University, Rawalpindi, Pakistan

Abstract

Among the highly fragrant rose species, *R. centifolia* and *R. gruss* an teplitz have high commercial importance and value added potential. Most of the modern roses are the result of hybridization, selection and spontaneous mutation. For floriculture trade, there is always demand and necessity for new varieties due to change in taste and fashion. Mutation breeding is an established method for crop improvement. Mutant lines were taken from the Plant Tissue Culture Laboratory of Department of Horticulture, PMAS-Arid Agriculture University, Rawalpindi. Rose genotypes mutants were sown in the field at similar conditions of irrigation, fertilizers and pest/disease management. Plants were treated with different levels of gamma rays and colchicine through solution. Data of various parameters like plant height shoot length, fresh leaf weight, dry leaf weight, flower diameter, rose water, number of shoots, number of flowers/plant/week, weight/10 flowers and number of petals were collected for different treatments. Gamma radiations show greater improvement in *R. centifolia* but colchicine impact was more on *R. gruss* an teplitz.

Keywords: *R. centifolia*; *R. gruss an teplitz*; Rose mutants; Colchicine application; Irradiated mutants

Introduction

Rose is one of the most important commercial flower crops belonging to family *Rosaceae* which is mostly used in perfumery, cosmetic industry and for medicinal purposes. Gamma radiations are basically the source of mutation and mutation is the sudden change in heredity material of the plant cells. While colchicine is a basically alkaloid which is obtain from colchicine aquatus tree. It actually doubles the chromosomes by stopping the spindle fiber growth by affecting the meiosis process. Mutation induction methods can largely increase the gene mutation frequency and produce new materials, germplasm and new cultivars in a normally short period. Rose species were bringing in the western world since ancient times and rigorous rose breeding was commenced since the 18th century. Introduction of new genotypes and replacement of species and cultivars by travelers significantly increases the genetic changes to the horticulturists and the propagation of the species: this importantly involved a decreasing of genetic changes available [1]. The establishment of cultivar of roses is mainly has three equal phases: the past (1876-1968) and the future (1966) moreover significant achievements are underlined [1] Considered that of the 250-350 identified species of rose, only 15-25 have added to the growth of some new cultivars of rose. Rose breeding majorly attempted by developed companies and their checked inherited awareness[1]. Different biotechnological methods are recently available for rose breeding. *R. centifolia* is commercially significant among the perfumed roses and yield highly fragrant important oil. Its petals have commercial significance and used in perfume industry, food stuff and medicines. More than 400 unstable compounds have been known in the floral bouquet of different rose cultivars. The flowers are commercially harvested for the manufacturing of rose oil which is normally used in perfumery [2]. The present research study was conducted to find out the performance of different rose genotypes for different morphological and yield parameters.

Materials and Methods

Present research work was carried out at the research area of PMAS-Arid Agriculture University, Rawalpindi, during the year 2011-2012. Design used for this purpose was randomized complete block design (RCBD) with 3 replications. The experimental material consist of plant of two varieties (*R. centifolia* and *R. gruss-an-teplitz*) treated *in vitro* with mutagens (Tables 1 and 2) Plants were treated with different level of gamma radiations (Table 1) and colchicine through solutions (Table 2). These treated plants were proliferated and rooted in Plant Tissue Culture Laboratory before acclimitization in green house. Now the mutant lines were taken from the Plant Tissue Culture Laboratory of Department of Horticulture, PMAS-Arid Agriculture University, Rawalpindi [3].

Field was prepared by plugging and hoeing the field followed by planking. Different plots were prepared for transplanting the rose genotypes mutants of two varieties (*R. centifolia* and *R. gruss an teplitz*). Rose genotypes mutants of two varieties (*R. centifolia* and *R. gruss an teplitz*) were sown in the field on October, 2011. The total length of the research area was 130 ft and width 33 ft. The plant to plant distance was 3 ft and row to row distance was also3 ft. All the cultural practices such as irrigation, weeding, hoeing, insects and pest control measures were given uniformly to all the treatments. The parameters such asplant

Rose species	Gamma radiation (Gy)						
	T_0	$T_{1...}$	T_2	T_3	T_4	$T_{5...}$	T_6
R. centifolia	00	10	20	30	40	50	60
R. gruss an teplitz	00	10	20	30	40	50	60

Table 1: Gamma radiations.

Rose species	Colchicine (mg L^{-1})				
	T_0	T_1	T_2	T_3	T_4
R. centifolia	00	100	300	500	700
R. gruss an teplitz	00	100	300	500	700

Table 2: Colchicine solutions.

---

*Corresponding author: Atif sarwar, Department of Horticulture, PMAS-Arid Agriculture University, Rawalpindi, Pakistan
E-mail: atif_malikuos@ymail.com; sbutt2@hotmail.com

height, shoot length, fresh leaf weight, dry leaf weight, flower diameter and rose water were collected. The data were collected from different treatments laid in RCBD which were statistically analyzed through the analysis of variance techniques and the tables of variance were constructed.

Results and Discussion

Effect of gamma radiations on rose mutant lines

Plant height of two rose lines (Figure 1) depicted significant difference under varying levels of gamma radiations. Results revealed the maximum (32.66 inches) was observed in *R. centifolia* at T_2 and (29.00 inches) in case of *R. gruss an teplitz* at T_1.[3]Martin observed that plant height increase by increasing the irradiation doses and the best result of irradiation was shown on sapota and blue blood plant height [4].

Shoot length of both the lines were depicted a great difference at different levels of gamma radiations. The maximum by *R. centifolia* (11.33 inches) at T_2 and *R. gruss an teplitz* shows (7.33 inches) at T_1. Gamma irradiation effects upon the shoot lengths constituents so some of aromatic herbs and spices were studied and their results shows significant results by Calucci et al [5].

There was a significant difference seen in fresh leaf weight of both the lines (Figure 2) at different levels of gamma radiations. *R. gruss an teplitz* depicted the maximum (74.56 mg) at T_1 and *R. centifolia* show (60.36 mg) at T_2. The decline in fresh weights coincided with the onset of flower wilting and desiccation [6]. According to our findings gamma rays show greater improvement.

Dry leaf weight of both the lines (Figure 3) was also significantly different from each other. Maximum was observed in (Figure 4) *R. gruss an teplitz* (27.16 mg) at T_1, *R. centifolia* shows (19.32 mg) at T_2. Similar results were found by Hong et al. [7]. Observed that dry weight of leaves of rose plant was significantly increased as result of gamma rays compared with control in the seasons.

Flower diameter of *R. gruss an teplitz* and *R. centifolia* showed a significant at various treatments of gamma radiations. Maximum was observed (6.26 cm) at T_1 in *R. gruss an teplitz* and in *R. centifolia* reveal (5.76 cm) at T_2. *R. gruss an teplitz* and *R. centifolia* were showing a significant difference of about (0.5 cm). Our results are not relevant to

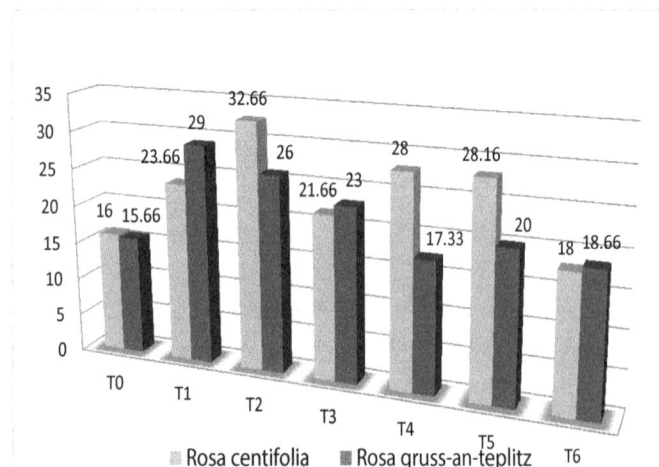

Figure 2: Bar chart graph showing the effect of colchicine solutions on Fresh leaf weight (mg) of *Rosa centifolia* and *Rosa gruss-an-teplitz*.

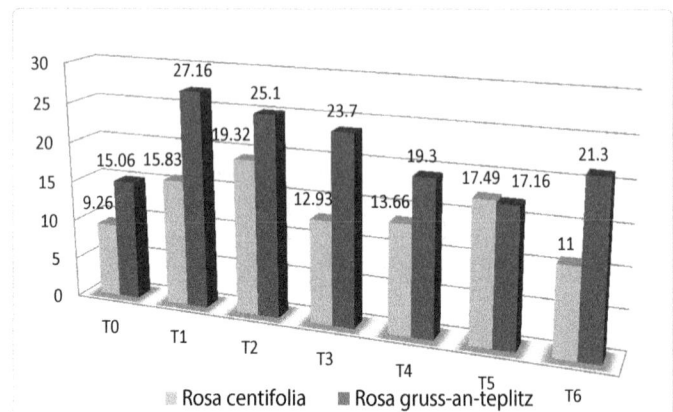

Figure 3: Bar chart graph showing the effect of gamma radiations on Dry leaf weight (mg) of *Rosa centifolia* and *Rosa gruss-an-teplitz*.

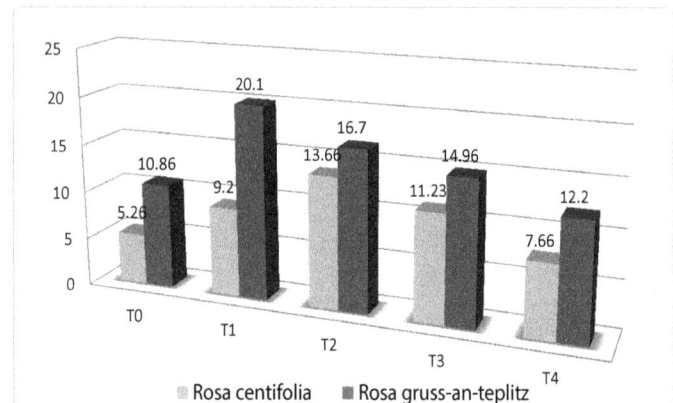

Figure 4: Bar chart graph showing the effect of colchicine solutions on dry leaf weight (mg) of *Rosa centifolia* and *Rosa gruss-an-teplitz*.

Figure 1: Bar chart graph showing the effect of gamma radiations on Plant height (inches) of *Rosa centifolia* and *Rosa gruss-an-teplitz*.

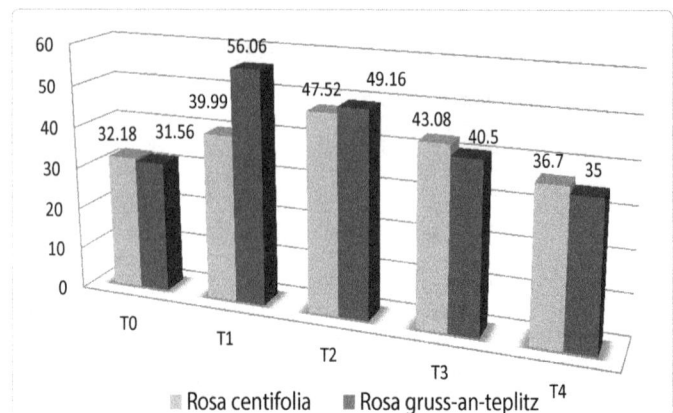

Bendini et al. [8] that flower diameter has no significant results with the application of gamma radiations. Rose water percentage (Figure 5) was also revealed a significant difference among both the lines at different levels of gamma radiations. *R. centifolia* shows the rose water percentage (3.70%) at T_3 and *R. gruss an teplitz* (1.36%) at T_1. Results are agreed with Hanson et al. [9]. As he also observed that gamma radiation show significant water rose % in case of *R. centifolia*.

Effect of gamma radiations on number of shoots in *Rosa centifolia* and *R. gruss an teplitz* showed a significant difference that *R. centifolia* with maximum (6.00) in T_2. While in *R. gruss an teplitz*, maximum (4.33) in T_1. The results supported with the observations of Muthuswamy and Pappiah (1976) conducted experiment on *J. auriculatum* under climatic conditions; it was found that gamma rays produced beneficial effect on quantum of new shoots. It shows that the application of gamma radiation at different ratios increased the number of branches compared to untreated plants in *J. sambac* and *J. auriculatum* [3].

Number of flowers/plant/week was significantly reveals that in *R. centifolia*, maximum (11.00) was found in T_2. But in *R. gruss an teplitz*, examined that maximum (7.66) was found in T_1 with minimum (5.00) in T_3. Similar findings are reported by Khattak et al. [10] he observed that maximum number of flower were 20.6.

Weight/10 flowers of *R. gruss an teplitz* and *R. centifolia* showed a significantly results at various treatments of gamma radiations. *R. centifolia*, maximum in T_2 treatment (31.99) as compared to lowest in T_0 (19.09). *R. gruss-an-teplitz*, the highest weight/10 flowers were observed in T_1 treatment (17.33) as compared to lowest in T_0 (12.30). The results are in consonance with the findings of Nikabakht [11] who observed that those of *Rosa gruss-an-teplitz* and *Rosa indica* showed the lowest values (1.358g and 1.388g, respectively) for flower weight.

Gamma radiations treatments showed significant results in both *Rosa* varieties. *R. centifolia*, maximum number of petals (36.00) were noted in T_2 treatment in comparison with lowest in T_0 (21.66). While in *R. gruss an teplitz*, maximum (34.33) were noted in T_1 treatment with lowest in T_0 (21.00). The present results are so much agreed with the findings of Kaul et al. [12] who found that number of petals were 38, 32 and 47, respectively. Similar findings were observed by Tabaei-Aghdaei et al. [13] that positive correlation was observed between number of petals and number of stamens.

Effect of colchicine solutions on rose mutant lines

Colchicine has been used for a long time as a polyploidizing agent. It has been used successfully to produce polyploids for cytogenetic research and for breeding programme in many plant species. Plant height of two rose lines depicted significant difference under varying colchicine treatments. Results revealed the maximum (25.33 inches) was observed in *R. gruss an teplitz* at T_1 and (25.00 inches) at T_2 in case of *R. centifolia* but the control shows the minimum plant height (17.00 inches) at T_0 in *R. centifolia*. Our findings show same results as mentioned by Mensah et al. [14] reported colchicine application revealed that increase in plant height at moderate application but decrease at high.

Shoot length of both the lines were reveal a great difference at various levels of colchicine treatments. The maximum (8.00 inches) at T_2 in *R. centifolia* and *R. gruss an teplitz* shows (7.33 inches) at T_1. Agreed with our results that data show the increase in shoot length (inches) over the application of colchicine [15].

There was a significant difference seen in fresh leaf weight of both the lines (Figure 4) at different levels of colchicine treatments. Data shows the increase in fresh leaf weight (mg) over the application of colchicine [15].

Dry leaf weight of both the lines was also significantly different from each other. Maximum was observed in *R. gruss an teplitz* (20.10 mg) at T_1 and the minimum dry leaf weight were recorded in *R. centifolia* (13.61 mg) at T_2. Same results were found by some of other scientist as our results that shoot dry weight increases by the application of colchicine [14].

Flower diameter of *R. gruss an teplitz* and *R. centifolia* show a significant results at various treatments of colchicine. Maximum (5.76 cm) at T_2 in *R. centifolia* and in *R. gruss an teplitz* (5.56 cm) at T_1. Hessayon [16] observed that varying flower diameters in different rose cultivars showed similar results as founded. Rose water percentage (Figure 6) showed a significant difference among both the lines at different levels of colchicine treatments. *R. centifolia* shows (2.83%) at T_2 and in *R. gruss an teplitz* (1.46%) at T_1.

Number of shoots was significantly different from each other. *R. centifolia* shows maximum (6.33) at T_1, while *R. gruss an teplitz* shows (4.33) in T_1 treatment. Senapati and Rout [17] observed that *R. gruss an teplitz* and *R. centifolia* showed significant results having 2.809 and 2.158 number of shoots, respectively. Shoot multiplication rate varied in different species and was specific to culture medium.

Maximum number of flowers/plant/week shows significant results at various colchicine treatments. *R. centifolia* (11.33) was found in T_2, while *R. gruss an teplitz* showed maximum (8.66) was found in T_1. The useful mutant lines isolated and treated with colchicine to establish any changes in locus for the increase of number of flower/plant/week reported by Biswas.

Weight/10 flowers (Figure 7) indicated significantly results from

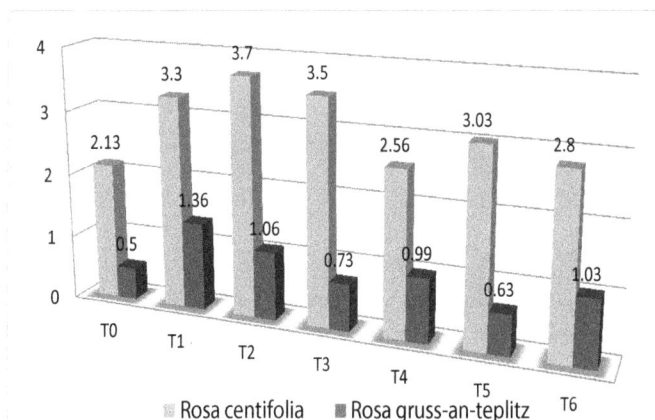

Figure 5: Bar chart graph showing the effect of gamma radiations on rose water % of *Rosa centifolia* and *Rosa gruss-an-teplitz*.

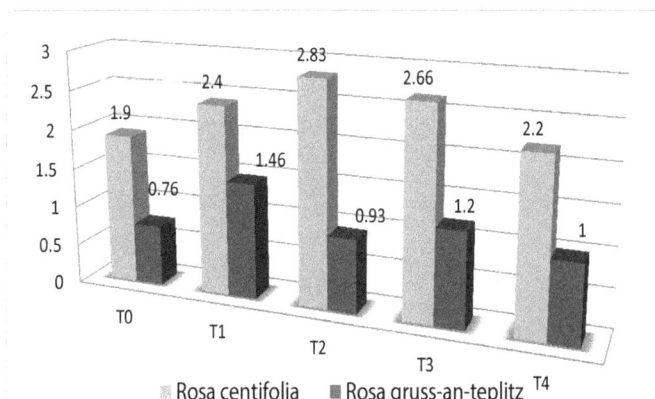

Figure 6: Bar chart graph showing the effect of colchicine solutions on Rose water (%) of *Rosa centifolia* and *Rosa gruss-an-teplitz*.

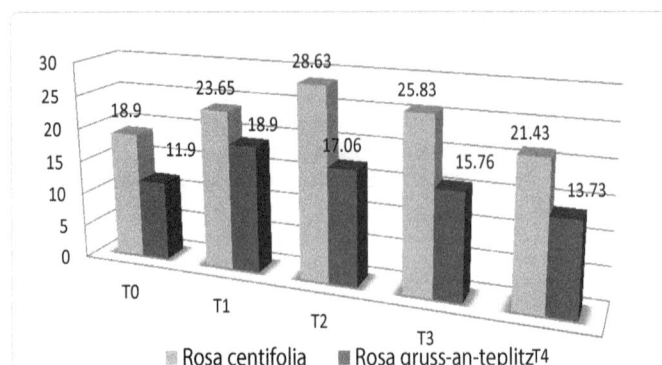

Figure 7: Bar chart graph showing the effect of colchicine solutions on weight / 10 flowers of *Rosa centifolia* and *Rosa gruss-an-teplitz*.

Sr.#	Parameters	Control	R. centifolia	R. gruss an teplitz
1.	Plant height (inches)	16.00± 2.73 15.66± 1.45	32.66 ± 1.86	29.00 ± 1.53
2.	Shoot length (inches)	3.00± 0.88 4.66 ± 0.33	11.33 ± 0.88	7.33 ± 0.33
3.	Fresh leaf weight (mg)	39.20± 4.20 56.40 ± 1.48	60.36 ± 4.93	74.56 ± 1.45
4.	Dry leaf weight (mg)	9.26± 1.22 15.06 ±0.69	19.32 ± 1.01	27.16 ± 0.45
5.	Flower diameter (cm)	2.33± 0.60 4.80 ± 0.15	5.76 ± 0.21	6.26 ± 0.12
6.	Rose water (%)	2.13± 0.20 0.50 ± 0.15	3.70 ± 0.35	1.36 ± 0.12
7.	No of shoots	1.66± 1.20 1.33 ±0.58	6.00 ± 0.58	4.33 ± 0.58
8.	No of flowers/plant/ week	4.33 ± 0.58 5.33 ± 0.33	11.00 ± 1.00	7.66 ± 0.33
9.	Weight / 10 flowers (g)	19.09 ± 1.92 12.30 ± 0.56	31.99 ± 2.36	17.33 ± 0.07
10.	No of petals	21.66± 0.88 21.00 ± 1.20	36.00 ± 3.52	34.33 ± 0.58

Values are not significantly different by LSD ($P < 0.05$)

Table 3: Effect of gamma radiations on both Rosa varieties.

Sr.#	Parameters	Control	R. centifolia	R. gruss an teplitz
1.	Plant height (inches)	15.66±1.45 22.33 ± 0.88	25.00 ± 0.58	25.33 ± 0.88
2.	Shoot length (inches)	4.66±0.33 4.33 ± 0.33	8.00 ± 0.58	7.33 ± 0.33
3.	Fresh leaf weight (mg)	32.18±1.48 31.56 ± 0.81	47.52 ± 0.46	56.06 ± 1.57
4.	Dry leaf weight (mg)	15.06±0.69 10.86 ± 0.65	13.66 ± 0.35	20.10 ± 0.87
5.	Flower diameter (cm)	4.80±0.15 3.70 ± 0.06	5.76 ± 0.15	5.56 ± 0.09
6.	Rose water (%)	1.90 ± 0.06 0.76 ± 0.13	2.83 ± 0.03	1.46 ± 0.15
7.	No of shoots	1.33 ± 0.56 2.33 ± 0.33	6.33 ± 0.24	4.33 ± 0.18
8.	No of flowers/plant/week	6.33 ± 0.55 3.66 ± 0.33	11.33 ± 0.33	8.66 ± 0.33
9.	Weight/10 flowers (g)	18.90 ± 0.58 11.90 ± 0.21	28.63 ± 0.56	18.90 ± 0.35
10.	No of petals	28.00 ± 0.58 19.33 ± 0.33	37.00 ± 0.58	25.66 ± 0.88

Values are significantly different by LSD ($P < 0.05$)

Table 4: Effect of colchicine solution on both *Rosa* varieties.

each other. *R. centifolia* shows maximum (28.63) in T_2 treatment, similarly *R. gruss an teplitz* shows (18.90) in T_1. Colchicine solution has no significant effect on increase in flower weight reported by Barnabas et al. [18].In another study it is revealed that *R. centifolia* showing great variation in its wait as compared to *R. gruss an teplitz* [19].

Numbers of petals create a significant difference in between both lines. *R. centifolia* illustrated that maximum (37.00) were noted in T_2 similarly *R. gruss an teplitz* found that (31.00) in T_1. However, the counting petals with large size which will be further conducted accurately [20].

Conclusion

On the basis of results as summarized above, it is concluded that the considerable difference for both the lines by application of different levels of gamma radiations and colchicine as compared to control. Gamma radiation showed great variation in its results as compared to colchicine treatment. Gamma radiations reveals significant improvement in mutant line *R. centifolia* instead of *R. gruss an teplitz* but the colchicine treatment show great variation in line *R. gruss an teplitz* as compared to *R. centifolia*. Also suggests that an extensive research work should be carried out to reach in a final conclusion for using such treatments in roses for plant height, flower color and size to increase in their commercial value.

References

1. Devries DP, Dubois AM, Morisot A, Ricci P (1996) Rose breeding: past, present prospects. Second Int symposium on roses, Antibes, France. Acta Hort 424: 241-248.

2. Lavid N, Wang J, Shalit M, Guterman I, Bar E, et al. (2002) O-Methyl transferases involved in the biosynthesis of volatile phenolic derivatives in rose petals. Plant Physiol 129: 1899-1907.

3. Martin G (1980) Use of thin layer and gas liquid chromatography for analysis of some essential oils metadysoverm. Moscow. USSR. Chem Abst 84: 19-90.

4. Hase Y, Shikazono N, Tano S, Watanae H (2002) Introduction to PlantPhysiology. Plant physiol 129: 60-64.

5. Calucci L, Pinzono C, Zandomeneghi M, Capocchi A (2003) Effects of γ-irradiation on the free radical and antioxidant contents in nine aromatic herbs and spices. J Agri and Food Chem 51: 927-934.

6. Reid MS, Evans RY, Dodge LL, Mor y (1989) Ethylene and silver thiosulfate influence opening of cut rose flowers. J Amer Soc Hort Sci 114: 436-440.

7. Hong V, Wrostlad O (1990) Use of HPLC separation/photodiode array detection for characterization of anthocyanin. J Agric Food Chem 38: 708-715.

8. Bendini A, Toschi TG, Lercker G (2002) Antioxidant activity of oregano leaves. Ital J Food Sci 14: 17-25.

9. Hanson RE, Zwick MS, Choi S, Islam-Faridi MN, McKnight TD, et al. (1995) Fluorescent in situ hybridization of a bacterial artificial chromosome. J Agri and Food Chem 38: 646-651.

10. Khattak AM, Dawar SH, Khan MA, Razaq A (2011) Effect of summer pruning on the quality and performance of rose cultivars. Sarhad J Agric 27: 1.

11. Nikabakht A, Kafi M (2008) A study on the relationship between Iranian people and damask rose (Rosa damascena) and its therapeutic and healing properties. Acta Hort 790: 251-254.

12. Kaur N, Sharma RK, Sharma M, Singh V, Ahuja PS (2007) Molecular evaluation and micro propagation of field selected elites of Rosa damascena. Gen Appl Plant physiol 33: 171-186.

13. Tabaei-Aghdaei, Babaei SRA, Khosh-Khui M, Jaimand K, Rezaee MB, et al. (2007) Morphological and oil content variations amongst Damask rose (Rosa damascene) landraces from different regions of Iran. Hort Sci 113: 44-48.

14. Mensah JK, Obadoni BO, Akomeah PA, Ikhajiagbe B, Ajibolu J (2006) The effects of sodium azide and colchicine treatments on morphological and yield traits of sesame seed (Sesame indicum). Afr J Biotechnol 6: 534-538.

15. Amiri S, Kazemitabaar SK, Ranjbar G, Azadbakht M (2010) the effect of trifluralin and cholchicine treatments on morphological characteristics of Jimsonweed (Datura stramonium). Turk J Sci 8: 47-61.

16. Hessayon DG (1988) The Rose expert. Pub, Britanica House, Waltham Cross, Herts, England.2003.

17. Senapati KA, Rout DC (2010) Gamma irradiation for insect disinfestations damages native Australian cut flowers. Hort Sci 52: 343-355.

18. Barnabas B, Obert B, Ovacs GK (2009) Colchicine an efficient genome-doubling agent for maize (Zea mays L.) microspores cultured. Plant cell Rep 18: 858-862.

19. Ojomo OA (2007) Pollination, fertilization and fruit characters in cowpea (Vigna unguiculata). Gha J Sci 10: 33-37.

20. Raufe S, Khan IA, Khan FA (2006) Colchicine-induced tetraploidy and changes in allele frequencies in colchicine-treated populations of diploids assessed with RAPD markers in (Gossypium arboreum). Turk J Biol 30: 93-100.

Faba Bean Gall; a New Threat for Faba Bean (*Vicia faba*) Production in Ethiopia

Endale Hailu[1]*, Gezahegne Getaneh[2], Tadesse Sefera[1], Negussie Tadesse[1], Beyene Bitew[3], Anteneh Boydom[1], Daniel Kassa[1] and Tamene Temesgen[1]

[1]Ethiopian Institute of Agricultural Research, P.O.Box 2003, Ethiopia
[2]Addis Ababa University, Salale Campus, P.O.Box 2003, Ethiopia
[3]Debre Birhan Research center, Ethiopia

Abstract

Faba bean ranks first in its production volume and cultivated land among pulse crops cultivated in Ethiopia and it is valuable as the cheap source of protein in most Ethiopian diet. It also plays a significant role in soil fertility restoration and in export market. Production of the crop is, however, constrained by several pests. The objective of this study was to find out the distribution and intensity of newly occurred epidemic faba bean gall disease and other diseases affecting faba bean in the major growing areas of central and northern part of Ethiopia. The survey was made in 2013 main cropping season (in September) along the main roads and accessible routes in each surveyed district at every 5-10 km intervals as per faba bean fields available. Five samples were taken in each faba bean field by moving "X" fashion. The mean prevalence of faba bean gall, Ascochyta blight, Chocolate spot and rust were 48.5%, 63.6%, 94.6% and 1.9%, respectively. The mean incidence of the formal all diseases were 15.4%, 30.3%, 42.4% and 0.1% in their previous order. Based on severity scale, mean disease severity of Ascochyta blight and chocolate spot were 1.9 and 1.5, respectively. Mean severity of faba bean gall and faba bean rust were 6.4% and 0.1% were recorded in their order. The disease was more sever in Amhara region with mean disease severity of 22.2% followed by Tigray and Oromiya region which had 11.3% and 7.8% severity value, respectively. The newly observed Faba bean gall disease was found the most devastating and widely disseminated in the area within a few years since its occurrence. The production of the crop is highly challenged and farmers are frustrated by the nature of the disease. Farmers witnessed the appearance of the disease three years ago in 2011 in Oromiya. They observed fast dissemination and increasing coverage of the pathogen in time and space. In fact the disease is epidemic and more serious from the record and the information from Faba bean growers in the surveyed areas. The epidemic conditions of the disease have significant implication on the production of Faba bean and on the country's economy. This survey information helps to consider the disease as serious pest in Ethiopia and in development of management options. In the future, ecology and biology of the causative agent, yield loss due to the diseases, breeding for disease resistance/tolerance and management strategies should get attention.

Keywords: Disease; Faba bean; Faba bean gall; Incidence; Prevalence; Severity

Introduction

Faba bean (*Vicia faba* L.) is believed to be originated in the Near East and is one of the earliest domesticated legumes after chickpea and pea. China has been the main producing country, followed by Ethiopia, Egypt, Italy and Morocco [1]. It is the first among pulse crops cultivated in Ethiopia and leading protein source for the rural people and used to make various traditional dishes. According to Central Statistics Agency of Ethiopia 2012/13, Faba bean takes over 30% (nearly half a million hectares) of cultivated land with an average national productivity of 1.5 tons ha^{-1}. Ethiopia is considered as the secondary center of diversity and also one of the nine major agro-geographical production regions of faba bean.

As the faba bean is familiar in Ethiopian feeding culture, the majority of the seed produced would be consumed domestically and only a smaller percentage of the crop is delivered to the export market. However, still this small portion of export volume put Ethiopia among the top broad bean exporting countries of the world [2]. Amhara and Oromia regions are the major faba bean producing regions. Within the regions some zones such as West Shoa, North Shoa, South Wello and East Gojjam are identified as major production areas of faba bean [2]. The growing importance of faba bean as an export crop in Ethiopia has led to a renewed interest by farmers to increase the area under production [3].

An average national productivity is 1.5 tons ha^{-1}, while world average grain yield of faba bean is around 1.8 t ha^{-1} (ICARDA, 2008). However, the productivity of faba bean in Ethiopia is still, far below its potential due to a biotic and biotic factor. Among which diseases are the most important biotic factors causing faba bean yield reduction [4]. More than 17 pathogens have been reported so far on faba bean from different parts of the country. Diseases that are economically most important in the major faba bean growing regions including chocolate spot (*Botrytis fabae*), faba bean rust (*Uromyces viciae-fabae*), black rot (*Fusarium solani*), Aschochyta blight (*Aschochyta fabae*) and faba bean necrotic yellow virus (FBNYV) (Dereje and Tesfaye, 1994). In recent years, in additional to the previous common diseases, the crop is threatening by new gall forming disease with typical symptoms of green and sunken on the upper side of the leaf and bulged to the back side of the leaf, and finally develops light brownish color lesion, chlorotic galls, and progressively broaden to become circular or elliptical uneven spots.

---

*Corresponding author: Endale Hailu, Ethiopian Institute of Agricultural Research, Ethiopia, E-mail: endalehailu@gmail.com

The disease affects leaves and stems. According to the information obtained from farmers and woreda agricultural office expertise, the disease existed for a long period in the region but well recognized after 2010. Locally, the disease is known by different names: such as *Qormid (North Shewa and South Wollo)*, *Kolsim and Kortim (North Gondar)*, *Aqorfid (East Gojam)*, *Chimid and Kurnchit (South Gondar)*; but in many places it is known by the name *Qormid* which is based on its symptoms on the leaf, in local Amharic language. The disease was highly expanding and distributing aggressively in the northern and central part of the country from year to year.

Even if the disease is disseminated at an alarming rate, there is no enough information on its status. Therefore, monitoring of the status of the disease is found to be crucial to draw management options. Hence, the present study was conducted to assess the distribution and intensity of the new Faba bean disease in central and northern Faba bean growing areas of Ethiopia.

Materials and Methods

Survey areas descriptions

The survey was conducted in 2013 main cropping season and covered three major Faba bean producing regional states (Amhara, Oromia and Tigray) of the central and northern parts of the country. A total of 278 faba bean fields in 66 Woredas (districts) of eleven zones from three regions were assessed of which 96 fields were in 18 districts, 192 fields in 44 districts and 10 fields in 4 districts of Oromia, Amhara and Tigray regional states, respectively. The altitude of study area was ranged from 1805 to 3332 m.a.s.l.

Survey methodology

In each surveyed regions, faba bean production fields along the main roads were randomly selected for observation and assessed at about 5-10 km using Vehicle odometer. Five stops were made in each faba bean field by moving in 'X' fashion of the fields using 1 X 1 meter square quadrants and data were collected from individual quadrants and the five samples per field were used as one site after averaged. All faba bean diseases in the surveyed areas were assessed and data on diseases prevalence, incidence and severity were recorded and evaluated.

The prevalence of a disease was calculated using the number of fields affected divided by the total number of field assessed and expressed in percentage. Incidence was calculated by using the number of plants infected and expressed as percentage of the total number of plants assessed. Severity was recorded by examining visually the whole plants using percent leaf area affected in the quadrants for faba bean gall and rust; but for chocolate spot and Ascohayta blight 0–9 scale were used.

The results of the survey were summarized by zones in the regions. The geographic coordinates (latitude and longitude), and altitude were recorded using Geographic Positioning System (GPS (ddd.WGS-84)) unit. The latitude and longitude coordinates were used to map the distribution of the new gall forming faba bean disease in the surveyed areas of the country.

Result and Discussions

The status of faba bean gall forming disease

The prevalence of the new disease was in the range of 0 and 100% (Figure 1). The overall average percent prevalence was about 48.5% in the surveyed areas. It was more prevalent in Awi zone of Amhara region with prevalence value of 100%. Prevalence percentage of 95.2, 91.7, 86.7 and 85.7 were recorded in north Shewa of Oromia region, north Shewa of Amhara region, south Wollo and south Tigray, respectively. Less

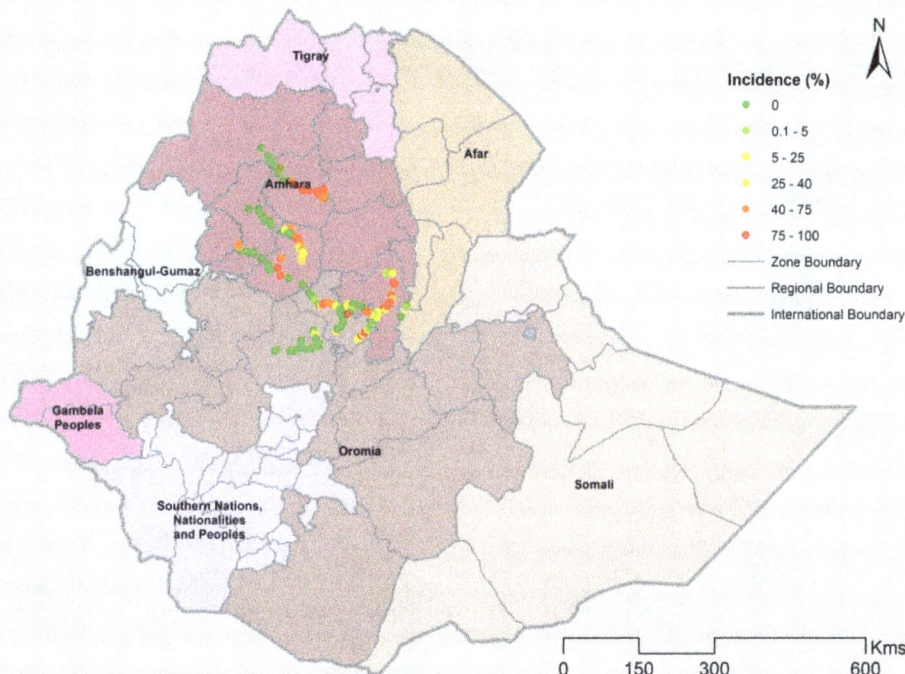

Figure 1: Distribution and incidence of faba bean gall disease in central and northern Ethiopia.

Region	Zone	Altitude	Faba bean gall					
		(m.a.s.l)	No of field assessed	Prevalence (%)	Incidence Range (%)	Mean Incidence (%)	Severity range (%)	Mean Severity (%)
Amhara	Awi	2501-2564	4	100.0	10-70	42.5	40-80	57.5
	East Gojam	2223-3318	36	69.4	0-100	37.4	0-100	39.9
	North Gondar	1873-1974	5	0.0	0	0	0	0.0
	North Shewa	2539-3324	60	91.7	0-100	32.6	0-80	13.6
	South Gondar	1805-3232	38	71.1	0-100	43.44	0-100	40.7
	South Wollo	1805-3232	15	86.7	0-100	41.5	0-80	14.6
	West Gojam	1825-2397	14	0.0	0	0	0	0.0
	Mean	1805-3332	172	72.1	0-100	29.9	0-100	22.2
Oromia	North Shewa	2528-3099	71	95.2	0-100	25	0-80	11.6
	West Shewa	2159-3090	25	36.0	0-40	3.0	0-50	4.0
	Mean	2159-3099	96	65.6	0-100	14.09	0-80	7.8
Tigray	East Tigray	2537-2724	3	66.7	0-60	26.7	0-5	8.3
	South Tigray	2247-2755	7	85.7	0-20	37.4	0-35	14.3
	Mean	2247-2755	10	80.0	0-60	32.05	0-35	11.3
Over all mean		1805-3332	278	48.5	0-100	15.4	0-100	6.4

Table 1: Distribution of faba bean gall disease in central and northern part of Ethiopia in 2013/14 cropping season.

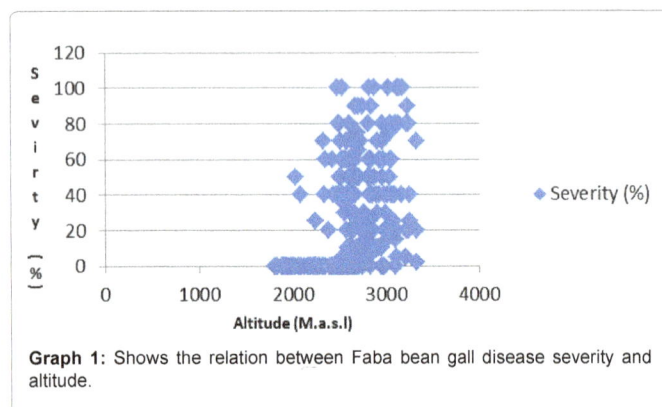

Graph 1: Shows the relation between Faba bean gall disease severity and altitude.

disease prevalence was observed in west Shewa zone with prevalence value of 36% but no new disease was observed in north Gondar and west Gojam zones (Table 1).

The overall mean incidence of faba bean gall in central and northern part of the country was 15.4% in which the disease incidence ranged from 0 to 100%. The maximum mean incidence of 43.4%, 42.5% and 41.5% were recorded in south Gondar, Awi and south Wollo zone, respectively. Among the three regions the highest mean incidence of 32.05% were recorded in Tigray.

The mean disease severity was also ranges from 0 to 57.5%. The maximum mean severity of 57.5% was observed in Awi zone followed by south Gondar zone with mean severity of 40.7%. The overall mean faba bean gall severity of 6.4% was observed in the surveyed areas of central and northern part of the country. In Amhara region, 22.2% mean severity was recorded and the disease was more severe in this region as compared to other regions (Table 1).

The faba bean gall disease incidence and severity was related to altitude. At altitude above 2400 m.a.s.l the disease became more Sevier. This showed that faba bean gall disease incidence has direct relation with altitude. The distribution and severity of the disease was high at higher altitude (Graph 1).

In general, faba bean production in Ethiopia is highly challenged by new faba bean gall forming disease. The disease was widely disseminated in the country and cause epidemic in short period of time.

Yet, the mechanism of introduction of the disease into the country and its transmission is not known. It was first reported in north Shewa zone of Oromiya region in 2011 in some pocket areas [5]. In 2013, the same problem with a significant magnitude was reported from Tigray Region [6]. In Chaina Xing reported the occurrence of gall forming disease on faba bean in Chaina and was identified as *Olpidium viciae*. In Ethiopia, it is not proved weather the causative agent similar to this pathogen.

The status of other faba bean diseases (chocolate spot, rust and ascohayta blight)

Chocolate spot (*Botrytis fabae*): Chocolate spot was widely distributed and the most prevalent in all Faba bean growing areas of the region. The recorded mean prevalence value ranged from 57.1 to 100%. The highest prevalence value of 100% was recorded in north Shewa, south Wollo, east Gojam and Awi zone of Amhara, north and west Shewa of Oromia and east and south Tigray zone of Tigray region. The minimum prevalence (57.1%) was recorded in west Gojam. The overall mean chocolate spot prevalence in all surveyed areas was 94.6%.

The overall mean chocolate spot incidence 42.4% was recorded in the surveyed areas of central and northern part of the country. The maximum incidence (100%) was recorded in Awi zone; whereas the minimum incidence (26.3) was recorded in north Gondar. Among the three regains surveyed, the highest mean incidence (70.8%) was recorded in Oromia followed by Amhara region with 70.6%.

Mean chocolate spot severity scale ranged from 1.1 to 5.4 in which the maximum was recorded in north Shewa zone of Oromia region and the minimum was in south Wollo zone (Table 2). Past study showed that Chocolate spot is the most important disease of faba bean and causes a significant yield loss of up to 61% on susceptible cultivars in Ethiopia [6,7]. It also indicated that the disease was the most important and widely prevalent in the faba bean growing regions of central and northern highlands of Ethiopia [8,9].

The status of faba bean rust (*Uromyces fabae*): The overall distribution of Faba bean rust disease in the surveyed area of the country was less with the mean prevalence of 1.9%. The disease prevalence was ranged from 0-50% in which the maximum rust prevalence was recorded in Awi zone. Mean rust disease prevalence of 15.3%, 1.4% and 0% was scored in Amhara, Oromia and Tigray regions (Table 2)

Faba bean rust mean disease incidence recorded in the range of 0

Region	Zone	Altitude range (m.a.s.l)	No of field assessed	Chocolate spot			Rust			Ascochyta blight		
				Prevalence (%)	Incidence (%)	Severity (1-9)	Prevalence (%)	Incidence (%)	Severity (%)	Prevalence (%)	Incidence (%)	Severity (1-9)
Amahara	Awi	2501-2564	4	100	100.0	4.0	50	0.0	0.0	25	17.5	1.5
	East Gojam	2223-3318	36	100	79.1	3.6	5.5	2.5	1.6	91.7	9.5	0.6
	North Gondar	1873-1974	5	60	26.3	2.1	20	26.1	28.5	0	3.8	0.3
	North Shewa	2539-3324	60	100	61.2	2.5	0	0.0	0.0	3.3	26.0	1.0
	South Gondar	1805-3232	38	78.9	64.0	2.8	31.6	22.6	15.2	5.3	1.3	0.1
	South Wollo	1805-3232	15	100	65.0	1.1	0	0.0	0.0	100	49.3	1.0
	West Gojam	1825-2397	14	57.1	88.6	4.0	0	1.4	0.7	0	12.1	0.8
	Mean	1805-3332	172	83.7	70.6	2.9	15.3	5.3	4.4	32.2	21.3	0.9
Oromia	North Shewa	2528-3099	71	100	70.7	5.4	2.8	0.7	0.5	97.1	31.3	1.9
	West Shewa	2159-3090	25	100	70.9	2.8	0	0.0	0.0	20	20.8	1.1
	Mean	2159-3099	96	100	70.8	4.1	1.4	0.4	0.3	58.5	26.1	1.5
Tigray	East Tigray	2537-2724	3	100	43.3	2.0	0	0.0	0.0	100	46.7	3.0
	South Tigray	2247-2755	7	100	69.3	1.3	0	0.0	0.0	100	82.8	3.0
	Mean	2247-2755	10	100	56.3	1.7	0	0.0	0.0	100	64.8	3.0
Over all mean		1805-3332	278	94.6	42.4	1.9	1.9	0.1	0.1	63.6	30.3	1.5

Table 2: Distribution of other common Faba bean diseases in northern and central part of Ethiopia, 2013 cropping season.

to 26.1% in all surveyed areas. The maximum incidence was recorded in north Gondar. Mean incidence of 5.3% and 0.4% of the disease was recorded in Amhara and Oromia regions, respectively. Insignificant result was observed in the surveyed areas of Tigray region. The overall mean incidence was less than 1%. Generally, the occurrences of the faba bean rust in the surveyed area were less (Table 2).

Faba bean rust disease severity was also ranged from 0 to 28.5% in which the maximum severity recorded in North Gondar zone. Mean rust disease severity of 4.4% and 0.3% were recorded in Amhara and Oromia regions, respectively. The overall disease severity in the surveyed areas of the country was 0.1% which was very low (Table 2). Past result also showed that the disease prevalence was less [6]. This might be due to unfavorable weather conditions to rust development in this cropping season as it is known to favor by high humidity, cloudy and warm weather conditions [10].

The status of ascochyta blight: The disease was distributed in all faba bean growing areas with over all mean prevalence value of 63.4%. The recorded prevalence ranges from 0-100 in which the maximum score recorded in east & south Tigray and south Wollo zones.

The mean incidence of Ascochyta blight in the surveyed regions ranged from 0 to 82.8%. Incidence of 21.3%, 26.1% and 64.8% was recorded in Amhara, Oromia and Tigray regions respectively. Tigray region was highly affected by Ascochyta blight as compared to the two regions.

The mean severity value of 1.5 was scored in the surveyed areas. Low severity was observed in the surveyed areas with a range of 0 to 3 in which the maximum was scored in Tigray region. The overall mean incidences of 30.3% were scored in the surveyed areas. Previous study indicated that Ascochyta blight was categorized as miner diseases [11]. Dereje and Tesfaye [8] indicated as Viruses and Ascochyta blight will be the potential treat for faba bean Production in Ethiopia. The recent studies indicated that the disease became among the major treats of faba bean production in the country [6].

Conclusions

The main purpose of this field inspection was to study the status of unidentified new gall forming faba bean disease which occurred in recent years and in the meantime observation of other most important

faba bean diseases. As a result, the new unidentified pest was found an epidemic and becoming a serious pest in the country within short period of time. Chocolate spot, Ascochyta blight and Faba bean rust are still the most common diseases of Faba bean in Ethiopia [12-14]. The mean prevalence of faba bean gall, Ascochyta blight, Chocolate spot and rust were 48.5%, 63.6%, 94.6% and 1.9% respectively. The mean incidence of all former diseases was 15.4%, 30.3%, 42.4% and 0.1% in their previous order. Based on severity scale, mean disease severity of Ascochyta blight and chocolate spot were 1.5 and 1.9, respectively. Mean severity of faba bean gall and faba bean rust were recorded as 6.4% and 0.1% in their order.

Faba bean gall was prevalent and the most challenging diseases and threatening faba bean production in the country. The severity of the disease has gone up to 57.5% (Awi zone). Among the three regions surveyed, it was more severe in Amhara region followed by Tigray and Oromiya regions. The mean disease severity of 22.2%, 7.8% and 11.3% were recorded in Amhara, Oromiya and Tigray regions, respectively. All improved and local faba bean varieties affected by the diseases indifferently.

The new gall forming disease recorded lower prevalence and incidence values, but in its area of destination it has got higher severity among all recorded common faba bean diseases, in which severity has direct impact on crop yield. The new disease is wider, fast coverage in time and space and so far the causal agent and the spread mechanism of this disease is unknown [15]. In fact the disease is epidemic and more serious from the record and the information of Faba bean growers in the surveyed areas. The epidemic conditions of the disease have significant implication on the production of Faba bean and on the country's economy. This survey information helps to consider the disease as serious pest in Ethiopia and in development of management options. Joint work on lab analysis and identification by all experts is crucial.

Acknowledgement

Without the support of some individuals and institutions the successful completion of this experiment would have not been realized. ICARDA is duly acknowledged for fully funding this work and EIAR for technical support of the study.

References

1. Salunkhe DK, Kadam SS (1989) Handbook of World Food Legumes: Nutrition Chemistry, Processing Technology, and Utilization. CRC press, Inc. Boca Rotan, Florida. 310.

2. Biruk Bereda (2009) Production & Marketing Activity of Broad Bean in Ethiopia. Ethiopia Commodity Exchange Authority study report, Addis Ababa, Ethiopia.

3. Samuel S, Fininsa C, Sakhuja PK, Ahmed S (2008) Survey of chocolate spot (*Botrytis fabae*) disease of faba bean (*Vicia faba* L.) and assessment of factors influencing disease epidemics in northern Ethiopia. *Crop Prot.* 27: 1457-1463.

4. Yohannes D (2000) Faba bean (*vicia fabae)* in Ethiopia. Institute of bio diversity, conservation and Research (IBCR), Adiss Ababa, Ethiopia, 43.

5. Dereje G, Wendafrash, Gemechu K (2012) Faba Bean Galls: a new disease of faba bean in Ethiopia. Available at Google.doc.com. 1-6.

6. Abebe, Tsehaye Birhane, Yemane Nega, Assefa Workineh (2014) The Prevalence and Importance of Faba Bean Diseases with Special Consideration to the Newly Emerging "Faba Bean Gall" in Tigray, Ethiopia. *JAFS,* 2: 33-38.

7. Teshome E, Tagegn A (2013) Integrated management of Chocolate spot (*Botrytis fabae* Sard.) of Faba bean (*Vicia faba* L.) at highlands of Bale, south eastern Ethiopia. Res J Agric Environ Manage 2: 011-014.

8. Dereje G, Tesfaye B (1994) Faba bean disease in Ethiopia. In: Asefaw T *et al.* (Eds.), cool-season food Legumes of Ethiopia. Proceedings of first National Cool-season Food legume Review Conference, Addis Ababa, Ethiopia, 16-20 December 1993. ICARDA/IAR.ICARDA. Syria. 328-345.

9. Dereje G, Beniwal SPS, Alem B (1988) Screaning of faba bean lines for chocolate spot in and rust resistance. Page 90-94 in IAR progress report on Faba Bean, 185-1987 crop season (S.P.S. Beniwal, Ed.). IAR, Addis Abeba.

10. Hawthore WA, Bretag T, Raynes M, Davidson JA, Kimber, et.al. (2004) Faba bean disease management stratege for south regions GRDC. Pulse Australia.

11. Nigussie T, Seid A, Derje G, Tesfaye B, Chemeda F et.al. (2008) Review of Research on Diseases Food Legumes. *In*: Abraham Tadesse (Eds). Increasing crop production through improved plant protection. 1: 85-124.

12. Agegnehu G, Gizaw A, Sinebo W (2006) Yield performance and land-use efficiency of barley and faba bean mixed cropping in Ethiopian highlands. Eur J Agron 25: 202-207.

13. Central Statistical Agency (CSA) 2013. Report on area and production of major crops (private peasent holdings, meher season). Stasistical bulletin 532:10-14.

14. ICARDA (2008) Drought and Broomrape-A threat to Faba Bean. Aleppo, Syria.

15. Xing Zhesheng (1984) Faba bean gall disease caused by *Oplidium* and its control. Acta Psychopathological Sinica 14:165-173.

3

Antennal and Behavioral Response of *Cydia pomonella* and *Lobesia botrana* Moths to Allyl Cinnamate

Marta Giner[1]*, Mercè Balcells[2] and Jesús Avilla[1]

[1]Department of Forest Science and Crop Production, University of Lleida, Lleida (Spain) and Department of Crop Protection, IRTA Center, Lleida, Spain
[2]Department of Chemistry, University of Lleida, Lleida, Spain

Abstract

Electroantenographical (EAG) response to *allyl cinnamate* were assessed on virgin and mated *Cydia pomonella* and *Lobesia botrana* adults to determine whether this compound could be used within integrated management programs (IMP). Adult behavioral reaction was later assessed in a wind tunnel, with and without the main compound of the corresponding female sex pheromone.

Allyl cinnamate elicited antennae responses of *C. pomonella* and *L. botrana*, both males and females. *Allyl cinnamate* EAG response was as high as pheromone response, and it was not reduced after mating.

In wind tunnel assays, allyl ester itself was not attractive to *C. pomonella* males, but its presence did not interfere with the pheromonal action when the number of contacts was compared. For females, a higher proportion of codling moths moved towards the source when *allyl cinnamate* was in the wind-tunnel plume. No differences were recorded depending on the mating status of codling moth adults. The same trend was observed in *L. botrana* males and females.

Results suggest that *allyl cinnamate* acts as a female behavioral modifier, but more assays are required to determine its role in insect communication in field conditions before inclusion in integrated pest management.

Keywords: *Allyl cinnamate*; Behavioral response; *Cydia pomonella*; EAG; *Lobesia botrana*; Wind-tunnel

Introduction

Cydia pomonella (L.) (codling moth) and *Lobesia botrana* (Dennis and Schiffermüller) (grapevine moth) (Lepidoptera: Tortricidae) are key pests in pome, pear and walnut orchards, and in vineyards, respectively [1]. Both species are pests of high-value crops and have low tolerance thresholds, leading to repeated insecticide treatments (with or without other pest control interventions) during the season. To solve negative effects of insecticide use [2,3], research has developed alternative pest control methods.

Mating disruption, based on the use of sex pheromones, is widely used in many countries to control codling and grapevine moths [4,5]. However, its application requires some specific field characteristics in order to succeed [6-8]. Other techniques based on the use of sex pheromones (mass-trapping and, attract and kill) are also available, but less used [9]. All these techniques focus in male moth control, but is also described an effect in female behavior by exposure to its own sex pheromone [10,11].

Tree fruit volatiles are usually added to pheromone traps to increase the number of moth captures [12-18]. Some of these plant volatiles synergize response to sex pheromone [19,20].

Our research group has been working on allyl ester synthesis using glycerol as starting material [21,22], which is produced in large amounts as a by-product in biodiesel production (http://www.biodiesel.org/). The insecticidal properties of some allyl esters have been assessed previously in order to give an added value to the surplus of glycerol [21,23,24]. As several volatile compounds from codling and grapevine moth host-plants are used by adults for host and mate localization [25-28], an effect in moth behavior due to allyl esters used as fruit aromas [29] was suspected. Moreover, some volatile compounds described as moth attractants [30-32] are chemically related to *allyl cinnamate*. This

fact drives to hypothesize that the latter may influence codling moth and grapevine moth adult behavior.

This paper describes for the first time the capacity of *allyl cinnamate* to elicit antennal response and to modify behavior from *C. pomonella* and *L. botrana* adults. Improved understanding of the role of this chemical compound may ultimately be incorporated into Tortricidae integrated pest management programmes (IPM).

Material and Methods

Insects

The experiments were conducted with a *C. pomonella* laboratory strain originated from a population collected in an unsprayed apple orchard in Lleida (north-east Spain), and with a *L. botrana* laboratory strain established in our laboratory from a mass-reared strain from INRA Bordeaux (France). Both strains were reared at the Crop Protection Laboratory of the UdL-IRTA Centre for Research and Development (Lleida, Spain) at 23 ± 2°C, with a photoperiod of 16:8 (L:D) and on agar-based semi-synthetic diet [33].

Cydia pomonella individuals were sexed as last instar larvae, whereas *L. botrana* individuals were sexed as pupae. The sexed individuals were kept separately by sex until adult emergence in the same conditions as

*Corresponding author: Marta Giner, Department of Forest Science and Crop Production, University of Lleida, Lleida (Spain) and Department of Crop Protection, IRTA Center, Lleida, Spain, E-mail: mginergil@gmail.com

the laboratory strains were. Next, four to six less than 24-h old adults of the same sex were transferred into plastic cages (diameter (d) = 15 cm, Height (H) = 5 cm), and maintained in the above mentioned conditions to obtain virgin individuals. To obtain mated adults, two or three couples were maintained together in the above mentioned plastic cages. Adults were used in electroantennogram (EAG) or wind-tunnel bioassays at second or third day after adult emergence. The mating status of females was ascertained by the presence of a spermatophore, by dissection of females after EAG recording or wind tunnel assay. Males were considered mated when females of their own group were ascertained to be mated.

Chemicals

Allyl cinnamate was purchased from Sigma-Aldrich (Madrid, Spain), and acetone for residue analysis was purchased from Panreac (Barcelona, Spain).

E,E-8,10-dodecadienol (codlemone) (≥ 99.5% purity), the main compound of *C. pomonella* sex pheromone, and *E,Z*-7,9-dodecadienyl acetate (EZ79Ac), the main compound of *L. botrana* sex pheromone (≥ 99.5% purity), were purchased from PheroBank (Wageningen, the Netherlands).

Electrophysiological assays

An EAG apparatus from Syntech (Hilversum, the Netherlands) was used to record electroantennographical responses of adults. Signals after stimulus application (mV) were amplified (100×) and filtered (DC to KHz) with an ID-2 interface (Syntech), digitized on a PC and analyzed with the EAG2000 program.

Each antenna was carefully cut from an insect that had previously been anesthetized with ice and immobilized using a fine needle. Another cut was done at the end of the antenna using a scalpel, and then the antenna was placed between EAG electrodes. Electrode gel (Parker, Orange, NJ) was used to facilitate connection between the antenna and electrodes.

Each chemical (stimulus) was presented by applying 0.1 µg of it to a piece of filter paper (2×2 cm). The piece of paper was then inserted into a Pasteur pipette, which was placed so that the tip of the pipette was 5 cm from the antenna. A puff of air (300 mL min⁻¹) through the pipette then carried the stimuli to the antenna.

Allyl cinnamate intrinsic activity

Twenty antennae per species (*C. pomonella* and *L. botrana*), sex (male and female) and mating status (virgin and mated) were used in the bioassay. Five consecutive puffs of allyl ester, pheromone and acetone (control puffs) were applied to each antenna in randomized order. Each puff was separated 30 s to minimize potential onset of antennae. No fatigue was observed in the antennae used in the bioassay.

The response to each stimulus was calculated as the mean response to the five puffs, and was compared to mean response to control puffs (acetone) by *t*-test ($P<0.05$). If significant greater response was observed to stimuli compared to control, mean response to stimuli was corrected as follow: corrected EAG response = mean EAG response to stimulus – mean EAG response to acetone. Corrected EAG responses were transformed [log (x + 1)] and analyzed by *t*-test or one-way ANOVA followed by Tukey-Kramer HSD test ($P<0.05$) for each species, sex and mating status. Statistical analysis was carried out with the JMP 8.0.1 program (SAS Institute, Cary, NC).

Synergism between *Codlemone* and *Allyl cinnamate*

Five consecutive puffs of codlemone (0.1 µg), a mixture of codlemone and allyl cinnamate (0.05 + 0.05 µg), allyl cinnamate alone (0.1 µg) and control puffs (acetone) were applied to each antenna (N = 20) of *C. pomonella* virgin males and females.

Mean corrected EAG response (calculated as described in the intrinsic activity assay) for each treatment (*codlemone, allyl cinnamate* and mixture) were compared for each sex by one-way ANOVA followed by Tukey-Kramer HSD test ($P < 0.05$) using the JMP 8.0.1 program (SAS Institute, Cary, NC).

Wind-tunnel assay

The assay was conducted in a glass wind tunnel (H=50 cm, Long (L)=200 cm and wide=50 cm) situated in a room maintained at 23 ± 2°C. Light was supplied by an incandescent light bulb situated on the ceiling of the room (2 lux for *C. pomonella* and 100 lux for *L. botrana, following* results from our research group). Two ventilators either side of the wind tunnel operated simultaneously, producing an air flow of 0.15 cm s⁻¹ through the tunnel (measured using an anemometer).

The assay was performed during the first two hours of scotophase, and at least 40 insects were used per treatment. Insects were tested individually and only once. Each individual was placed into a two-side open glass tube (d=2.5 cm, L=15 cm), which was oriented to the stimulus source and situated on a 20 cm high metal stand (the insect starting point). One 8 mm red rubber septum (Sigma-Aldrich, Spain) loaded with the main compound of the sex pheromone, allyl ester, or a mixture of pheromone and allyl ester in acetone (Table 1) was used as source of the stimulus. The source was situated on a 20 cm high metal stand located at 150 cm from the insect starting point. For solvent (acetone, code 0:0), 10 µL were added to the septum.

The effect of the *allyl cinnamate* alone or mixed to the pheromone at different proportions (Table 1) was tested on virgin males and females of *C. pomonella* and *L. botrana*. Moth behavior (activation, non-oriented flight, oriented flight and contact with source) in response to the stimulus was recorded for three minutes. The wind tunnel was cleaned with acetone after each experimental day and used material (glass tubes and metal stands) was washed with acetone and oven-dried at 200°C overnight. The percentage of insects that showed a specific behavior (activation, non-oriented flight, oriented flight and contact) for each species, sex and stimulus were compared by X^2 test ($P<0.05$) using the GraphPad program (Graph Pad Software Inc., La Jolla, CA).

Stimulus	Code	Species	Dose (µg / septum)	
			Pheromone	Allyl cinnamate
Pheromone*	1:0	C. pomonella	10.0	-
		L. botrana	2.5	-
Allyl cinnamate	0:1	C. pomonella	-	10.0
		L. botrana	-	2.5
Mixture of pheromone and allyl cinnamate	1:0.1	C. pomonella	10.0	1.0
		L. botrana	2.5	0.25
	1:1	C. pomonella	10.0	10.0
		L. botrana	2.5	2.5
	1:5	C. pomonella	10.0	50.0
		L. botrana	2.5	12.5

*Pheromone dose was fixed according to previous bioassays in our wind tunnel conditions (Giner et al., (2009) for *C. pomonella* [41] and Cruz -personal communication- for *L. botrana*).

Table 1: Composition of each stimulus used for each species in the wind tunnel assay.

Comparison between results obtained per sex on *C. pomonella* for the combination *codlemone+allyl cinnamate* and the sum of results for *codlemone* and *allyl cinnamate* separately were compared by *t*- test (*P*<0.05) using the GraphPad program to ascertain synergism.

Results

Electrophysiological assays

Allyl ester intrinsic activity: *Allyl cinnamate* elicited virgin *C. pomonella* female antenna and EAG responses as great as the one produced by codlemone. The EAG response was maintained after mating (Table 2).

For *C. pomonella* males, *allyl cinnamate* also caused a significant EAG response (1.070 ± 0.460 mV), not significantly different from the response to codlemone (0.753 ± 0.068 mV). The EAG response to *allyl cinnamate* did not reduced after mating (0.756 ± 0.207 mV) (*P*<0.05).

For *L. botrana* virgin males, EAG values regarding *allyl cinnamate* were not significantly different than the pheromone ones (0.635 ± 0.141 and 0.892 ± 0.403, respectively). A significant reduction of response was recorded to pheromone after mating (0.323 ± 0.086; *P*>0.05), but not to *allyl cinnamate* (0.237 ± 0.059; *P*<0.05), as observed in *C. pomonella*. In the case of females, only virgins elicited a response followed by *allyl cinnamate* stimulation (0.305 ± 0.019; *P*<0.05).

Synergism among *Codlemone* and *Allyl cinnamate*: No synergism was observed among *codlemone* and *allyl cinnamate* in *C. pomonella* antennae. No differences in EAG response were observed between *codlemone* - *allyl cinnamate* mixture and *codlemone* alone ($P_{males} = 0.91$; $P_{females} = 0.57$) or *allyl cinnamate* alone ($P_{males} = 0.76$; $P_{females} = 0.55$).

Wind tunnel: In the wind-tunnel, the codlemone caused the complete range of behaviors from *C. pomonella* males (Figure 1). The addition of *allyl cinnamate* to codlemone blend at different doses did not have a significant effect on activation and non-oriented flight behavioral steps when compared to the ones produced by codlemone alone. The addition of *allyl cinnamate* to codlemone at 1: 0.1 and 1: 5 caused some reduction on oriented flight and contact behavior, but not at 1:1 (Figure 1).

When *allyl cinnamate* was presented alone a significant reduction in percent of males that reach each behavioral step was recorded. If these values were compared to the solvent alone, a greater percent of males were activated and started the flight in the presence of the allyl ester, but few contacts were scored (Figure 1).

In the case of *C. pomonella* females, codlemone did not cause significant behavioral effect but the presence of *allyl cinnamate* (alone or mixed with codlemone) did. A higher percent of virgin females were activated and flew to the source when *allyl cinnamate* was present (Figure 2). The same was observed in the case of mated females (data not shown). No differences were recorded in the response to *allyl cinnamate* between virgin or mated *C. pomonella* females (*P*>0.05). No increase of attraction was observed in the case of mated males (data not shown), compared to virgins.

In *L. botrana*, *allyl cinnamate* did not produce an effect in males, and the addition to pheromone causes a significant reduction in percent of contacts with the source at 1:0.1 or 1:5 proportions, but not at 1:1 (Figure 3), as observed in *C. pomonella*. For females, *allyl cinnamate* caused a significant activation, non-oriented flight and oriented flight – compared to solvent – (Figure 4), but no contacts were observed.

Chemical	C. pomonella Virgin		C. pomonella Mated
Codlemone	0.520 ± 0.090 a		-
Allyl cinnamate	0.580 ± 0.036 a	ns	0.642 ± 0.166

Mean and SE, n = 20.
Values followed by the same letter in the same column are not significantly different (*t*-test, *P*<0.05).
ns =not significant differences between virgin and mated responses (*t*-test, *P*<0.05).

Table 2: Corrected (Response to chemical – response to acetone) EAG response (mV) of antennae of virgin and mated females of *Cydia pomonella* to 0.1 µg of the main component of female sex pheromone (codlemone) and allyl cinnamate.

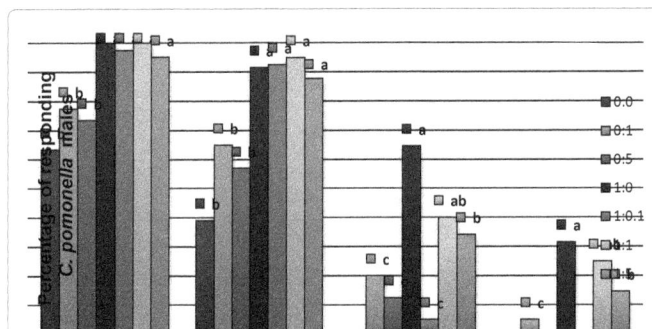

Different letters into each behavioral step indicate significant differences in moth response to different treatments (X^2, $P < 0.05$).

Figure 1: Behavioral response of *C. pomonella* virgin males (n = 40, minimum) in the wind tunnel flying towards a source baited with pheromone (1:0) (10 µg) and different proportions of pheromone: allyl cinnamate.

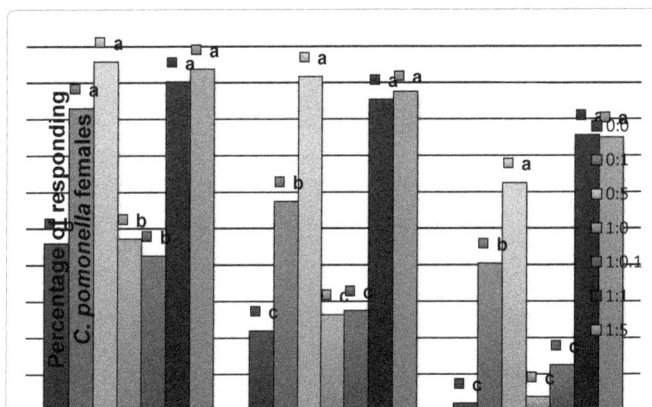

Different letters into each behavioral step indicate significant differences in moth response to different treatments (X^2, $P < 0.05$).

Figure 2: *C. pomonella* virgin female responses in the wind tunnel to pheromone (1:0) (10 µg) and different proportions of pheromone:allyl cinnamate.

Discussion

To our knowledge, this is the first report of *allyl cinnamate* eliciting antennal response of *C. pomonella* and *L. botrana* moths and (or) causing a behavioral reaction in a wind tunnel.

Allyl cinnamate elicited an antenna response in both males and females of *C. pomonella* and *L. botrana*, independently of the state of mating. This indicates that its action would not be strictly related with the mating process but has to be involved in general activity (feeding, host-localization). This observation contrasts with the one produced by several plant volatiles, which are more attractive after mating [34,35].

It is interesting to note that the same EAG recordings were scored

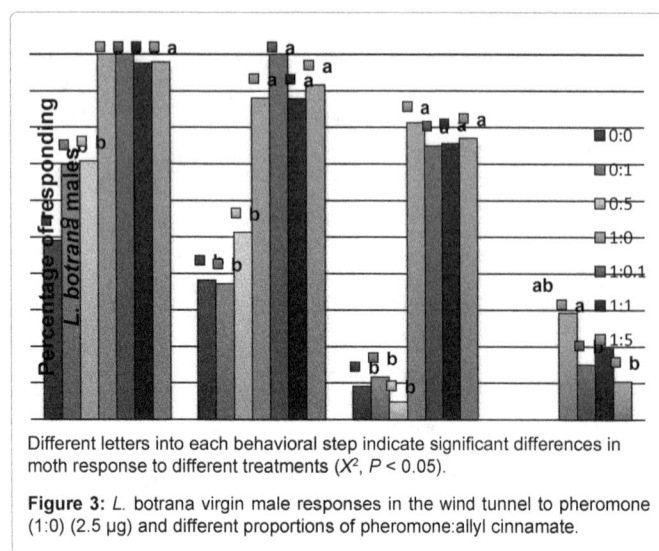

Different letters into each behavioral step indicate significant differences in moth response to different treatments (X^2, $P < 0.05$).

Figure 3: *L. botrana* virgin male responses in the wind tunnel to pheromone (1:0) (2.5 µg) and different proportions of pheromone:allyl cinnamate.

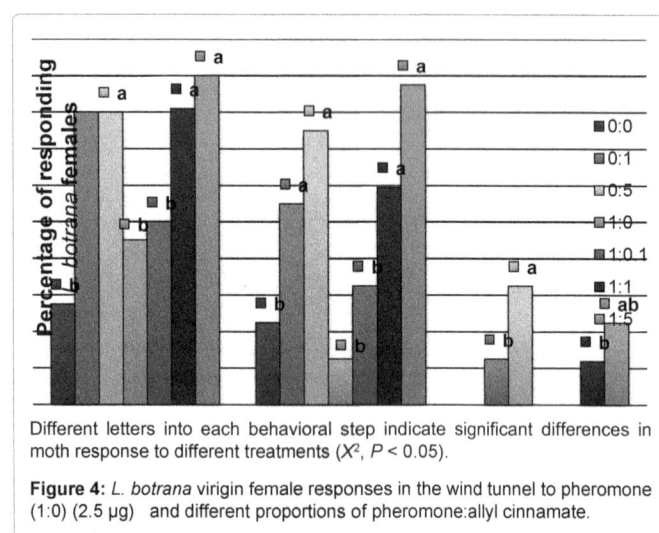

Different letters into each behavioral step indicate significant differences in moth response to different treatments (X^2, $P < 0.05$).

Figure 4: *L. botrana* virigin female responses in the wind tunnel to pheromone (1:0) (2.5 µg) and different proportions of pheromone:allyl cinnamate.

by both codlemone and *allyl cinnamate* alone, or when blended, suggesting that the same receptors could be involved in the perception process. This possibility was previously described for plant volatiles [25,36-38]. In the case of *allyl cinnamate* this fact should be more deeply studied to be confirmed.

Although an effect of allyl esters of fatty acids could be suspected from similarities in chemical structure to butyl hexanoate (moth attractant) [15], no effect of alkyl allyl esters were recorded in *C. pomonella* or *L. botrana* (unpublished data).

It is known that the size of EAG response does not always indicate a behavioral response [34]. Even though male and female antenna elicitation was recorded, only females were attracted to sources baited with allyl cinnamate in the wind-tunnel. An increase of fluttering was observed in males and *allyl cinnamate* alone does not clearly produce a behavior of source contact in any case. This lack of contact could be related with the fact that the behavior was observed for three minutes. More assays should be done with increased observation time or semi-field assays to reaffirm the properties of *allyl cinnamate* as *Tortricidae* moth attractant.

No increase of attraction was observed in the wind-tunnel assay

after mating, reassuring the results from the EAG recordings and indicating the lack of *allyl cinnamate* effect on the mating process.

Only on *C. pomonella* females a synergism between pheromone and *allyl cinnamate* was suggested in the wind-tunnel. Other authors have described a similar synergism to plant volatiles [21,25]. This could be extremely interesting if confirmed in a field situation.

Allyl cinnamate is suggested as a candidate to bait pheromonal traps with the aim of increasing moth captures in pest monitoring, or in attract & kill and mass-trapping strategies in Tortricidae pest control.

The fact that allyl cinnamate is authorized for use as food aroma (http://eur-lex.europa.eu/) indicates its availability to use in ethologic control. Moreover, allyl esters can be synthesized from glycerol or fat wastes [21,22], what fosters the re-use of industrial sub products. However, more assays need to be done to fix the proportion of *allyl cinnamate* into the blend to improve the attractiveness of moths [39,40] and asses it in field conditions, as well as the fate of *allyl cinnamate* in field conditions.

Acknowledgements

M.G. was financed by fellowship n° BES-2008-004779, and M.G., M.B. and J.A. were supported by Spanish Ministry of Economy and Competitiveness research grant AGL 2010-17486.

References

1. International Society of Pest Information (ISPI) (2009) Pest directory (CD-ROM for Windows 98 & up).

2. Devine GJ, Furlong MJ (2007) Insecticidal use: context and ecological disturbances. Agric Human Values 24: 281-306.

3. Rodríguez MA, Marques T, Bosch D, Avilla J (2011) Assessment of insecticide resistance in eggs and neonate larvae of Cydia pomonella (Lepidoptera: Tortricidae). Pesticide Biochemistry and Physiology 100: 151-159.

4. Angeli G, Anfora G, Baldessari M, Germinara GS, Rama F, et al. (2007) Mating disruption of codling moth Cydia pomonella with high densities of Ecodian sex pheromone dispensers. J Appl Entomol 131: 311-318.

5. Dunkelblum E (2007) Pest management programs in vineyards using male mating disruption. Pest Manag Sci 63: 769-775.

6. Sauer AE, Karg G (1999) Variables affecting pheromone concentration in vineyards treated for mating disruption of grapes vine moth Lobesia botrana. J Chem Ecol 24: 289-302.

7. Witzgall P, Bäckman AC, Svensson M, Koch U, Rama F, et al. (1999) Behavioural observations of codling moth, Cydia pomonella, in orchards permeated with synthetic pheromone. Biocontrol 44: 211- 237.

8. Louis F, Schirra K-J (2001) Mating disruption of Lobesia botrana (Lepidotpera: Tortricidae) in vineyards with very high population densities. Pheromones for Insect Control in Orchards and Vineyards. IOBC wprs Bull 24: 75-79.

9. Mansour M (2010) Attract and kill for codling moth Cydia pomonella (Linneeus) (Lepidoptera: Tortricidae) control in Syria. J Appl Entomol 134: 234-242.

10. Stelinki LL, Il'ichev AL, Gut LJ (2006) Antennal and behavioral responses of virgin and mated oriental fruit moth (Lepidoptera: Tortricidae) females to their sex pheromone. Ann Entomol Soc Am 99: 898-904.

11. Kuhns EH, Pelz-Stelinki K, Stelinski LL (2012) Reduced mating success of female tortricid moths following intense pheromone auto-exposure varies with sophistication of mating system. J Chem Ecol 38: 168-175.

12. Knight AL, Light DM (2004) Use of (E,Z)-2,4-decadienoic acid in codling moth management: improved monitoring in Barlett pear with high dose lures. J Entomol Soc British Columbia 101: 59-66.

13. Bengtsson M, Bäckman A-C, Liblikas I, Ramirez MI, Borg-Karlsson AK, et al. (2001) Plant odor analysis of apple: Antennal response of codling moth females to apple volatiles during phenological development. J Agric Food Chem 49: 3736-3741.

14. Reed HC, Landolt PJ (2002) Attraction of mated females codling moths

(Lepidoptera: Tortricidae) to apples and apple odor in a flight tunnel. Fla Entomol 85: 324-329.

15. Hern A, Dorn S (2004) A female-specific attractant for the codling moth, Cydia pomonella, from apple fruit volatiles. Naturwissenschaften 91: 77-80.

16. Fernández DE, Cichón L, Garrido S, Ribes-Dasi M, Avilla J (2010) Comparison of lures loaded with codlemone and pear ester for capturing codling moths, Cydia pomonella, in apple and pear orchards using mating disruption. J Insect Sci 10: 139.

17. Knight AL (2010) Improved monitoring of female codling moth (Lepidoptera: Tortricidae) with pear ester plus acetic acid in sex pheromone-treated orchards. Environ Entomol 39: 1293-1290.

18. Knight AL, Light DM (2014) Combinal approaches using sex pheromone and pear ester for behavioural disruption of codling moth (Lepidoptera: Tortricidae). J Appl Entomol 138: 96-108.

19. Yang Z, Bengtsson M, Witzgall P (2004) Host plant volatiles synergize response to sex pheromone in codling moth, Cydia pomonella. J Chem Ecol 30: 619-629.

20. Varela N, Avilla J, Anton S, Gemeno C (2011) Synergism of pheromone and host-plant volatile blends in the attraction of Grapholita molesta males. Entomol Exp Appl 141: 114-122.

21. Escribà M, Barbut M, Eras J, Balcells M, Avilla J (2009) Synthesis of allyl esters of fatty acids and their ovicidal effect on C. pomonella (L.). J Agric Food Chem 12: 345-358.

22. Escribà M, Eras J, Villorbina G, Balcells M, Blanch C, et al. (2011) Use of crude glycerol from biodiesel producers and fatty materials to prepare allyl esters. Waste Biomass Valorization 2: 285-290.

23. Giner M, Balcells M, Avilla J (2012) Insecticidal action of five allyl esters on eggs and larvae of three tortricid pests: laboratory tests. Bull Insectol 65: 63-70.

24. Giner M, Avilla J, Balcells M, Caccia S, Smagghe G (2012) Toxicity of allyl esters in insect cell lines and in Spodoptera littoralis larvae. Arch Insect Physiol Biochem 79: 18-30.

25. Yan F, Bengtsson M, Witzgall P (1999) Behavioural response of female codling moths, Cydia pomonella, to apple volatiles. J Chem Ecol 25: 1343-1351.

26. Reddy GVP, Guerrero A (2004) Interactions of insect pheromones and plant semiochemicals. Trends Plant Sci 9: 253-261.

27. Tasin M, Anfora G, Ioriatti C, Carlin S, de Cristofaro A, et al. (2005) Antennal and behavioural response of grapevine moth Lobesia botrana females to volatiles from grapevine. J Chem Ecol 31: 77-87.

28. Trona F, Anfora G, Bengtsson M, Witzgall P, Ignell R (2010) Coding and interaction of sex pheromone and plant volatile signals in the antennal lobe of the codling moth Cydia pomonella. J Exp Biol 213: 4291-4303.

29. Burdock GA (2010) Fenaroli's handbook of flavour ingredients. 10th edition. Taylor and Francis group. CC Press. Orlando, Florida, USA. 2159.

30. Casado D, Gemeno C, Avilla J, Riba M (2006) Day-night and phenological variation of apple tree volatiles and electroantennogram responses in Cydia pomonella (Lepidoptera: Tortricidae). Environ Entomol 35: 258-267.

31. Massa MJ, Robacker DC, Patt J (2008) Identification of grape juice aroma volatiles and attractiveness to the Mexican fruit fly (Diptera: Tephritidae). Fla Entomol 91: 266-276.

32. Ortiz A, Echeverría G, Graell J, Lara I (2010) The emission of flavour-containing volatile esters by "Golden Reinders" apples is improved after mid-term storage by postharvest calcium treatment. Postharvest Biol Technol 57: 114-123.

33. Pons S, Eizaguirre M, Sarasúa MJ, Avilla J (1994) Influencia del fotoperiodo sobre la inducción de diapausa de Cydia pomonella (Lepidoptera: Tortricidae) en laboratorio y campo. Investigaciones Agrarias Producción Protección Vegetal 9: 477-491.

34. Ansebo L, Coracini MDA, Bengtsson M, Liblikas I, Ramirez M, et al. (2004) Antennal and behavioural response of codling moth Cydia pomonella to plant volatiles. J Econ Entomol 128: 488-493.

35. Landolt PJ, Guédot C (2008) Field attraction of codling moth (Lepidoptera: Tortricidae) to apple and pear fruit, and quantification of kairomones from attractive fruits. Ann Entomol Soc Am 101: 675- 681.

36. Shields VAC, Hildebrand JG (2001) Responses of a population of antennal olfactory receptor cells in the female moth Manduca sexta to plant-associated volatile organic compounds. J Compr Physiol 186: 1135-1151.

37. De Cristofaro A, Ioriatti C, Molinari F, Pasqualini E, Anfora G, et al. (2004) Electrophysiological responses of Cydia pomonella to codlemone and pear ester ethyl (E,Z)-2,4-dodecadienote: peripheral interactions in their perception and evidences for cells responding to both components. B Insectol 57: 137-144.

38. Ansebo L, Ignell R, Löfqvist J, Hansson BS (2005) Responses to sex pheromone and plant odours by olfactory receptor neurons housed in sensilla auricillica of the codling moth, Cydia pomonella (Lepidoptera: Tortricidae). J Insect Physiol 51: 1066-1074.

39. Tasin F, Casado D, Coracini M, Bengtsson M, Ioriatti C, et al. (2010) Flight tunnel response of codling moth Cydia pomonella to blends of codlemone, codlemone antagonists and pear ester. Physiol Entomol 35: 249-254.

40. Solé J, Sans A, Riba M, Guerrero A (2010) Behavioural and electrophysiological responses of the European corn borer Ostrinia nubilalis to host-plant volatiles and related chemicals. Physiol Entomol 35: 354-363.

41. Giner M, Sans A, Riba M, Bosch D, Gago R, et al. (2009) Development and biological activity of a new antagonist of the pheromone of codling moth Cydia pomonella. J Agric Food Chem 57: 8514-8519.

4

Effect of Seed Variety and Cutting Age on Dry Matter Yield, Nutritive Values and *In Vitro* Digestibility of Teff Grass

Benjamin SA[1]* and Bradford BJ[1]

[1]*Department of Animal Sciences and Industry, Kansas State University, Manhattan, KS 66506, USA*
[2]*Department of Agronomy, Kansas State University, Manhattan, KS 66506, USA*

Abstract

While there is substantial ongoing work to improve the drought tolerance of grain crops, less effort has been made to decrease the water needs for forage crops. Water-efficient warm-season forage crops, with acceptable nutritional value, could prove an attractive alternative to traditional forages like alfalfa and corn silage. Teff (Eragrostis tef) is a and to 5 cutting ages (40, 45, 50, 55, or 60 d after planting [DAP]). Samples were dried, weighed, and analyzed for crude protein (CP), neutral detergent fiber (aNDFom), and 24 h *in vitro* NDF digestibility (IVNDFD). It was found that seed variety had no effect on dry matter (DM) yield, CP, aNDFom, or IVNDFD. DM yield increased linearly as cutting age increased from 40 to 60 DAP. Similarly, aNDFom concentration increased quadratically with increasing cutting age. CP and IVNDFD decreased linearly as cutting age increased from 40 to 60 DAP. To assess carryover effects of cutting age on yield and nutritive values, 2 additional cuttings were taken from each pot. It was found that increasing the age at first cutting from 40 to 60 DAP significantly decreased CP concentration in the second cutting. Additionally, increasing DAP significantly reduced DM yield in subsequent cuttings. Across all seed varieties and cutting ages, CP decreased and aNDFom increased linearly with each additional cutting. Results indicate that, under greenhouse conditions, the first cutting of teff should be taken between 45 and 50 DAP to optimize nutritive values and digestibility in that cutting and additional cuttings.

Keywords: Drought; Teff grass; Dry matter yield; Nutritive value; Dairy cattle

Introduction

One of the most pressing issues facing the dairy industry is drought. In the Southwestern and High Plains regions of the United States, where annual precipitation is low, irrigation for growing feed presents the greatest water-utilization challenge for dairy producers. More than 90% of the water used to support a dairy farm is devoted to producing crops that feed the cattle [1]. While the dairy industry has seen impressive growth in states like Kansas, New Mexico, and Texas, ground water levels in these areas have been decreasing at an alarming rate [2]. As ground water levels drop, some wells are no longer able to provide fields with the intended volume of water. Given the high water demands of crops like alfalfa and corn, and that alfalfa hay and corn silage are the most commonly fed forages in the dairy industry, the sustainability of the dairy industry in the Southwest and High Plains is questionable without an intentional shift toward water conservation.

While there is substantial ongoing work to improve the drought tolerance of grain crops, less effort has been made to decrease the water needs for forage crops. Water-efficient warm-season forage crops, with acceptable nutritional value, could prove an attractive alternative to traditional forages like alfalfa and corn silage. Teff (*Eragrostis tef*) is a warm-season annual grass (C4 physiology) native to Ethiopia that is well-adapted to arid conditions. For thousands of years, teff has been used as a grain crop for human consumption [3]. Once introduced to the United States, however, researchers began evaluating teff as a forage crop [4].

While teff grass has potential to fit the needs for forage production in water-stressed regions, very little is currently known about its nutritional characteristics and whether it can support high levels of milk production by dairy cattle. In Ethiopia, because teff is primarily grown as a grain crop, most feeding trials have aimed at improving the nutritive value of low-quality teff straw [5-7]. Additionally, studies that have investigated the quality of teff grass before it reaches full maturity have reported nutritive values that are highly variable. The crude protein (CP) concentration of teff has been reported to range anywhere from 8.5 to 21.5% [4,8-10]. The neutral detergent fiber (NDF) concentration, a predictor of intake in ruminants, has been reported to range from 52.5 to 72.5% [4,8,10]. Due to the extreme variation in reported nutritive values for teff, it is difficult to know at this point if teff grass is a suitable forage source for high producing dairy cows. Given that the productivity of a dairy cow is highly dependent on forage quality and digestibility [11], standardized quality and digestibility values for teff must be established before the productivity of cows fed teff grass can be investigated. Because both variety and age at harvest play a crucial role in dictating the quality of a given forage, the objective of this study was to investigate the effect of variety and cutting age on dry matter yield, nutritive values, and *in vitro* digestibility of teff grass.

Materials and Methods

Design and treatments

This experiment was conducted in a climate-controlled greenhouse

---

*Corresponding author: Benjamin Saylor, Department of Animal Sciences and Industry, Kansas State University, Manhattan, KS 66506, USA
E-mail: bsaylor318@gmail.com

space at Kansas State University (Manhattan, KS). The designated space averaged 24.6°C with 14 h of light/d as a combination of both natural and artificial light. Eighty plastic pots (3.78 L) were blocked by location and randomly assigned to 4 teff varieties and 5 cutting ages. The 20 treatment combinations were assigned in replicates of 4. The 4 varieties of teff seed used in this study were Corvallis, Dessie, Moxie, and Tiffany. All 4 varieties were commercially available at the start of the study and coated. Although the exact coating used on the seeds is proprietary, most seed coatings consist of a combination of lime to regulate soil pH, fertilizer to direct specific nutrients to the site of seed-soil contact, as well as insecticides and fungicides, all held together by a binding agent. Coating grass seeds can both enhance germination and add weight to the seeds for easier and more uniform sowing [12].

Seeds were planted in Metro Mix 360 (Sungro Horticulture, Agawam, MA) at a rate of 30 seeds per pot (equivalent to 16.81 kg/ha) and to an average depth of 0.48 cm. At planting, 0.15 g of urea (equivalent to 56 kg N/ha) was applied to each pot and the pots were lightly watered with a spray bottle. Pots were watered with a spray bottle until the seedlings were strong enough to withstand watering with a hose. Mature plants were watered to maintain "well-watered" conditions. An additional 0.15 g of urea (equivalent to 56 kg N/ha) was applied to all pots at d 60 after planting. The 5 cutting ages were 40, 45, 50, 55, and 60 d after planting (DAP).

Data and sample collection

Each pot was harvested at the assigned cutting age. Entire plants were cut with gardening clippers to a height of 10 cm and top biomass was collected and weighed. To assess the carryover effects of first-cutting harvest age on DM yield and nutritive values, a second cutting was taken from each pot 30 d after the first cutting. A third cutting was taken 30 d after the second cutting. After the third cutting, regrowth was insufficient to justify a fourth cutting.

Analytical techniques

Harvested samples were placed in paper bags and dried at 55°C in a forced-air oven for 72 h. After 24 h of air equilibration, dried samples were weighed to determine dry matter (DM) yield. Samples were then ground through a 1-mm screen using a Cyclone Sample Mill (UDY Corporation, Fort Collins, CO). Concentrations of amylase-treated, ash-free neutral detergent fiber (aNDFom) were determined in the presence of sodium sulfite [13] using an Ankom Fiber Analyzer (ANKOM Technology, Macedon, NY). Crude protein (CP) was determined by oxidation and detection of N_2 (LECO Analyzer, LECO Corp., St. Joseph, MI), multiplied by 6.25. Concentrations of all nutrients except for DM were expressed as percentages of DM determined by drying at 105°C in a forced air oven for more than 8 h. In vitro NDF digestibility (IVNDFD) was analyzed using a DAISY Incubator (ANKOM Technology, Macedon, NY). Ground grass samples were placed in filter bags with 25 μm porosity (ANKOM Technology, Macedon, NY) and incubated for 24 h in rumen fluid collected from a mature Holstein steer fed a 50:50 forage:concentrate diet. Once removed from incubation, samples were dried at 55°C and transferred to an Ankom apparatus to determine NDF concentration of the residue. Second and third cutting samples were analyzed by Dairy One Forage Testing Laboratory (Dairy One Inc., Ithaca, NY) using identical analytical techniques.

Statistical analysis

The data were analyzed using JMP (version 10.0, SAS Institute, Cary, NC). An analysis of variance was conducted to analyze how the fixed effects of teff seed variety, cutting age, and their interaction influenced

dependent variables. Independent variables were declared significant at P<0.05 and means were separated by Tukey's HSD test.

Results and Discussion

Cutting 1

Plant maturity at harvest is one of the principal factors influencing forage quality and digestibility [13]. With the development of higher quality and more digestible varieties, however, plant genetics are playing an increasingly crucial role in determining the overall quality of a given forage. Researchers worldwide have investigated the effect of seed variety on the quality and digestibility of a number of forage types including alfalfa [14], corn silage [15], sorghum [16], tall fescue [17], oats, and vetch [18] to name a few. There are multiple varieties of teff seed on the market today; some are better for grain production, others for forage production. Grain types tend to mature earlier than forage types, resulting in lower DM yields and forage quality [4]. In this experiment, all 4 teff varieties evaluated were bred for forage production. In Cutting 1, seed variety had no effect (P>0.30) on DM yield, aNDFom, CP, or IVNDFD (Table 1).

Cutting age, however, had significant impacts on first cutting forage yield, quality, and digestibility (Figure 1). As expected, DM yield increased linearly (P<0.001) from 4.1 to 26.4 ± 0.45 g/pot as cutting age increased from 40 to 60 DAP. Additionally, aNDFom concentration increased (P<0.001) from 51.7 to 63.5 ± 0.81% of DM with increasing DAP and CP decreased linearly (P<0.001) from 28.7 to 11.2 ± 0.49% of DM. As forages mature, quality decreases as photosynthetic products are converted to fibrous, structural components [19]. Grasses like teff, as opposed to legumes, have structural components in both their leaves and stems. Therefore, the forage quality of grasses tends to decline more rapidly with age than that of legumes [19]. In this study, the CP concentration of first cutting teff decreased linearly at a rate of 0.88% per d (Figure 1). Similar trends have been seen with brome grass [20] and sorghum-sudan grass [21]. The average greenhouse temperature could explain the higher-than-expected CP concentration of teff cut at 40 and 45 DAP. Lower temperatures slow the maturation process and the subsequent production of fibrous structural compounds thus improving CP concentration and overall forage quality [19].

Cutting age also had a significant effect on the IVNDFD of first-cutting teff (Figure 1). As cutting age increased from 40 to 60 DAP, IVNDFD decreased linearly (P<0.001) at a rate of 0.95% per day (60.8 to 41.2 ± 1.0%). The NDF component of teff, like all forages, is composed primarily of cellulose, hemicellulose and lignin. Lignin represents the indigestible fraction of NDF [19]. As a plant ages, lignin concentration increases, ultimately decreasing the overall digestibility of the fiber [22]. Other studies have confirmed this trend [20-21]. While the nutrient composition and digestibility of forages grown in a greenhouse are not always the same as those grown in the field, other studies [14,23,24] have used quality and digestibility values of greenhouse grown forages as initial estimates of what could be expected in a more practical cultivation scenario.

Cuttings 2 and 3

In Cutting 2, teff variety had no effect (P=0.47) on DM yield, aNDFom concentration (P=0.13), or CP concentration (P=0.84, Table 1). Additionally, there was no effect (P=0.30) of variety on the cumulative DM yielded from the 2 cuttings. First-cutting harvest age, however, had a significant effect (P<0.001) on second-cutting DM yield as well as second-cutting aNDFom and CP concentrations (Figure 2). Dry matter yield from Cutting 2 decreased from 23.68 to 11.59 ± 0.91

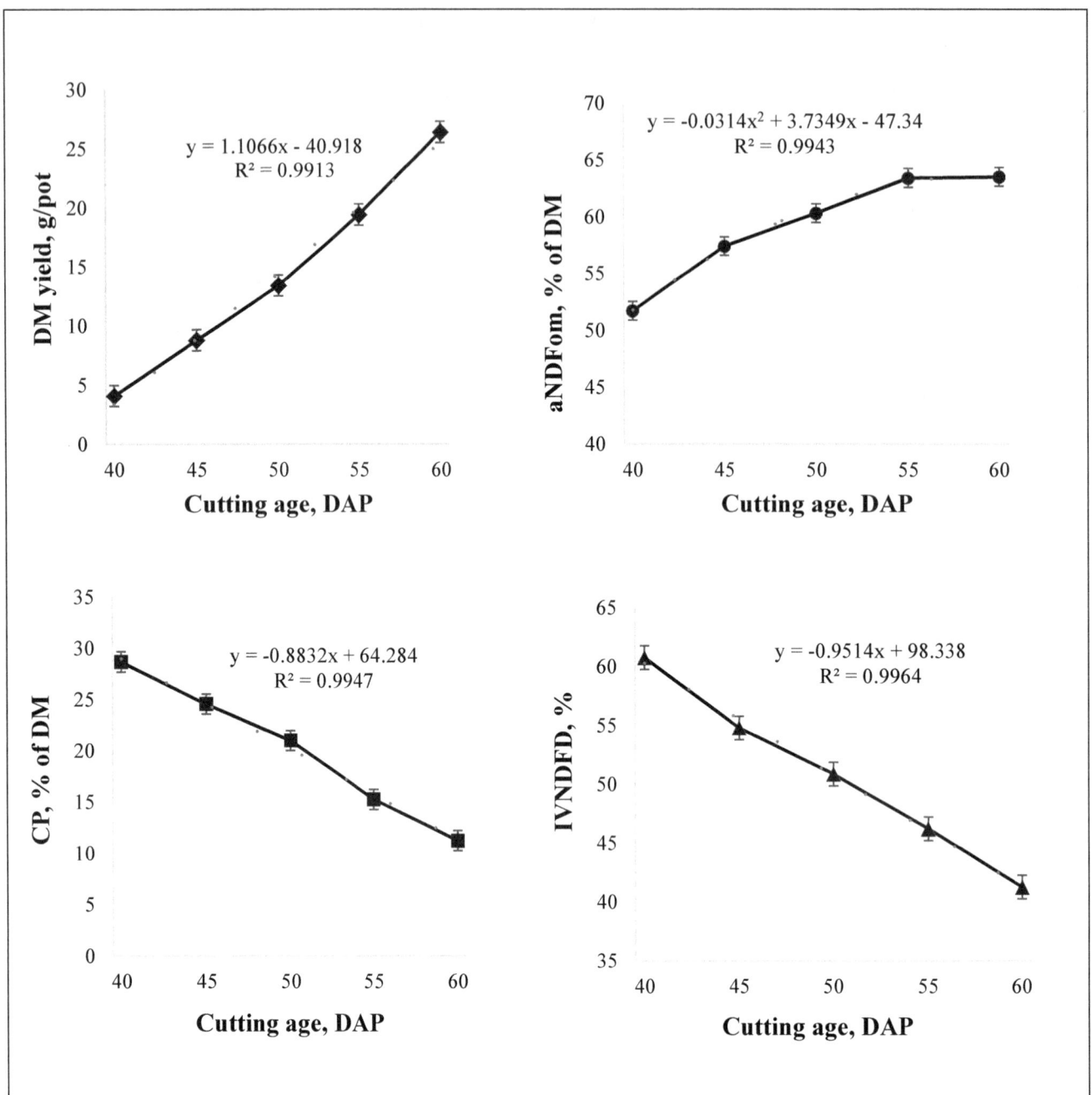

Figure 1: Effect of cutting age on yield, nutritive values, and digestibility of first-cutting teff grass. Increasing cutting age from 40 to 60 DAP significantly increased DM yield and aNDFom concentration (P<0.001) but significantly decreased CP concentration and IVNDFD (P<0.001).

g/pot when first-cutting harvest age increased from 40 to 60 DAP. We found that second-cutting aNDFom concentration was greatest (P<0.001) in those samples that were first cut at 45 and 50 DAP. Crude protein concentration of the second-cutting teff decreased dramatically, from 11.94 to 6.43 ± 0.32% of DM, when first-cutting harvest age was increased from 40 to 60 DAP.

In Cutting 3, again, teff variety had no effect (P=0.40) on DM yield, aNDFom concentration (P=0.10), or CP concentration (P=0.48, Table 1). Additionally, seed variety had no effect (P=0.49) on the cumulative DM yielded from the 3 cuttings. Like what was seen with Cutting 2,

first-cutting harvest age had a significant effect (P<0.001) on third-cutting DM yield, aNDFom concentration, and CP concentration (Figure 2). DM yield decreased from 18.70 to 5.24 ± 0.30 g/pot when first-cutting harvest age increased from 40 to 60 DAP. Third-cutting aNDFom concentration was greatest in samples originally cut at 45 DAP and least in those cut at 55 DAP (P<0.001). CP was greatest in samples originally cut at 45 DAP and least in those cut at 55 DAP.

Whereas seed variety had no effect on the agronomic characteristics of teff, first-cutting harvest age played a critical role in influencing yield and nutritive values in Cuttings 2 and 3. According to Van Soest,

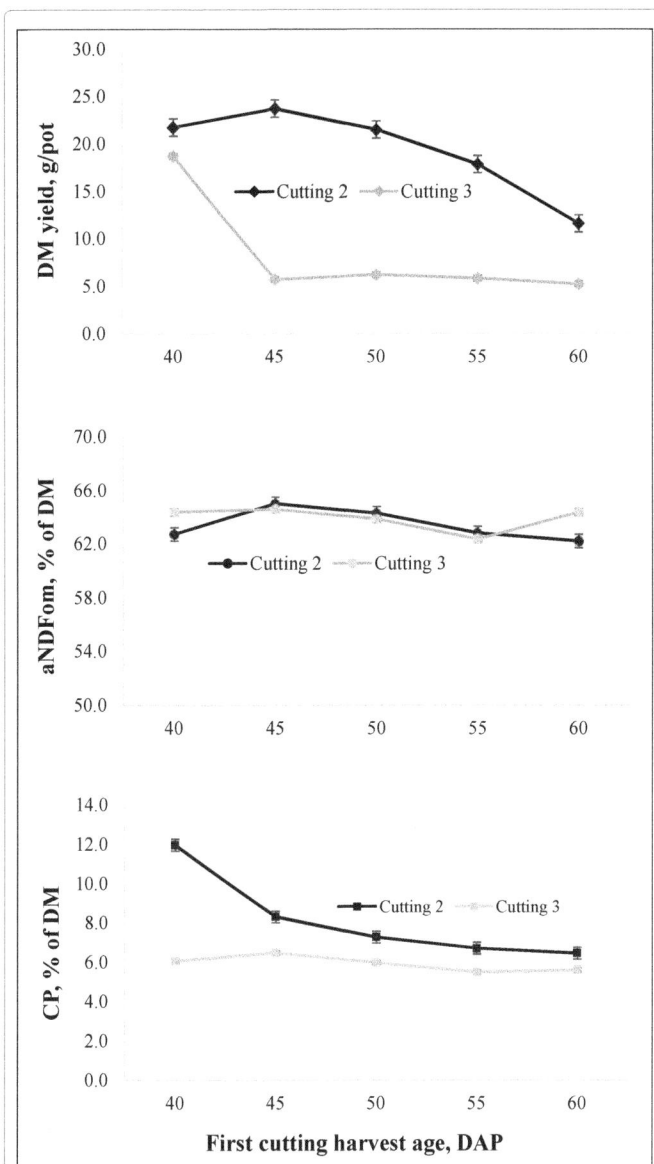

Figure 2: Effect of first-cutting harvest age on yield, nutritive values and digestibility of second- and third-cutting teff grass. For all pots, Cutting 2 was taken 30 d after Cutting 1. Cutting 3 was taken 30 d after Cutting 2. For Cuttings 2 and 3, first-cutting harvest age was a significant predictor (P<0.001) of DM yield and concentrations of aNDFom and CP.

Item	Variety				SEM	P-value
	Tiffany	Moxie	Corvallis	Dessie		
Cutting 1						
DM[1] yield, g/pot	14.77	14.56	14.48	13.83	0.40	0.38
aNDFom[2,3]	60.27	58.61	59.13	59.13	0.73	0.43
CP[2,4]	20.39	19.89	19.58	20.63	0.44	0.32
IVNDFD[5], %	51.21	49.66	50.39	51.81	0.85	0.32
Cutting 2						
DM yield, g/pot	19.48	20.12	19.12	18.30	0.82	0.47
aNDFom	62.70	63.33	64.21	63.37	0.44	0.13
CP	7.99	8.27	8.23	8.44	0.35	0.84
Cumulative DM yield (g/pot)	34.26	34.68	33.60	32.13	0.99	0.30
Cutting 3						
DM yield, g/pot	8.35	8.38	7.97	8.63	0.27	0.40
aNDFom	63.82	64.44	63.68	63.66	0.24	0.10
CP	5.87	5.83	5.96	5.98	0.08	0.48
Cumulative DM yield (g/pot)	42.61	43.06	41.57	40.76	1.15	0.49

[1]Dry matter

[2]Nutrients expressed as a percent of DM

[3]Ash-free neutral detergent fiber with amylase

[4]Crude protein

[5]In-vitro neutral detergent fiber digestibility

Table 1: Effect of teff variety on yield, nutritive values, and in-vitro digestibility of teff grass.

(Figure 3). After 2 cuttings, delaying the first cutting from 40 to 60 DAP significantly increased (P<0.001) total DM yield from 25.76 to 38.00 ± 1.11 g/pot. This was most likely due to the fact that the first-cutting yield from plants harvested at 40 and 45 DAP was so low that the cumulative yield for the early-cut plants was still less than that of the late-cut plants after 2 cuttings, despite having a relatively higher second-cutting yield. After 3 cuttings, however, an initial cutting age of 40 DAP yielded significantly more (P<0.01) DM than an initial cutting age of 45 DAP (44.47 vs. 38.15 ± 1.29 g/pot, or roughly 26 vs. 22 tons DM/ha) and numerically more DM than original cutting ages of 50, 55, and 60 DAP. After 3 cuttings, the advantage of harvesting a plant at an earlier maturity at Cutting 1 significantly outweighed the greater first cutting yield of a more mature plant. It is important to note that, although yield data collected from the greenhouse is useful for detecting differences among seed varieties and first-cutting harvest dates, yields observed in field trials do not typically match those observed in a controlled greenhouse setting.

Finally, across all teff varieties and cutting ages, Cutting 2 yielded significantly more DM (P<0.01) than Cuttings 1 and 3 and Cutting 1 yielded significantly more DM (P<0.001) than Cutting 3 (Table 2). Additionally, aNDFom concentration increased (P=0.01) and CP decreased (P<0.001) when cutting number increased from 1 to 3. Van Soest describes lignification as one of a plant's protective mechanisms against predatory attack or, in this case, a harvest event. As cutting number increases, then, it is expected that the concentration of the protective, fibrous component of teff would increase. This is supported by the fact that, as cutting number increased from 1 to 3, forage DM concentration, at harvest, increased (P<0.001) from 19.96 to 31.37 ± 0.92% (Table 2). Reid et al. reported a similar trend with smooth brome grass. As cutting number increased from 1 to 4, yield and digestibility

photosynthetic compounds are either stored or converted to structural material in plants. In a young plant, most of these photosynthetic compounds are stored. Stored nutrients are crucial for regrowth. When grasses are harvested during the late vegetative to early boot stage (40 to 45 DAP), these stored nutrients assist in the regrowth process and improve overall nutritive values. Grasses harvested during the boot to early heading stage (55 to 60 DAP), however, have already converted a large portion of these photosynthetic compounds to structural compounds. These structural compounds are mostly unavailable to the plant [19]. Therefore, after harvesting, mature plants have less nutrients available for regrowth, ultimately reducing subsequent yield and protein concentration while increasing the fiber concentration.

Delaying the first cutting from 40 to 60 DAP had a significant impact on the cumulative DM yielded over the course of the trial

Item	Cutting Number			SEM	P-values
	1	**2**	**3**		
DM yield, g/pot	14.58[b]	19.26[a]	8.33[c]	0.72	0.001
DM %	19.96[c]	26.72[b]	31.37[a]	0.92	<0.001
aNDFom[1]	59.40[b]	63.40[a]	63.80[a]	0.40	<0.001
CP[1]	19.97[a]	8.23[b]	5.90[b]	0.45	<0.001

[1]Expressed as a percent of DM

[a,b]Means with different superscripts are significantly different (P<0.05) by Tukey's HSD

Table 2: Effect of cutting number on yield and nutritive values of teff across all varieties and first-cutting harvest ages.

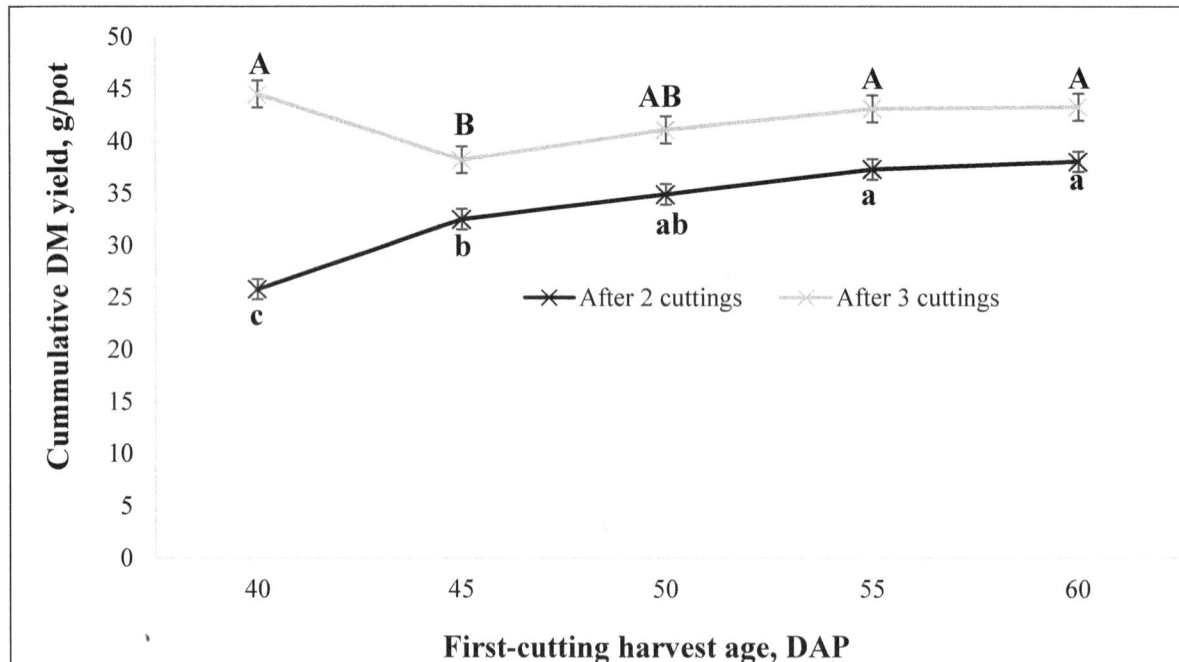

^{a,b}Means with different superscripts differ (P<0.05)

^{A,B}Means with different superscripts differ (P<0.05)

Figure 3: Effect of first-cutting harvest age on cumulative DM yielded from 3 cuttings.

tended to decrease while lignin concentration increased. The decrease in the CP concentration as cutting number increased could be due to both the increase in the fiber portion of the plant as well as the overall depletion of N and other key nutrients from the soil over time. While additional N (0.15 g of urea) was applied at d 60, N was not applied between Cuttings 2 and 3.

Conclusion

Results from this study indicate that, under greenhouse conditions, the first cutting of teff grass should be harvested at 45 to 50 DAP to optimize forage yield, quality, and digestibility in that cutting and in subsequent cuttings. For best results, N should be applied at planting and at every cutting to optimize regrowth and CP concentrations. Overall, the agronomic characteristics and nutrient profile of teff grass are similar those of other commonly fed forages like timothy [4], smooth brome grass, and sorghum-sudan grass. To use teff grass in the diets of a high producing dairy cow, maturity at first cutting and soil fertility must be well managed to ensure that the forage provided in the diet is of the highest quality and as valuable as possible to the animal.

Acknowledgements

Contribution no. 17-332-J from the Kansas Agricultural Experiment Station. Partial funding for this work was provided from the Kansas EPSCoR Program supported by National Science Foundation Grant No. EPS-0903806 and the State of Kansas. The authors would like to thank Byron Seeds (Rockville, IN) and Allied Seed (Macon, MO) for the donation of teff seed used in this trial. Additionally, we would like to thank Michael Stamm of Kansas State University for donating a section of his greenhouse space to this project.

References

1. Innovation Center for US Dairy (2013) US Dairy's Environmental Footprint.

2. Cross JA (2015) Change and Sustainability Issues in America's Dairyland. Focus on Geography 58: 173-183.

3. Mengesha MH (1966) Chemical composition of teff (*Eragrostis teff*) compared with that of wheat, barley and grain sorghum. Econ Botany 20: 268-273.

4. Miller D (2011) Teff grass: crop overview and forage production guide.

5. Bonsi MLK, Osuji PO, Tuah AK, Umunna NN (1995) Intake, digestibility, nitrogen balance, and certain rumen characteristics of Ethiopian Menz sheep

fed teff straw supplemented with cotton seed cake, dry sesbania, dry leucaena, or fresh leucaena. Agrofor Syst 31: 243-256.

6. Bonsi MLK, Tuah AK, Osuji PO, Nsahlai VI, Umunna NN (1996) The effect of protein supplement source or supply pattern on the intake, digestibility, rumen kinetics, nitrogen utilization, and growth of Ethiopian Menz sheep fed teff straw. Anim Feed Sci Technol 64: 1996.

7. Mesfin R, Ledin I (2004) Comparison of feeding urea-treated teff and barley straw based diets with hay based diet to crossbred dairy cows on feed intake, milk yield, milk composition and economic benefits. Livest Res for Rural Dev, p: 16.

8. Roseberg RJ, Norberg S, Smith J, Charlton B, Rykbost K, et al. (2005) Yield and quality of teff forage as a function of varying rates of applied irrigation and nitrogen. Klamath Exp Stn 2005 Annu Rep 119-136.

9. Roseberg RJ, Charlton BA, Norberg S, Kugler J (2006) Yield and forage quality of six teff seed brands grown in three Pacific Northwest environments. Klamath Exp Stn 2006 Annu Rep 31-45.

10. Young A, Creech E, ZoBell D, Israelsen C, Eun J (2014) Integrating Teff into Livestock Operations. Utah State Univ Febr 2014 Ext Rep.

11. Allen MS (1996) Relationship between forage quality and dairy cattle production. Anim Feed Sci Technol 59: 51-60.

12. Burns J, Bennett B, Rooney K, Walsh J, Hensley J (2002) Coatings for legume and grass seed. Proc 57th South Pasture Forage Crop Improv Conf Athens, GA April, pp: 23-25.

13. Van Soest PJ, Robertson JB, Lewis BA (1991) Methods for dietary fiber, neutral detergent fiber and nonstarch polysaccharides in relation to animal nutrition. J Dairy Sci 74: 3583-3597.

14. Guo D, Chen F, Wheeler J, Winder J, Selman S, et al. (2001) Improvement of in-rumen digestibility of alfalfa forage by genetic manipulation of lignin O-methyltransferases. Transgenic Res 10: 457-464.

15. Ballard CS, Thomas ED, Tsang DS, Mandebvu P, Sniffen CJ, et al. (2001) Effect of corn silage hybrid on dry matter yield, nutrient composition, in vitro digestion, intake by dairy heifers, and milk production by dairy cows. J Dairy Sci 84: 442-452.

16. Carmi A, Aharoni Y, Edelstein M, Umiel N, Hagiladi A, et al. (2006) Effects of irrigation and plant density on yield, composition and *in vitro* digestibility of a new forage sorghum variety, Tal, at two maturity stages. Anim Feed Sci Technol 131: 120-132.

17. Chen L, Auh C, Dowling P, Bell J, Chen F, et al. (2003) Improved forage digestibility of tall fescue (Festuca arundinacea) by transgenic down-regulation of cinnamyl alcohol dehydrogenase. Plant Biotechnol J 1: 437-449.

18. Assefa G, Ledin I (2001) Effect of variety, soil type and fertiliser on the establishment, growth, forage yield, quality and voluntary intake by cattle of oats and vetches cultivated in pure stands and mixtures. Anim Feed Sci Technol 92: 95-111.

19. Van Soest PJ (1982) Nutritional Ecology of the Ruminant. Corvallis: O & B Books Inc., Print.

20. Kilcher MR, Troelsen JE (1973) Contribution of stems and leaves to the composition and nutrient concentration of irrigated brome grass at different stages of development. Can J Plant Sci 53: 767-771.

21. Ademosum AA, Baumgardt BR, Scholl JM (1968) Evaluation of a sorghum-sudan grass hybrid at varying stages of maturity on the basis of intake, digestibility and chemical composition. J Anim Sci 27: 818-823.

22. Jung HG (1987) Forage lignins and their effect on fiber digestibility. Agron J 81: 33-38.

23. Mir Z, Acharya SN, Mir PS, Taylor WG, Zaman MS, et al. (1997) Nutrient composition, *in vitro* gas production and digestibility of fenugreek (*Trigonella foenum-graecum*) and alfalfa forages. Can J Anim Sci 77: 119-124.

24. Reid RL (1962) Investigation of plant species and maturity stage on forage nutritive value as determined by in vitro techniques. Final Report USDA Contract 12-14-100-4524. W Va Agr Exp Sta, Morgan Town.

5

Effects of Nitrogen and Phosphorous Fertilization on Western Flower Thrips Population Level and Quality of Susceptible and Resistant Impatiens

Yan Chen[1]*, Richard Story[2], and Michelle Samuel-Foo[3]

[1]*LSU Agricultural Center Hammond Research Station, 21549 Old Covington Highway, Hammond, LA 70403, USA*
[2]*LSU Agricultural Center Department of Entomology, 404 Life Sciences Building, LSU, Baton Rouge, LA 70803, USA*
[3]*University of Florida, IR-4 Southern Region Program, Gainesville, FL 3261, USA*

Abstract

Impatiens (Impatiens wallerana) cultivars 'Super Elfin Red' and 'Dazzler Violet', resistant or susceptible, respectively, to western flower thrips (Frankliniella occidentalis) were grown under 112 or 336 mg·L-1 N in combination with 10, 20, or 40 mg·L-1 P to investigate the effect of fertilization and host plant resistance on thrips population level and plant quality. Half of the plants were inoculated with thrips at two weeks after fertigation treatments began (WAT) and sampled at 4 and 8 WAT. For thrips-free plants, both N rates and the two higher P rates resulted in market-quality plants with various tissue N and P concentrations. Plant quality was lower in thrips-infested plants due to thrips damage to foliage as distortion on expending leaves and browning on edges of fully expended leaves. Higher numbers of adult and immature thrips were found in 'Dazzler Violet' than 'Super Elfin Red'. However, distortion index, which represented degree of distortion on young leaves, was higher in the resistant cultivar, and the two cultivars had similar quality ratings at 8 WAT. For both cultivars, N had no effect on thrips population, and plants fertilized with 20 or 40 mg·L-1 had higher number of thrips than the low rate. However, percentage of damaged leaf area, which represented the severity of browning, was found higher in plants fertilized at the low P rate. As a result, both cultivars fertilized with higher P rates had better plant quality although these plants had more thrips. Therefore, when infestation level is moderately low, i.e. 10 thrips per plant, plant nutrient status favoring thrips development may not necessarily result in lower plant quality. The final outcome of plant marketability is a combination of plant growth, thrips damage, and the ability of plants to compensate for pest damage.

Introduction

The western flower thrips (WFT), *Frankliniella occidentalis* Pergande, is one of the most serious pests of ornamental crops as well as many other crops throughout the world [1]. Both adult and immature feed on plants by piercing and rasping leaf or petal surface and withdrawing sap that exudes from injured cells, causing aesthetic injuries including lesions, discoloration, distortion, and dropping of leaves and floral buds [2]. In addition, WFT vectors impatiens necrotic spot virus (INSV) and several strains of tomato spotted wilt virus (TSWV) [3]. Thrips management in greenhouse productions has traditionally relied on a limited number of insecticides, and thrips resistance to carbamates, organophosphates, pyrethroids, and spinosad has been well documented [4-6]. Alternative approaches that incorporate available cultural, biological, and chemical measures are needed for effective crop protection as well as managing pest resistance development.

Host plant resistance to WFT have been identified in ornamental crops, i.e., chrysanthemum, eustoma [7], impatiens [8], miniature rose [9], cut rose [10], and verbena [11]. Warnock [8] reported significant lower level of WFT damage in resistant impatiens after being inoculated with 30 thrips and grown for eight weeks. Of these cultivars, some had high thrips population levels, indicating tolerance, while others had low thrips population levels, an indication of antibiosis. One of the possible mechanisms for host plant resistant is the content of secondary metabolic chemicals (i.e. phenols) in the plant tissue that might be altered by fertilization. However, whether or not production practices such as nitrogen (N) and phosphorous (P) fertilization rates would alter host plant resistance has not been studied for ornamental crops [12].

Previous research suggests that manipulating fertilization levels can impact thrips populations in row crops through changing hosts nutrient content [13]. Either positive or no effects of N fertilization has been reported for thrips. However, many of these reports are based on experiments where insect responses to plants deficient in N are compared to plants with sufficient or luxurious N concentrations [14,15]. Such results have limited application to commercially produced crops because N and P are more likely to be over-applied for a faster crop cycle and better economic returns [16]. Fewer studies have investigated insect population growth in response to P fertilization. Chen et al. [17] reported that a slight population increase was observed when tissue P concentration increased and became luxurious in impatiens plants.

Impatiens is one of the top three warm season bedding plants, contributing a wholesale value of $62 million to the floriculture industry in 2011 (USDA NASS 2011 Floriculture Summary, 15-state data). Feeding and oviposition injuries caused by WFT on impatiens include distorted young leaves and brown areas on leaf edges. Level of injury is affected by cultivar and also growing stage because of WFT's preference for nectar and pollen [17] thus flowering plants may have less damage on their foliage than plants at vegetative stages. If plant marketability is affected by cultivar, fertilization, and pest level, interactions among these factors need to be determined before they can

*Corresponding author: Yan Chen, LSU Agricultural Center Hammond Research Station, 21549 Old Covington Highway, Hammond, Los Angeles, USA
E-mail: yachen@agcenter.lsu.edu

be manipulated for thrips management. Therefore, the objective of this study was to determine the effects of N and P fertilization on WFT population level and damage in resistant vs. susceptible impatiens.

Materials and Methods

The study was conducted at Louisiana State University Agricultural Center Hammond Research Station, Hammond, LA in 2010. 'Super Elfin Red' and 'Dazzler Violet' seeds were sown in MetroMix300 potting mix (mixture of sphagnum peat moss and perlite, SunGro Horticulture Canada Ltd.) on 5 Mar. Seeds were not covered by potting mix and exposed to a natural photoperiod of about 14:10 day/night to germinate. Seedlings were grown in a greenhouse covered with 60% AlumiNet (Green-Tek, Dinuba CA) for three weeks and then individually transplanted to 4 inch (8.5-cm) square pots (280 ml) on 2 April using MetroMix360, which was formulated with sphagnum peat moss, coarse perlite, and fine bark. The potting mix was amended with 4.7 kg dolomitic limestone per m^3 to adjust pH. Plants were placed into thrips-proof cages, one plant per cage to avoid natural infestation by thrips and other pests. Cages were constructed from 5-gal (19 L) plastic buckets with four windows, each 6 x 12 inch, cut into the sidewall then covered with no-thrips screen (GreenTek, Janesville, WI) and sealed with silicone. The top of the bucket was covered by a muslin cloth and held in place by rubber bands. The study was conducted in a research greenhouse with temperatures set at 26°C/20°C day/night. Actual temperature and relative humidity in the greenhouse were recorded by HOBO sensors (Onset Computer Corp., Bourne, MA). During the 8-week experiment, temperatures ranged from 18.8 to 31.4°C with an average daily temperature of 24.5 ± SD of 3.1°C; and the relative humidity ranged from 23.8% to 93.6%, with an average of 56.2% ± 18.4%.

Application of six experimental nutrient solutions started on the day of transplant (week 1) as combinations of 112 or 336 mg·L^{-1} N by 10, 20, or 40 mg·L^{-1} P. These rates were chosen because previous experiments had shown that applications of N and P within these ranges produced commercial quality impatiens [17]. The selected N rates are also close to those recommended by Dole and Wilkins [18] for impatiens. Nitrogen sources were NO_3^- and NH_4^+ at a rate of 3:2. Phosphorous source was KH_2PO_4 and other essential and micro nutrients were held constant among solutions at the following rates: K at 6 mM, Ca at 0.65 mM, Mg at 1.5 mM, Fe at 1.5 mg·L^{-1}, Cu at 0.24 mg·L^{-1}, Zn at 0.19 mg·L^{-1}, and Mn at 0.43 mg·L^{-1}.

Water source for the nutrient solution was municipal (Hammond LA) with a pH at 8.3 and alkalinity at 31 mg·L^{-1}. Fertigation was delivered by pumping single-strength nutrient solutions to individual pots inside sealed cages through spaghetti tubing. A submersible pump (8.8 L per min, Little Giant Pump Co., Oklahoma City, Ok) was housed in 120 L containers filled with one of the six nutrient solutions. Timing of fertigation was determined by weighing six non-experimental plants that were grown under the same cultural practices. Benches were fertigated when the weight of sample pots dropped by 20% ± 5% from their container capacity (weight after irrigation and leaching) due to water loss. This procedure resulted in plants receiving fertilizer solution about twice per week at the beginning to almost daily as plants reached marketable sizes. Leaching fraction was maintained at ~15% at each fertigation.

A virus-free WFT colony was obtained from the Department of Entomology, Kansas State University and reared on green beans in plastic containers (20×14×10 cm) at 24°C and relative humidity of 50%. Adult and immature stages were maintained in separate containers to produce large numbers of even-aged thrips. Prior to inoculation, containers with thrips were placed in a refrigerator at 6°C for 20 minutes to slow their movement. Plants were inoculated with ten female adults at two weeks after fertigation began. Cages were resealed after inoculation.

The experimental design was a completely randomized block design (CRBD) with a total of 24 treatment combinations [2 N × 3 P × 2 cultivar × 2 thrips densities (0 and 10)] and 8 replications (blocks). The 8 replications were arranged on eight benches in a research greenhouse. The benches were arranged parallel to the cooling pads so that blocking accounted for the temperature gradient in the greenhouse. A total of 24 cages with 12 plants of each cultivar were arranged on each bench, with six nutrient treatments and two inoculation densities randomly assigned to them.

Four replications were each sampled on 4 May and 2 June (weeks 4 and 8). At the time of sampling, a plant was removed from its cage and placed in a plastic container (34 × 25 × 15 cm). Plant heights, widest width, perpendicular width to the widest width, and number of flowers were recorded. An overall quality rating (QR) was given to the plant using a scale from 1 to 10 by taking into consideration of plant size and form, leaf greenness and glossiness, and number and size of flowers. Plants meeting commercial size were assigned a rating between 7 and 10 and then classified into one of three market ratings (MR): "premium" that plants with highest market quality (MR ≥2, QR between 9 and 10), "Discounted" plants that likely to be marketed at a lower price (MR ≥ 1, QR between 8 and 9), and "unmarketable" plants that likely will be discarded by growers (MR ≤ 1, QR between 7 and 8).

Another typical injury symptom caused by WFT on many ornamental plants is distortion on young leaves, and a distortion index was developed and used on ivy geraniums to describe this injury [19]. Leaf distortion in impatiens caused by thrips feeding is usually observed on young leaves that cannot expand normally because of feeding scars. Each plant was assessed for the severity of distortion on three stems using a rating scale of 0 to 3 to represent no distortion (0), and slight (1), moderate (2) and severe (3) distortion. Total distortion was then calculated as distortion index (DI) which was the average from the ratings of three stems.

Plants were then destructively sampled for thrips by cutting each plant at its base and laying it horizontally in a container that was divided into three sections by inserting two plastic boards across the width of the container. Each stem was then cut based on these leaf categories: stem with young leaves (Y), stem with fully-expanded leaves (FE), and stem with old leaves (O) and put into the three sections in the container. The container was shaken for ten times and thrips dislodged into each section was counted by a tally counter. The Y stratum with the meristems was then washed with a 48% ethanol stream and then filtered and checked under a 10x microscope (Micro Master, Fisher, USA) for thrips. Thrips in flowers and buds were individually counted by destructively peeling back petals. Numbers of adult and immature (nymph, pre-pupae, and pupae) stages were recorded separately, and total WFT number was the sum of adult and immature.

Leaves were then detached from stems and scanned by a HP ScanJet 3100C (Hewlett-Packard Co., Palo Alto, CA) and the acquired images were analyzed for Percent Damaged Area (PDA) as described by Chen and Williams [20] that can be distinguished from normal leaf tissue by digital imaging and computer software [20]. Damaged area of a plant is quantified and better correlated with thrips population than subjective visual damage ratings. The three strata were scanned and analyzed separately. The PDA for the entire plant was computed as total damaged area divided by total leaf area.

After the images were acquired, leaf tissue samples were collected from FE leaves, which can best represent plant nutrient status at the time of sampling. Tissue samples were washed and dried at 70°C for 48 h, then ground in a stainless steel Wiley mill to a particle size<1 mm. Tissue analyses were conducted by Louisiana State University Soil Testing and Plant Analysis Laboratory (Baton Rouge, LA). Tissue N concentration was determined by an LECO TruSpec CN nitrogen analyzer (LECO Corp., St. Joseph, MI), and tissue P was determined by ICP-emission spectroscopy (Fisons Instruments, Dearborn, MI).

All data were subjected to normality check and those that failed were transformed using appropriate means to improve normality based on suggestions from Hartwig and Dearing [22]. LSMEANS were back-transferred after analysis. Effects of treatment factors on thrips population levels were tested by PROC MIXED and the LSMEANS statement was used to compute means for each effect (SAS software v 12.1, SAS Institute, 2001). The MIXED model for analysis included treatment factors and their interactions; block, block x N x P, and block x cultivar x N x P were included as random effects.

Results and Discussions

For all dependable variables, interactions among treatment factors were not significant or significant but did not affect the main effects of treatment factors.

Plant growth and tissue N% and P%: When comparing the two cultivars, although 'Super Elfin Red' had greater DW than 'Dazzler Violet' at 4 weeks after fertigation treatment (WAT), they had similar DW at 8 WAT and were similar in plant size (SI) at 4 and 8 WAT (Table 1). 'Dazzler Violet' was higher in tissue P% than 'Super Elfin Red' at 4 WAT and was higher in both N% and P% at 8 WAT. 'Super Elfin Red' had more flowers than 'Dazzler Violet' at 8 WAT (13 vs. 3.2).

Nitrogen rate at 336 mg·L^{-1} resulted in similar plant SI and DW but higher tissue N% at both sample dates compared to 112 mg·L^{-1} and (Table 1). Tissue P% was affected by N rate at 8 WAT that plants fertilized with the higher N rate had slightly higher tissue P concentration than those fertilized at the low N rate. This is possibly a result of more update of both N and P at higher N rate. Similarly, planted fertilized at higher P rates had higher tissue N% than those fertilized at lower P rates, i.e. 10 mg·L^{-1}.

Phosphorus at the low rate (10 mg·L^{-1}) resulted in smaller plants, lower tissue N% and P%, and fewer flowers compared with plants fertigated at higher P rates at both sample dates (Table 1). Therefore, the 2N x 3 P nutrient treatments resulted in similar marketable plants (except those at the low P rate) but varying tissue N% (ranging from 3.5% to 4.3% at 4 WAT and 3.3% to 4.2% at 8 WAT) and P% (from 0.21% to 0.52% at 4 WAT and 0.16% to 0.48% at 8 WAT).

Inoculating 10 thrips at 2 WAT did not affect plant growth and flowering at 4 or 8 WAT, suggesting that thrips infestation at this level does not adversely affect growth and flowering of both cultivars.

More thrips were found on 'Dazzler Violet' than on 'Super Elfin Red' especially the number of immature at 4 WAT (20.2 vs. 9.2) and both adults and immature at 8 WAT (22.8 vs. 9.1 and 30.8 vs. 12.4, Table 2). This confirmed that 'Super Elfin' impatiens is more resistant to thrips. Evaluation of onion thrips on onion cultivars grown in the field showed low population level on resistant cultivars, indicating a combination of antibiosis and/or antixenosis [22]. In the current

study, because thrips were inoculated onto plants held in individual cages and was not given a choice between the two cultivars, the low population level found on 'Super Elfin' indicated antibiosis as the possible mechanism for resistance, meaning that 'Super Elfin Red' was not as suitable of a host as 'Dazzler Violet'. A coincidence was that tissue P% in Dazzler Violet' was higher than 'Super Elfin Red' at both sample dates (0.42 vs. 0.32% and 0.38 vs. 0.24%; (Table 1). Similar results was reported by Zhi et al. [22] that resistant impatiens 'Cajun Carmine' had significant lower number of WFT and two spotted spider mite (*Tetranychus urticae* Koch) compared to a susceptible cultivar 'Impulse Orange', which also indicated antibiosis. However, tissue nutrient concentrations were not measured in their study.

Percent Damaged Area was an indication of thrips damage. 'Dazzler Violet' had higher PDA than 'Super Elfin Red' at 8 WAT, possibly due to the higher number of thrips on this cultivar. 'Super Elfin Red' had higher DI in both inoculated and thrips-free plants comparing with inoculated and thrips-free 'Dazzler Violet', respectively (data not shown). As a result, across thrips inoculation rate and fertilization levels, 'Super Elfin Red' and 'Dazzler Violet' had similar visual QR and MR at both sample dates (Table 2).

Across cultivar and inoculation rate, N fertilization had no effect on thrips population at 4 WAT, however, number of immature was higher in plants fertilized with the low rate than those with 336 mg·L^{-1} N (25 vs. 18, Table 2). Hunt et al. [12] reported that N ranging from 50 to 100 mg·L^{-1} was suitable for WFT development than lower or higher rates. A moderate N rate is more favorable for pest population growth because N is not a limiting factor, while at relatively high N rates; accumulation of certain secondary metabolites, i.e., phenolic acids, can negatively impact insect growth and reproduction.

Effect of P on WFT population level was not significant at 4 WAT, however, at 8 WAT, for both cultivars, number of adult were significantly higher in plants fertilized with 20 or 40 mg·L^{-1} than those fertilized with the low P rate (19 and 18 vs. 12, Table 2). Numbers of immature were higher in plants fertilized with 20 mg·L^{-1} than those at 10 mg·L^{-1} P (25 vs. 18) but similar to those fertilized at 40 mg·L^{-1}. Chen et al. [17] reported similar P effects with 'Dazzler Violet', where plants fertilized with 40 mg·L^{-1} P had marginally more adult and immature WFT than plants fertilized at 10 mg·L^{-1}. Our results indicate that this trend was consistent in both resistant and susceptible cultivars.

The percentage of damaged leaf area (PDA) of thrips-free plants was about 1.1% at both sample dates indicating that there are other factors affecting PDA in addition to thrips damage. Plants inoculated with 10 adult thrips developed higher PDA than thrips-free plants, and increased from 3.8% at 4 WAT to 5% at 8 WAT (Table 2). These PDA values represented minor but noticeable browning on leaf edges. 'Dazzler Violet' had higher PDA than 'Super Elfin Red' at 8 WAT possibly because of higher numbers of thrips on this cultivar. Fertilization had no effect on PDA at both sample dates except that plants fertilized with 20 mg·L^{-1} P had lower PDA than those at 10 mg·L^{-1} at 4 WAT despite similar population levels at these rates. A possible explanation is that plants fertilized at higher P rates may provide higher nutrients per feeding area thus reduced the overall leaf area needed for supporting similar population levels.

The distortion index is a measure of how easily distorted young leaves can be noticed by bare eyes. Inoculation with 10 thrips did not affect distortion index (DI) in young leaves until 8 WAT (Table 2). Because DI was also found in thrips-free plants (i.e., 0.49 vs. 0.68 in thrips-free and inoculated plants, respectively), apparently, some

Treatment[z]	4 WAT				8 WAT				
	SI cm	DW g	N%	P%	SI cm	DW g	Flower no.	N%	P%
Cultivar									
Super Elfin Red	20.3	4.1 [a]	3.9	0.32 [b]	24.3	7.9	13 [a]	3.6 [b]	0.24 [b]
Dazzler Violet	19.3	3.6 [b]	3.9	0.42 [a]	23.7	7.4	3.2 [b]	3.8 [a]	0.38 [a]
LSD$_{05}$[y]	NS	0.5	NS	0.03	NS	NS	1.2	0.1	0.03
N									
112	19.9	3.7	3.5 [b]	0.38	24.2	7.7	7.6	3.3 [b]	0.29 [b]
336	19.8	3.9	4.3 [a]	0.36	23.9	7.6	7.2	4.2 [a]	0.33 [a]
LSD$_{05}$[x]	NS	NS	0.1	NS	NS	NS	NS	0.1	0.03
P									
10	17.9 [b]	3.2 [b]	3.7 [b]	0.21 [c]	22.0 [b]	6.4 [b]	6.6 [b]	3.5 [b]	0.16 [c]
20	20.8 [a]	4.2 [a]	3.9 [a]	0.38 [b]	25.5 [a]	8.7 [a]	9.0 [a]	3.8 [a]	0.28 [b]
40	20.7 [a]	4.0 [a]	4.0 [a]	0.52 [a]	24.6 [a]	7.8 [a]	8.4 [a]	3.9 [a]	0.48 [a]
LSD$_{05}$[w]	1.7	0.6	0.1	0.04	1.7	1.0	1.9	0.1	0.03
Thrips									
0	19.8	3.8	3.8	0.33	24.1	7.8	7.6	3.7	0.28
10	19.8	3.8	3.9	0.38	24.0	7.6	6.4	3.8	0.31
LSD$_{05}$[v]	NS	NS	NS	NS	NS	NS	NS	NS	NS

Table 1: Plant size index (SI), dry weight (DW), number of flowers, and tissue N and P concentrations in the youngest fully expended leaves of 'Super Elfin Red' and 'Dazzler Violet' impatiens at 4 or 8 weeks after transplant (WFT) and being fertigated with 2N x 3P nutrient solutions and inoculated with 10 thrips or kept thrips-free.

[z]Interactions among treatment factors were not significant for all dependent variables.
[y]Least significant difference of the least square means of each cultivar, α=0.05, n=48.
[x]Least significant difference of the least square means of each N rate, n=32.
[w]Least significant difference of the least square means of each P rate, n=48.
[v]Least significant difference of the least square means of each thrips inoculation rate, n=48.

Treatment[z]	4 WAT						8 WAT					
	Adult	Immature	PDA%	DI (0 to 3)	QR (1 to 10)	MR (1 to 3)	Adult	Immature	PDA%	DI (0 to 3)	QR (1 to 10)	MR (1 to 3)
Cultivar												
Super Elfin Red	2.0 [b]	9.2 [b]	3.0	0.99 [a]	7.9	2.0	9.1 [b]	12.4 [b]	3.7 [b]	1.03 [a]	7.8	2.1
Dazzler Violet	3.3 [a]	20.2 [a]	3.3	0.31 [b]	7.7	2.0	22.8 [a]	30.8 [a]	4.7 [a]	0.30 [b]	7.7	2.0
LSD$_{05}$[y]	0.7	3.3	NS	0.14	NS	NS	3.6	5.1	0.7	0.14	NS	NS
N												
112	2.8	16.0	3.30	0.70	7.7	2.1	17.7	25.0 [a]	425	0.68	7.6	2.1
336	2.4	13.4	3.00	0.57	7.9	1.9	14.2	18.2 [b]	4.11	0.60	7.9	1.9
LSD$_{05}$[x]	NS	NS	NS	NS	NS	NS	NS	5.1	NS	NS	NS	NS
P												
10	2.4	13.0	3.49 [a]	0.66	7.4 [b]	1.9 [b]	11.5 [b]	17.5 [b]	4.37	0.65	7.2 [c]	1.8 [b]
20	2.3	13.5	2.78 [b]	0.55	8.0 [a]	1.9 [b]	18.5 [a]	24.7 [a]	3.84	0.64	8.2 [a]	2.0 [b]
40	3.1	17.8	3.19 [ab]	0.69	8.0 [a]	2.3 [a]	17.9 [a]	22.5 [ab]	4.35	0.63	7.8 [b]	2.3 [a]
LSD$_{05}$[w]	NS	NS	0.54	NS	0.3	0.2	4.4	6.2	NS	NS	0.4	0.25
Thrips												
0	0.2 [b]	1.5 [b]	1.1 [b]	0.54	8.1 [a]	2.1 [a]	0.5 [b]	0.5 [b]	1.1 [b]	0.49 [b]	8.2 [a]	2.1 [a]
10	3.4 [a]	19.1 [a]	3.8 [a]	0.66	7.7 [b]	1.6 [b]	19.8 [a]	26.5 [a]	5.0 [a]	0.68 [a]	7.7 [b]	1.6 [b]
LSD$_{05}$[v]	0.8	3.9	0.5	NS	0.3	0.2	4.5	6.4	0.8	0.18	0.4	0.3

Table 2: Number of adult and immature thrips per plant, distortion index (DI), percent damage area (PDA), visual quality rating (QR) and market rating (MR) of 'Super Elfin Red' and 'Dazzler Violet' impatiens at 4 or 8 weeks after transplant (WFT) and being fertigated with 2N x 3P nutrient solutions and inoculated with 10 thrips or kept thrips-free.

[z]Interactions among treatment factors were not significant for all dependent variables.
[y]Least significant difference of the least square means of each cultivar, α=0.05, n=48.
[x]Least significant difference of the least square means of each N rate, n=32.
[w]Least significant difference of the least square means of each P rate, n=48.
[v]Least significant difference of the least square means of each thrips inoculation rate, n=48.

distortion was caused by environmental conditions. For example, we observed that 'Super Elfin Red' tends to have more distorted young leaves when grown at low temperatures. Correlation analyses indicated that there was no relationship between DI and thrips numbers (p=0.3822) with a data set including both cultivars and N/P rates. Therefore, PDA, which was significantly correlated to total number of thrips (p=0.0325), is a more relevant measurement for quantifying thrips damage.

Thrips feeding negatively affected plant visual quality rating (QR) compared with thrips-free plants at 4 and 8 WAT (~7.7 vs. 8.2, Table 2). Plant marketability was also affected by thrips injury in that, thrips-infested plants were of lower MR than thrips-free plants and had to be marketed at a discounted price (MR<2). Therefore, to avoid economic loss in both cultivars, control action is needed within two weeks of infestation with 10 adults per plant. This is consistent with the economic injury threshold recommended for impatiens by Frey [23].

Nitrogen fertilization had no effect on QR and MR, which was expected because of similar plant growth, PDA, and DI between the two N rates. Phosphorus rate at 40 mg·L^{-1} resulted in higher MR than 10 or 20 mg·L^{-1} although plants fertilized with 40 mg·L^{-1} had more thrips than those fertilized at 10 mg·L^{-1}, suggesting that plants having higher tissue P% may be able to sustain higher pest pressure than those with lower tissue P% without showing more severe damage. Another possibility is that plants having higher tissue P% may be able to compensate for thrips damage better than those having low tissue P%. Thus This manipulate P fertilization might provide an advantage to IPM programs where it takes longer for bio pesticides to be effective, or when a low pest number is needed to maintain biological control agents on crops to be protected from WFT damage.

Population distribution of WFT on both cultivars varied between the two sample dates because plants were at vegetative growth at 4 WAT, and both cultivars began to flower at 8 WAT with the resistant 'Super Elfin Red' having more flowers (Table 1). Therefore, a larger portion of the population (both adult and immature) was found in 'Super Elfin Red' flowers than 'Dazzler Violet' at 8 WAT (Figure 1). Preference for flowers (nectar and pollen) was reported in impatiens [17], tomato [24], and cucumber [25]. Among the three leaf strata, more adult and immature thrips were found in FE than in Y or O (Figure 1). Therefore, thrips sampling plan should consider these distribution patterns for various growing stages of the crop. For example, FE leaves can be sampled for thrips monitoring at vegetative growing stage, and flowers can be sampled to determine population level when plants are in full bloom.

In summary, fertilizer applications with N from 112 to 336 mg·L^{-1} and P from 20 to 40 mg·L^{-1} can be used to produce quality 'Super Elfin Red' and 'Dazzler Violet' impatiens. Nitrogen had no positive effect on thrips population at these rates, while P rates resulted in more numbers of thrips compared to a lower rate at 10 mg·L^{-1} which did not grow quality plants. Resistant cultivar 'Super Elfin Red' had significantly less number of thrips and less leaf browning than the susceptible cultivar 'Dazzler Violet'. Plant quality however, was an outcome from the combined effects of plant nutrient status and pest damage and was similar between cultivars and N rates, and higher in thrips-free and plants fertilized at 20 mg·L^{-1} [26-28]. Based on these results, we would recommend using resistant cultivars in combination with P rates at about 20 mg·L^{-1} for an impatiens IPM program.

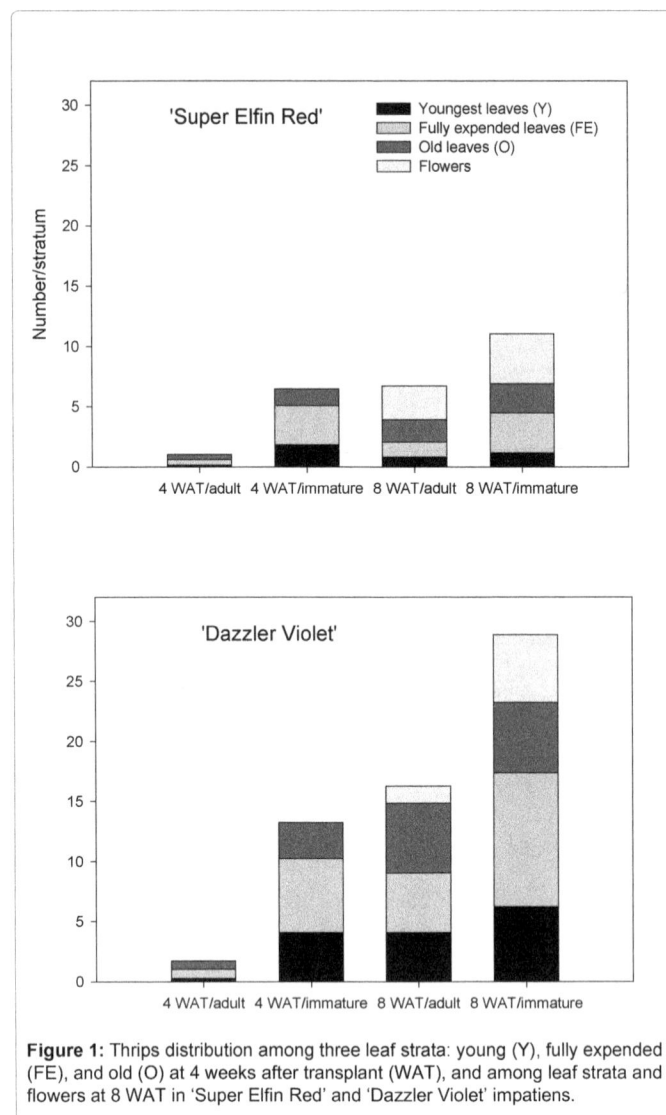

Figure 1: Thrips distribution among three leaf strata: young (Y), fully expended (FE), and old (O) at 4 weeks after transplant (WAT), and among leaf strata and flowers at 8 WAT in 'Super Elfin Red' and 'Dazzler Violet' impatiens.

Acknowledgement

We thank Joey Quebedeaux and Ashley Edwards for technical assistance, and Ball Horticultural Company for providing impatiens seeds. This work was financially supported by the USDA NIFA Pest Management Alternative Program Grant no: 2009-34381-20019.

References

1. Lewis T (1997) Pest thrips in perspective. 1-15.

2. Childers CC, Achor DS (1995) Thrips feeding and oviposition injuries to economic plants, subsequent damage and host response to infestation. 276: 31-52.

3. Ullman DE, Casey CA, Whitfield AE, Campbell LR, Robb KL et al. (1998) Thrips and tospoviruses: present and future strategies for management. Proc. Brighton Crop Protection Conference: Pest and Disease 2: 391-400.

4. Immaraju JA, Paine TD, Bethke JA, Robb KL, Newman JP (1992) Western flower thrips (Thysanoptera: Thripidae) resistance to insecticides in coastal California greenhouses. J Econ Entom 85: 9-14.

5. Loughner RL, Warnock DF, Cloyd RA (2005) Resistance of greenhouse, laboratory, and native populations of western flower thrips to spinosad. HortScience 40: 146-149.

6. Robb KL, Newman J, Virzi JK, Parrella MP (1995) Insecticide resistance in western flower thrips. 276: 341-346.

7. Morishita M (2003) Varietal difference in preference exhibited by western flower thrips Frankliniella occidentalis (Pergande) in Eustoma russelianum. Proc. Kansai Plant Protection Soc. 45: 1-4.

8. Warnock DF (2003) Resistance to western flower thrips feeding damage in impatiens populations from Costa Rica. HortScience 38: 1424-1427.

9. Bergh JC, Le Blanc JPR (1997) Performance of western flower thrips (Thysanoptera: Thripidae) on cultivars of miniature rose. J Econ Entom. 90: 679-688.

10. Bolkan L, Casey C, Newman JP, Robb K, Tjosvaold S (2001) IPM works for rose growers. Grower Talks. 65: 52-58.

11. Warnock DF, Loughner R, Bertschinger L, Anderson JD (2004) Verbena cultivars differentially attract adult western flower thrips. Acta Horticulturae 638: 89-94.

12. Hunt D, Carter N, Drury C (1999) The influence of nitrogen on seedless cucumber resistance and susceptibility to western flower thrips. Acta Horticulturae 481: 561-567.

13. Waring GL, Cobb NS (1989) The impact of plant stress on herbivore population dynamics. VI: 167-225. In: E.A. Bernays (edn.) Insect-plant interaction. CRC Press, Boca Raton, FL.

14. Mollema C, Cole RA (1996) Low aromatic amino acid concentrations in leaf proteins determines resistance to Frankliniella occidentalis in four vegetable crops. Entomol Expt et Applicata. 78: 325-333.

15. Schuch UK, Redak RA, Bethke JA (1998) Cultivar, fertilizer, and irrigation affect vegetative growth and susceptibility of chrysanthemum to western flower thrips. J Amer Soc Hort Sci. 123: 727-733.

16. Nelson PV (1989) Developing root management strategies to minimize water and fertilizer waste: the United States perspective with emphasis on surface applied non recirculated systems. Acta Hort 272:175-184.

17. Chen Y, Williams KA, Harbaugh BK, Bell ML (2004) Effects of tissue phosphorus and nitrogen in Impatiens wallerana on western flower thrips (Frankliniella occidentalis) population levels and plant damage. HortScience 39: 545-550.

18. Dole JM, Wilkins HF (1999) Floriculture – Principles and Species. Prentice-Hall, Inc., Simon & Scuster/A Viacom Company. Upper Saddle River, New Jersey.

19. Opit GP, Chen Y, Williams KA, Nechols JR, Margolies DC (2005) Plant age, fertilization, and biological control affect damage caused by twospotted spider mites on ivy geranium: development of an action threshold. J Amer Soc Hort Sci 130: 159-166.

20. Chen Y, Williams KA (2006) Short communication: Quantifying western flower thrips (Frankliniella occidentalis Pergande) (Thysanoptera: Thripidae) damage on ivy geranium (Pelargonium peltatum L.) (Geraniaceae Juss.) with Adobe Photoshop™ and Scion® Image software. J Kansas Entom Soc 79: 83-87.

21. Hartwig F, Dearing BE (1979) Exploratory Data Analysis. Newberry Park, CA. Sage Publications, Inc.

22. Diaz-Montano J, Fuchs M, Nault BA, Shelton AM (2010) Evaluation of onion cultivars for resistance to onion thrips (Thysanoptera: Thripidae) and Iris yellow spot virus. J Econ Entomol 103: 925-937.

23. Frey JE (1993) Damage threshold levels for western flower thrips, Frankliniella occidentalis (Perg.) (Thysanoptera: Thripidae) on ornamentals. International Organization for Biological Control of Noxious Animals and Plants (IOBC/OILB), West Palaearctic Regional Section (WPRS/SROP), Bulletin OILB/SROP 16: 78-81.

24. Reitz SR (2002) Seasonal and within plant distribution of Frankliniella thrips (Thysanoptera: Thripidae) in North Florida tomatoes. Florida Entomologist 85: 431-439.

25. De Kogel WJ, Van der Hoek, Dik MTA, Gebala B (1997) Seasonal variation in resistance of chrysanthemum cultivars to Frankliniella occidentalis (Thysanoptera: Thripidae). Euphytica 97: 283-288.

26. Zhi J, Margolies DC, Nechols JR, Boyer Jr JE (2006) Host-plant-mediated interaction between populations of a true omnivore and its herbivorous prey. Entomologia Experimentalis Applicata 121: 59-66.

27. Yang T, Stoopen G, Thoen M, Wiegers G, Jongsma MA (2013) Chrysanthemum expressing a linalool synthase gene 'smells good', but 'tastes bad' to western flower thrips. Plant Biotechnol J 11: 875-882.

28. USDA National Agricultural Statistics Service (2011) United States Department of Agriculture Floriculture Crops 2011 Summary. Washington, D. C.

6

Agropastoralist Evaluations of Integrated Sorghum Crop Management Packages in Eastern Ethiopia

Tadeos Shiferaw*, Fano Dargo and Abdurhman Osman

College of Dry Land Agriculture, Department of Dryland Crop Sciences, Jigjiga University, P.O. Box, 1020, Jigjiga, Ethiopia

Abstract

On farm participatory field experiment was conducted in 2013 cropping season in districts located in eastern Ethiopia to evaluate effect of different sorghum crop management packages on grain and fodder yield of improved variety *Teshale* and local check *Elmijam*. Six different sorghum management alternatives along with agro-pastoralist indigenous management practice were evaluated in six randomly selected agro-pastoralist fields in Ethiopian Somali province Fafen administrative Zone. The result reviled that compared to agro pastoralist indigenous practices on both varieties improved production practice had significantly increased fodder and grain yield of sorghum by 60-70%. The result also showed significant varietal differences between improved variety *Teshsale* and the local check *Elimjema* in all aspects. Therefore based on agropastoralist interest and rating production packages composed of improved sorghum and local check variety, tide-ridge planting, fertilizer (urea and DAP application at 50 kg/ha and 100 kg/ha), hand weeding once at 45 days after emergence and with recommended seed rate 10 kg/ha were selected as best management package because it balanced both grain and fodder yield with the production costs.

Keywords: Participatory; Evaluations; Sorghum; Integrated; Management; Packages

Introduction

Sorghum (*Sorghum bicolor* (L.) Moench) is one of the most important cereal crops in the semi-arid tropics Catherine et al. [1]. It is the most widely grown cereal crops in Ethiopia is a staple food crop on which the lives of millions of poor Ethiopians depend. It has tremendous uses for the Ethiopian farmer and no part of this plant is ignored Asfaw [2]. Sorghum grows in a wide range of agro ecologies most importantly in the moisture stressed parts where other crops can least survive and food insecurity is rampant. Ethiopian national average yield is 1.302 ton/ha it far lower than other developing country and the world. The low productivity of sorghum in Ethiopia could be attributed to biotic and edaphic factors affecting directly and indirectly sorghum production and productivity Tekle and Zemach [3].

Sorghum is drought tolerant, widely produced and popular cereal crop in pastoralist and agro-pastoralist communities in eastern Ethiopia in Somali regional sate Mahdi et al. [4]. The target area is characterized by erratic rainfall and recurrent drought therefore the yield obtained is significantly lower than the other parts of the county. Mekbib F [5] reported that the low productivity of sorghum in this area could be due to the low level of sorghum research and low input production systems. Beside its popularity in Eastern part of the country its productivity is constrained by numerous factors. Among the factors including unspecified planting density, unfavorable rainfall patterns, soil infertility, and low yield potential of available varieties at farmers hand, poor crop management skills and low extensions systems contributed a lot in yield reduction and resulting food insecurity in the study area.

Traditional sorghum management practice in the study area involves seed broadcasting densely, no fertilizer application no weed management practices are employed moreover farmer don't use early maturing sorghum variety and don't know how to conserve soil moisture. This had resulted in reduced sorghum both grain and fodder yield. Currently no recommended integrated crop management practice for sorghum crop is available for the agro-pastoralist. Most research conducted in sorghum only focused on varietal improvement and agronomic practice limited only on fertility and planting density

and the recommendations were also presented separately. Such kind of findings may not help the growers unless they integrated in suitable manner. Livelihood of current study area is characterized as pastoralism and agro-pastoralism this indicates that there is poor skill and knowledge of integrated crop management practices that led to the lower crop yield per unit of area there by resulting food insecurity Mahdi et al. [4].

According to Bellon [6] farmer participation in agricultural research is the most efficient way of technology transfer and adoption of new technology. Many innovations have spread from farmer to farmer without the intervention of any formal agricultural extension services Bellon, and Reeves [7]. It is crucial to involve agropastoralist or farmers in every step of technology development and transfer Obaa et al. Freeman [8,9]. Therefore current study was undertaken to evaluate effect of different sorghum crop management practices on grain and fodder yield of tow sorghum cultivars through participatory farmers' researchers' group approach in agro-pastoralist community in the eastern Ethiopia.

Description of study area

On-farm field experiment was conducted in Ethiopian Somali region Fafen zone Gursum district on randomly selected agropastoralist field. The area is located between 9° 15'N 43°00'E at an altitude of 1746 meter above sea respectively. It experiences a bimodal type of rainfall classified as a short from March to April rainy season and a main rainy season is from June to September with mean annual precipitation of

---

*Corresponding author: Tadeos Shiferaw, College of Dry Land Agriculture, Department of Dryland Crop Sciences, Jigjiga University, P.O. Box, 1020, Jigjiga, Ethiopia, E-mail: tadu1352@gmail.com

890.55 mm. The mean annual temperature of the area is 28.21°C the soil of the area is characterized as light sandy to alluvial soil. (Figure 1) The farming system of area is mainly agro-pastoralism and they mostly produce sorghum, maize, wheat, and chat through traditional agronomic practices Mahdi et al. [4].

Establishing researchers and agropastoralist research group

A total of 56 participant farmers composed of female 16 and 30 male were selected from six villages in Gursum districts to form agropastoralist and researchers group. The researchers were composed of multi disciplinary and multi institutes including three local university researchers, invited two local research institute researchers and two district agriculture department experts with different fields of specializations.

Treatments and methods of evaluations

Six different sorghum management alternatives along with farmer's indigenous management practice were evaluated in six randomly selected agro pastoralist fields' in villages in Fafen administrative zone. Improved and local check sorghum verities namely *Teshale* and *Elimjema* respectively planted in experimental plot threatened with different production packages. The size of each plots were 7 m by 10 m each plots were separated by 2 m distance there were a total of 7 plots replicated in six different agropastoralist field. Participant agropastoralist evaluated production packages presented on trails on selected group of agropastoralist field. They evaluated sorghum management alternatives from different perspectives. They set their own six important treats including days to emergence, days to heading, and days to maturity, fodder and grain yield. .

Methods of data collections and analysis

Participant agro pastoralist evaluation data was collected using farmers ranking methods modified 1-4 scale was developed from Bellon MR [6] Bellon and Reeves [7], Obaa et al. [8] and Freeman HA [9]. Scale was developed after farmers picked five different types of livestock based on their importance to them (Table 1). Cards were prepared using livestock symbols selected by participant agro pastoralist. Then the cards with four types of live stock symbols were distributed among participant agro pastoralist finally they were allowed to cast the cards that contain livestock symbol in the basket prepared in front of the treatment plots for five types of quality parameters including days to emergence, heading, maturity, fodder and grain yield. After evaluations coded data was summarized ANOVA and descriptive statistics were performed using SPSS version 17 software (Table 2).

Results

Days to emergence, heading and maturity

Based on the agro-pastoralist perception the maximum (Good) rate was given to T6, T7 and T2 respectively for early emergence, they also rated T5 and T 4 as fair where T1 ranked as poor. Similarly participant agro-pastoralist evaluated days to heading and maturity of treatments

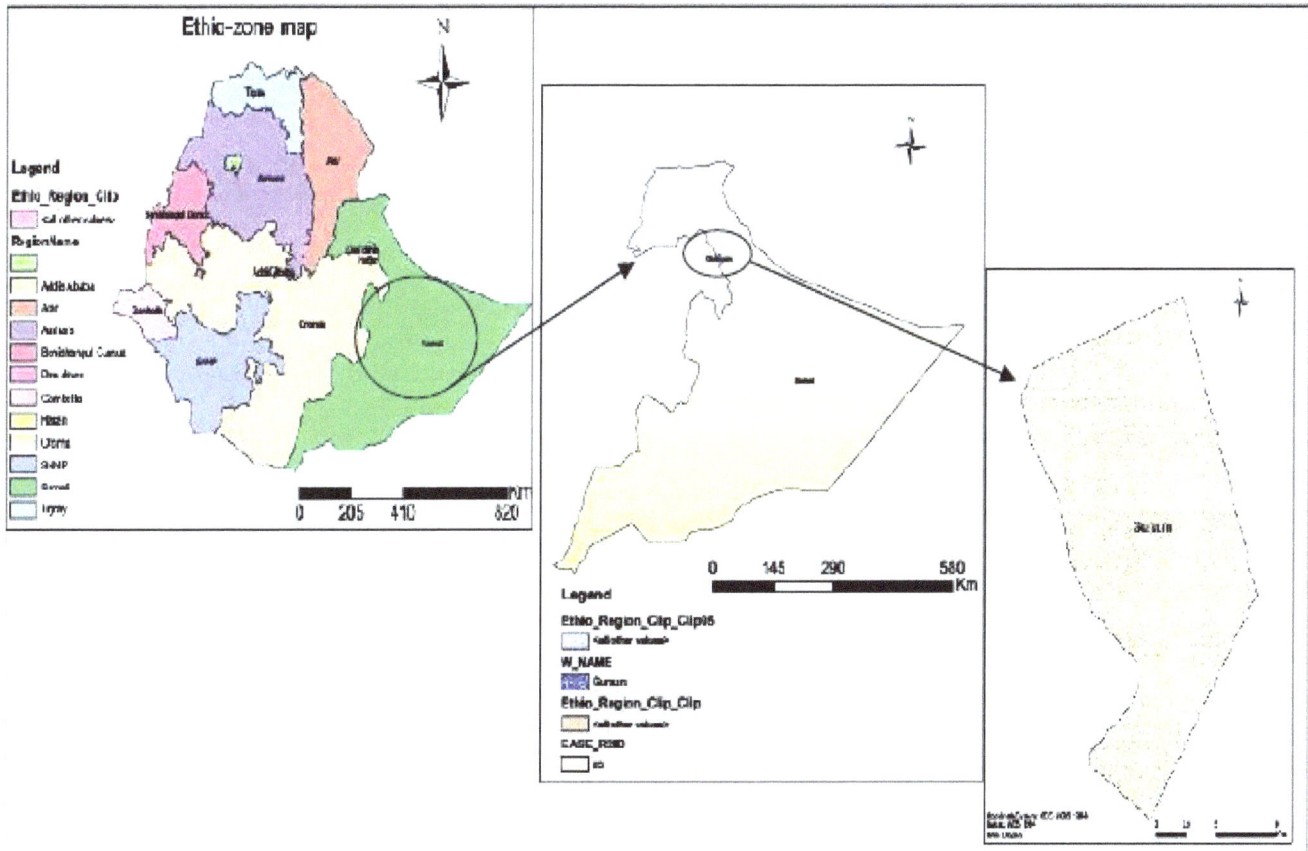

Figure 1: Shows map of Ethiopia, Ethiopian Somali province and Fafen administrative zone districts.

accordingly they gave maximum (Good) grade for T2, T4, T6 and T7 fair for T1 and poor for T5. In general participant agro pastoralist perceived that improved variety *Teshale* had advantage over their local one on days to emergence, heading and maturity. Agropastoralist observed that days to emergence of T1 and T3 was delayed compared to the other treatment this due that general crop management practices and seed quality in which participant agropastoralist used local seed and placed it deeper and coved it with traditional moldboard plough to the depth of at least 10 cm in average. The other treatments including T2, T4, T5, T6 and T7 showed early emergence.

They also differentiated effect of management practices on crop days to heading and maturity beside the varietal differences. Participant agropastoralist indicted that compared to their local management practices (T1) management practice involving tide-ridge planting, fertilizer (N at 50 kg/ha and P 100 kg/ha) and hand weeding once at (T6) resulted early heading and maturity. In this study variety *Tehsale* was significantly earlier than the local check both in days to heading and maturity regardless of management intensity. It was observed that compared to control treatment (T1 and T2 agro-pastoralist indigenous practice) application of fertilizes, weeding twice, straw mulch application at 2 ton/ha and tide-ridge planting for both varieties resulted in slight delay in heading and in maturity. Moreover when local agropastoralist variety was treated with in the same management at (T5) it became let for heading and maturity. Agropastoralist concluded that early matured improved sorghum variety *Teshale* was excellent type of variety when the rainfall condition is unreliable and short. They also indicated that it could be planted twice in a single season during long rainy season (from June to October) in the area (Table 3).

Fodder yield

The other participant agro-pastoralist criteria was fodder yield in this parameter the highest rank excellent(4) and good was give to treatment (T5) and followed by (T4) and where they categorizes T6 and T7 as fair and T2 as poor. According to their evaluations when

local check sorghum variety was tasted in intense managements in T4 and T5 it produced maximum amount of fodder at this level improved variety (*Teshale*) showed low fodder yield under the same management it was categorized as poor to fair. Participant agro-pastoralist view of this parameter of the treatments revealed that better fodder yield could be obtained from the local variety through management intensification. They also compared fodder yield of both local check and improved variety under the same management they found that the local check was superior over the improved one. All treatments composed of *Teshale* showed reduced fodder yield (T2, T6 and T7) it was because *Teshale* was significantly shorter than that of local and it had also low number of leaf/plant and stem was also thin.

Grain yield and yield component

Finally participant agropastoralist compared and ranked total grain yield obtained from of each treatment. Based on their evaluations T6 and T7 ranked as good and the rest T5, T4, T3 and T2 evaluated as fair and T1 poor respectively. At this stage participant agropastoralist able to compare yield obtained from different treatments they also compared grain yield difference of both varieties at similar management. According to participant agropastoralist view grain yield obtained from improved variety *Teshale* was very high compared to their local check under similar management intensity. They indicted that grain yield obtained at T6 and T7 was far greater than other treatments. They perceived crop management options in T6 and in T7 had contributed in yield increment beside the nature of the variety.

Participant agropastoralist clearly demonstrated their understanding toward the use of early maturing variety and improved cultural practices. They explained early maturing sorghum variety not only excellent in crop yield but also it can be planted twice in one season as it matures early. During group discussion agropastoralist pointed out that tide-ridge planting and straw mulch application with fertilizer as most advantages compared with their tradition management practices (T1 using their own variety and T2 using improved variety respectively). They also noted that flat planting as not effective method for moisture conserving as tide-ridge they explained tide-ridge planting was an excellent way to conserve moisture and applied nutrient.

Discussion

Participant agro-pastoralist concluded that early days to emergence observed on both varieties were due to treatments especially on tide-

Livestock Symbol	Scale	Definitions
Camel	4	Excellent
Cow	3	Good
Goat	2	Fair
Chicken	1	Poor

Table 1: Modified agropastoralist treatment rating scale.

Code	Treatments detail
T1	Agropastoralist seed and indigenous management practice (local sorghum variety *Elmijama* was planted broadcasted planting pattern in flat bed densely and fertilizer was not applied. When the plant was 45 days old it was cultivated by locally produced animal drown cultivator.
T2	Improves variety *Teshale* with Agro-pastoralist practice. The seed was planted in broadcasted planting pattern in flat bed densely and fertilizer was not applied. When the plant was 45 days old it was cultivated by locally produced animal drown cultivator
T3	Agropastoralist variety *Elmijama* was planted in flat bed with row spacing (70 cm × 30 cm) with recommended seed rate at 8 kg/ha. N and P fertilizers were applied in the form of urea and DAP at 50 and 100 kg/ha respectively. When the plants were 45 days old it was hand weeding once.
T4	Agro-pastoralist variety *Elmijama* was planted in tied-ridge with row spacing (70 cm × 30 cm) with recommended seed rate at 8 kg/ha. N and P fertilizers were applied in the form of urea and DAP at 50 and 100 kg/ha respectively. When the plants were 45 days old it was hand weeding once
T5	Agro-pastoralist variety *Elmijama* was planted in tied-ridge with row spacing (70 cm × 30 cm) with recommended seed rate at 8 kg/ha. N and P fertilizers were applied in the form of urea and DAP at 50 and 100 kg/ha respectively. Dried straw mulch was applied at the rate of 2 ton/ha. The crops were weeded twice when the plants were 45 and 82 days old.
T6	Improved sorghum variety *Teshale* was planted in tied-ridge with row spacing (70 cm × 30 cm) with recommended seed rate at 8 kg/ha. N and P Fertilizers were applied in the form of urea and DAP at 50 and 100 kg/ha respectively. When the plants were 45 days old it was hand weeding once
T7	Improved sorghum variety *Teshale* was planted in tied-ridge with row spacing (70 cm × 30 cm) with recommended seed rate at 8 kg/ha. N and P Fertilizers were applied in the form of urea and DAP at 50 and 100 kg/ha respectively. Dried straw mulch was applied at the rate of 2 ton/ha. The crops were weeded twice when the plants were 45 and 82 days old.

Recommended rate of fertilizers application and seeding rate was used from the recommendations of Ethiopian Ministry of Agriculture (MoARAD, [10]

Table 2: Different sorghum management packages and agropastoralist indigenous production practices.

Treatment	Days to emergence	Days to heading	Maturity period	Fodder yield	Panicle size	Seed size	Grain yield
T1	1.40[a]	2.40 [ab]	2.33[a]	2.67[a]	2.10[a]	1.37a	1.00[a]
T2	3.20[b]	3.80[b]	1.52[b]	1.32[b]	3.00[b]	3.45[b]	2.02[a]
T3	2.40[ab]	1.60[a]	3.55[ac]	3.90[c]	2.47[ab]	2.55[bc]	2.75[a]
T4	2.60 [ab]	3.40[b]	3.90[c]	4.75[d]	2.22[a]	2.50[bc]	2.95[ab]
T5	2.80[ab]	1.60[a]	3.00[ac]	4.88[d]	2.15[b]	2.92[bc]	3.55[b]
T6	3.40[b]	3.40[b]	4.70[d]	2.75[ac]	4.10[c]	4.35[d]	4.40[c]
T7	3.40[b]	3.40[b]	3.22[ac]	2.35[ac]	4.85[c]	4.20[d]	4.40[b]
LSD $_{0.05}$	0.91	1.81	1.071	0.761	0.54	0.82	0.93
CV%	18.55	20.52	26.76	16.93	13.27	19.57	22.01

Rating scale 4=excellent; 3=Good; 2=Fair; 1=poor
Means followed by the letter are not statistically significant at p<0.05

Table 3: Agro-pastoralist evaluations of treatments for days to emergence maturity, heading and maturity.

ridge planting, straw mulch and fertilizes application promoted early emergence. Similar report was made by Ravender et al. [11] who reported that mulching is known to influence water use efficiency of crops by affecting the hydrothermal regime of soil, which may enhance root and shoot growth. Also Heluf [12] indicted that tide-ridge sorghum planting was advantageous over the flat for moisture conservation and applied nutrient management. Mulch application and tide-ridge planting resulted in better soil water condition and provided the plants with good moisture in stressed area positively affecting plant growth and yield Hari et al.[13], Sunday et al. [14]. In addition to the tide-ridge planting method straw mulch application and fertilize availability tends to affect days to emergence positively for both varieties. Despite management effect there was also varietal effect contributed for early emergence, heading and maturity. Similar result was reported by Tekle and Zemach, [3] who indicted that compared to local check improved variety Tehsale had germinated and matured earlier. It has been seen that when both verities were treated with (T5 and T7) with tide-ridge planting, fertilizer application (N and P), straw mulch application and hand weeding showed slight delay in both days to heading and maturity compared to the control. The delay to heading maturity in these treatments was due to management intensity in which resulted longer vegetative growth periods. Sorghum growth and yield could be affected by management practices like planting pattern, seed bed type, fertilizer application and weed removal [12,15-17].

Availability of fertilizer and weed removal along with tide-ridge planting method at (T4 and T5) resulted in higher fodder production compared to the control and other treatments (T1 andT3). It was due to treatment and varietal effect. Similar report Chiroma et al. [18] whom revealed that sorghum fodder yield was significantly influenced by straw mulch, land planting patterns and nutrient management and other essential cultural practices. In our study improving management practice in both local and improved variety showed increments in plant height, leaf number and stem thickness the result was in agreement with other previous findings who reported that sorghum fodder yield could be affected by management practices [15,16]. For both varieties fertilize application, weeding and tide-ridge planting gave better fodder yield. Similar findings of previous works indicated that fodder yields of rabi sorghum were significantly influenced by moisture conservation and fertilizer management practices Mudalagiriyappa et al. [19], Jehan et al. [20] also reported that tide-ridge sowing gave maximum stalk yield when compared with maize planted in other sowing methods. According to the result of this study and participant agro pastoralist discussions, fodder yield was significantly influenced by varietal and management differences. They noted that fodder yield of local variety was significantly greater than that of the improved Teshale regardless of treatment effect. However when management was intensified at (T6

andT7) the fodder yield tends to increase compared to untreated Teshale (T2). Current finding was in the contrary to that of Tekle and Zemach [3] who reported high biomass yield of improved variety Teshale compared to local check. In current study participant agro-pastoralist identified that when local variety planted in flat bed produced lower amount of fodder while those treatments treated with tide-ridge with straw mulch along with fertilizer produced maximum fodder yield.

Grain yield for both varieties were significantly affected by management practices (Table 4). Hence improved production packages regardless of varietal difference increased grain yield. Similar findings have also been reported that good production practices resulted in higher mean grain yield and net returns compared to farmers' practice Pandey et al. [21]. Tekle and Zemach [3] indicated that improved variety Teshale have grain yield superiority over the local check. Dinesh et al. [22] also reported that grain as well as straw yield of wheat increased significantly when the improved variety was sown in rows provided with balanced fertilizations. Moreover Heluf G [12] reported that tied ridge planting with appropriate fertilization gave the highest yield of sorghum in areas with low and erratic rainfall.

Conclusion

Improved variety showed best performance in yield and maturity duration in T6 and T7 local variety also produced significantly higher quantity of fodder and grain yield at T4 and T5. Participant agropastoralist understood that by practicing appropriate and recommended good cultural practice both grain and fodder yield can be increased to greatest extent. Finally both participant agropastoralist and the research team concluded that both grain and fodder yield of sorghum in current study area can be increased through the adoption of improved production packages. They also argued that improved sorghum variety Teshale can be used for grain production during uncertain and unreliable rainfall condition but also it coul be produced twice during longer rainy season in the area from June to October. Participant agropastoralist need their let maturing variety for its quality in fodder and stalk yield they decided to keep using it through adoption of improved cultural practice. Therefore based on agro-pastoralist interest and rating T6 and T4 were selected as best management practices because it balances the both grain and fodder yield with the production costs. Even though both grain and fodder yield was higher in T5 and T7 framers didn't interested to pick the as best management practices because of straw mulch application on T5 and 7. Therefore based on the interest of participant agropastoralist T6 and T4 were recommended for the current study areas and other similar places.

Treatments	Panicle length(cm)	100 seed weigh(gm)	Grain yield (ton/ha)	Biomass yield (ton/ha)
T1	21.80[a]	1.22[a]	0.72[a]	20.00[a]
T2	17.00[b]	2.25[b]	1.14[ab]	14.33[b]
T3	24.00[c]	1.89[a]	0.96[a]	29.33[c]
T4	26.66[d]	1.99[a]	1.48[ab]	32.00[d]
T5	27.00[d]	2.24[b]	1.91[ab]	39.00[e]
T6	21.00[a]	3.55[bc]	2.45[c]	22.47[af]
T7	20.67[ab]	3.32[bc]	2.39[c]	21.12[af]
LSD $_{0.05}$	1.67	0.59	0.85	1.60
CV%	13.59	20.61	6.94	20.03

Means followed by the same later are not statistical significant at $P<0.05$.

Table 4: Effect of different crop management practice on yield and yield components of sorghum.

References

1. Catherine W, Muui R, Muasya M, Duncan TK (2013) Participatory identification and evaluation of sorghum (Sorghum bicolor (L.) Moench) landraces from lower eastern Kenya. International Research Journal of Agricultural Science and Soil Science 3: 283-290.

2. Asfaw A (2007) the role of introduced sorghum and millets in Ethiopian agriculture. Journal of SAT Agricultural Research 3: 1-4.

3. Tekle Y, S Zemach (2014) Evaluation of sorghum (Sorghum bicolor (L.) Moench) varieties, for yield and yield components at Kako, Southern Ethiopia. Journal of Plant Sciences 2: 129-133.

4. Mahdi E, T Pichai, Savitree R, T Sayan (2012) Factors Affecting the Adoption of Improved Sorghum Varieties in Awbare District of Somali Regional State, Ethiopia. Kasetsart Journal of Social Science 33: 152-160.

5. Mekbib F (2007) Genetic erosion of Sorghum bicolor L. Moench) in the centre of diversity, Ethiopia. Genet. Resour. Crop Evol 55: 351-364.

6. Bellon MR (2001) Participatory Research Methods for Technology Evaluation: A Manual for Scientists Working with Farmers. Mexico, DF. CIMMYT.

7. Bellon MR, Reeves J (eds) (2002) Quantitative Analysis of Data from Participatory Methods in Plant Breeding. Mexico, DF: CIMMYT.

8. Obaa B, M Chanpacho, Agea J (2007) Participatory farmers' evaluation of maize varieties: A case study from Nebbi District, Uganda. African Crop Science Conference Proceedings 7: 1389-1393.

9. Freeman HA (2001) Comparison of farmer-participatory research methodologies: case studies in Malawi and Zimbabwe. Working Paper Series no. 10. PO Box 39063, Nairobi, Kenya: Socioeconomics and Policy Program. International Crops Research Institute for the Semi-Arid Tropics 28.

10. Mo MARD (2009) Ministry of Agriculture and Rural Development Animal and Plant Health Regulatory Directorate Crop Variety Register June, 2009 Addis Ababa, Ethiopia 12: 38-44.

11. Ravender SD, Kundu K, Bandyopadhyay KK (2010) Enhancing Agricultural Productivity through Enhanced Water Use Efficiency. Journal of Agricultural Physics 10: 1-15.

12. Heluf G (2003) Grain Yield Response of Sorghum (Sorghum bicolor) to Tied Ridges and Planting Methods on Entisols and Vertisols of Alemaya Area,

Eastern Ethiopian Highlands. Journal of Agriculture and Rural Development in the Tropics and Subtropics 104: 113-128.

13. Hari R, Dadhwal V, Kumar V, Harinderjit K (2013) Grain yield and water use efficiency of wheat (Triticum aestivum L.) in relation to irrigation levels and rice straw mulching in North West India. Agricultural Water Management 128: 92-101.

14. Sunday EO, Okpara IM, Martin EO, Toshiyuki W (2011) Short Term Effects Of Tillage- Mulch Practices Under Sorghum And Soybean on Organic Carbon and Eutrophic Status Of A Degraded Ultisol In Southeastern Nigeria. Tropical and Subtropical Agroecosystems 14: 393-403.

15. Adil BK, IM Hassan, Haitham RE, Amir BS, El Atif (2012) Effect of Planting Geometry on Soil Moisture Content, Yield and Yield Component of Sorghum (Sorghum bicolor L. Moench). Global Journal of Plant Ecophysiology 2: 23-30.

16. Afzal M, Ahmad A, Ahmad AH (2012) Effect Of Nitrogen On Growth And Yield Of Sorghum Forage (Sorghum Bicolor (L.) Moench Cv.) Under Three Cuttings System. Cercetări Agronomice În Moldova 45: 57-64.

17. Reddya BVS, Sanjana Reddya P, Bidingera F, Blumme M (2003) Crop management factors influencing yield and quality of crop residues. Field Crops Research 84: 57-77.

18. Chiroma AM, Alhassan AB, Yakubu H (2006) Growth, Nutrient Composition and Straw Yield of Sorghum as Affected by Land Configuration and Wood-chips Mulch on Sandy Loam Soil in Northeast Nigeria. International Journal of Agriculture & Biology 6: 770-773.

19. Mudalagiriyappa BK, Ramachandrappa H, Nanjappa V (2012) Moisture conservation practices and nutrient management on growth and yield of rabi sorghum (Sorghum bicolor) in the vertisols of peninsular India. Agricultural Sciences 3: 588-593.

20. Jehan B, Faisal M, Siddique M, Shafi H, Akbar MK, et al. (2007) Effect of Planting Methods and Nitrogen Levels on the Yield and Yield Components of Maize. Sarhad Journal of Agriculture 23: 3.

21. Pandey AK, Prakash V, Singh RD, Gupta HS (2001) Contribution and impact of production factors on growth, yield attributes, yield and economics of rainfed wheat (Triticum aestivum). Indian Journal of Agronomy 46: 674-681.

22. Dinesh KS, Purushottam K, Bhardwaj AK (2014) Evaluation of Agronomic Management Practices on Farmers' Fields under Rice-Wheat Cropping System in Northern India. International Journal of Agronomy.

7

Field Management of Anthracnose (*Colletotrichum lindemuthianum*) in Common Bean through Fungicides and Bioagents

Mohammed Amin*, Sileshi Fitsum, Thangavel Selvaraj and Negeri Mulugeta

Ambo University, Ethiopia

Abstract

Common bean anthracnose is a major production constraint in bean growing regions of Ethiopia. This study aimed to determine whether foliar sprays of mancozeb, folpan and mancolaxyl or antagonistic bioagents; *Trichoderma harzianum, Trichoderma viride* and *Pseudomonas fluorescens* could reduce anthracnose symptoms and consequently, increase yield and yield components. A total of seven treatments were arranged in a randomized complete block design with three replications. Statistical analysis showed significant differences among treatments. Anthracnose incidence, severity, infected pods per plant and the area under disease progress curve were highest in the control plots compared to the fungicide sprayed and bioagent treated seed plots. The highest percentage of infected pods per plant of 78.9 and 55.0 recorded on the control and mancozeb sprayed plots, respectively. The highest AUDPC value resulted in the lowest yield of 1.01 t/ha in the control plots compared to a highest yield of 3.3 t/ha from the sprayed plots with folpan and 1.8 t/ha from plots treated with Pseudomonas fluorescens. Relative yield losses of 69.7, 46.3 and 22.8% were recorded from the control, seed treated plots with P. fluorescens and sprayed plots with mancolaxyl, respectively. Economic analysis revealed that the highest rate of return of 8,740 was obtained from *Pseudomonas fluorescens* seed treatment and the highest net benefit value of 43,154 calculated on folpan foliar spray treatment. The results of the present study support the novel possibility of using folpan foliar spray and *Pseudomonas fluorescens* seed treatments to decrease anthracnose symptoms in common bean and consequently, achieve greater yield.

Keywords: Bioagents, Folpan; *Pseudomonas fluorescens*; Mancolaxyl; Severity; Seed treatment

Introduction

Common bean (*Phaseolus vulgaris* L.) is a key grain legume crop and a vital source of nutrition worldwide. The FAO reports that half of the world's common bean production occurs in low income, food deficit countries where this staple crop contributes to food security. The other half is produced in countries like the U.S., where common bean is an important economic crop with 769 thousand hectares of dry and snap beans planted in 2012, and with a farm gate value of $1.5 billion [1]. The value of the common bean crop exceeds that of all other legumes combined, including chickpea, lentil, pea, and cowpea, thus indicating the current and potential future economic role of this crop.

In Ethiopia, common bean is mainly cultivated in the Eastern, Southern, South-western and Rift valley regions of the country [2]. Despite its economic significance and wide area of production, the national annual yield is low, ranging from 0.615-1.487 tons/ha between the years 2004 and 2010 [3]. The low national yield could be attributed to various constraints. But a recent study revealed that pests and diseases are ranked as the second important production constraints in the Central Rift Valley region, next to drought and third in the Southern parts of the country, next to drought and shortage of land [4]. Moreover, Yesuf [5] emphasized that diseases are known to be the major factors which threaten the productivity of common beans in general and common bean in particular. Anthracnose (*Colletotrichum lindemuthianum*), rust (*Uromysis appendiculatus*), angular leaf spot (*Phaeoisariopsis griseola*) and common bacterial blight (*Xanthomonas compestris pv. phaseoli*) are common diseases of bean in Ethiopia [2]. Among these, anthracnose caused by the fungus *Colletotrichum lindemuthianum* (Sacc. and Magnus) Lams-Scrib. is the most wide spread and economically important seed borne disease, mainly in the tropical and sub-tropical bean growing regions of the world including Ethiopia [2,6]. It has been confirmed that infection of susceptible cultivars like Mexican-142 and Awash-1 in favorable environmental conditions leading to an epidemic could result in 100% yield loss [7]. A study by Tesfaye [8] stated that yield loss up to 62.8% due to anthracnose was recorded in Ethiopia on susceptible cultivars of common bean.

The efficacies of seed dressing and foliar fungicides like benlate, difenoconazole, mancozeb and carbendazim have been carried out in Ethiopia [9,10]. However, due to the increased risk of developing resistance, there is a growing need for the evaluation of the efficacies of new alternative fungicides and other cost effective and eco-friendly management options. Folpan, a protective fungicide which is commonly applied for the control of a number of fungal diseases including anthracnose of cucumbers, melons, pumpkins, squash and tomatoes, has a multi-site activity, which provides excellent resistance management option [11]. Similarly, the co-formulation of metalaxyl and mancozeb (Mancolaxyl 72 WP), which resulted in satisfactory control of Colletotrichum coccodes on tomatoes, was found to be a good option to prevent fungicide resistance [12]. It has been reported that seed dressing or soil application of bioagents like *Trichoderma viride, Trichoderma harzianum* and *Gliocladium virens* caused significant inhibition of mycelial growth of C. lindemuthianum, thereby effectively controlled the seed borne infection and increased the seed germination of common bean [13]. Extracellular metabolites like siderophores, antibiotics, lytic enzymes and volatile compounds produced by rhizobacteria (*Pseudomonas fluorescens and Bacillus cepacia*) have also been reported to effectively reduce lesions and damages caused by C. lindemuthianum on bean plants [14]. Generally, recent innovations showed that biological control of crop diseases is getting increased attention as economic and environmentally sound approach. But in Ethiopia, the method has received comparatively

---

*Corresponding author: Professor Mohammed Amin, Department of Plant Science, Ambo University, Ethiopia, E-mail: yonias_1986@yahoo.com

little attention. Apparently, the management of common bean anthracnose through bioagents, particularly, P. *fluorescens*, *T. viride* and *T. harzianum* have not been studied so far in Ethiopia. Much work still needs to be done in the management of bean anthracnose through alternative fungicides and bio agents, since the disease is still causing devastation on the crop. Therefore, the present work was carried out with the objective of evaluating the efficacies of alternative foliar spray fungicides and seed treatment bioagents for the management of common bean anthracnose and also to determine the economics of fungicide sprays as well as bioagents seed treatments.

Materials and Methods

Description of the study area

The field experiment for the management of common bean anthracnose through bioagents and fungicides was conducted at Ambo University's research farm, during the main cropping season of 2013. Ambo is located 120 km west of Addis Ababa at 8°98' South latitude and 37°83' North longitude. It has a total geographical area of 83,598.69 sq. km., with elevation ranging from 1380-3300 meter above sea level. Annual rainfall ranged from 900-1100 mm and temperature ranged from 10-27°C, with an average of 18°C. The soil type of the study site is vertisol with a pH value of 6.8.

Experimental materials used

The highly susceptible common bean variety to bean anthracnose, Mexican-142 [15], was used in the field experiment. Discolored seeds with typical anthracnose lesions were obtained from the Ethiopian Seed Enterprise, Hawassa. Three fungicide schedules; mancozeb (Unizeb 80% WP), folpan 80 WDG and mancolaxyl 72 WP (Mancozeb with Metalaxyl) were evaluated separately against common bean anthracnose disease. Three antagonistic bioagents viz., *Trichoderma harzianum*, *T. viride* and *Pseudomonas fluorescens* were also evaluated separately against common bean anthracnose disease. All the three bioagents were obtained from the Department of Plant Sciences, Ambo University, Ethiopia.

Experimental design and treatments

The treatments were subjected to RCBD with three replications. Three fungicides, three antagonistic bioagents and one control, total 7 treatments were evaluated against common bean anthracnose. Spacing between plants was maintained at 10 cm and between rows 40 cm. There were 21 plots, each consisting of 6 rows. Plots had a width of 2.4 m and length 2m. Each row had 20 plants. In general, there were 120 plants per plot in which thirteen of them were pre-tagged from the central four rows. The total experimental area landed on 182.4 m² of land. Seeds were planted at the rate of two seeds per hole and thinned to one plant, 15 days after sowing (DAS) to insure 120 plants per plot. All agronomic practices were kept uniform for all plots of each treatment. Mancozeb (Unizeb 80% WP) was applied at the rate of 2 kg/ha, folpan 80 WDG at the rate of 2.6 kg/ha and mancolaxy 72 WP (mancozeb + metalaxyl) at the rate of 3.5 Kg/ha. All the fungicides were applied as of the disease onset at 7 days interval for 4 times using a Knap-sack sprayer. Bioagents were applied as seed treatments by the method of [16]. Talc based formulations (28x10⁻⁶ cfu/g product) of *T. viride* and *T. harzianum* were used at the rate of 40 g/Kg of seeds, soaked in 1L of water for 24 hrs [17]. Similarly, a talc based formulation of *P. fluorescens* by the method of [18] was used at the rate of 10 g/Kg to soak the seeds in 1L of water for 24 hrs. The treated seeds were dried overnight before sowing.

Anthracnose disease assessment

Thirteen plants were selected and pre-tagged from each plot using W-shaped sampling after the plants emerged. Disease epidemic data were collected from pre-tagged plants starting from the onset of the first anthracnose symptoms at 14-days intervals. Plants that showed symptoms of anthracnose were counted from the pre-tagged plants and the percentage of disease incidence was calculated according to the formula by Wheeler [19].

The severity of anthracnose on the leaves of pre-tagged common bean plants were graded using standard disease scales of: 1-9, where, 1= no visible disease symptoms; 3= presence of very small lesions, mostly on the primary vein of leaf's lower side or on the pod, that covers approximately 1% of surface area; 5= presence of several small lesions on the petiole or on the primary and secondary veins of the leaf's lower side or small round lesions on the pods, with or without reduced sporulation, that covers approximately 5% of the pods surface area; 7= presence of enlarged lesions on the lower side of the leaf. Necrotic lesions can also be observed on the upper leaf surface and on the petioles. On the pods, the presence of medium lesions are evident but also some small and large lesions generally with sporulation and that cover approximately 10% of the pod's surface area may be found and 9= more than 25% of the leaf surface area covered with large coalescing and generally necrotic lesions resulting in defoliation [20].

Assessment of crop growth, seed yield and yield components

The Plant height, number of pods per plant, infected pods per plant and seeds per pod were recorded from the 13 pre-tagged plants. The harvested pods were sun dried and the respective seed yields of the different treatments were measured. Bean yield data was adjusted at 10% moisture content after measuring using a moisture tester. Seed yield per plot was converted into tons per hectare. The weight of 100 seeds was taken randomly from harvested seed lots.

The area under the disease progressive curve and disease progress rates

The area under the disease progressive curve (AUDPC) was computed from the PSI data recorded at each date of assessment as described by Campbell and Madden [21].

$$AUDPC = \sum_{i=1}^{n-1} 0.5\left(x_{i+1} + x_i\right)\left(t_{i+1} - t_i\right)$$

Where n is the total number of assessments, t_i is the time of the ith assessment in days from the first assessment date, xi is percentage of disease severity at i[th] assessment. AUDPC was expressed in percent-days because the severity (x) was expressed in percent and time (t) in days. The rates of disease progress were obtained from the regression of the PSI data fit to the Gompertz model [-ln (lnY)] with dates of assessments.

Statistical analysis

Analysis of variance was performed for disease parameters (incidence, PSI, AUDPC, disease progress rates and infected pods per plant) and yield related parameters (seed yield, pods per plant and 100 seed weight) using the Statistical Analysis System version 9.1.3 software [22]. Least significance difference was used to separate treatment means (P<0.05). Correlation analysis was used to examine the relationship between epidemic data and seed yield.

Cost and benefit analysis

The price of common bean seeds (Birr/Kg) was assessed from the local market and the total price of the commodity obtained from each treatment was computed on hectare basis. Input costs like fungicides, bioagents and labor were converted into hectare basis according to their frequencies used. Since there were significant differences between mean yields of treatments, the obtained data were analyzed using the partial budget analysis method [23].

Results and Discussion

Bean anthracnose incidence and severity

The severity and incidence of anthracnose increased with time from the 39 DAS onwards. Disease incidence data showed highly significant differences (P<0.01) among treatments at 39 and 67 DAS and significant differences (P<0.05) at 53 and 81 DAS. There were no significant differences in disease incidences among treatments at the final (95 DAS) date of disease assessment (Table 1). Maximum disease incidences were recorded from the untreated control, 74.4% at the initial (39 DAS) and 100% at the final (95 DAS) date of disease assessment. The least disease incidences were recorded from plots treated with folpan foliar spray fungicide, 5.1% and 87.2%, at the initial (39 DAS) and final (95 DAS) respectively, dates of disease assessment. Among the bioagents, plots treated with T. *viride and P. fluorescens* showed the least disease incidence (38.5%) at the initial date (39 DAS) of disease assessment. *P. fluorescens* showed the least disease incidence at 39, 53, 67 and 81 DAS. Disease incidence reached its maximum at the podding stage on the control plots, which could be due to the accumulation of secondary inoculums, susceptibility of the crop's stage and or the occurrence of favorable environmental condition. As stated by [24] when no heavy seedling infection is observed, another phase of marked susceptibility will be encountered at the early stage of pod formation. Mean severity of 59.3 at the podding stage was recorded at Ambo by Tesfaye [8].

The percent of severity index (PSI) data revealed that the severity of anthracnose on the control plot was higher than the protected plots. Highly significant differences (P<0.001) among treatments were recorded at all dates of assessment. Maximum disease severity, 32.2% at the initial (39 DAS) and 96.0% at the final (95 DAS) dates of disease assessment were recorded from the untreated control. The least disease severities were recorded from plots spayed with folpan foliar spray fungicide, 1.1% and 34.2%, at the initial (39 DAS) and final (95 DAS) respectively, followed by plots treated with mancozeb foliar spray fungicide, 7.7% at the initial (39 DAS) dates of disease assessment (Table 1). Spraying foliage at flowering initiation, late flowering, and pod fill

to achieve satisfactory bean anthracnose disease control [25]. Among the bioagents, the least disease severities were recorded from plots treated with *P. fluorescens*, 5.4% and 55.6%, followed by plots treated with *T. viride*, 18.52% and 56.98% at the initial (39 DAS) and final (95 DAS) respectively, dates of disease assessment. High disease severity on the control plots indicated that all the treatments significantly reduced the severity of anthracnose at both dates of assessment. The fungicide sprays particularly permitted the crop to reach physiological maturity without being under severe anthracnose infection.

Area under diseases progress curve (AUDPC)

For evaluating practical disease management strategies, the two most commonly used tools are comparing disease progress curves and AUDPC between treatments [26]. The AUDPC analysis showed the overall disease development was significantly affected by the management program applied. The increase in incidence throughout the assessment days indicated the spread of the disease in space. The data showed highly significant differences (P<0.01) among treatments. Maximum AUDPC value (3197.5 %-days) was computed from the untreated control, whereas the minimum value (835.8 %-days) from plots treated with folpan spray fungicide followed by mancozeb (1552.3 %-days). Among the bioagents, the least AUDPC value was recorded on plots treated with *P. fluorescens* seed treatment (Table 1).

Infected pods per plant due to bean anthracnose

Data on the percentage of infected pods per plant showed highly significant differences (P<0.01) among treatments. The highest percentage of pod infection (78.9%) was recorded from the control plots, whereas the least (31.0%) from plots treated with *P. fluorescens*, followed by (31.8%), from folpan treated plots (Figure 1). Frequent rainfall with optimum temperature encountered in August and September of the growing season might have contributed much for the observed severe pod infection. As reported by [27] if the pod filling stage of beans coincides with frequent rainfall and moderate temperature in areas like Bako and Ambo, the risk of severe pod and seed infection must be anticipated.

Field management of bean anthracnose using fungicidal and bioagents on yield components and seed yield

Data on yield parameters showed highly significant differences (P<0.01) among treatments in the number of pods per plant, seeds per pod and seed yield, whereas, no significant differences were observed in the 100 seed weight. Plots treated with mancolaxyl foliar spray fungicide gave the highest number of pods per plant (29.4) followed by folpan (28.9) and mancozeb (25.6) foliar spray fungicides. Among

Treatments	Incidence (PDI)		Severity (PSI)		
	Initial (39 DAS)	Final (95 DAS)	Initial (39 DAS)	Final (95 DAS)	AUDPC
Mancolaxyl	28.2(26.80cd)	100(99.9a)	18.8(25.7b)	60.7(51.22b)	2098.7b
Folpan	5.1(10.74d)	87.18(76.7a)	1.1(4.8d)	34.2(35.44c)	835.8d
Mancozeb	28.2(30.9bcd)	97.4(91.3a)	7.7(15.9c)	65.5(54.23b)	1552.3c
T .harzianum	61.5(51.9ab)	97.4(91.3a)	15.1(22.8b)	65.2(54.01b)	2143.7b
T. viride	38.5(38.3abc)	97.4(91.3a)	18.5 (25.5b)	56.9 (49.05b)	1867.7bc
P. fluorescens	38.4(38.1bc)	97.4(91.3a)	5.4(13.2c)	55.6(48.24b)	1664.0c
Control	74.5(59.8a)	100(99.9a)	32.2(34.5a)	96.0 (78.77a)	3197.5a
CV (%)	33.3	13.9	16.4	12.8	9.6
LSD (0.05)	21.71	Ns	5.9	12.1	327.4

Table 1: Incidence, severity and AUDPC of common bean anthracnose as influenced by different fungicides and bioagents during 2013.
Figures in parenthesis are arc sine transformed values.DAS = Days after sowing, AUDPC = Area under disease progress curve, Ns = Non-significant, Mean values within columns followed by the same letter are not significantly different (P ≤ 0.05), PDI =percent of disease incidence, PSI= percent of severity index

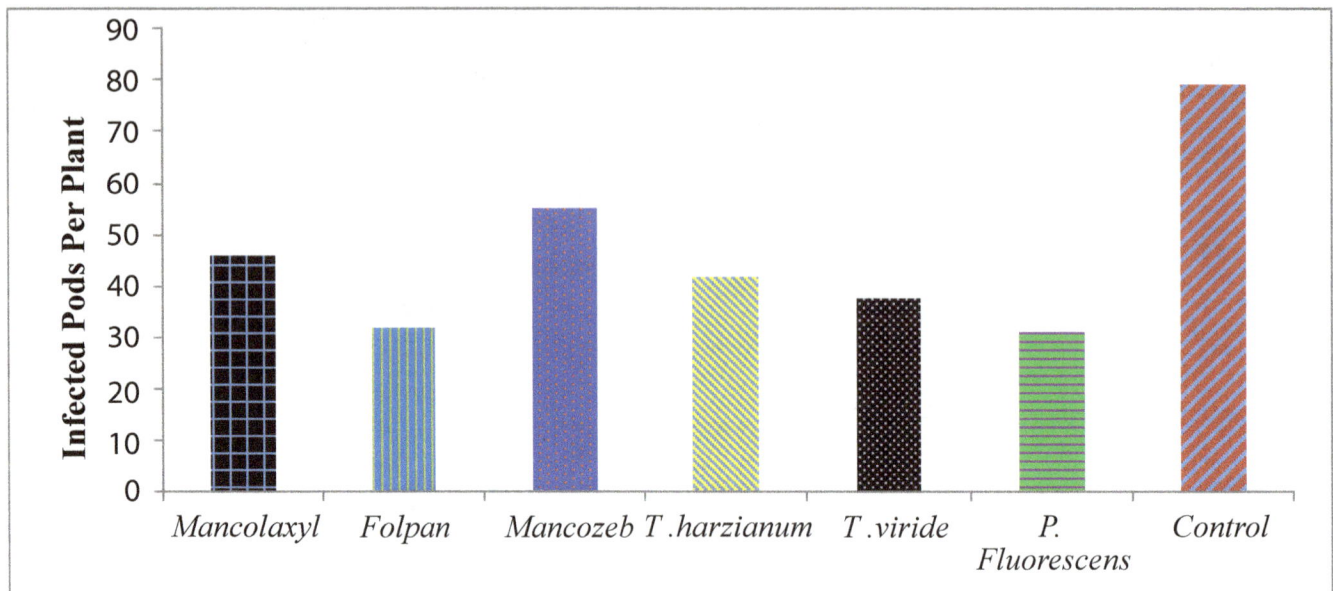

Figure 1: Infected pods per plant as influenced by different fungicides sprays and bioagents seed treatment during 2013

Treatments	PH(cm)	Pods/plant	Seeds/pod	HSW (g)	Yield (t/ha)	RYL (%)
Mancolaxyl	68.7a	29.4a	4.6b	17.0a	2.6b	22.8
Folpan	64.9a	28.9ab	5.1a	17.3a	3.3a	-
Mancozeb	67.1a	25.6abc	4.5b	18.0a	2.4b	27.3
T .harzianum	65.2a	20.5c	4.9ab	16.8a	1.5cd	54.1
T. viride	57.5a	21.4c	4.5b	16.4a	1.5cd	54.1
P. fluorescens	51.3a	22.4bc	4.7ab	16.4a	1.8c	46.3
Control	57.8a	13.1d	3.2c	16.1a	1.0d	69.7
CV (%)	11.2	16.9	6.1	6.5	14.9	
LSD (0.05)	Ns	6.9	0.5	Ns	0.5	

Table 2: Plant height, pods per plant, seeds per pod, hundred seed weight, seed yield and relative yield loss of common bean as influenced by different fungicides and bioagents during 2013
RYL = Relative yield loss, Ns = Non-significant, Mean values within columns followed by the same letter are not significantly different (P ≤ 0.05), HSW=hundred seed weight, PH=plant height

Treatments	Rate (r)	R2	SEE	Significance
Mancolaxyl	0.96	0.90	3.14	(P<0.05)
Folpan	0.93	0.82	5.10	Ns
Mancozeb	0.91	0.79	7.23	Ns
T. harzianum	0.95	0.89	4.60	Ns
T. viride	0.93	0.83	3.87	0.001
P. fluorescens	0.95	0.87	4.89	0.001
Control	0.98	0.95	3.65	0.05

Table 3: Effect of fungicides foliar sprays and bioagents seed treatments on disease progressive rate of bean anthracnose during the 2013 main cropping season
SEE = Standard Error of Estimate, R2 = Coefficient of determination, Ns = Non-significant

Parameters	Yield	PSI	AUDPC	IP (%)	IR
Yield	1				
PSI	-0.76288*	1			
AUDPC	-0.81961*	0.94256**	1		
IP (%)	-0.48984ns	0.92583**	0.79772*	1	
IR	-0.57513ns	0.66890ns	0.81998*	0.49103ns	1

Table 4: Correlation coefficient (r) between PSI, AUDPC, pod infection, infection rate and crop yield in different fungicides and bioagents under field condition during 2013

*Correlation is significant at 0.05 probability level, ** Correlation is highly significant at 0.05 probability level, ns = None Significant, IP=infected pods, IR=infection rate, AUDPC=area under disease progress curve, PSI=percent of severity index

the bioagents, *P. fluorescens* gave the highest number of pods per plant (22.4). The least number of pods per plant was recorded from the untreated control (13.1). The highest number of seeds per pod was obtained from plots treated with folpan foliar spray (5.1) followed by plots treated with the bioagents, *T. harzianum* (4.9) and *P. fluorescens* (4.7). The untreated control gave the least number of seeds per pod (3.2). The seed yield data showed highly significant differences among treatments. The maximum yield (3.3t ha-1) was obtained from plots treated with folpan foliar fungicide, followed by 2.6 t ha-1 and 2.4t ha-1, from plots treated with the foliar fungicides, mancolaxyl and mancozeb respectively. Among the bioagents, *P. fluorescens* gave the highest yield (1.8 t ha^{-1}) (Table 2).

Common bean yield loss due to anthracnose

The computed relative yield losses showed notable differences among treatments. Yield losses were highly reduced by fungicide chemicals; mancolaxyl (22.8%) and mancozeb (27.3%), compared to the untreated control and the bioagents. The highest yield loss was calculated from the untreated control (69.7%). Among the bioagents, plots treated with *P. fluorescens* gave the least relative yield loss (46.3%), whereas both *T. viride and T .harzianum* showed a relative yield loss of (54.1%). The results are in agreement with the findings of [8], which evaluated the severity of bean anthracnose and its effect on yield and found that high disease severity between 17.2%-76.6% resulted in mean yield loss of 67.2%. Similarly, Sharma et al. [28] reported that pod infection had direct effect on seed yield and stressed that the pod development stage is the most vulnerable stage of common beans for quick disease progress. Apparently, the severe pod infection (78.9%) which was recorded on the untreated control plots, could have contributed much for the estimated yield loss. As reported by Conner et al. [29], application of the foliar fungicide headline could reduce losses in seed yield and quality from bean anthracnose disease.

Disease progress rate (r)

Comparisons of the rates of development of disease among the treatments were subsequently made based on the Gompertz model by fitting the PSI data with dates of assessment (Table 3). The highest disease progress rate (0.98 unit-days), was computed from the untreated control, whereas the least (0.91 unit-days), from plots treated with mancozeb foliar spray. Among the bioagents, the least disease progress rate was attained in plots treated with *T. viride* (0.93 unit-days). But generally, high disease progress rates were observed in all the treatments. This could be due to high density of initial inoculum from the infected seeds.

Experimental studies have shown that the rates of disease increase were considerably influenced by the number of initial disease foci [26,30]. In an experiment with southern blight of processing carrot, the rate of disease increase generally increased as the number of initial foci increased [31].

Correlation between yield and disease parameters

Correlations among the disease parameters and with the yield parameters revealed highly significant (P<0.01) positive correlations between PSI and AUDPC and also between PSI and the percentage of pod infection. Significant (P<0.05) positive correlations were observed between AUDPC and the percentage pod infection and also with the disease progressive rate. Disease parameters, AUDPC and the terminal disease severity (PSI) showed significant (P<0.05) negative correlations with the seed yield. As reported by Sharma et al. [28], highly significant correlations between anthracnose severity and percentage reductions

in the number of seeds per pod and seed weight. Marcinkowska and Borucka [32] found significant positive correlation between the incidences of *C. lindemuthianum* in *P. vulgaris* seeds and leaf, pod and stem infection by the pathogen under natural field conditions. The disease progressive rates and percentage of pod infection showed non-significant negative correlations with seed yield. Similarly, the disease progressive rates showed non-significant positive correlations with the percentage of pod infection and terminal disease severity. The results suggest that reliable yield loss estimates could be made on the basis of the severity level by employing regression equations. Especially, for crops growing under epidemic conditions, PSI value recorded during the podding stage of common bean could be a good indication of the expected yield (Table 4).

Cost benefit analysis

Cost benefit analysis was computed for all the treatments using the partial budget analysis method (Table 5). Due to the seasonal pattern of production and marketing of common bean, fluctuations in prices are very common. Meanwhile, the price of common bean from November to December was assessed and an average of 17 Birr/kg was used to compute the total sale revenue and net benefit of the total produce obtained.

From the data analyzed, the highest variable cost (input and labor cost) was computed for mancolaxyl (13,939 Birr/ha). The highest net benefit (43,154 Birr/ha) was obtained from folpan and the least (7,503 Birr/ha) from the control. Folpan also gave the highest marginal benefit and cost benefit ratio. Input cost of the bioagents was found to be cheaper than the fungicidal treatments. Among them, *P. fluorescens* gave the highest marginal benefit. This bioagent also gave the highest marginal rate of return (8,740%) among all the treatments which indicated its economic advantage, especially for the resource poor farmers.

Conclusions

Common bean anthracnose is serious threat to bean production in the major common bean growing regions of Ethiopia, especially in areas like Ambo, where frequent rainfall and moderate temperature that prevail during the main cropping season, predispose the crop to attack by various pathogens including *C. lindemuthianum*. An alternative fungicide, folpan, with multi-site activities could be an important option for the management of the anthracnose disease. Similarly, the bioagents evaluated in the study were found to be economically important options that need to be further investigated, where *P. fluorescens* showed promising results in the control of common bean anthracnose.

References

1. National Agricultural Statistics Service (2013) Crops and Plants USDA-NASS: Washington DC, USA

2. Habtu A, Ivan S, Zadoks JC (1996) Survey of cropping practices and foliar disease of common beans in Ethiopia. Crop Protection 15: 179-186.

3. Central Statistical Authority (2010) Area under production of major crops. Statistical Bulletin.245 Addis Ababa, Ethiopia.

4. Kutangi E, Farrow A, Mutuoki T, Gebeyehu S, Karanja D, et al. (2010) Improving common bean productivity: An Analysis of socioeconomic factors in Ethiopia and Eastern Kenya. Baseline Report Tropical Legumes II. Centro Internacional de Agricultura Tropical-CIAT. Cali, Colombia.

5. Yesuf M (2005) Seed Borne Nature of *Colletotrichum lindemuthianum* and its Epidemic on Common Beans in the Major Bean Growing Areas of Ethiopia.A PHD Thesis in Tropical Agriculture. Graduate School, Kasetsart University.

6. Tu JC (1983) Epidemiology of anthracnose caused by *Colletotrichum lindemuthianum* on white bean (Phaseolus vulgaris) in Southern Ontario: Survival of the pathogen. Plant Disease 69: 402-404.

7. Fernandez M, Casares A, Rodr´ıguez R, Fueyo M (2000) Bean germplasm evaluation for anthracnose resistance and characterization of agronomic traits: a new physiological strain of *Colletotrichum lindemuthianum* infecting *Phaseolus vulgaris L.* in Spain. Euphytica 114: 143-149.

8. Tesfaye B (1997) Loss assessment study on haricot bean due to Anthracnose. Pest Management Journal of Ethiopia 1: 69-72.

9. Tesfaye B, Pretorius ZA (2005) Seed treatment and foliar application of fungicide for the control of bean anthracnose. Pest Management Journal of Ethiopia. 9: 57-62.

10. Amin M, Ayalew A, Dechassa N, Negeri M (2013). Effect of Integrated Management of Anthracnose (Colletotrichum lindemuthianum) on Plant and Seed Health of Common Bean in Hararge Highlands, Ethiopia. Journal of Science and Sustainable Development 1: pp.1-19

11. Pest Management Regulatory Agency (2008) Folpan 50 WP (Folpate) Fungicide Commercial. Makhteshim Agan of North America Inc.

12. Johnston SA (1986) Fungicide and Nematicide Update: Current Status of Fungicide Testing on Vegetable Crops. The American Phytopathological Society 70.

13. Padder BA, Sharma PN, Kapil R, Pathania A, Sharma OP (2010) Evaluation of Bioagents and Biopesticides against *Colletotrichum lindemuthianum* and its Integrated Management in Common Bean. Notulae Scientia Biologicae. 2: 72-76.

14. Hernandez A, Hernandez AN, Velazquez MG, Bigiramana Y, Audenaert K, et al. (2004) Rhizobacteria application to induce resistance in the plant-pathogen systems, bean-*Colletotrichum lindemuthianum* and tomato-Botrytis cinerea. Mex. J. Phytopathol. 22: 100-106.

15. Tesfaye B (2003) Biology and control of bean anthracnose (Colletotrichum lindemuthianum) in Ethiopia. Ph.D. Thesis, South Africa, Free State University.

16. El-Mohamedy RSR, El-Samad A, Hoda AM, Habib TSH, El-Bab (2011) Effect of using biocontrol agents on growth, yield, head quality and root rot control in Broccoli plants. International Journal of Academic Research Vol. 3: 71-80.

17. Raja U (2011) Green Max Agro Tech. Trichoderma Species Manufacturer and Whole Sale. Coimbatore, Tamil Nadu, India.

18. Kloepper JW, Schroth MN (1981) Development of a powder formulation of rhizobacteria for inoculation of potato seed pieces. Phytopathology. 71: 590-592.

19. Wheeler JBEJ (1969) An Introduction to Plant Diseases. Wiley, London 347.

20. Sharma P, Pathania K (2010) Evaluation of Bioagents and Biopesticides against *Colletotrichum lindemuthianum* and its Integrated Management in Common Bean. CSK HP Agricultural University, Molecular Plant Pathology Laboratory, Department of Plant Pathology, India

21. Campbell CL, Madden LV (1990) Introduction to Plant Disease Epidemiology. Jhon Wiley and Sons, New York.

22. SAS institute Inc. (2002) SAS/ stat guide for Personal Computer. Version, 9.1.3 editions. SAS Institute Inc. Carry, NC.

23. CIMMYT (1988) Farm agronomic data to farmer's recommendations: a training manual completely revised edition. International maize and wheat center, Mexico.

24. Araujo E, Zambolim L, Vieira C, Chaves GM, Arauio GA (1994) Reaction of seedlings and seeds of bean (Phaseolus vulgaris L.) to six physiological races of Colletotrichum lindemuthianum (Sacc.and Magn.) Scrib. Revista – Ceres 41: 584-594.

25. Schwartz HF Pastor-Corrales MA, eds. Bean production problems in the Tropics. 2nd ed. CIAT, Cali, Colombia.

26. Jerger MJ, Termorshuizen AJ, Nagtzaam MPM, Van den Bosch F (2004) The effects of spatial distributions of mycoparasites on biocontrol efficacy: a modeling approach. Biocontrol Science and Technology. 14: 359-373.

27. Yesuf M, Sangchote S (2005) Seed Transmission and Epidemics *Colletotrichum lindemuthianum* in the major common bean growing areas of Ethiopia. Kasetsart J. (Nat. Sci.) 39: 34-45.

28. Sharma PN, Sharma OP, Padder BA, Kapil R (2008) Yield loss assessment in common bean due to bean anthracnose (Colletotrichum lindemuthianum) under sub temperate conditions of north-western Himalayas. Indian Phytopathology 61: 323-330.

29. Conner FA, Kiehn SJ, Park HH, Muendel (2001) Crop Development Centre, University of Saskatchewan, Saskatoon

30. Xu XM, Ridout MS (1998) Effects of initial epidemic conditions, sporulation rate and spore dispersal gradient on the spatio-temporal dynamics of plant disease epidemics. Phytopathology 88: 1000-1012.

31. Smith VL, Campbell CL, Jenkins SF, Benson DM (1988) Effects of host density and number of disease foci on epidemics of southern blight of processing carrot. Phytopathology 78: 595-600.

32. Marcinkowska JZ, Borucka K (2001) *Colletotrichum lindemuthianum* in *Phaseolus vulgaris* seeds. Plant Breeding and Seed Sciences, 45: 59-64.

8

Effects of Fe Deficiency on Organic Acid Metabolism in *Pisum sativum* Roots

Nahida Jelali[1]*, Marc El Beyrouthy[2], Marta Dell'orto[3], Mohamed Gharsalli[1] and Wissem Mnif[4]

[1]*Laboratory of Plant Adaptation to Abiotic Stress, CBBC, Borj-Cedria Technopark , BP 901, 2050 Hammam-Lif, Tunisia*
[2]*Faculty of Agricultural and Food Sciences, Holy Spirit University of Kaslik, P.O. Box 446, Jounieh, Lebanon*
[3]*Dipartimento di Scienze Agrarie e Ambientali - Produzione, Territorio, Agroenergia, Università degli Studi di Milano, Via Celoria 2, I-20133 Milano, Italy*
[4]*LR11-ES31 Biotechnology and Bio-Enhancement of Geo Resources, Higher Institute of Biotechnology of Sidi Thabet Sidi Thabet BiotechPole, 2020, Tunisia Manouba University, Tunisia*

Abstract

Iron deficiency induces several responses to iron shortage in plants. Metabolic changes occur to sustain the increased iron uptake capacity of Fe-deficient plants. The aim of this work was to investigate the impact of Fe deficiency on the organic acid metabolism in *Pisum sativum* roots. For this purpose, seedlings of *Pisum sativum* (cv. Douce) were grown under controlled conditions, in the presence of iron sufficient (C) or deficient (D) mediums. Our results showed that PEPC activity increased by 290% in root extracts of Fe deficient plants, compared to the control. Citrate concentration increased in Fe deficient *Pisum* roots (114% of the control). As well, MDH, CS and ICDH activities showed a marked increase in roots subjected to D treatment. However, the extent of stimulation was especially important for MDH and ICDH activities (99% and 150% of the control, respectively). These data suggest that the capacity of Douce cultivar to increase organic acid metabolism enzyme activities under iron deficiency is related to its better Fe-use efficiency, which indicate the tolerance level of this cultivar to iron chlorosis.

Keywords: Iron deficiency; Pea; PEPC activity, Organic acids, Metabolic responses, *Pisum sativum*

Introduction

Iron deficiency is a yield-limiting factor, and a worldwide problem in crop production of many agricultural regions, particularly in calcareous soils [1]. Theoretically, total soil-Fe content would be sufficient to meet Fe needs of plants; however, most of the Fe in the soil is present as inorganic forms, poorly available for root, thus exposing the plant to severe deficiency of this nutrient [2,3]. Since Fe is a vital element for plants, as it is essential for the proper functioning of metabolic processes related to electron transport, such as respiration and photosynthesis, and for chlorophyll biosynthesis [4,5]; plants have evolved different adaptive mechanisms to mobilize and increase the availability of Fe from soil [6].

Under such tricky environmental conditions, dicots and non-graminaceous monocots termed Strategy I plants induce a set of physiological and biochemical responses [7]. As anticipated by Zocchi [8], metabolic responses were shown to be deeply involved in the response to iron deficiency, with particular regard to the central role played by Phosphoenolpyruvate Carboxylase (PEPC). As well, several enzymes of Krebs cycle, especially those involved in the organic acid metabolism, such as Citrate Synthase (CS), Isocitrate Dehydrogenase (ICDH), Fumarase and Aconitase has been assumed to be important for the whole response to Fe deficiency [9], and it has been shown to occur in roots of Strategy I plant species [10]. For this reason, research on carboxylate exudation (e.g. citrate, malate, etc.) has gained interest globally, as carboxylate release into the rhizosphere enhances the availability of Fe, by accessing soluble forms of this nutrient to the plant [11]. It has been shown that root PEPC activity increased in Fe-deficient plants, and correlates with an enhancement of organic compounds accumulation in roots (e.g. organic acids, phenols and amino acids) [10]. In this perspective, organic acids were shown to be the source for H+ released by the roots [12]. Citrate and malate are the main carboxylate anions accumulating in the root tissues and exudates, in response to Fe deficiency [13,14].

Previous studies investigating the morpho-physiological responses to Fe deficiency in pea showed that Douce cultivar has a tolerant behavior to Fe deficiency [15]. In the present work, using 18-d-old Fe-deficient pea plants (*Pisum sativum* L., cv 'Douce'), we studied the effect of Fe-deficiency on organic acids and phenols concentrations in roots and exudates, together with the activities of some root enzymes related to the organic acids metabolism (PEPC, MDH, CS and ICDH).

Materials and Methods

Plant material and growth conditions

Pea seeds (cv. Douce) obtained from the Tunisian National Institute of Agronomic Research (INRAT) were sterilized in a saturated solution of calcium hypochlorite (30%) for 2 min, and then abundantly rinsed in distilled water. Seeds were germinated for 6 days on filter paper constantly moistened with 0.1 mM CaSO4 and then grown for 12 days in a continuously aerated nutrient solution (pH adjusted at 6.0 with 1 M KOH), being exposed to 30 µM Fe (Fe-EDTA); thereafter, the plants were transferred for a further week to a Fe-free nutrient solution (Fe-deficient). The nutrient solution was renewed every three days. The composition of the nutrient solution was: (mM) 1.25 Ca (NO$_3$)$_2$, 1.25 KNO$_3$, 0.5 MgSO$_4$, 0.25 KH$_2$PO$_4$ and (µM) 10 H$_3$BO$_3$, 1 MnSO$_4$, 0.5 ZnSO$_4$, 0.05 (NH4)$_6$Mo$_7$O$_{24}$ and 0.4 CuSO$_4$. Two treatments were established as follows: control (presence of 30 µM Fe(III)-EDTA: C); Fe

*Corresponding author: Dr. Nahida Jelali, Laboratory of Plant Adaptation to Abiotic Stress, CBBC, Borj-Cedria Technopark , BP 901, 2050 Hammam-Lif, Tunisia. E-mail: nahidajelali@yahoo.fr

deficiency (D). The pH was adjusted to 6.0 with NaOH, for both C and D treatments. $NaHCO_3$ and $CaCO_3$ were added to the nutrient solution to simulate the effect of a calcareous soil on Fe availability. Aerated hydroponic cultures were maintained in a growth chamber, with a day/night regime of 16/8 h, 24C/18C regime, PPFD of 200 $\mu mol\ m^{-2}\ s^{-1}$ at the plant level, and a relative humidity of 60% in the dark and 80% in the light.

Determination of plant dry weight

To determine the dry weight (DW) of roots, twelve plants from each treatment (n=12) were sampled at the end of the treatment. Roots were dried at 65°C, and dry weight was recorded as grams per plant.

Soluble protein extraction and cytosolic enzyme assays

Soluble protein extraction for measuring enzyme activities was performed, as reported by De Nisi and Zocchi [16]. Roots were homogenized in a mortar 2-4°C in one volume of a buffer containing 50 mM Tris–HCl (pH 7.5), 10 mM $MgCl_2$, 10% (v/v) glycerol, 1 mM EDTA, 14 mM μ-mercaptoethanol and 1 mM PMSF. To avoid or minimize proteolysis, 10 μg Ml^{-1} leupeptin were added. The homogenate was centrifuged at 13000×g for 15 min, and the supernatant was again centrifuged at 100,000×g for 30 min. The extracted soluble proteins were dialysed against the same homogenization buffer and used for enzyme activity assays directly, or after storing in liquid N_2.

Phosphoenolpyruvate carboxylase (PEPC; EC 4.1.1.31) was determined, as reported by De Nisi and Zocchi [16]. Reaction was started by adding aliquots of protein extracts, and the enzymatic assay was performed at 25°C in 1.5 ml final volume. Oxidation of NADH was followed spectrophotometrically at 340 nm.

Malate dehydrogenase (MDH; EC 1.1.1.37) activity was determined with oxaloacetate as a substrate, by measuring the decrease in absorbance at 340 nm due to the enzymatic oxidation of NADH, as reported by López-Millán et al. [17], with some modifications. The reaction was carried out with 5 μL of extract in 0.2 mM NADH, 0.5 mM oxaloacetate and 94 mM phosphate buffer (pH 7.5). The enzyme assay was performed at 26°C and in 1 mL final volume.

Citrate synthase (CS; EC 4.1.3.7) was assayed spectrophotometrically, by monitoring the reduction of acetyl CoA to CoA with 5-50-dithio-bis-2-nitrobenzoic (DTNB) acid at 412 nm. The reaction was carried out with 50 μl of extract in 0.1 mM DTNB, 0.36 mM acetyl CoA, 0.5 mM oxalacetate and 100 mM Tris–HCl, pH 8.1 [17].

Isocitrate dehydrogenase (ICDH; EC 1.1.1.42) activity was determined by monitoring the reduction of $NADP^+$ at 340 nm, with 50 μL of extract in a reaction mixture containing 3.5 mM $MgCl_2$, 0.41 mM $NADP^+$, 0.55 mM isocitrate and 88 mM imidazole buffer, pH 8.0 [17].

Root exudate collection

Root exudates were collected, as described by Zocchi et al. [7]. Five plants from each treatment, with four replicates, were transferred to vessels containing 250 ml of distilled water, and root exudates were collected for a period of 4 h. After collection, Micropur was added to the exudates to prevent decomposition of organic matter by microorganisms. The samples were then cooled to 0°C and filtered through filter paper. Solutions containing root exudates were thereafter concentrated by evaporation, and again filtered on a 0.22 μm cellulose acetate filters. This final solution was used for quantification of citrate, malate, and phenols.

Quantification of malate and citrate

At the end of the treatment, entire root systems of 18-day-old plants were excised, rinsed in distilled water, homogenized in the presence of 5 mL of 10% (v/v) perchloric acid and centrifuged for 15 min at 10,000×g. Supernatant pH was brought to 7.5 with 0.5 M K_2CO_3, to neutralize the acidity and to precipitate the perchlorate. The extract was clarified with another centrifugation at 15,000×g for 15 min. Citric and malic acid contents were determined enzymatically, according to Rabotti et al. [18]. The recovery of both organic acids was more than 90%, as determined by the use of an internal standard.

Statistical analysis

Variance analysis of data (TWO-WAY ANOVA) was performed using the SPSS 10.0 program, and means were separated, according to Duncan's test at $p \leq 0.05$. Data shown are means of twelve (DW), and five (organic acids concentrations and enzyme activities) replicates for each treatment.

Results

Root dry weight

Although grown under Fe-deficient conditions for 7 days, *Pisum sativum* root DW was not affected by such constraint (Figure 1); no significant decrease was recorded for this plant organ.

PEPC, MDH, CS and ICDH activities

Since the accumulation of organic acids in Fe-deficient plants is attributable to the stimulation of PEPC activity [10], this metabolic enzyme and others with a key role in the organic acid synthesis, were assayed in Douce cultivar roots grown under Fe deficiency conditions.

As shown in table 1, the (D) treatment resulted in a significant increase of PEPC activity in root extracts of Douce cultivar (1.9-fold increase), compared to the control. In addition, the D treatment led to a significant increase of root MDH, CS and ICDH activities in *P. sativum* plants. The extent of stimulation was especially important for MDH and ICDH activities (99% and 150% of the control, respectively).

Concentration of organic acids in roots and exudates

Under Fe deficiency, the root concentration of citrate and malate underwent a significant increase in Douce cultivar (Table 2). The values recorded were larger for citrate (114% of the control) than malate (90% of the control). As well, citrate and malate concentrations were stimulated in root exudates in Fe deficient *P. sativum* roots (Table 2).

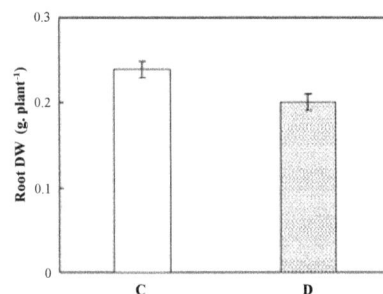

Figure 1: Effect of Fe deficiency on root dry weight (DW) in Pisum sativum plants (cv. Douce). Values are means ± SE (n=12), and differences between means were compared using Duncan's test at (P<0.05).

	Treatments	
	C	D
PEPC activity (nmol NADH min⁻¹ mg⁻¹ prot)	180 ± 12^b	530 ± 10^a
MDH activity (nmol NADH min⁻¹ mg⁻¹ prot)	4020 ± 60^b	8000 ± 40^a
CS activity (nmol g⁻¹ FW min⁻¹)	618 ± 30^b	860 ± 50^a
ICDH activity (nmol g⁻¹ FW min⁻¹)	200 ± 12^b	500 ± 21^a

Values are means ± s.e. and differences between means were compared using Duncan's test (P=0.05).
Different letters correspond to significantly different values.

Table 1: PEPC, MDH, CS and ICDH activities in roots of *P. sativum* plants (cv. Douce) grown on iron-sufficient (C), or iron-deficient medium (D).

	Treatments	
	C	D
Citrate in roots (µg g⁻¹ FW)	135 ± 10^b	290 ± 18^a
Malate in roots (µg g⁻¹ FW)	304 ± 19^b	578 ± 22^a
Citrate in exudates (µg g⁻¹ FW)	3 ± 0.5^a	6 ± 1^a
Malate in exudates (µg g⁻¹ FW)	1.5 ± 0.1^b	2.1 ± 0.2^a

Values are means ± s.e. and differences between means were compared using Duncan's test (P=0.05). Different letters correspond to significantly different values.

Table 2: Amounts of citrate and malate in roots and exudates of *P. sativum* plants (cv. Douce) grown on iron-sufficient (C), or iron-deficient medium (D).

Discussion

As shown above in our results, Fe deficiency elicited an increase in PEPC activity in *P. sativum* plants. These results suggest an important role of this enzyme in the adaptation of plant to environmental changes. Similar results were described in other plant species, such as pepper, kiwifruit and sugar beet [13,19,20]. In addition, all the organic acid related enzymes determined in our work (MDH, CS and ICDH) showed a significant increase in their activities, under Fe deficiency conditions in *P. sativum*.

The extent of stimulation was especially important for MDH and ICDH activities (Table 1). In accordance with our findings, López-Millán et al. [17] reported that the increase in PEPC activity in Fe-deficient roots of sugar beet would lead to carbon fixation, accumulation of organic acids, and in turn, to increased activities of ICDH, MDH, and G6PDH enzymes. The extent of stimulation of all enzymes measured was markedly lower than that reported in sugar beet [17], and tomato [21]. These observations could be ascribed to the metabolic differences of these species that may be due to their different Fe efficiency. Iron deficiency has been generally shown to cause increase in the organic acid concentrations in roots, leaves and exudates of different plant species [10]. The current study revealed that Fe deficiency led to an increase of citrate and malate concentration in *P. sativum* roots (Table 2). Similar results were found in other species, such as barley, sorghum and maize [22]. It has been shown that citrate could be used as a biochemical marker of Fe-chlorosis tolerance, as suggested by Ollat et al. [14] and Rombolà et al. [13].

Moreover, in the present study, we found small amounts of citrate and malate exuded from *P. sativum* roots under Fe deficiency conditions (Table 2), compared to the concentration of these organic compounds in roots. Based on model calculations, Jones et al. [23] suggested that even the low exudation of citrate in roots might be sufficient for Fe(III) solubilization, at a rate which would be high enough to cover the plant Fe requirements. On the other hand, another possible explanation could be that pea plants prevent organic acids decrease from the tissue pool, thus increasing the availability of these compounds for translocation through the xylem.

Taken together, the observed changes, mainly the increase in PEPC activity and organic acid concentrations of *Pisum* roots, may contribute to the iron deficiency stress response. In fact, increased PEPC activity would sustain the carbon replenishment in the tricarboxylic acid cycle, the enhanced synthesis of malic, citric and amino acids [8-24]. Furthermore, since iron deficiency can induce the accumulation of other metals such as Mn, Zn and Cu [25], organic acids could be involved in metal binding to avoid oxidative damage due to other catalytic ions [26].

Conclusion

In summary, we conclude that Douce cultivar showed several responses characteristics of Strategy I-efficient plants under iron deficiency conditions, such as enhanced root PEPC activity and malic, citric concentrations, in addition to some enzyme activities involved in the organic acid metabolism. These physiological responses could be partly related to tolerance in biochemical and molecular studies, would be helpful to gain more knowledge concerning the regulation of intracellular ion transport, when plants are under Fe deficiency conditions.

Acknowledgments

This work was funded by the Tunisian Ministry of Higher Education, Research and Technology (LR02CB02). Thanks are also to the anonymous reviewers for additional valuable comments.

References

1. Mengel K, Kirk E, Kosegarten H (2001) Iron in mineral nutrition. Kluwer, Dordrecht 553-571.

2. Lindsay WL, Schwab AP (1982) The chemistry of iron in soils and its availability to plants. J Plant Nutr 5: 821-840.

3. Bavaresco L, Poni S (2003) Effect of calcareous soil on photosynthesis rate, mineral nutrition and source-sink ratio of table grape. J Plant Nutr 26: 2123-2135.

4. Kader JC, Delseny M (2007) Advances in Botanical Research. Incorporating Advances in Plant Pathology.

5. Briat JF, Vert G (2004) Acquisition and management of iron by plants Workbook French Studies and Research. Agriculture 13: 183-201.

6. Deng H, Ye ZH, Wong MH (2009) Lead, zinc and iron (Fe²⁺) tolerances in wetland plants and relation to root anatomy and spatial pattern of ROL. Environ Exp Bot 65: 353-362.

7. Zocchi G, De Nisi P, Dell'Orto M, Espen L, Gallina PM (2007) Iron deficiency differently affects metabolic responses in soybean roots. J Exp Bot 58: 993-1000.

8. Zocchi G (2006) Metabolic changes in iron-stressed dicotyledonous plants. Iron Nutrition in Plants and Rhizospheric Microorganisms 359-370.

9. Espen L, Dell'Orto M, De Nisi P, Zocchi G (2000) Metabolic responses in cucumber (*Cucumis sativus L*) roots under Fe deficiency: a 31P-nuclear magnetic resonance *in vivo* study. Planta 210: 985-992.

10. Abadía J, López-Millán AF, Rombolà A, Abadía A (2002) Organic acids and Fe deficiency: a review. Plant Soil 241: 75-86.

11. Ryan PR, Delhaize E, Jones DL (2001) Function and mechanism of organic anion exudation from plant roots. Annu Rev Plant Physiol Plant Mol Biol 52: 527-560.

12. Landsberg EC (1986) Function of rhizodermal transfer cells in the Fe stress response mechanism of Capsicum annuum L. Plant Physiol 82: 511-517.

13. Rombolà AD, Brüggemann W, López-Millán AF, Tagliavini M, Abadía J, et al. (2002) Biochemical responses to iron deficiency in kiwifruit (*Actinidia deliciosa*). Tree Physiol 22: 869-875.

14. Ollat N, Laborde B, Neveux M, Diakou-Verdin P, Renaud C, et al. (2003) Organic acid metabolism in roots of various grapevine (*Vitis*) rootstocks submitted to iron deficiency and bicarbonate nutrition. J Plant Nutr 26: 2165-2176.

15. Jelali N, Salah BI, M'sehli W, Donnini S, Zocchi G, et al. (2011) Comparison of three pea cultivars (*Pisum sativum*) regarding their responses to direct and bicarbonate-induced iron deficiency. Sci Hortic 129: 548-553.

16. De Nisi P, Zocchi G (2000) Phosphoenolpyruvate carboxylase in cucumber (*Cucumis sativus L.*) roots under iron deficiency: activity and kinetic characterization. J Exp Bot 51: 1903-1909.

17. López-Millán AF, Morales F, Andaluz S, Gogorcena Y, Abadía A, et al. (2000) Responses of sugar beet roots to iron deficiency Changes in carbon assimilation and oxygen use. Plant Physiol 124: 885-898.

18. Rabotti G, De Nisi P, Zocchi G (1995) Metabolic implications in the biochemical responses to iron deficiency in cucumber (*Cucumis sativus L.*) roots. Plant Physiol 107: 1195-1199.

19. Landsberg EC (1981) Organic acid synthesis and release of hydrogen ions in response to Fe deficiency stress of mono- and dicotyledonous plant species. J Plant Nutr 3: 579-591.

20. Jiménez S, Ollat N, Deborde C, Maucourt M, Rellán-Álvareze R, et al. (2011) Metabolic response in roots of Prunus rootstocks submitted to iron chlorosis. J Plant Physiol 168: 415-423.

21. López-Millán AF, Morales F, Gogorcena Y, Abadía A, Abadía J (2009) Metabolic responses in iron deficient tomato plants. J Plant Physiol 166: 375-384.

22. Alhendawi RA, Römheld V, Kirkby EA, Marschner H (1997) Influence of increasing bicarbonate concentrations on plant growth, organic acid accumulation in roots and iron uptake by barley, sorghum, and maize. J Plant Nutr 20: 1731-1753.

23. Jones DL, Darrah PR, Kochian LV (1996) Critical evaluation of organic acid mediated iron dissolution in the rhizosphere and its potential role in root iron uptake. Plant Soil 180: 57-66.

24. Jiménez S, Gogorcena Y, Hévin C, Rombolà AD, Ollat N (2007) Nitrogen nutrition influ- ences some biochemical responses to iron deficiency in tolerant and sensitive genotypes of *Vitis*. Plant Soil 290: 343-55.

25. Jiménez S, Morales F, Abadía A, Abadía J, Moreno MA, et al. (2009) Elemental 2-D mapping and changes in leaf iron and chlorophyll in response to iron re-supply in iron-deficient GF 677 peach-almond hybrid. Plant Soil 315: 93-106.

26. Sharma SS, Dietz KJ (2006) The significance of amino acids and amino acid-derived molecules in plant responses and adaptation to heavy metal stress. J Exp Bot 57: 711-726.

9

Genetic and Correlation Studies in Double Genotypes of Tuberose (*Polianthes tuberosa*) for Assessing the Genetic Variability

Ranchana P*, Kannan M and Jawaharlal M

Department of Floriculture and Landscaping, HC& RI, TNAU, Coimbatore-3, India

Abstract

Five genotypes of tuberose (double) were evaluated for twelve different parameters to ascertain the genetic variability and association among the characters during the year 2011-12 at Tamil Nadu Agricultural University, Coimbatore. The results of the experiment revealed that 'Suvasini' showed its superiority for certain parameters *viz.*, plant height, number of leaves per plant, number of florets/ spike, length of the floret, weight of florets per spike, number of spikes/m² and yield of florets/ plot (2*2 m). The phenotypic coefficient of variation (PCV) was higher than genotypic coefficient of variation (GCV) for all twelve characters studied. The higher PCV and GCV estimates were found for number of florets/spike. High heritability with high genetic advance was observed for number of florets per spike, number of spikes/m² rachis length and yield of florets per plot (2×2 m). The correlation studies revealed that plant height exhibited positive correlation with spike length, yield of florets/plot (2×2 m), number of florets per spike, flowering duration, number of leaves per plant, weight of florets per spike, number of spikes/m², rachis length and length of the floret. There exists a positive relationship of number of leaves per plant with weight of florets per spike, yield of florets/plot (2×2 m), spike length, flowering duration, and number of florets per spike, length of the floret, number of spikes/m² and rachis length. Spike length exhibited positive and significant association with yield of florets/ plot (2×2 m), number of florets per spike, weight of florets per spike, number of spikes per m², length of the floret and rachis length.

Keywords: Tuberose; Double types; Heritability; Genetic advance; Correlation

Introduction

The top cut flowers like rose, carnation, gladiolus, tuberose, chrysanthemum, etc., are commonly and frequently demanded in both local as well as international market. Among them, tuberose (*Polianthes tuberosa*) is one of the most important flowers used for both cut and loose flower purpose. There are only two types of tuberose (Single and Double) cultivated in the world. Among these two types, double flowered types are highly preferred for cut flower purpose. As of commercial importance, the genotypes *viz.* Calcutta Double, Hyderabad Double, Pearl Double, Suvasini and Vaibhav are cultivated throughout India. It is an ornamental bulbous plant, native of Mexico and belongs to family *Amaryllidaceae*. Waxy white flowering spikes of single as well as double flower tuberose impregnate the atmosphere with their sweet fragrance and longer keeping quality of flower spikes [1,2], and are in great demand for making floral arrangement and bouquets in major cities of India. A huge quantum of variability exists in this crop with respect to growth habit, flowering behavior, etc. Inspite of such variability, very few are having desirable characters in terms of yield and quality. Considering the fact, there is a need for selection as well as maintenance of good germplasm. The study of interrelationship of various characters in the form of correlation is an important aspect in crop breeding. Knowledge of correlation studies helps the plant breeder to ascertain the real components of yield and provide an effective basis of selection. The characters contributing significantly to desirable traits can be significantly identified, and can be used as alternate selection criteria in crop improvement programme. Very little work on this aspect has been reported so far, hence the present study to find out the association among important quantitative characters in tuberose.

Materials and Methods

The study was carried out at Botanical gardens, Tamil Nadu Agricultural University, Coimbatore, during the year 2011-2012. It is situated at 11°02" N latitude, 76°57" E longitude and 426.76 m above mean sea level. Experimental material consists of five genotypes of tuberose *viz.*, Calcutta Double, Hyderabad Double, Pearl Double, Suvasini and Vaibhav. The experiment was laid out in randomized block design (RBD) with three replications. The soil was brought to a fine tilth by giving four deep ploughings. Weeds, stubbles, roots, etc. were removed. At the time of last ploughing, FYM was applied at the rate of 25 t ha⁻¹. After levelling, raised beds of 1 m width and convenient length were formed and the medium sized bulbs (3.0-3.5 cm diameter) of about 25 grams were planted, with a spacing of 45×20 m which accommodates 11 plants per m². Uniform cultural practices were followed throughout the experimentation. The data were recorded on five plants from each genotype in each replication for 12 characters *viz.*, days taken for sprouting of bulb (days), plant height (cm), number of leaves per clump, days to spike emergence, flowering duration, spike length (cm), rachis length (cm), number of florets/spike, length of the floret, weight of the florets/spike, number of spikes/m², yield of florets/plot. Data were analysed and presented in tabular form. Data were put to statistical analysis as per [3] Genetic parameters like genotypic coefficient of variation (GCV) and phenotypic coefficient of variation (PCV) were estimated according to Burton and Vane [4], and heritability as suggested by Weber and Moorthy [5] Correlation analysis was carried out as per the formulae suggested by Fisher [6]. The significance of phenotypic and genotypic correlation coefficients

*Corresponding author: Ranchana P, Department of Floriculture and Landscaping, HC& RI, TNAU, Coimbatore-3, India, E-mail: ranchanahorti@gmail.com

was tested against 'r' value given in Fisher and Yates table [7] at (n-2) degrees of freedom.

Results and Discussion

A significant variation in growth was observed among the genotypes with regard to days taken for sprouting of bulb, number of leaves per plant and plant height under tropical condition (Table 1 and 2). The variations among the growth parameters may be due to their diversified origin, and also evolution of the particular genotype as a morphotype in their specific geographical location. This offers scope for selecting genotypes with better performance under tropical condition.

Mean performance of the cultivars for growth parameters reflected the variation among the cultivars. Among the genotypes, significantly less number of days taken for sprouting of bulbs (12.32) was recorded in 'Suvasini', followed by 'Vaibhav' (12.67), and more number of days was taken by 'Hyderabad Double' (16.15 days). Maximum plant height (86.25 cm) was noticed in 'Suvasini'. This is in accordance with the results of Gudi [8]. 'Suvasini' also produced maximum number of leaves/ plant (270) followed by 'Vaibhav' (250), while minimum number of leaves was recorded in 'Hyderabad Double' (235). The differences among the varieties for vegetative characters are attributed to their variation in their genetic makeup [9]. The number of days taken for spike emergence was less (84 days) in 'Suvasini', followed by 'Vaibhav' (85 days), while it was more in 'Hyderabad Double' (89 days). Similarly, the duration of flowering was also significantly more in Suvasini (12.40 days), followed by Vaibhav (11.43 days). This is in line with the findings of Patil et al. [10].

Suvasini produced spike with maximum length of 71.25 cm followed by 'Vaibhav' 66.38 cm, and it was minimum in 'Hyderabad Double' (53.87 cm). The rachis length was significantly higher in 'Vaibhav' (54.00 cm), followed by 'Suvasini' (44.00 cm), and it was minimum in 'Hyderabad Double' (33.95 cm). The variation in spike length and rachis length in different genotypes might be due to variation in their intrinsic factor. 'Suvasini' showed its superiority for number of florets/ spike (54.00), followed by 'Vaibhav' (44), and it was minimum in 'Pearl Double' (30). The increased floret length was noticed in 'Suvasini'

(7.50), and it was lowest in 'Hyderabad Double' (6.70). This finding is in consonance with the findings of Patil et al. [10]. Weight of florets/spike was maximum in 'Suvasini' (146.88 g), followed by 'Vaibhav' (119.24 g). This might be due to the increased number of florets/spike. The increased number of spikes/m² and yield of florets/plot (4*1 m) were noticed in Suvasini (34.10 and 3.42 kg). The maximum yield may be accorded due to its capacity to produce more number of florets per spike, increased floret length and weight of florets/spike.

The phenotypic coefficient of variation (PCV) and genotypic coefficient of variation (GCV) (Table 3) was the highest for number of florets/spike (24.59, 24.38), suggesting that this character is under genetic control. Hence, these characters can be relied upon selection for further improvement. The phenotypic coefficient of variation (PCV) was higher than genotypic coefficient of variation (GCV) for all the characters under study, indicating the role of environment in expression of genotype. Similar results were also reported by Misra et al. (1987) in dahlia and Sheela et al. [11] in *heliconia*. Minimum values of phenotypic coefficient of variation (PCV) and genotypic coefficient of variation (GCV) were recorded for days to spike emergence (3.58, 1.49), length of the floret (5.04, 3.93) and number of leaves per plant (6.14, 5.25). This type of findings indicated that very minimum variation existed among the genotypes with respect to these characters.

High heritability, coupled with high genetic advance, was observed for number of florets per spike (98.35, 69.81), number of spikes/m² (96.93, 65.78), rachis length (96.91, 45.18) and yield of florets per plot (95.78, 40.20). This indicates the lesser influence of environment in expression of these characters and prevalence of additive gene action in their inheritance. Hence, these traits are found suitable for selection. High heritability with moderate genetic advance was recorded for weight of florets per spike (94.67, 46.64), spike length (93.79, 44.46), flowering duration (92.83, 32.39), days taken for sprouting of bulb (91.91, 22.26) and plant height (90.52, 19.17), suggesting the presence of both additive and non-additive gene actions, and simple selection offers best possibility of improvement of this trait. The estimate of heritability was high with low genetic advance as percentage of mean for length of the floret (60.68, 6.30), days taken for spike emergence (77.38,

S.NO	Genotypes	Days taken for sprouting of bulb (days)	Plant height (cm)	No. of leaves per plant (Nos.)	Days to spike emergence	Flowering duration (days)
1.	Calcutta Double	13.25	73.92	246.00	88.00	10.48
2.	Hyderabad Double	16.15	68.87	235.00	89.00	9.12
3.	Pearl Double	14.62	69.70	238.00	86.00	10.39
4.	Suvasini	12.32	86.25	270.00	84.00	12.40
5.	Vaibhav	12.67	81.38	250.00	85.00	11.43
	SE(D)	0.38	1.96	6.45	0.62	0.28
	CD (0.5)	0.87	4.53	14.88	1.26	0.64

Table 1: Performance of tuberose genotypes for growth parameters (2011- 2012).

S.No.	Genotypes	Spike length (cm)	Rachis length (cm)	Number of florets/ spike (Nos.)	Length of the floret (cm)	Weight of florets per spike (g)	Number of spikes/m² (Nos.)	Yield of florets/ plot (2*2 m) (kg)
1.	Calcutta Double	58.92	38.75	35.00	7.10	112.36	32.00	2.57
2.	Hyderabad Double	53.87	33.95	34.00	6.70	108.56	21.01	2.48
3.	Pearl Double	54.70	42.67	30.00	7.40	109.32	31.50	2.42
4.	Suvasini	71.25	44.00	54.00	7.50	146.88	34.10	3.42
5.	Vaibhav	66.38	54.00	44.00	7.20	119.24	33.75	3.26
SE(D)		1.57	5.08	6.02	0.08	6.07	0.25	0.07
CD (0.5)		3.62	11.12	12.34	0.19	14.08	1.52	0.16

Table 2: Performance of tuberose genotypes for spike and yield parameters (2011- 2012).

1.28) and number of leaves per plant (73.06, 9.25), which indicated that high heritability were due to non-additive gene effects and influence of environment. Hence, there is a limited scope for selection. These results are in accordance with the findings of Sheikh and John [12] in Iris.

The genotypic and phenotypic correlation coefficients were computed in all possible combinations for twelve characters, and are presented in Table 4 and 5. Correlation coefficient analysis measures the mutual relationship between various plant characters, and determines the component characters on which selection is based for genetic improvement for a particular character [13]. A positive correlation between desirable characters is favorable to the plant breeder, because it helps in simultaneous improvement of both the characters. In the present study, genotypic correlation coefficients were found to be higher than phenotypic correlation coefficients for most of the characters, indicating a strong inherent association between various characters, and were masked by environmental component with regard to phenotypic expression

Correlation coefficient analysis measures the mutual relationship between various plant characters and determines the component characters on which selection is based for genetic improvement for a particular character [13]. A positive correlation between desirable

characters is favorable to the plant breeder because it helps in simultaneous improvement of both the characters. In the present study, genotypic correlation coefficients were found to be higher than phenotypic correlation coefficients for most of the characters, indicating a strong inherent association between various characters, and were masked by environmental component with regard to phenotypic expression. Similar results were obtained by Singh [14] in antirrhinum. The study showed a highly significant and positive correlation between days taken for sprouting of bulb with days to spike emergence (0.671). This trait however, showed negative correlation with flowering duration (-0.978), number of leaves per plant (-0.954), number of spikes per m^2 (-0.952), plant height (-0.939), spike length (-0.927), weight of florets per spike (-0.874), yield of florets per plot (2*2 m) (-0.803), rachis length (-0.802), length of the floret (-0.761) and number of florets per spike (-0.736).

Highly significant and positive correlations for plant height was observed with spike length (1.001), yield of florets per plot (2*2 m) (0.975), number of florets per spike (0.971), number of leaves per plant (0.958), flowering duration (0.944), weight of florets per spike (0.904), number of spikes per m^2 (0.695), rachis length (0.652) and length of the floret (0.623). The trait however, showed negative correlation with

S.NO.	Characters	GCV	PCV	HERT	GA (%) of mean
1	Days taken for sprouting of bulb	11.27	11.76	91.91	22.26
2	Plant height	9.78	10.28	90.52	19.17
3	Number of leaves per plant	5.25	6.14	73.06	9.25
4	Days to spike emergence	1.49	3.58	77.38	1.28
5	Flowering duration	11.28	11.71	92.83	32.39
6	Spike length	12.26	12.66	93.79	44.46
7	Rachis length	17.35	17.62	96.91	45.18
8	Number of florets/ spike	24.38	24.59	98.35	69.81
9	Length of the floret	3.93	5.04	60.68	6.30
10	Weight of florets per spike	13.29	13.66	94.67	46.64
11	Number of spikes/ m^2	17.64	17.92	96.93	65.78
12	Yield of florets/ plot	14.98	15.31	95.78	40.20

Table 3: Estimates of variability and genetic parameters for flower yield and its components.

S.NO	1	2	3	4	5	6	7	8	9	10	11	12
1.	1.000	-0.772	-0.637	1.036*	-0.830	-0.792	-0.646	-0.690	-0.477	-0.623	-0.852	-0.698
2.		1.000	0.939**	-0.347	0.940**	0.999**	0.639**	0.955**	0.581**	0.900**	0.685**	0.971**
3.			1.000	-0.157	0.914**	0.928**	0.413**	0.897**	0.608**	0.946**	0.649**	0.931**
4.				1.000	-0.445**	-0.389**	-0.411**	-0.416**	-0.125**	-0.364**	-0.427**	-0.401**
5.					1.000	0.940*	0.683*	0.854*	0.806*	0.868*	0.859*	0.896*
6.						1.000	0.639**	0.961**	0.595**	0.899**	0.685**	0.974**
7.							1.000	0.470*	0.544*	0.320*	0.734*	0.500*
8.								1.000	0.456**	0.946**	0.490**	0.993**
9.									1.000	0.621**	0.812**	0.554**
10.										1.000	0.516*	0.963*
11.											1.000	0.550*
12.												1.000

1. Days taken for sprouting of bulb	7. Rachis length
2. Plant height	8. Number of florets/ spike
3. No. of leaves per plant	9. Length of the floret
4. Days to spike emergence	10. Weight of florets per spike
5. Flowering duration	11. Number of spikes/m^2
6. Spike length	12. Yield of florets/ plot (2 * 2 m)

Note:

*Significant at 5% level
**Significant at 1% level

Table 4: Phenotypic correlation coefficient among different characters in tuberose (double).

S.NO	1	2	3	4	5	6	7	8	9	10	11	12
1.	-0.588	-0.939	-0.954	0.671**	-0.978	-0.927	-0.803	-0.736	-0.761	-0.874	-0.952	-0.803
2.		1.000	0.958**	-1.574	0.944**	1.001**	0.652**	0.971**	0.623**	0.904**	0.695**	0.975**
3.			1.000	-1.759	0.941**	0.965**	0.438**	0.981**	0.730**	0.994**	0.665**	0.987**
4.				1.000	-1.706	-1.518	-1.384	-1.276	-2.136	-1.406	-1.418	-1.430
5.					1.000	0.944*	0.695*	0.859*	0.851*	0.875*	0.862*	0.899*
6.						1.000	0.653**	0.967**	0.604**	0.905**	0.696**	0.974**
7.							1.000	0.481*	0.566*	0.347*	0.742*	0.517*
8.								1.000	0.489**	0.950**	0.500**	0.996**
9.									1.000	0.630**	0.918**	0.561**
10.										1.000	0.534*	0.965*
11.											1.000	0.565*
12.												1.000

1. Days taken for sprouting of bulb	7. Rachis length
2. Plant height	8. Number of florets/ spike
3. No. of leaves per plant	9. Length of the floret
4. Days to spike emergence	10. Weight of florets per spike
5. Flowering duration	11. Number of spikes/m^2
6. Spike length	12. Yield of florets/ plot (2 * 2 m)

Note:
*Significant at 5% level
** Significant at 1% level

Table 5: Genotypic correlation coefficient among different characters in tuberose (double).

days to spike emergence (-1.574). Prabhat et al. [15] did similar studies and reported significant and positive association of plant height with spike length in gladiolus. Further, the number of leaves per plant had highly significant relationship with weight of the florets per spike (0.994), yield of florets per plot (2*2 m) (0.987), number of florets per spike (0.981), spike length (0.965), flowering duration (0.941), length of the floret (0.730), number of spikes per m^2 (0.665) and rachis length (0.438). The trait however, showed negative correlation with days to spike emergence (-1.759). Similar findings were also reported by Vetrivel [16] and Prabhat et al. [15] in gladiolus.

Days to spike emergence showed highly significant, but negative correlation with flowering duration (-1.706), spike length (-1.518), yield of florets per plot (2*2 m) (-1.430), number of spikes per m^2 (-1.418), weight of florets per spike (-1.406), rachis length (-1.384), length of the floret (-2.136) and number of florets per spike (-1.276). This is line with the findings of Prabhat et al. [15] in gladiolus for spike length and weight of florets per spike.

Positive and significant association for flowering duration was observed for spike length (0.944), yield of florets per plot (2*2 m) (0.899), weight of florets per spike (0.875), number of spikes per m^2 (0.862), number of florets per spike (0.859), length of the floret (0.851) and rachis length (0.695). This is in line with the findings of Rakesh et al. [17] in snapdragon. The spike length showed highly significant and positive correlation with yield of florets per plot (2*2 m) (0.974), number of florets per spike (0.967), weight of florets per spike (0.905), number of spikes per m^2 (0.696), rachis length (0.653) and length of the floret (0.604). This is in consonance with the findings of Rakesh et al. [17] in snapdragon. There exists a positive and highly significant relationship of rachis length with number of spikes per m^2 (0.742), length of the floret (0.566), yield of florets per plot (2s*2 m), number of florets per spike (0.481) and weight of florets per spike (0.347). Similar such findings were reported by Rakesh et al. [17] in snapdragon. Positive and significant association was observed for number of florets/ spike with yield of florets per plot (2*2 m) (0.996), weight of florets per spike (0.950), number of spikes per m^2 (0.500) and length of the floret (0.489). These results are in conformity with the findings of Prabhat

et al. [15] in gladiolus. Length of the floret exhibited positive and significant association with number of spikes per m^2 (0.918), weight of florets per spike (0.630) and yield of florets per plot (2*2 m) (0.561). In genotypic and phenotypic levels, weight of florets per spike exhibited positive relationship with yield of florets per plot (2*2 m) (0.965) and number of spikes per m^2 (0.534). Positive and significant association was also observed for number of spikes/ m^2 with yield of florets per plot (2*2 m) (0.565). This is in consonance with the findings of Prabhat et al. [15] in gladiolus for number of spikes/ m^2.

References

1. Sadhu MR, Bose TK (1973) Tuberose for most artistic garlands. Indian Hort 18: 17-20.

2. Benschop M (1993) Polianthes. In: De Hertogh A, Le Nard M (Ed.), The physiology of flower bulbs, Elsevier, Amsterdam, the Netherlands 589-601.

3. Panse VG, Sukhutme PV (1967) Statistical methods for agricultural workers. (2nd Edn.), Indian Council of Agricultural Research, New Delhi, India 381.

4. Burton GW, Vane D (1953) Estimating heritability in late fescue from replicated clonal material. Agronomy Journal 45: 478-479.

5. Weber CR, Moorthy BR (1952) Heritable and nonheritable relationship and variability of soil content and agronomic characters in F2 generation of soybean crosses. Agronomy Journal 44: 202-209.

6. Fisher RA (1954) Statistical methods for research workers. Din Oliver and Boyd Ltd, London, United Kingdom.

7. Fisher RA, Yates F (1963) Statistical table for biological, agricultural and medical research. Oliver and Boyd Ltd, Edinburgh, Scotland 146.

8. Gudi G (2006) Evaluation of tuberose varieties. Thesis submitted to University of Agricultural Sciences, Dharwad, Karnataka, India.

9. Swaroop K (2010) Morphological variation and evaluation of gladiolus germplasm. Ind J Agric Sci 80: 742-745.

10. Patil VS, Munikrishnappa PM, Tirakannanavar S (2009) Performance of growth and yield of different genotypes of tuberose under transitional tract of north Karnataka. J Ecobiol 24: 327-333.

11. Sheela VL, Rakhi R, Jayachandran Nair CS, Sabina George T (2005) Genetic variability in heliconia. J Ornamental Hort 8: 284-286.

12. Sheikh MQ, John AQ (2005) Genetic variability in Iris (Iris japonica Thumb). J Ornamental Hort 8: 75-76.

13. Robinson HF, Comstock RE, Harvey PH (1949) Estimates of heritability and degree of dominance in corn. Agronomy Journal 41: 353-359

14. Singh AK (2011) Assessment of snapdragon germplasm for various traits. Indian Journal of Ornamental Horticulture 76.

15. Prabhat K, Kumar MR, Binayak C, Rakesh M, Mishra DS (2011) Genetic variability and correlation studies in Gladiolus hybrida L. under tarai condition of Uttarakhand. Progressive Horticulture 43: 323- 327.

16. Vetrivel T (2010) Evaluation of gladiolus (Gladiolus spp l.) varieties suitable for shevaroys conditions. M.Sc. (Hort.) Thesis, Tamil Nadu Agricultural University, Coimbatore, India.

17. Rakesh Kumar, Santosh Kumar, Awani Kumar Singh (2012) Genetic variability and diversity studies in snapdragon (Antirrhinum majus) under tarai conditions of Uttarakhand. Indian Journal of Ornamental Horticulture 82.

10

Assessment of Combining Ability in Pearl Millet Using Line x Tester Analysis

Jagendra Singh[1] and Ravi Sharma[2*]

[1]Research Scholar Senior Research Fellow Directorate of Research Services, RVSKVV Race Course Road, Gwalior 474002, MP, India
[2]Ex-Principal ESS College, Dr B.R. Ambedkar University (formerly Agra University), Dayalbagh, Agra, India

Abstract

Present experiment was conducted at Raya, Mathura (U.P.) during the kharif season of 2008, 2009 and 2010 with four male sterile lines (female parents) and nine inbreds used as testers (male parents) of pearl millet in line x tester fashion. In general combining ability analysis GIB 144 found maximum g.c.a. effects for yield, stem thickness, leaf area, panicle length, panicle-girth, and 1000-grain weight, dry weight per plant and harvest index followed by ICMA 93222, GIB 3346 and ICMA 95333. None of the parents showed significant positive g.c.a. effects for number of nodes per main stem and number of leaves per main stem. In specific combining ability analysis seven crosses viz., ICMA 93222 x GIB 78, ICMA 96111 x GIB 129, ICMA 93222 x GIB 144, ICMA 93222 x GIB 129, ICMA 97333 x GIB 157, ICMA 97333 x GIB 135 and ICMA 95333 x GIB 157 were identified as the best specific combiners for yield and major yield components. Analysis of s.c.a. effects revealed that good combining parents yield better hybrids, because parents with significant positive g.c.a. effects were involved more in selected crosses than those with non-significant g.c.a. effects and negative g.c.a. effects. In the present study, the involvement of at least one good general combiner was found essential for obtaining combinations with high specific effects. Combining ability studies revealed that both general and specific combining ability variances were important but the estimates of s.c.a. variance were higher in magnitude for all the characters. Thus, indicating the predominance of non-additive gene action.

Keywords: Kharif; Inbreds; Pearl; Millet; Line x Tester Analysis

Introduction

Pearl millet is being grown in arid and semi-arid regions of the world including West Africa, India and Pakistan with the rainfall ranging from 150-700 mm. India is a major pearl millet producing country with 43.3 per cent of the world area and 42 per cent of world production. It is mainly cultivated in the states of Rajasthan, Maharashtra, Gujarat, Madhya Pradesh, Karnataka, Andhra Pradesh, Uttar Pradesh and Tamil Nadu on a total area of 9.16 million hectare with the production of 8.01 million tonnes. The national average productivity is 850 kg/ha [1]. Pearl millet (*Pennisetum glaucum*) belonging to family Poaceae is a major crop of semi-arid tropics and possesses tetraploid (2n=4x=28) chromosomes. It is a monocotyledonous and cross-pollinated annual C_4 crop species. Its protogynous nature of flowering can be used to make hybrids. The principal aim of any breeding Programme is to increase the yield potential. The yield is a complex character comprising of a number of components each of which is genetically controlled and susceptible to environmental fluctuations. The concept of combining ability is gaining importance in plant breeding as it provides valuable genetic information about the parents and the characters under study. It helps in assessing the breeding value of parental lines in terms of their superiority in hybrid combinations and also provides the information regarding the nature and extent of gene action involved in controlling the inheritance of characters in question, like yield and yield attributing characters, thus helps in deciding upon the future breeding strategy. Hence the present investigation based on '*line* x *tester*' analysis was designed, to collect the information regarding the genetic composition of various quantitatively inherited yield contributing traits including grain yield in pearl millet.

Materials and Methods

The present study was brought from author's Ph.D. thesis. The experiment was conducted during the *kharif* season of 2008, 2009 and 2010 and recommended package of practices were applied. The material comprised of four male sterile lines (female parents) and nine inbreds used as testers (male parents) of pearl millet. During the year

2008 the experimental material was generated by crossing four male sterile lines to nine testers in line x tester fashion resulting in 36 F_1s. During the year 2009, the 36 F_1 crosses along with four lines and nine testers constituting a total of 49 treatments were grown in a randomized block design (RBD) with three replications, each entry represented by a single row of four meter length, with row to row and plant to plant spacing being 50 cm and 15 cm respectively. During the year 2010 the experiments were again repeated as in the year 2009. The observations for 14 quantitative characters [*viz.*, Plant height, Stem thickness, Number of nodes per main stem, Number of leaves per main stem, Leaf area, Flag leaf length, Number of productive tillers per plant, Panicle length, Panicle-girth, Grain density, 1000-grain weight, Grain weight per plant (economical yield), Dry weight per plant (Biological yield), Harvest Index (%)] were recorded on ten competitive plants, selected randomly from each row in each replication during 2009 and 2010 and averaged. Combining ability analysis was computed using line x tester procedure developed by Kempthorne [2].

Result and Discussion

Analysis of variance for combining ability

The results of analysis of variance for combining ability indicated that the mean squares due to lines were found to be highly significant for plant height , leaf area, flag leaf length, panicle length, panicle-girth,

---

*Corresponding author: Ravi Sharma, Ex-Principal ESS College, Dr B.R. Ambedkar University (formerly Agra University), Dayalbagh, Agra, India
E-mail: drravisharma327@yahoo.com

grain density, 1000- grain weight, grain yield per plant, dry weight per plant and harvest index (Table 1). In case of testers significant values were obtained for all the characters except for number of nodes per main stem, number of leaves per main stem and flag leaf length, whereas the mean squares due to line x tester were found highly significant for all the characters under study.

Estimates of general combining ability effects

The estimates of general combining ability (g.c.a.) effects of parents for all the characters have been given in Table 2. General combining ability effects suggested that GIB 144, ICMA 93222, GIB 3346 and ICMA 95333 were found to be the best general combiners for yield and some of its attributes. GIB 144 showed maximum g.c.a. effects for yield, stem thickness, leaf area, panicle length, panicle-girth, 1000-grain weight, dry weight per plant and harvest index, hence was considered most desirable. ICMA 93222 was fond to be good general combines for grain yield, number of productive tillers per plant, grain density and dry weight per plant while GIB 3346 proved to be good general combiner for yield, stem thickness, leaf area, flag leaf length and dry weight per plant. Similarly ICMA 95333 was identified as good general combiner for yield, stem thickness, leaf area, flag leaf length, panicle length and dry weight per plant. None of the parents showed significant positive g.c.a. effects for number of nodes per main stem and number of leaves

per main stem.

Estimates of specific combining ability effects

The specific combining ability estimates revealed that no cross combination was consistently superior for all the characters under study as reported by Upadhyay and Murthy [3], Pokhriyal et al. [4] and Basavraju et al. [5]. Seven crosses viz., ICMA 93222 x GIB 78, ICMA 96111 x GIB 129, ICMA 93222 x GIB 144, ICMA 93222 x GIB 129, ICMA 97333 x GIB 157, ICMA 97333 x GIB 135 and ICMA 95333 x GIB 157 were identified as the best specific combiners for yield and major yield components (Table 3). Analysis of s.c.a. effects revealed that good combining parents yield better hybrids, because parents with significant positive g.c.a. effects were involved more in selected crosses than those with non-significant g.c.a. effects and negative g.c.a. effects. In the present study, the involvement of at least one good general combiner was found essential for obtaining combinations with high specific effects. For example in the case of hybrid ICMA 93222 x GIB 78, parent ICMA 93222 was a good general combiner for most of the characters while parent GIB 78 was a low combiner, whereas in the hybrid ICMA 93222 x GIB 144 both the parents were high combiners. Several workers Singh et al., Mathur and Mathur and Dass et al. [6-8] have also made similar observations in pearl millet.

S.N.	Source	d.f.	Plant height (cm)	Stem thick-ness (mm)	No. of nodes / main stem	No. of leaves / main stem	Leaf area (cm²)	Flag leaf length (cm)	No. of product-ive tillers / plant	Panicle length (cm)	Panicle-girth (mm)	Grain density / cm²	1000-grain weight (gm)	Grain weight / plant (gm)	Dry weight / plant (gm)	Harvest index (%)
1.	Replication	2	14.25	130.01	1.14	0.41	3420.00	36.42	0.00	1.21	3.98	4.59	0.02	0.48	5.89	4.04
2.	Lines	3	829.16**	5.27	0.42	0.28	5654.63**	94.84**	0.09	9.89**	233.18**	8.50**	3.94**	188.28**	2047.72**	36.66**
3.	Testers	8	205.42**	8.54**	0.38	0.37	1678.58**	27.75	0.18**	14.68**	78.28**	5.61**	3.46**	117.09**	366.06**	44.69**
4.	Line x Tester	24	439.50**	8.82**	1.55**	1.44**	2706.51**	86.08**	0.21**	29.48**	210.07**	13.63**	3.17**	79.28**	576.67**	51.42**
5.	Error	70	22.01	2.15	0.38	0.32	561.59	13.46	0.06	1.65	23.53	1.22	0.01	13.56	41.29	5.72

** Significant at 1% level.

Table 1: Analysis of Variance for combining ability for 14 characters in pearl millet (Mean sum of squares).

S. N.	Parents	Plant height (cm)	Stem thick-ness (mm)	No. of nodes / main stem	No. of leaves /main stem	Leaf area (cm²)	Flag leaf length (cm)	No. of productive tillers / plant	Panicle length (cm)	Panicle-girth (mm)	Grain density / cm²	1000-grain weight (gm)	Grain weight / plant (gm)	Dry weight / plant (gm)	Harvest index (%)
	Lines (Females)														
1.	ICMA 93222	1.78**	0.24	0.10	0.03	1.03	- 0.05	0.07**	0.23	-3.64**	0.64**	- 0.26	2.76**	7.62**	-0.20
2.	ICMA 95333	2.78**	0.47*	0.11	0.12	19.83**	2.56**	- 0.00	0.38*	0.10	0.01	- 0.23**	1.21	6.77**	-1.44**
3.	ICMA 96111	3.64	-0.21	- 0.10	-0.10	- 7.56*	-0.62	- 0.00	0.28	-0.00	0.05	- 0.06	- 0.60	- 4.07**	1.36**
4.	ICMA 97333	-8.23	-0.50*	- 0.11	-0.05	-13.29*	-1.88**	-0.06*	-0.90**	3.55**	-0.72**	0.55**	- 3.37**	-10.33**	0.27
5.	SE (g_i)	0.67	0.20	0.08	0.08	3.38	0.52	0.03	0.18	0.69	0.15	0.01	0.52	0.97	0.34
	SE (g_i - g_j)	0.94	0.282	0.112	0.112	4.765	0.733	0.042	0.253	0.972	0.211	0.014	0.733	1.283	0.479
	Testers (Males)														
1.	GIB 1	- 5.89**	- 1.06**	- 0.24	0.31	- 4.95	- 0.28	0.23**	-1.45**	-2.60*	-0.92**	-0.49**	1.59	6.46**	- 0.54
2.	GIB 77	4.42**	0.10	0.10	0.12	- 4.70	0.66	- 0.02	1.66**	-0.01	0.15	0.07**	- 0.40	- 3.53*	0.81
3.	GIB 78	5.23**	- 0.06**	- 0.10	-0.17	2.77	1.31	- 0.02	-0.29	-3.52**	-0.42	-0.72**	- 2.90**	- 5.03**	- 1.50**
4.	GIB 129	- 2.68*	- 0.16	0.10	0.10	-19.56**	- 2.23*	0.05	0.74*	-0.01	0.90**	-0.13	-3.82**	- 6.78**	- 1.78**
5.	GIB 135	- 3.33**	0.17	0.07	0.17	- 4.09	-1.20	- 0.14**	0.45	-1.06	0.99**	-0.34**	- 1.99**	- 3.87*	0.51
6.	GIB 144	5.13**	1.52**	- 0.15	0.07	19.95**	1.61	- 0.18**	1.04**	5.23**	0.07	1.03**	6.25**	8.54**	4.06**
7.	GIB 157	- 2.28*	- 0.56	0.27	0.12	- 2.57	-0.76	0.03	-1.58**	0.49	-0.59*	0.13**	-1.82*	0.37	- 1.94**
8.	GIB 8436	- 1.90	0.26	-0.19	- 0.19	- 2.72	-1.12	0.08	-0.65*	-0.50	0.40	0.53**	0.84	- 1.45	1.67**
9.	GIB 3346	1.30	0.80*	0.13	0.07	15.89**	2.01*	- 0.03	0.07	1.99	-0.59*	- 0.07*	2.25*	5.29**	- 0.27
	SE (g_i)	1.09	0.34	0.14	0.13	5.52	0.85	0.05	0.30	1.13	0.25	0.03	0.85	1.49	0.55
	SE (g_i - g_j)	1.53	0.479	0.197	0.183	7.783	1.198	0.070	0.423	1.593	0.352	0.042	1.198	2.100	0.775

* Significant at 5% level. ** Significant at 1% level.

Table 2: Estimates of General combining ability effects of Parents for 14 characters in Pearl millet.

S. N.	Crosses	Plant height (cm)	Stem thick-ness (mm)	No. of nodes / main stem	No. of leaves / main stem	Leaf area (cm²)	Flag leaf length (cm)	No. of productive tillers / Plant	Panicle length (cm)	Panicle-girth (mm)	Grain density / cm²	1000-grain weight (gm)	Grain weight / plant (gm)	Dry weight / plant (gm)	Harvest index (%)
1.	ICMA 93222 x GIB 1	- 11.75**	- 0.68	0.67	0.64**	- 6.04	- 0.79	- 0.24	0.68	- 7.42**	1.85**	-0.82**	-1.51	-15.12**	4.23**
2.	ICMA 95333 x GIB 1	5.72**	- 0.39	- 0.53	- 0.51	34.04**	8.67**	- 0.11	1.58**	12.00**	-1.23**	0.50**	1.14	-2.79	2.52**
3.	ICMA 96111 x GIB 1	- 0.42	0.20	- 0.32	- 0.35	- 11.78	- 3.48*	- 0.05	- 0.80	3.18	0.35	-0.09	-4.35**	-11.29**	-0.39
4.	ICMA 97333 x GIB 1	- 20.98**	0.08	- 0.33	- 0.30	- 7.80	- 3.84*	0.07	- 5.07**	- 1.12	0.01	0.81**	-4.76**	-15.21**	0.60
5.	ICMA 93222 x GIB 77	10.14**	1.20*	0.49*	0.49*	12.41	- 0.36	- 0.12	1.72**	- 6.67**	2.26**	- 0.54**	-0.26	8.20**	-2.98**
6.	ICMA 95333 x GIB 77	8.74**	- 0.67	- 0.93**	- 0.80**	14.32	7.77**	- 0.01	2.99**	8.29**	-1.14*	0.37**	-1.18	5.45**	- 3.73**
7.	ICMA 96111 x GIB 77	- 5.23**	0.68	- 0.23	- 0.32	- 17.45	- 4.40**	- 0.17	- 0.95	- 0.50	0.18	0.04	-3.43*	-0.71	-3.52**
8.	ICMA 97333 x GIB 77	- 4.41*	- 2.31**	- 0.10	- 0.06	-19.44**	- 2.41	0.23*	- 1.17*	- 2.77	-1.14*	0.07	1.56	12.45**	-2.72**
9.	ICMA 93222 x GIB 78	18.20**	1.87**	1.30**	1.22**	1.75	- 1.14	0.42**	1.02	- 4.93*	-1.14*	-0.34**	12.81**	19.03**	5.98**
10.	ICMA 95333 x GIB 78	13.76**	- 0.05	0.20	-0.04	46.88**	7.51**	- 0.15	0.71	6.11**	-1.51**	-0.19**	-2.96*	-10.27**	0.61
11.	ICMA 96111 x GIB 78	- 3.35	1.57**	0.05	0.05	-24.28*	- 5.82**	- 0.40	-2.02**	- 1.23	-2.26	-0.02	-5.96**	-4.94	- 4.76**
12.	ICMA 97333 x GIB 78	- 21.76**	- 2.96**	- 0.39	- 0.31	-47.78**	- 4.31**	0.43**	-5.61**	- 1.82	-0.01	0.50**	-2.12	0.22	-2.14*
13.	ICMA 93222 x GIB 129	16.18	- 0.55	1.12**	1.07**	- 16.95	-1.58	0.22*	0.19	-11.40**	2.98**	0.91**	5.12**	16.63**	-0.01
14.	ICMA 95333 x GIB 129	-10.09**	0.83	-0.51*	-0.46*	17.55	2.10	-0.04	1.02	5.64**	-0.76	-0.17**	0.28	- 4.94	2.03*
15.	ICMA 96111 x GIB 129	0.10	2.28**	0.38	0.30	29.41**	-1.18	-0.06	3.70**	5.88**	-2.51**	-1.53**	9.70**	26.63**	-0.63
16.	ICMA 97333 x GIB 129	- 2.64	- 0.75	-0.44	-0.21	-28.13**	-0.33	0.37**	-2.04**	-8.78**	-0.51	0.14**	1.45	- 4.19	3.05**
17.	ICMA 93222 x GIB 135	7.61**	0.08	0.42	0.43	-11.15	-2.78	-0.00	3.02**	-12.85**	5.14**	-2.36**	-4.21**	- 11.36**	-0.51
18.	ICMA 95333 x GIB 135	0.19	- 0.46	-0.83**	-0.83**	34.44**	6.35**	-0.35**	1.01	12.44**	-0.51	1.48**	-1.29	- 7.77**	2.37**
19.	ICMA 96111 x GIB 135	- 2.98	0.37	-0.84**	-0.61**	-23.21*	- 4.72**	0.10	-0.17	2.75	0.11	-0.06	-0.48	15.90**	-6.38**
20.	ICMA 97333 x GIB 135	- 13.18**	- 1.80**	-0.46	-0.38	-27.83**	-3.95**	0.49**	-2.30	-0.88	0.36	1.10**	4.51**	2.24	4.12**
21.	ICMA 93222 x GIB 144	12.88**	1.23*	0.81**	0.91**	11.85	-0.66	-0.07	0.83	-9.00**	0.94*	-0.59**	5.35**	10.40**	1.55
22.	ICMA 95333 x GIB 144	1.50	-1.19*	-0.59*	-0.56*	24.08*	6.12**	-0.17	2.26**	7.77**	-1.38**	0.21**	-2.06	0.15	-3.33*
23.	ICMA 96111 x GIB 144	- 1.08	0.73	-0.03	-0.16	-17.00	2.86	-0.04	-0.23	1.42	-1.80**	-0.17**	-1.56	-13.09**	3.58**
24.	ICMA 97333 x GIB 144	- 4.34*	- 0.48	-0.06	-0.19	-23.12*	-4.12**	-0.03	-4.75**	-0.20	3.44**	-1.06**	-6.14**	-26.50**	5.06**
25.	ICMA 93222 x GIB 157	15.90**	1.90**	1.30**	1.34**	27.50**	3.32*	-0.22*	4.23**	-4.34	0.11	-0.01	0.93	15.32**	-5.07**
26.	ICMA 95333 x GIB 157	5.35**	1.83**	0.03	-0.13	46.93**	9.92**	-0.01	1.48**	7.99**	-1.55**	0.80**	3.60*	6.49*	0.72
27.	ICMA 96111 x GIB 157	-14.05**	-2.60**	-0.15	-0.20	-19.19*	-3.08	-0.02	-1.35*	-5.50**	-0.22	-0.21**	-4.14**	-10.92**	-0.25
28.	ICMA 97333 x GIB 157	0.96	0.36	-0.03	0.00	-17.62	-2.00	0.29**	-1.22**	-1.40	-0.44	1.08**	4.96**	9.50**	1.53
29.	ICMA 93222 x GIB 8436	10.81**	0.62	0.94**	0.84**	13.07	1.08	0.02	2.73**	-9.88**	3.13**	-1.58**	0.29	5.50*	-1.88
30.	ICMA 95333 x GIB 8436	9.30**	1.52**	-0.10	-0.23	47.71**	8.44**	-0.31**	5.53**	7.65**	-1.27**	0.18**	1.12	0.66	0.98
31.	ICMA 96111 x GIB 8436	3.30	1.66**	-0.18	-0.20	0.67	-0.75	-0.11	2.61**	4.74*	-1.61**	-0.10*	1.71	-1.58	2.73**

32.	ICMA 97333 x GIB 8436	1.03	-2.77**	0.04	0.12	-12.96	1.12	0.21*	-2.51**	-0.40	0.30	0.90**	1.54	9.83**	-2.62**
33.	ICMA 93222 x GIB 3346	-4.49*	-1.12	0.61*	0.69**	-20.61*	-2.46	0.12	-1.94**	-13.96**	0.22	-0.85**	-2.37	-5.58**	-0.69
34.	ICMA 95333 x GIB 3346	-8.01**	-1.83**	-0.61*	-0.80**	18.08	1.41	0.03	-1.23*	7.63**	0.22	-0.17**	1.04	-10.41**	5.54**
35.	ICMA 96111 x GIB 3346	-8.56**	0.39	-0.35	-0.24	-16.33	-4.72**	-0.21*	-3.33**	7.63**	-0.44**	1.48**	-0.95	-7.58**	2.51*
36.	ICMA 97333 x GIB 3346	-4.34*	1.18*	-0.31	-0.18	-17.00	-2.12	-0.03	-0.68	-2.00	1.88**	-0.93	-7.37**	-0.33	-8.10**
	SE (Sij)	1.89	0.59	0.25	0.22	9.57	1.48	0.10	0.52	1.96	0.44	0.05	1.48	2.59	0.96
	SE (Sij-Sij)	2.664	0.831	0.352	0.310	13.49	2.086	0.141	0.733	2.763	0.620	0.070	2.086	3.651	1.353

* Significant at 5% level. ** Significant at 1% level.

Table 3: Estimates of specific combining Ability effects of crosses for 14 characters in Pearl millet.

Combining ability studies revealed that both general and specific combining ability variances were important but the estimates of s.c.a. variance were higher in magnitude for all the characters. Thus, indicating the predominance of non-additive gene action.

Acknowledgements

First author is thankful to the Head Department of Botany and the Principal K.R. College, Mathura for permission to work for his PhD in the Department of Botany K.R. College, Mathura 281001 (Dr B.R. Ambedkar University formerly Agra University, Agra) UP India.

References

1. Anonymous (2007) Annu. Rep. (2006-07). AICPMIP, Jodhpur, Rajasthan.

2. Kempthorne O (1957) An introduction to general statistics. John Wiley & Sons. Inc. New York. 545.

3. Upadhyaya MK, Murthy BR (1971) Genetic diversity and combining ability in pearl millet. Indian J Genet 31: 63-71.

4. Pokhariyal SC, Patil RR, Rama Das, Singh B (1974) Combining ability for new male sterile lines in pearl millet. Indian J Genet 34: 208-215.

5. Basavaraju R, Safeeulla KM, Murthy BR (1980) Combining ability in pearl millet. Indian J Genet, 40: 528-536.

6. Singh YP, Kumar S, Tiwari SN, Chauhan BPS (1980) Combining ability analysis for yield and its components in pearl millet. Indian J Genet Pl Br 40: 276-280.

7. Mathur PN, Mathur JR (1983) Combining ability for yield and its components in pearl millet. Indian J Genet & Pl Br 43: 299-303.

8. Dass S, Kapoor RL, Chandra S, Jatasra DS, Yadav HP (1985) Combining ability analysis for yield components of pearl millet in different environments. Indian J Genet Pl Br, 45: 70-74.

11

Adoption of IPM Approach-An Ideal Module against Thrips (*Thrips tabaci Linderman*) in Onion

P. Tripathy*, Sahoo BB, Das SK, Priyadarshini A, Patel D and Dash DK

All India Network Research Project on Onion and Garlic, College of Horticulture, (OUAT), Chiplima, Sambalpur-768025, Odisha, India

Abstract

Onion (*Allium cepa* L.) is an important export oriented vegetable among the cultivated Allium crops in India. Onion Thrips (*Thrips tabaci* Linderman) is the key biotic factor for reducing yield loses in both onion as vegetable crop as well as seed crops. Besides direct damage to both foliage and bulbs, thrips can also indirectly aggravate purple blotch and act as a vector for viral disease such as Iris yellow spot. In absence of high levels of host plant resistance to *Thrips tabaci* and development of resistance towards number of pesticide of late, there is an urgent need to look at other IPM options for effective management. A field study was conducted under the All India Network Research Project on Onion and Garlic, at the College of Horticulture (OUAT), Sambalpur, Odisha, India during the winter season 2010-11 to 2012-13 to find out the most effective eco-friendly IPM modules for management of thrips in onion. The treatment consists of M_1: IPM module, M_2: Farmers' Practices and M_3: Control, laid out in RBD. The results obtained over three years indicated that both M_1 and M_2 not only significantly reduced the thrips population (21.68 and 21.02 thrips plant^{-1}) but also increased total marketable yield (25.86 and 25.70 tha^{-1}), respectively over the control, M_3 (39.13 thrips plant^{-1} and 20.58 tha^{-1}) . Higher BC Ratio was recorded in M_1 (3.26) than M_2 (2.70). It is concluded that adoption of the IPM module approach consisting of planting of border crop of two rows of wheat and one row of maize, 10-15 days prior to planting of onion seedling, dip treatment with Carbosulfan and need based insecticides spray, when thrips population exceed ETL (30 thrips/plant^{-1}) not only reduces the thrips infestation but also increases the bulb yield with quality of onion bulbs.

Keywords: BC ratio; IPM; Marketable bulb yield; Onion Thrips; *Thrips tabaci* infestation

Introduction

Onion (*Allium cepa* L.) is an important export oriented vegetable among the cultivated *Allium crops* in India. It is grown in Rabi, *Kharif* as well as late *Kharif* seasons in India with maximum area of 60% in Rabi (winter) season alone. Onion Thrips (*Thrips tabaci* Linderman) is the key biotic factor for reducing yield loses in both bulb as well as seed crops in all seasons in onion. Besides direct damage to both foliage and bulbs, *Thrips tabaci* can indirectly aggravate purple blotch and vector for viral diseases, as well as Iris yellow spot [1] *Thrips tabaci* is now treated as the measure production constraints of onion at the national level in India [2] . Abundant literature is available on various aspects on onion thrips and their management in onion. Though host plant resistance is a crucial component of IPM, in absence of high levels of host plant resistance to *Thrips* and development of resistance towards number of pesticide of late. There is an urgent need to look at other IPM options for effective management of thrips. Keeping this in mind, the present study was carried out to study the effectiveness of IPM modules in onion crops against onion thrips.

Materials and Methods

The experiment was carried out under the All India Network Research Project on Onion and Garlic, at the College of Horticulture (OUAT), Sambalpur, Odisha, India during the winter season of 2010-11 to 2012-13. The three treatments used were: T_1: IPM, T_2: conventional farmer practices and T_3: Control with 8 replication per treatment by adopting Randomized Block Design. The details of each treatment are given in Table 1.

Onion seedlings variety Bhima Super of about 45 days old were transplanted in plot size of 250 m^{-2} for each module (40 beds of 3 m×2 m) with a spacing of 15 cm×10 cm on 20.10.2010 to 23.10.2010

; 7.12.2011 to 9.12.2011 and 28.11.12 to 30.11.12, respectively . All recommended practices recommended by Directorate of Onion and Garlic Research (ICAR), Rajgurunagar, Pune, India, was adapted uniformly to all the modules except the insecticidal treatments which were adopted on the basis of the treatments. Insecticidal treatments were given as soon as thrips were recorded in treatment plots. The observations on thrips population were recorded at 30, 45, 60, 90 days after planting (DAP). The marketable bulb yield was recorded for each replication including only A+, A, B & C grade bulbs. The bolters, double, small size bulbs and rotten bulbs were excluded while calculating the marketable bulbs for this study. All data generated were subjected to statistical analysis and the efficacy different module was assessed P value at 5% level [3].

Result and Discussion

In the field study on efficacy of IPM module, the population of *Thrips tabaci* was significantly low in T_1, (9.67 and 20.52 thrips/plant^{-1}) as compared to T_2, conventional Farmers' Practice (13.91 and 31.36 thripsplant^{-1}) up to 60 DAP. Subsequently, the thrips plant^{-1} was significantly lower in farmers' practice (29.27 and 25.42) than IPM plots (38.12 and 35.53), respectively. However, at all stages of observations both IPM and farmers' practice significantly reduced

---

*Corresponding author: P. Tripathy, All India Network Research Project on Onion and Garlic, College of Horticulture, (OUAT),Chiplima, Sambalpur-768025, Odisha, India, E-mail: ptripathy_ouat05@rediffmail.com

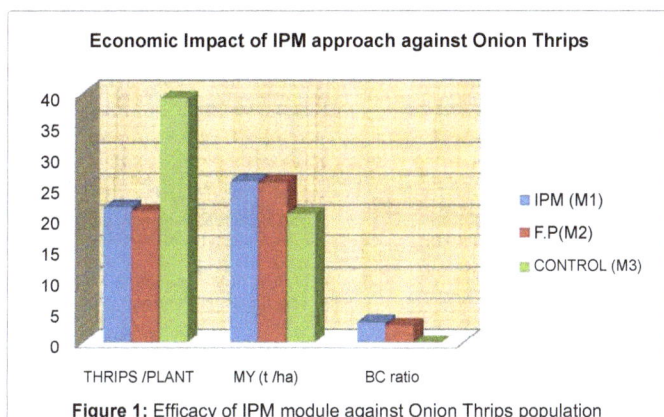

Figure 1: Efficacy of IPM module against Onion Thrips population

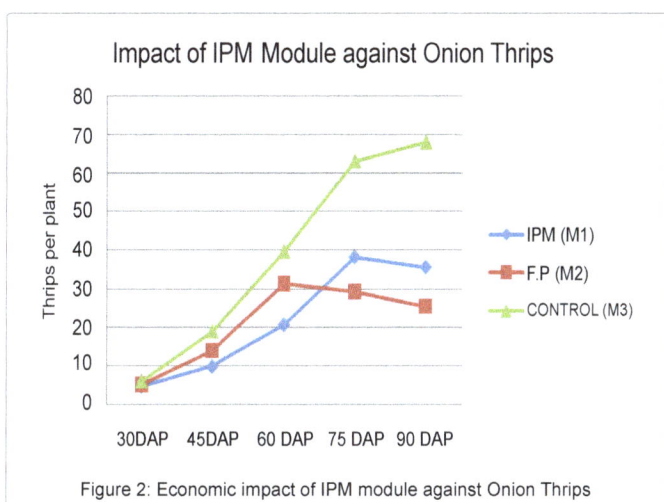

Figure 2: Economic impact of IPM module against Onion Thrips

thrips population up to 90 DAP and pooled results, except 30 DAP over the control plot (Table 2 and Figure 1). The non-significant effect on thrips population during the initial 30 DAP stage might be due to the initiation of thrips infestation stage. Similarly, the better efficacy of reducing the thrips population upto 60 DAP in IPM module might be due to border crop effect of both wheat and maize. Thrips are weak fliers and can be carried by the wind. Therefore, planting live barriers like maize and wheat could effectively block or reduce adult thrips reaching onion plants. Similar report of efficacy of border crop – maize and wheat to block thrips in onion has been reported by [4]. On the other hand, significant reduction in thrips population after 75 and 90 DAP in farmers' practice over IPM might be due to senescence of boarder crop effect, both at wheat and maize. On the contrary, due to repeated spraying of insecticides, the thrips population was well under control in T_2, the farmer's practice (Figure 1).

The thrips population counts pooled over all the stages of observations from 30-90 DAP, indicated significant reduction in thrips population in both IPM and farmers' practice over control plots. Significantly lowest thrips/plant^{-1} was recorded in farmers' practice (21.02) over control plot (39.13). However *statistical parity* was observed between IPM and farmers' practice indicating that both the treatments were equally effective in reducing thrips population in onion crops. The pooled results over 2010-11 to 2012-13 on marketable bulb yield revealed significant variations ranging from 20.58 tha^{-1} in control to 25.86 tha^{-1} in IPM plot with a mean value of 24.05 tha^{-1} (Table 2 and Figure 2). Significantly highest marketable bulb yield was recorded in IPM plots (25.86 tha^{-1}) than control (20.58 tha^{-1}). However, IPM and farmers' practice and (25.86 and 25.70 tha^{-1}) was non-significant statistically indicating the better efficacy of IPM modules.

The economics of marketable bulb yield in onion mean over three years revealed that adoption of IPM treatments was best

Sl.no.	Module	Details
01.	T₁: IPM module	• Planting barrier crops–outer row of maize+inner row of wheat on all 4 sides of the plot at least 7-10 days before onion planting. Wheat may be planted closely and maize at 25 cm interval. Avoid tall maize variety. No gaps should be left between maize plants. • Seedling root dip: Dip the seedlings (bottom 1/3rd) in Carbosulfan (2 ml/l) solution for 2 hours before transplanting. • Monitor thrips population twice a week regularly. Spray insecticides based on ETL-whenever thrips population (nymphs) crosses 30/plant. • Whenever thrips cross ETL Spray - Methomyl at the rate of 240 g active ingredient per ha or neem oil (3 ml/l)+Profenofos (0.5 ml/l), • or Fipronil (1 ml/l), or neem (3 ml/l)+carbosulfan (1 ml/l).
02.	T₂: Farmers' practice	• Insecticides spray at 15-days interval: Rogor, Monocrotophos, cypermethrin, chlorpyriphos, L-cyhalothrin. • Start spraying as soon as thrips appear in the field.
03.	T₃: Control	• Without any spraying schedule.

Table 1: Treatment Details of Evaluation of IPM Module for Onion Thrips

Module	Thrips/plant⁻¹ after						MY (t/ha)	BCR
	30 DAP	45 DAP	60 DAP	75 DAP	90 DAP	Pooled		
M1 (IPM Module)	4.54 (2.24)	9.67 (3.18)	20.52 (4.58)	38.12 (6.21)	35.53 (6.00)	21.68 (4.71)	25.86	3.26
M2 (Farmer's practice)	5.13 (2.37)	13.91 (3.78)	31.36 (5.64)	29.27 (5.46)	25.42 (5.09)	21.02 (4.64)	25.70	2.70
M3 (untreated control)	5.97 (2.54)	18.81 (4.39)	39.65 (6.34)	63.16 (7.98)	68.05 (8.28)	39.13 (6.29)	20.58	
Grand mean	5.21 (2.38)	14.13 (3.79)	30.51 (5.62)	43.52 (6.55)	43.00 (6.45)	27.28 (5.21)	24.05	
Sem ±	0.09	0.16	0.12	0.14	0.20	0.10	0.62	
LSD 5%	0.23	0.39	0.29	0.35	0.50	0.23	1.52	

MY=Marketable Yield, BCR= Benefit Cost Ratio Figures in the parentheses indicate the corresponding square root of (x+0.5) values*
Table 2: Efficacy of IPM Modules against Onion Thrips under Odisha Condition (mean value of 2010-11 to 2012-13)

with highest Benefit Cost ratio of 3.26 than farmers' practices (2.70). The higher BC ratio in IPM plot is primarily due to less insecticidal application [1] and additional income from the border crop – maize & wheat as compared to farmers' practice. Similar report of highest BC ratio was estimated in control plots between 45-75 DAP in onion [5].

Conclusion

It is concluded that adoption of IPM module approach consisting of planting border crops of two rows of wheat and maize, 10-15 days prior to planting, seedling dip treatment with Carbosulfan and need based insecticides spray, when thrips population exceed threshold levels (30 thrips/plant^{-1}) not only reduces thrips infestation but also increases the bulb yield with quality bulbs in onion.

Acknowledgements

We gratefully acknowledge the Orissa University of Agriculture & Technology, Odisha, India for the research facilities provided and to the Director, Directorate of Onion & Garlic Research (ICAR), Rajgurunagar, Pune, India for providing the financial & other facilities to carry out this study under AINRP on Onion and Garlic

References

1. Krishna Kumar NK, Srinivas PS, Rebjit KB, Asokan R, Ranganath HR (2011) Onion thrips Thrips tabaci Linderman : a prospective. In: National Symposium on Allium: Current Scenario and Emerging Trends, Pune 68-76.

2. Srinivas PS, Lawande KE (2004) Impact of planting dates on Thrips tabaci Linderman infestation and yield loss in onion (Allium cepa L.) Pest management in Horticultural Eco- system, 10:11-183.

3. Sukhatame, PV, Amble VN (1995) Statistical Methods for Agricultural workers. ICAR, New Delhi 145-156.

4. Srinivas PS, Lawande KE (2006) Maize border as a cultural method for management of thrips in onion (Allium cepa L.). Indian journal of Agricultural Science 76: 167-171.

5. Srinivas PS, Lawande KE (2008) Growth stage susceptibility of onion (Allium cepa L.) and its role in thrips management. Indian journal of Agricultural Science 78: 98-101.

12

Genetic Divergence in Ethiopian Coriander (*Coriandrum sativum* l.) Accessions

Miheretu Fufa*

Adami Tullu Agricultural Research Center; Plant Biotechnology Team; P.O. Box 35, Zeway, Ethiopia

Abstract

Although Ethiopia is a center of diversity for many crops including Coriander little is known on the genetic divergence of this crop due to its negligence in the research program of the country in the past. The genetic divergence of 25 land races was assessed using principal component and cluster analysis based on 8 characters. The accessions were grouped into five clusters. Cluster I was the largest consisting 19 accessions. High inter cluster distance (47.42) was observed between cluster IV, cluster II and IV (47.33) and cluster I and IV (41.47) indicating the presence of substantial genetic diversity in genetic makeup of the accessions included in these clusters. Accessions 16 and 8 having positive values for principal component 1 and 2 were of considerable breeding interest because of their good combination for the studied yield related traits.

Keywords: Coriander accessions; Genetic divergence; Cluster analysis; Principal component analysis

Introduction

Coriander (*Coriandrum sativum* L.) is an annual spice herb that belongs to the family of *Umbelliferae/Apiaceae*. It is used as a spice, medicine and a raw material in food, beverage and pharmaceutical industries. Its green foliage, rich in vitamins and other minerals, is used in vegetables and salads while its seeds contain essential oils rich in linalool [1]. Although coriander is one of the several plant species for which Ethiopia is known as a center of origin and diversity [2], there is little information on its genetic divergence which in turn hinders the exploitation of the wealth of its diversity. The only work so far done on genetic divergence on Ethiopian coriander is that of Mengesha et al. [3], that focused on collections from different agro-ecological and geographical areas of the country. However, ecological and geographical diversifications are not the only causes of genetic divergence. For the changing of genetic material, genetic drift, natural variation and artificial selection also contribute to the genetic divergence [4] coriander accessions were diversified in different agro-ecologies of Ethiopia. Therefore, intensive collection focusing on the desired traits will benefit breeders by large for effective improvement in coriander [3]. Accordingly, target collection of the present coriander accessions was made from the potential growing areas of Arsi and Bale zones that were not well covered before and thus not well addressed by work of Mengesha et al. [3].

Genetic divergence is an essential prerequisite factor in any crop improvement programme to identify potential parents for hybridization and to obtain high yielding variety [5]. Therefore, having precise information and knowledge on the nature and degree of genetic divergence is helpful and fundamental to identify and organize the available genetic resources aiming at the production of promising cultivars [6]. For the selection of parents based on the extent of genetic divergence in different crop species multivariate methods have successfully utilized [7]. Target collection of coriander accession was made from the present study was undertaken with the following objectives: a) to assess and evaluate genetic diversity of coriander accessions, b) to identify characters which contribute at maximum to genetic diversity and c) to identify accessions for future use in breeding programs in coriander.

Materials and Methods

The experiment was conducted at Sinana Agricultural Research

Center in 2012 which is located at an altitude of 2400 m.a.s.l. Sinana has a range of mean annual rainfall of 563-1018 mm with minimum and maximum temperature of 7.9 °C and 24.3°C, respectively. The soil type is dark-brown with slightly acidic reaction [8].

Twenty five coriander accessions collected from Arsi and Bale potential growing areas were sown in RCBD with three replications on a plot of 2 meter length with spacing of 15 and 30 cm between plants and rows, respectively. The experiment was conducted under rain-fed condition. Three times hoeing and weeding were carried out without the application of chemicals and fertilizers. Five plants were randomly selected for the measurement of the characters. A total of 8 characters were recorded according to the descriptors of International Plant Genetic Resource Institute (IPGRI) as given by Diederichsen [2]. These are number of basal leaves, length of basal leave, length of the longest basal leaves, habitus of the basal leaves, blade shape of the upper stem leaves, blade shape of the longest basal leaves, foliation and branching.

Recorded descriptors were subjected to principal component analysis and average linage hierarchical method of cluster analysis to determine the common pattern of variation among the accessions using SAS version 9.2 (2008) [8].Genetic divergences between clusters were calculated using mahalanobi's [9] and clustering of accessions was done according to Tocher's method as described by Rao [10].

Result and Discussion

The principal component analysis revealed that the majority of the total variation was contributed by component one and two (Figure 1). Principal component one and two contributed 35% and 19% of the total variation (Table 1) respectively. Maximum genetic variance was contributed by length of basal leaves (0.51) and foliation of the plant (0.49) to principal component 1 while number of basal leaf (0.62) and

*Corresponding author: Miheretu Fufa, Adami Tullu Agricultural Research Center; Plant Biotechnology Team, P.O. Box 35, Zeway, Ethiopia
E-mail: miheretufufag@gmail.com

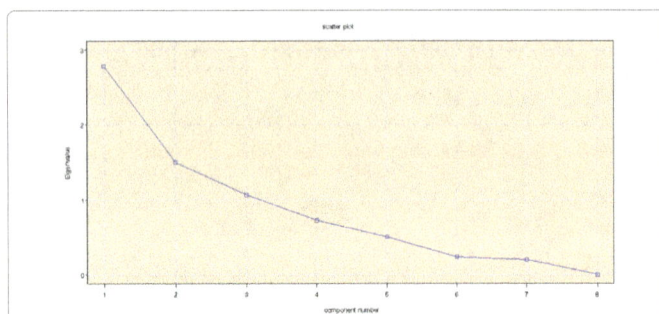

Figure 1: The contribution of the principal components to the total variation.

Characters	PC1	PC2
Eigenvalues	2.79	
Proportion of variance	0.400.21	
Cumulative variance	0.400.61	
	Factor loadings	
Number of Basal Leaf	0.22	0.60
Length of Basal Leaves	0.51	0.22
Length of the Longest Basal Leaf	0.32	0.28
Habitus of the Basal Leaves	-0.24-0.22	
Blade Shape of the Longest Basal Leaves	-0.510.25	
Blade Shape of the Upper Stem Leaves	-0.200.54	
Foliation of the plant	0.49-0.32	
Branching of the plant	0.00 0.00	

Table 1: Principal component analysis of 25 coriander accessions.

blade shape of the upper stem leaves (0.54) contributed to principal component two.

Figure 2 displays a biplot in the dimension of the first and second principal components of the 25 accessions of coriander. Accession 16 and 8 had positive values for PC1 and PC2 and were of considerable breeding interest because of their good combination for the studied traits. Hybridization between the accessions with high positive values for PC1 and close to 0 values for PC2 as accession 24 and 25 and accessions with high positive values for PC2 and near to 0 values for PC1 as accession 22 is expected to give promising and desirable segregates in subsequent generations.

The intra- and inter-cluster distance (D2) values are presented in Table 2. The intra-cluster distance was lower than the inter-cluster distances implying that the accessions included within a cluster had less diversity among themselves. The highest intra-cluster distance (5.67) was observed in cluster I followed by cluster II (1.33). The intra-cluster distance of cluster III, IV and V was zero for they contained only one accession. The intra-cluster distance in this study is similar to intra-cluster distance (2.81 to 5.59) recorded by Mengesha et al. [3], but relatively lower than the values between 13.8 and 28.25 reported by Singh et al. [1]. The magnitudes of inter-cluster distance (D2) were generally high and were indicator for the presence of substantial genetic diversity in Ethiopian coriander accessions. The highest inter-cluster distance (47.42) was observed between cluster IV and V followed by cluster II and IV (47.33) and cluster I and IV (41.47) suggesting more variability in genetic makeup of the accessions included in these clusters. The accessions belonging to the clusters separated by high statistical distance could be used in hybridization programme for obtaining a wide spectrum of variation among the segregates. In the

present study, the range of inter-cluster D2 values from 13.22 (cluster I and III) to 47.42 (cluster IV and V) was obtained. It is comparable to range of inter-cluster distance (13.8 to 91.3) reported by Singh et al. [1]. The 25 accessions were grouped into five clusters based on the 8 characters (Table 3 and Figure 3). Cluster I was the largest having 19 accessions. Cluster II and III had three and two accessions respectively. Cluster IV and V had one accession each.

The cluster means of the investigated traits are presented in Table 4. Comparison of means of various traits in different clusters revealed that cluster IV produced the highest mean value (55) for number of basal leaf, length of basal leaf (45.8cm) and length of the longest basal leaf (52 cm) than the other clusters. The same branching habit per plant was observed in all clusters. Cluster I (19 accessions), cluster III (single accessions) and IV (single accession) showed similar blade shape of the longest basal leaf and foliation per plant. That means, most of the accessions (22) evaluated had very many leaves. On the other hand, cluster II, with 3 accessions, had produced middle number of leaves while cluster V, with single accession, had very few leaves. It can be concluded from the finding of the foliation that the clusters with very many leaves can be produced for fresh herb purposes on top of their production for fruit.

On the other and Cluster II and IV and cluster III and V showed similar blade shape of the upper stem leaf. Cluster IV and V, each with single accession, gave the maximum length (52cm) of the longest basal leaf. Cluster II, with three accessions, gave the lowest length (31.33cm) of the longest basal leaf while cluster V, with single accession, produced

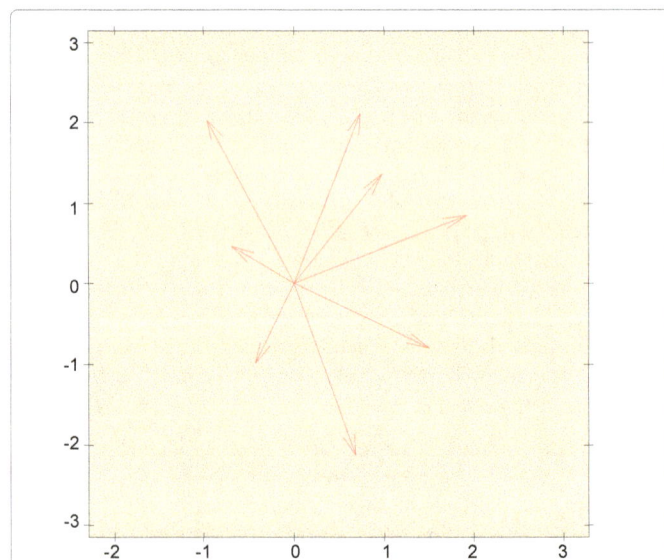

Figure 2: Scatter diagram for PC1 and PC2 in 25 coriander accessions based on the 8 traits.

	I	II	III	IV	V
:I	5.67	15.22	13.22	41.47	18.04
II		1.33	13.72	47.33	21.74
III			0.00	39.60	28.18
IV				0.00	47.42
V					0.00

Table 2: Intra-cluster (bolded diagonal) and inter-cluster (off diagonal) distance (D2) values among 25 Ethiopian coriander.

The SAS System

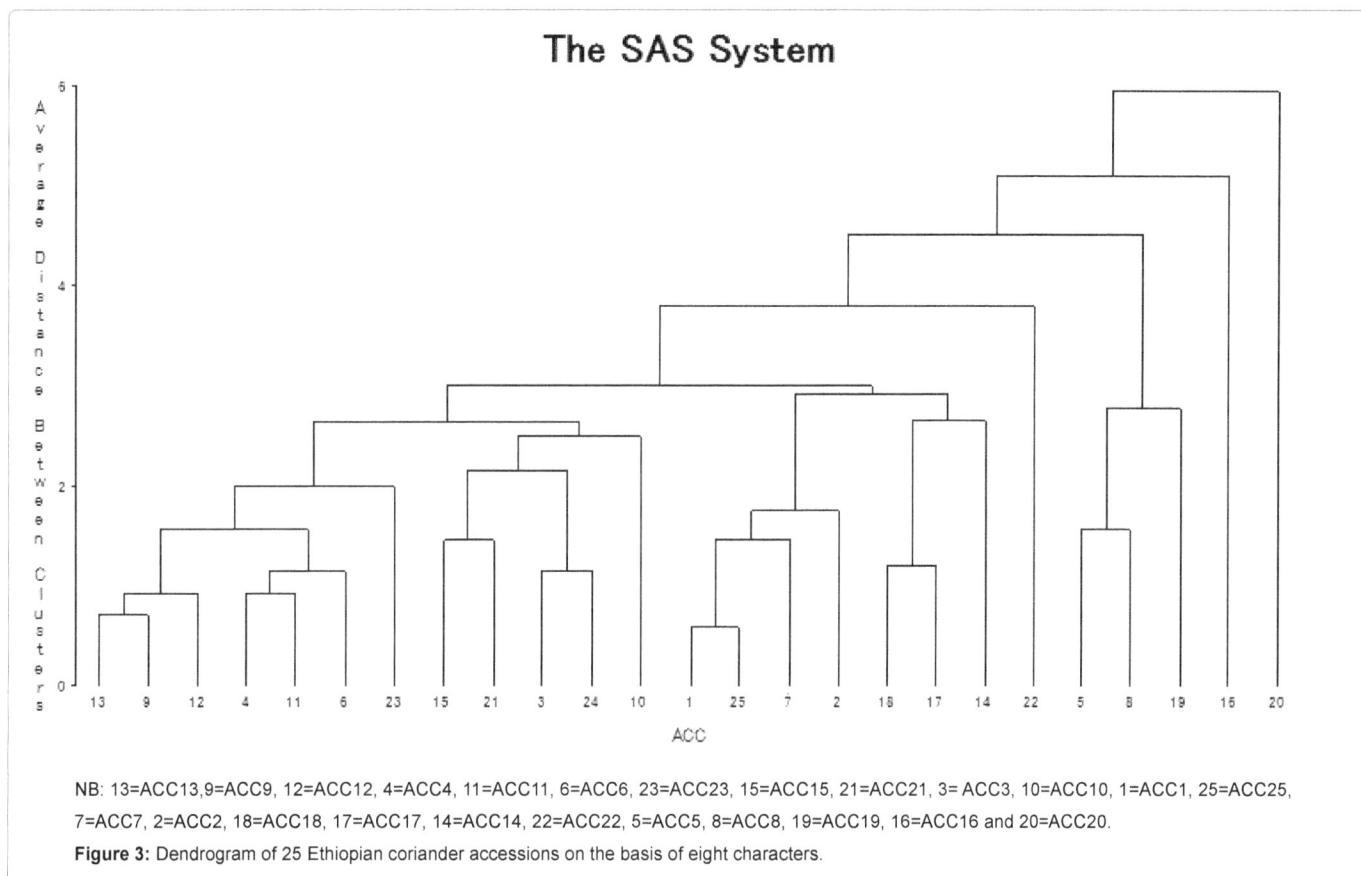

NB: 13=ACC13,9=ACC9, 12=ACC12, 4=ACC4, 11=ACC11, 6=ACC6, 23=ACC23, 15=ACC15, 21=ACC21, 3= ACC3, 10=ACC10, 1=ACC1, 25=ACC25, 7=ACC7, 2=ACC2, 18=ACC18, 17=ACC17, 14=ACC14, 22=ACC22, 5=ACC5, 8=ACC8, 19=ACC19, 16=ACC16 and 20=ACC20.

Figure 3: Dendrogram of 25 Ethiopian coriander accessions on the basis of eight characters.

Cluster	Number of accessions	Accessions
I	19	ACC13, ACC9, ACC12, ACC4,ACC11, ACC6, ACC23, ACC15,ACC21, ACC3, ACC10,ACC1, ACC25,ACC7,ACC2, ACC18, ACC17 and ACC1.
II	3 (5,8,19)	ACC5, ACC8 and ACC19
III	2 (22)	ACC22
IV	1 (16)	ACC16
V	1 (20)	ACC20

Table 3: Clustering pattern of 25 Ethiopian coriander accessions on the basis of 8 traits.

Cluster	NBL	LBL	LLBL	HBL	BSLBL	BSUSL	FOLN	BR
I	15.63	36.55	42.95	2	4	4	9	1
II	16.67	27.75	31.33	2	5	5	5	1
III	22	39.6	32	1	4	6	9	1
IV	55	45.8	52	1	4	5	9	1
V	14	23.6	52	4	5	6	1	1

Note: NBL=number of basal leaf per pant, LBL=length of basal leaf, LLBL=length of the longest basal leaf, HBL=habitus of the basal leaf, BSLBL=blade shape of the longest basal leaf, BSUSL=blade shape of the upper stem leaf, FOLN=foliation and BR=branching of the plant.
Table 4: Cluster mean values of 8 traits of 25 Ethiopian coriander accessions.

the lowest length of the basal leaf. Cluster III and IV had very flat (1) while cluster I and II showed flat (2) habitus of the basal leaf. On the other hand, cluster V that had only one accession showed a different leaf habitus that is raised with an arcus of about 45°. Cluster V produced the lowest number of basal leaf (14) per plant.

References

1. Singh SP, Katiyar RS, Rai SK, Tripayhi SM, Srivastava JP (2005) Genetic divergence and its implication in breeding of desired plant type in coriander (*Coriandrum sativum* L.). Genetika 37: 155-163.

2. Diederichsen A (1996) Coriander (*Coriandrum sativum* L.): Promoting the conservation and use of underutilized and neglected crops. Institute of Plant Genetics and Crop Plant Research, Gatersleben/ International Plant Genetic Resources Institute, Rome.

3. Mengesha B, Alemaw G, Tesfaye B (2011) Genetic divergence in Ethiopian coriander accessions and its implication in breeding of desired plant types. Afr Crop Sci J 19: 39-47.

4. Sirohi SPS, Dar AN (2009) Genetic divergence in soybean (Glycine max L. Mrrill). SKUAST Journal of Research 11: 200-203.

5. Marker S, Krupakar A (2009) Genetic divergence in exotic maize germplasm (Zeamays L.). Journal of Agricultural and Biological Science 4: 44-47.

6. Palomino EC, Mori ES, Zimback L, Tambarussi EV, Moracs CB (2005) Genetic diversity of common bean genotypes of Carioca commercial group using RADP markers. Crop Breeding and Applied Biotechnology 5: 80-85.

7. Dyulgerov N, Dyulgerova B (2013) Variation of yield components in coriander (*Coriandrum sativum* L.). Agricultural science and technology 5: 160-163.

8. SARC (1998) A decade of research experience. Geremew E Tilahun G Aliy H (eds.). Bulletin No. 4. Sinana agricultural research center, agricultural research coordination service, Oromia agricultural development bureau.

9. Mahalanobi's PC (1936). On the generalized distance in statistics. Proceedings of National Institute of Science, India 2: 49-55.

10. Rao CR (1952) Advance statistical methods in biometrical research. John Wiley and Sons, New York, 390.

DNA Fingerprinting Based Decoding of Indica Rice (*Oryza sativa* L) Via Molecular Marker (SSR, ISSR, & RAPD) in Aerobic Condition

Ashu Singh* and Sengar RS

Tissue culture Lab, College of Biotechnology, University of Agriculture & Technology, Meerut, India

Abstract

Genetic improvement mainly depends on the extent of genetic variability present in the population. The molecular marker is a useful tool for assessing genetic variations and resolving cultivar identities. The objective of this study was to evaluate the genetic divergence of 30 rice varieties (Basmati, Non-Basmati, Aerobic) using 10 ISSR, RAPD markers each. The diversity or similarities and dissimilarities between all thirty rice varieties were calculated using 0 1 sheets. SSR primers RM-263 is highly informative since it recorded high PIC value (0.995). The resolving power varies between 0.132(RM-256) to 4.662(RM-222) with an average value of 2.7502. In RAPD analysis PIC values varies from 0.811(OPD-08) to 0.9925(OPF-13) with average of 0.9635 and resolving power varies from 1.32(OPJ-08) to 2.066(OPJ-13) with average of 1.8256. In ISSR analysis, PIC value ranged from 0.8791(ISSR6) to 0.9916(ISSR5) with an average value of 0.9482. The resolving power varies between 1.6(ISSR3) and 8.366(ISSR2) with an average value of 5.2708. The PIC values and Resolving power were calculated for individual primers. The analysis indicated that ISSR expressed maximum resolving power of 8.336 and RAPD gave maximum PIC values of 0.9925. RAPD primer OPF-13 gave the maximum accessions coverage (depending on the value of PIC) in the rice genome. Out of 52 amplified bands, 49 bands were polymorphic and 3 bands were monomorphic. The cluster analysis using the marker systems could distinguish the different genotypes. The dendogram generated on the principle of Unweight Pair Wise Method using Arithemetic Average (UPGMA) was constructed by Jaccard's Coefficient and the genotypes were grouped in to clusters. The dendogram developed for aroma and quality traits showed that the genotypes with common phylogeny and geographical orientation tend to cluster together thus marker based molecular fingerprinting could serve as a sound basis in the identification of genetically distant accessions as well as in the duplicate sorting of the morphologically close accessions as the case is common in differentiating Basmati and non-basmati.

Keywords: *Phylogen; Polymorphism; Indica rice; Luster analysis; cTAB method*; Genome coverage; Genetic diversity

Introduction

Rice (*Oryza sativa* L.) is one of the leading cereal crops of the world and is the principal food crop of about half of the world's population. It is a major source of calories for them [1]. In many regions, it is eaten with every meal and provides more calories than any other single food. It can also be used in the manufacture of cosmetics and textiles; beer and wine are also made from it [2]. Besides its food value, it has high cultural and social values in rice consuming societies. Rice is the staple food of more than 50% of the world's population [3]. By the year 2025, 21% increase in rice production will be needed over that of year 2000 [4]. It is one of the most important crops that provide food for more than half of the world population [5].

This implies that thousands of valuable allelic variations of traits of economic significance remain unutilized [6]. The first step towards determining the magnitude of these risks is to evaluate the genetic diversity in improved rice genotypes as the success of a crop improvement program depends on the magnitude of genetic variability and the extent to which the desirable characters are heritable [7]. Hence assessment of genetic diversity becomes important in establishing relationships among different cultivars [8,9]. Therefore, different rice varieties of distinct genetic structure are a good promise for the future rice crop improvement. Thus, identification of genotypes and their inter-relationships is important. Development of new biotechnological techniques provides increased support to evaluate genetic variation in both phenotypic and genotypic levels and the results derived from analyses of genetic diversity at the DNA level could be used for designing effective breeding programs aiming to broaden the genetic basis of commercially grown varieties.

DNA fingerprinting/profiling is used to describe the combined use of several single locus detection systems and is being used as versatile tools for investigating various aspects of plant genomes including characterization of genetic variability, genome fingerprinting, genome mapping, gene localization, analysis of genome evolution, population genetics, taxonomy, plant breeding [10]. Genetic diversity can be evaluated with morphological traits, seed proteins, isozymes and DNA markers. Molecular marker technology is the powerful tool for determining genetic variation in rice varieties. In contrast to morphological traits, molecular markers can reveal abundant difference among genotypes at the DNA level, providing a more direct, reliable and efficient tool for germplasm characterization, conservation, and management and untouched by environmental influence. In the present study SSR, ISSR and RAPD markers were used, as these are dominant marker systems and are less costly and easier to be developed and used. The genetic diversity analysis can be extended to characters like salt tolerance/ other abiotic stress, which are controlled by large number of QTLs which may share homology between genes responsible for other abiotic stresses like temperature, drought, flood, submergence etc. Refinement in primer sequences and increasing their

*Corresponding author: Ashu Singh, Tissue culture Lab, College of Biotechnology, Sardar Vallabh Bhai Patel University of Agriculture and Technology, Meerut-250110, India, E-mail: ashubiot25@gmail.com

specificity in relationship to characters under study can enhance the efficacy of microsatellite markers as a tool for tagging specific gene throughout the concerned genome [11]. In the present study, 30 rice varieties were analyzed for genetic variation using SSR, ISSR & RAPD markers. Specially, the objective of the study was DNA fingerprinting and genetic diversity analysis of different varieties (Basmati, non-basmati & aerobic) to measure the extent of genotypic differences, genetic relationship and to assist in broadening the germplasm base of future aromatic rice breeding programs.

Materials and Methods

The field trial involving thirty Basmati and non-Basmati rice varieties adapted to traditional irrigated and aerobic agro-eco systems of rice cultivation was conducted at Sardar Vallabhbhai Patel University of Agriculture and Technology, Meerut during Kharif (Rainy) crop season in 2012-2013. For molecular studies, genomic DNA was isolated from single leaf taken from each of the 30 varieties/genotype of the rice following CTAB (Cetyl Trim ethyl Ammonium Bromide) method [12]. The 30 varieties/genotypes were subjected to classify for genetic diversity of rice with the help of 10 SSR, 10 ISSR, 10 RAPD primers. Agarose gel electrophoresis was used to quantify DNA on the basis of molecular weight. The purified DNA was amplified in PCR with different SSR, ISSR, RAPD primers (10 each). The 30 varieties/genotypes were subjected to screen for diversity of Basmati and non-Basmati genotypes adapted to irrigated and aerobic conditions.

For evaluating marker efficiency, PIC (Polymorphism Information Content) value and Resolving powers were estimated for each primer.

Data analysis and detection of genetic diversity for SSR, ISSR & RAPD markers

Thirty one rice varieties were used to estimate genetic diversity. Polymorphic products from all the marker system were assayed for presence (1) or absence (0). The proportion of bands that have been shared between any of the two varieties averaged over loci SSRs ISSRs & RAPDs were used as the measure of similarity. Genetic diversity was calculated using formula given by Chakravarthi et al. [12]. It refers to the value of a marker for detecting poly- morphism within a population, depending on the number of detectable alleles and the distribution of their frequency.

The power of each primer to distinguish among the studied genotypes was evaluated by the resolving power (Rp) [12]. Resolving power is the capacity of any primer to distinguish among different varieties. It is defined per primer as $Rp=\Sigma\ Ib$ where lb is the band informativeness, that takes the values of: $1-(2x\ [0.5-p])$, being p the proportion of the rice varieties containing the band.

The calculation was based on the number of bands in SSRs, ISSRs, & RAPD primers. Clustering pattern was based on distance matrices by using the Un- weighted Pair Group Method Analysis (UPGMA) program in NTSYS-pc version 2.2 [13]. Thirty rice varieties were analyzed for Genetic diversity and molecular characterization. Numbers of polymorphic as well as monomorphic bands were obtained for determination of PIC value and resolving power of 30 markers (Table-1). Gel photographs are given in Figures 1-3 for SSR, RAPD & ISSR respectively. The Resolving power of RAPD, ISSR & SSR molecular markers are presented by Pie diagrams (Figure 4) and PIC of combined markers values are given as bar diagram (Figure 5). Comparative analysis of RAPD, ISSR & SSR is given in Table 1. Clustering of genotypes based on different marker assays and joint assays is given in Figure 6 onwards.

Discussion

Traditionally used morphological and chemical parameters have not been found to be discriminative enough, warranting more precise techniques. Presently several molecular techniques are available for fingerprinting different cultivars of crops involving differences within and among cultivars. Among these, the DNA Markers RAPD, simple sequence repeat (SSR) and ISSR markers are considered effective and cost-efficient which could detect higher degree of polymorphism in rice [14].

In this present study, the detailed use of molecular markers for the assessment of genetic diversity and identification of economically important traits were evaluated. Genome analysis based, molecular markers have generated epitome of information and a number of databases. The availability of new techniques and new equipment such as invention of PCR technology has revolutionized progress of research in molecular biology. PCR-based methods during the last more than 20 years became the routine work of molecular laboratories. Presently publications related to the methodology and applications of PCR-based DNA fingerprinting behave like the DNA in a PCR, i.e., they multiply exponentially. DNA based molecular markers are the most powerful diagnostic tools to detect DNA polymorphisms both at the level of specific loci and at the genome level [15].

DNA based molecular markers are the most powerful novel tools to detect variation in rice genotypes on the basis of DNA polymorphisms both at the level of specific loci and at the whole genome level [16]. Polymorphisms at the DNA level can be studied by numerous approaches like polymorphism information content etc [17]. Direct strategy is the determination of the nucleotide sequence of a defined region [18], the establishment of lineage of this sequence to an orthologous region in the genome of related organisms. The extent of homology between various sequences can be deduced from the alignment, and phylogenies reconstructed by a variety of approaches and algorithms. DNA sequencing provides highly robust, reproducible, and informative data sets that can be utilized to different analyses for discrimination or mapping of targeted regions of a genome [19]. On the other hand, DNA sequencing can be tedious and expensive when very large number of individuals has to be assayed (e.g., in population genetics and marker-assisted plant breeding programs). In specific areas of research, it is not suitable for estimation of genetic diversity. PCR-based molecular markers SSR (Simple sequence repeat) and ISSR (Inter simple sequence repeat) instead provide a measure of genome wide genetic variation [20]. The analysis of genetic diversity and relatedness between or within different populations, species, and individuals is very important for many disciplines of biological science. Marker technology based on polymorphisms in DNA has catalyzed research in a variety of disciplines such as phylogeny, taxonomy, ecology, genetics, and plant and animal breeding [21].

Despite such large number of varieties developed using diverse germplasm, molecular marker based diversity analysis has shown the genetic base of Indian rice gene pool to be surprisingly narrow [22,23]. Moreover, with regard to trends of genetic diversity in major Indian rice cultivars, however, little work has been done, recently hypervariable microsatellite markers evenly distributed in rice genome have been demonstrated to be quite effective in estimating genetic diversity [24] and during last three decades, a classical strategy for estimation of genetic variability has been complemented by molecular techniques. These include, for example, the analysis of chemical constituents, but most importantly relate to the development of molecular markers.

Sr.	Primer	Molecular wt. range (bp)	Chr. No.	Motif	No. of Polymor. Bands	No. of mono. Band	Polymor-phism %	Diversity in value of PIC	Marker Index
SSR									
1	RM-235	700-3000	12	(CT)24	6	0	100%	0.958	5.875
2	RM-222	200-8000	10	(CT)18	2	0	100%	0.962	4.662
3	RM-236	400-700	2	(CT)18	2	0	100%	0.97	3.862
4	RM-242	500-1000	9	(CT)26	2	0	100%	0.976	3.394
5	RM-247	400-800	12	(CT)16	3	0	100%	0.982	2.728
6	RM-254	200-700	11	TC)6ATT(CT)11	1	0	100%	0.987	1.464
7	RM-253	100-200	6	(GA)25	2	0	100%	0.982	1.932
8	RM-256	100-280	8	(CT)21	2	0	100%	0.911	0.132
9	RM-263	100-240	2	(CT)34	2	0	100%	0.915	1.131
10	RM-264	100-200	8	(GA)27	1	1	50%	0.961	4.198
								Avg.=0.9718	

S.No.	Primer	Molecular wt. range (bp)	No. of Polymor. bands	No. of monomor. bands	Polymor-phism %	PIC Value	Marker Index/ Resolving Power
RAPD							
1	OPF-13	300-1400	4	2	66.6%	0.9925	1.43
2	OPC-15	500-900	7	0	100%	0.982	1.8
3	OPD-08	200-1700	7	0	100%	0.811	1.866
4	OPF-14	300-700	6	0	100%	0.979	1.93
5	OPF06	400-1400	6	0	100%	0.986	1.732
6	OPJ-08	100-900	3	0	100%	0.985	1.32
7	OPC-07	500-1800	4	0	100%	0.975	2.052
8	OPF-17	100-1200	3	0	100%	0.981	2.061
9	OPJ-13	400-900	4	1	80%	0.977	2.066
10	OPK-11	400-900	5	0	100%	0.967	2
ISSR						Avg.=.9635	
1	ISSR1	200-1100	6	0	100%	0.951	5.24
2	ISSR2	200-1000	7	1	87.5%	0.9167	8.366
3	ISSR3	200-900	2	0	100%	0.971	1.6
4	ISSR4	200-900	8	2	80%	0.912	7.926
5	ISSR5	200-1000	6	2	75%	0.8791	4.06
6	ISSR6	100-900	9	0	100%	0.9916	7.786
7	ISSR7	200-900	5	1	83.33%	0.966	3.326
8	ISSR8	200-1000	7	0	100%	0.941	6.088
9	ISSR9	200-1500	8	0	100%	0.975	4.722
10	ISSR10	200-1000	6	1	85.7%	0.979	3.594
						Avg=0.9482	

Table 1: Primer code, annealing temperature, total no. of alleles, no. of Polymorphic alleles, no. of monomorphic alleles and PIC (Polymorphism Information Content) value of 30 rice genotypes.

SSR marker have some merits such a quickness, simplicity, rich polymorphism and stability, thus being widely applied in genetic diversity analysis, molecular map construction and gene mapping [25,26] construction of fingerprints [27], genetic purity test [26], analysis of germplasm diversity [27-29] utilization of heterosis, especially in identification of species with closer genetic relationship. Availability of a number of marker assays provides great opportunities for exercising choice of efficient and robust marker system based on well-defined objectives, convenience and costs. Microsatellite (SSR) markers, a type of variable numbers of tandem repeats, containing generally two or three nucleotide repeats were introduced during 1990s. Such markers are simple, PCR based, locus specific, more reliable i.e. reproducible and typically co-dominant markers. Therefore, SSR markers and their alternative method inter-SSR assay available in public domain were presently used to fingerprint 30 different varieties of basmati and non-basmati rice and also to asses DNA based genetic diversity/similarity of the same varieties [28-29].

DNA fingerprint database has been prepared using the three different PCR-based marker (SSR, RAPD and ISSR) systems for 30 rice genotypes. All the above three molecular markers used in this study were able to generate sufficient polymorphisms and unique DNA fingerprints to identify each of the 30 rice varieties. The level of polymorphism generated by ISSR markers (maximum no. of polymorphic bands amplified by a primer was 9 for ISSR) was higher compared to the SSR (maximum no. of polymorphic bands amplified by a primer was 6) and RAPD (maximum no. of polymorphic bands amplified by a primer was 7) markers. Genetic relationships as determined by cluster of SSR, RAPD, ISSR and/or pooled allelic diversity data of 30 rice genotypes. The SSR polymorphism and diversity could likely be attributed to pedigree [30]. The dendrograms obtained using SSR, RAPD and ISSR data (Figures 7 and 8) were quite similar and most of the varieties were placed in their respective groups, which also match their known pedigrees. Salient features of finger print database obtained using different markers are given below:

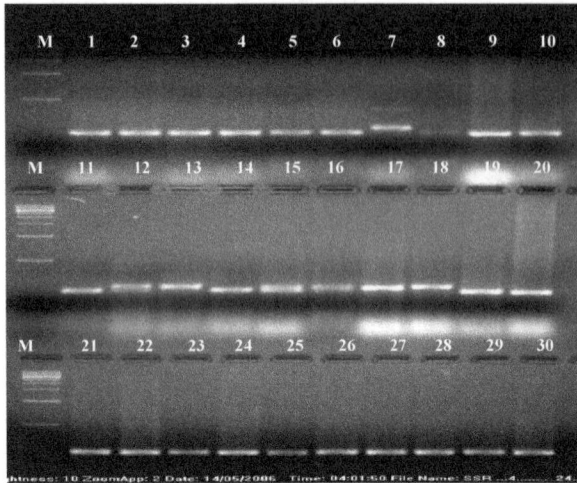

Figure 1: Banding Patterns of various primers of SSR, ISSR & RAPD, with all 30 varieties/genotypes.

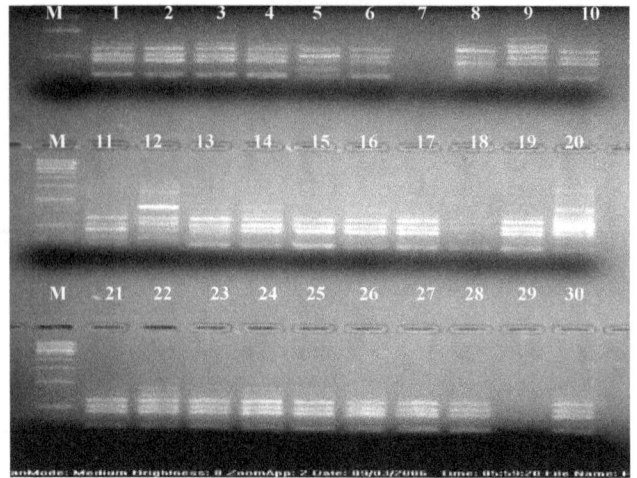

M=500 bp molecular marker.
Figure 2: SSR profiling pattern of 30 rice varieties with RM-242 primer.

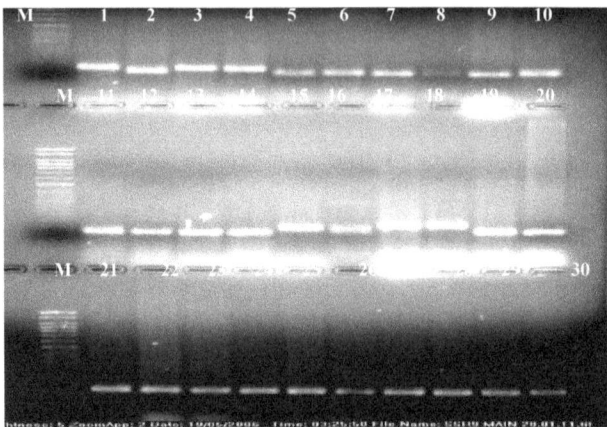

M=20 bp molecular marker .
Figure 3: SSR profiling pattern of 30 rice varieties with RM-263 primer.

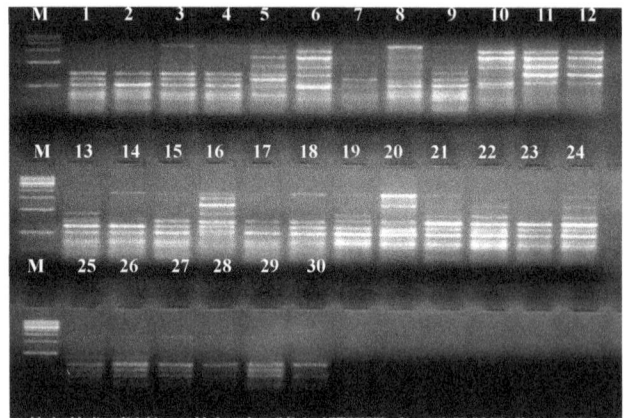

M=100 bp molecular marker.
Figure 4: RAPD profiling pattern of 30 rice varieties with OPD-08 primer.

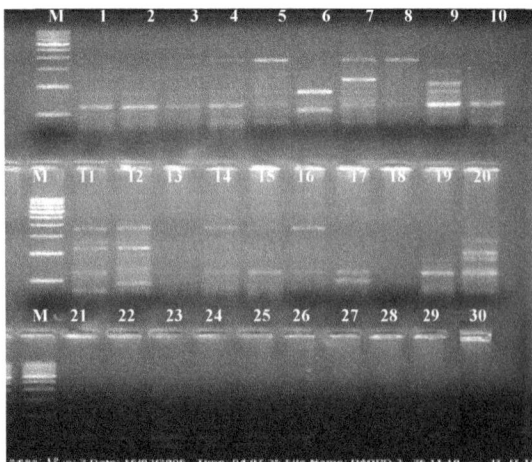

M=100bp molecular marker.
Figure 5: RAPD profiling pattern of 30 rice varieties with OPF-13F primer.

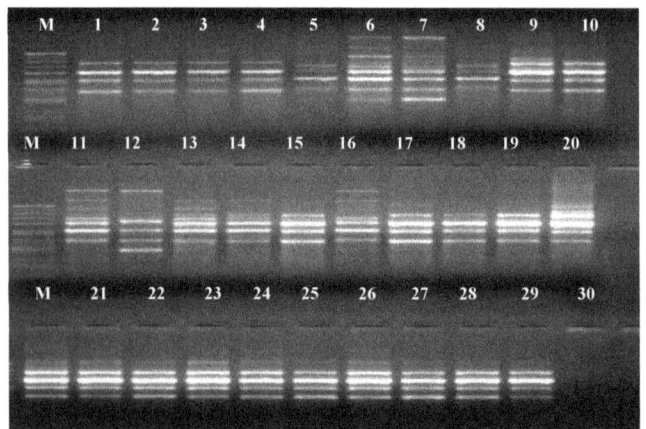

Figure 6: ISSR profiling pattern of 30 rice varieties with ISSR-2 primer (M=100 bp molecular marker).

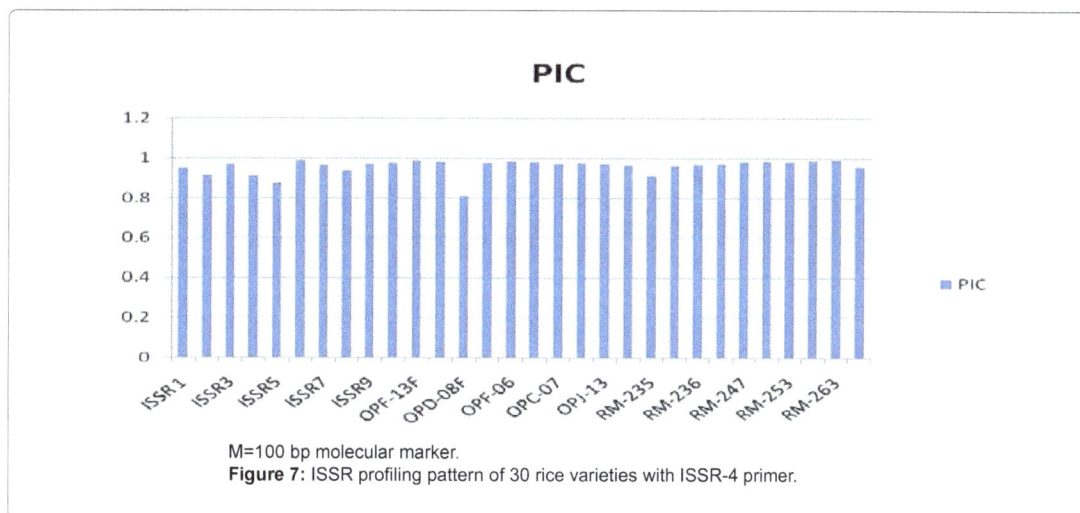

M=100 bp molecular marker.
Figure 7: ISSR profiling pattern of 30 rice varieties with ISSR-4 primer.

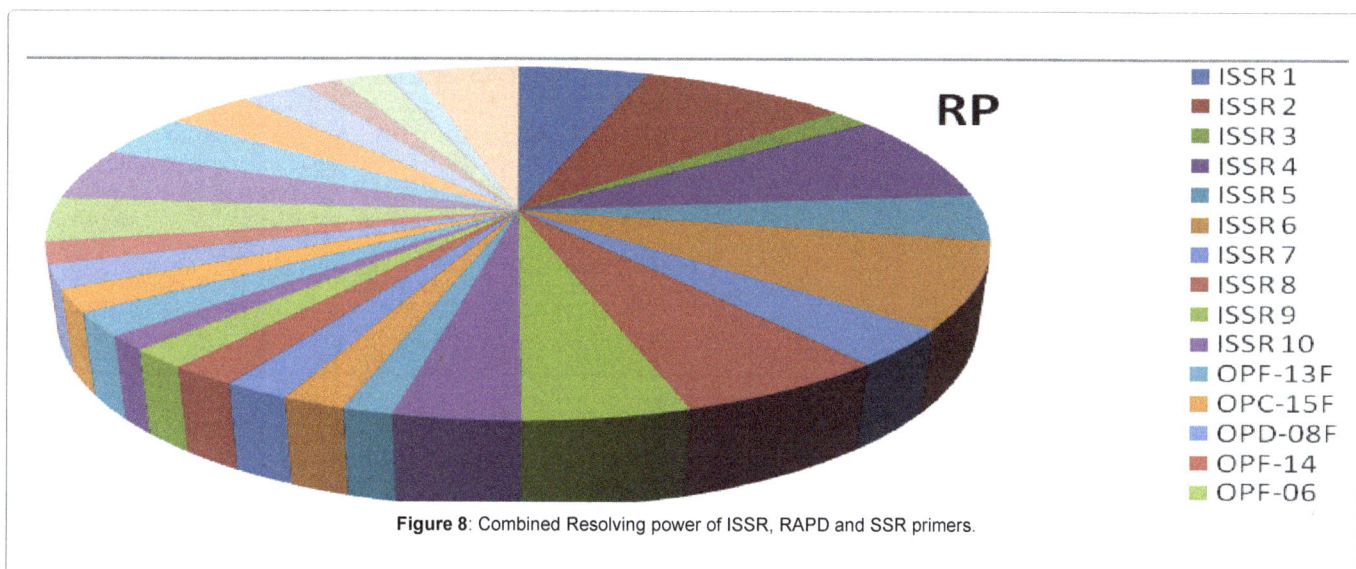

Figure 8: Combined Resolving power of ISSR, RAPD and SSR primers.

SSR analysis

The term SSR (simple sequence repeat) also known as microsatellites was coined by Litt & Luty. These are Co-dominant markers [31]. They are a class of repetitive DNA sequences usually 2.6 bp that are distributed throughout whole genome and are flanked by highly conserved region [32]. The objective of this present study was to evaluate these markers essentially belonging to the repetitive DNA family. Simple sequence repeats (SSRs) consist of 1 to 6 bp long monomer sequences which are repeated several times. A microsatellite fingerprint database, has been generated with 10 SSR markers for 30 rice genotypes, and used for diversity analysis and determination of genetic relationships, as we know the high level of polymorphism associated with microsatellites represents just one component of their rapid rise to become the "genetic tool of choice" for mappers working with all animal as well as plant species.

In SSR analysis, a total of 24 amplified bands were detected using 10 SSR primers in 30 rice varieties, out of 24 amplified bands, only 1 was monomorphic and 23 were polymorphic (Table 2). The maximum numbers of polymorphic bands (4 bands) were obtained using RM-235 primer with 98% polymorphism. RM-235 observed minimum

polymorphism with PIC Value of 0.912, Lower PIC value may be the result of closely related genotypes and higher PIC values might be the result of diverse genotypes. Low PIC values for some other primers were earlier reported by Ma et al. [33]. Among the primers used in the present study, RM-263 is highly informative since it recorded high PIC value (0.995) value of. The resolving power varies between 0.132(RM-256) to 4.662(RM-222) with an average value of 2.7502. On the basis of PIC values & RP (resolving power) cleared that SSR primers have important position in this analysis. In the analysis of SSR assay, all the 30 rice varieties were classified in four main clusters (Figure 3). All the ten Basmati varieties were clustered together. In addition, MAUB 13 and PS 2511 both the non-Basmati genotypes were grouped in the same cluster. The reason, being that these varieties had common parentage. MAUB-13 a non-Basmati and a Basmati variety Vallabh Basmati-21(Similarity 100%) had the same parentage. All varieties/genotypes adapted in aerobic conditions in Srilanka were also included in the same cluster along with Sarbati a variety of Indian origin. Two genotypes shahpasan (Chhattisgarh origin) and Sathi, a farmers' variety (Traditionally conserved in Northern India), both suitable for aerobic condition clustered together in a main cluster. The rest all non-Basmati rice genotypes were clustered together. DG154 and

Components	RAPD	ISSR	SSR
Total no. of primers used	10	10	10
Polymorphic markers	all	all	all
Total no of bands amplified	52	71	24
Average no of bands per primer	5.2	7.1	2.4
Maximum no. of bands amplified by a single primer	7	10	6
No. of polymorphic bands	49	64	23
Average no. of polymorphic bands per primer	4.9	6.4	2.3
Maximum no. of polymorphic bands amplified by a primer	7	9	6
Genetic similarity coefficient for all possible pairs of genotype			
maximum	0.9925	0.9916	0.995
minimum	0.811	0.8796	0.912
average	0.9635	0.9482	0.9718
Resolving power			
maximum	2.006	8.336	4.662
minimum	1.32	1.6	0.132
average	1.8256	5.2708	2.7502

Table 2: Comparative analysis of RAPD, ISSR and SSR markers in this study.

DG138 expressed maximum similarity (approximately 100%) perhaps they had same line of origin or they have been selected on the basis of phenotypic differences those were governed by small DNA sequences with a difference of very few base pairs. Also, DG234 and DG296 had maximum similarity (approximately 100%).

RAPD analysis

Williams et al. used simplest version of molecular markers, the Random Amplified Polymorphic DNA (RAPD). These are mainly Dominant markers [31]. Such primers are simple arbitrary sequences of decamer nucleotides with a GC content of at least 50%. Such primers are used under relaxed stringent conditions and in such cases no prior knowledge of DNA sequence is required. Such primers are still being used despite their low reproducibility. This is due to the simplicity of this technique, as an only very small amount of DNA is required and information on template DNA sequence is not needed [34].

This study indicates that RAPD is a sensitive and powerful technique to distinguish among rice cultivars and to detect genetic variation at DNA level. The results can complement classical morphological identification. The RAPD technique requires a small amount of DNA extracted from a single rice seed, and is not affected by environmental factors. In contrast, morphological identification involves the evaluation of many morphological parameters of the whole plant, and can be affected by environmental factors. DNA-based analysis could be used to identify Indica rice cultivars to prevent fraudulent commercial activity. However, the RAPD is not suitable for the analysis of DNA extracted from processed products, because of the high degradation rate of DNA in products.

RAPD analysis revealed a large number of distinct, scorable fragments per primer pair. A total of 52 bands were amplified using 10 RAPD primers in 30 rice gentypes. Out of 52 amplified bands, 49 bands were polymorphic and 3 bands were monomorphic (Table 2). The number of amplified fragments varied from 3 to 49, with an average of 5.2 (i.e. Average no of bands per primer). The PIC values and resolving power were calculated for individual primers. Polymorphic Information Content (PIC) refers to the value of a marker for detecting polymorphism within a population, depending on the number of detectable alleles and the distribution of their frequency whereas

resolving power is the capacity of any primer to distinguish among different varieties.

In RAPD analysis, PIC values varies from 0.811(OPD-08) to 0.9925(OPF-13) with average of 0.9635 and resolving power varies from 1.32(OPJ-08) to 2.066(OPJ-13) with average of 1.8256. RAPD markers could be employed both for estimating the relationships between varieties and for variety identification [35,36]. All the 30 rice varieties of two quality groups (Basmati, Non-Basmati) adapted in different agro-eco systems were grouped into three distinct major clusters (Figure 3). The Basmati type varieties/genotypes were not clustered together and distributed over three sub clusters in different major clusters. MAUB-13, MAUB-15 and MAUB-64 were clustered with Taroari Basmati, a Basmati variety suitable for normal as well saline/alkaline conditions. Pusa Basmati-1 was clustered with Basmati-370. Another variety of quality rice PS-1121 was clustered with PS-2511. While MAUB-57 the elite Basmati genotype could not be clustered with any variety. Also, NDR118 and Sathi varieties could not be grouped with other varieties. Therefore, these varieties were put individually in two separate clusters. The major cluster was subdivided into 10 sub clusters in case of RAPD. The genotypes having far distant place of origin were clustered together reflecting that the distribution of varieties over clusters was independent of their place of origin/development. It indicated the uniqueness of the DNA sequences represented by RAPD markers in the whole genome. Maximum varieties of Srilanka origin were grouped in one major cluster; perhaps they had same line of origin or were selected on the basis of minor differences governed by few numbers of base pairs.

ISSR analysis

ISSR (Inter-simple sequence repeat) primers are an alternative form of SSR and can be utilized to amplify inter-SSR DNA sequences [28]. These are mostly dominant markers though occasionally a few of them exhibit co-dominance. A total of 71 bands were amplified using 10 ISSR primers in 30 rice varieties. Out of 71 amplified bands, 7 bands were monomorphic and 64 were polymorphic (Table 2). In ISSR analysis, PIC value ranged from 0.8791(ISSR6) to 0.9916(ISSR5) with an average value of 0.9482. The resolving power varies between 1.6(ISSR3) and 8.366(ISSR2) with an average value of 5.2708.

In ISSR analysis, all 30 rice varieties classified in two main clusters (Figure 3). Sathi separately clustered with the Srilanka rice varieties. The export quality variety Pusa1121 was clustered with varieties of coarse rice Pusa677 and Vallabh Bangani. Furthermore, N22 a non-Basmati variety was clustered with two traditional varieties Basmati-370 and Ranbir Basmati. The situation could be explained that the ISSR molecular markers could not represent DNA sequences of Basmati characteristics adequately. It also reflected inefficiency of ISSR markers. Govind and shahpasan were clustered together reflecting that pattern of clustering was independent of place of origin. The genotypes MAUB 13, MAUB 15 & MAUB 164 of quality rice developed at SVPUA&T Meerut were included in a single cluster, reason may be the common DNA sequences received from common parents.

Conclusion

The use of more number of markers would be efficient to characterize the three varieties i.e basmati, non-basmati & aerobic used for the present study, which highlighted the presence of diversity at genomic level among the genotypes studied. India harbors a huge resource of rice cultivars that are lesser known at the market front but hold great significance not only for farmers but also for the local consumers. An effort was made to collect a set of 30 cultivars including

10 basmati varieties, 10 non-basmati and 10 aerobic varieties of rice and to assess their genetic diversity. Their genetic diversity with molecular markers (ISSR RAPD markers) were according to their genotypes having far distant place of origin were clustered together reflecting that the distribution of varieties over clusters was independent of their place of origin/development like MAUB-57 the elite Basmati genotype could not be clustered with any variety. Also, NDR118 and Sathi varieties could not be grouped with other varieties. Some genotypes recently developed (Sathi, MAUB-13, MAUB-21 [Vallabh Basmati-21], MAUB-15, MAUB-64 and MAUB-57) were also characterized and offered promise in their use in the genetic improvement of rice cultivars for grain quality even in case of SSR, out of 24 amplified bands, only 1 was monomorphic and 23 were polymorphic. In the RAPD assay, MAUB-13, MAUB-15 and MAUB-64 were clustered with Taroari Basmati. Pusa Basmati-1 was clustered with Basmati-370. Another variety of quality rice PS-1121 was clustered with PS-2511. While MAUB-57 the elite Basmati genotype could not be clustered with any other variety. Also, NDR118 and Sathi varieties could not be grouped with other varieties. In SSR marker assay, all the Basmati varieties were clustered together. In addition, MAUB 13 and PS 2511, the non-Basmati genotypes were grouped in the same cluster. All varieties/genotypes adapted to aerobic conditions in Srilanka were also included in the same cluster along with Sarbati. Two genotypes shahpasan and Sathi were clustered together. ISSR+RAPD+SSR assay showed that all the genotypes of Basmati rice MAUB-13, MAUB-21 (Vallabh Basmati-21), MAUB-15, MAUB-64 and MAUB-57 developed at SVPUA&T Meerut were clustered in cluster with Taroari Basmati. Pusa Basmati-1 was clustered with Basmati 370 and Ranbir Basmati. However, the most prominent variety of Basmati, PS 1121 was clustered with non- Basmati variety PS-2511 and could not be grouped in any cluster having traditional variety of Basmati. The dendogram also showed that Sathi and all genotypes of Srilanka origin well adapted to aerobic conditions were clustered together.

The dendogram also showed that Sathi and all genotypes of Srilanka origin well adapted to aerobic conditions, which is a newly developed water-saving rice system in which rice grows in nonflooded and unsaturated soil according to 2001 report of the International Rice Research Institute in Philippines, this system has been monitored to identify potentially promising varieties of rice able to grow as an irrigated upland crop and quantify yield potential and water use efficiency, thus these aerobic varieties clustered together, which can be used for proper identification and selection of appropriate parents for breeding programs, including gene mapping, and ultimately for emphasizing the importance of marker-assisted selection (MAS) in aromatic/ non-aromatic rice improvement worldwide.

References

1. Sasaki T, Burr B (2000) International Rice Genome Sequencing Project: the effort to completely sequence the rice genome. Curr Opin Plant Biol 3: 138-141.

2. Onwueme IC, Sinha TO (1991) Field crop production in Tropical Africa, principles and practice CTA (Technical Centre for Agriculture and Rural Cooperation) Ede, The Netherlands. pp. 267-275.

3. Aggarwal RK, Shenoy V, Ramadevi J, Rajkumar R, Singh L (2002) Molecular characterization of some Indian Basmati and other elite rice genotypes using fluorescent – AFLP. Theor Appl Genet 105: 680-690.

4. Bhuiyan MAR (2005) Efficiency in evaluating salt tolerance in rice using phenotypic and marker assisted selection. M.Sc. dissertation, Department of Genetics and Plant Breeding, Bangladesh Agricultural University, Mymen singh, Bangladesh. pp. 96.

5. Malik RK, Gupta RK, Singh CM, Brar SS, Singh SS, et al. (2008) Accelerating the adoption of resource conservation technologies for farm level impact on

6. Hossain MM, Islam MM, Hossain H, Ali MS, da Silva T, et al. (2012) Genetic diversity analysis of aromatic landraces of rice (Oryza sativa L.) by microsatellite markers. Genes, Genomes and Genomics 6(SI1): 42-47.

7. Ravi M, Geethanjali S, Sameeyafarheen F, Maheswaran M (2003) Molecular marler based genetic diversity analysis in rice (Oryza sativa L.) using RAPD and SSR markers. Euphytica 133: 243-252.

8. Sivaranjani AKP, Pandey MK, Sudharshan I, Kumar GR, Madhav MS, et al. (2010) Assessment of genetic diversity among Basmati and non-Basmati aromatic rices of India using SSR markers. Current Sci 99: 221-226.

9. Kibria K, Nur F, Begum SN, Islam MM, Paul SK, et al. (2009) Molecular marker based genetic diversity analysis in aromatic rice genotypes using SSR and RAPD markers. Int J Sustain Crop Prod 4: 23-34.

10. Singh Y (2011) Molecular approaches to assess genetic divergence in rice. GERF Bulletin of Biosciences. 2: 41-48.

11. Chakraborty S, Vhora Z, Trivedi R, Ravikiran R, Sasidharan N (2013) Molecular studies of aromatic and non aromatic rice (oryza sativa l.) Genotypes for quality traits using microsatellite markers. The Bioscan 8: 359-362.

12. Chakravarthi BK, Naravaneni R (2006) SSR marker based DNA fingerprinting and diversity study in rice (Oryza sativa. L). African J Biotech 5: 684-688.

13. Wei X, Yuan X, Yu H, Wang Y, Xu Q, et al. (2009) Temporal changes in SSR allelic diversity of major rice cultivars in China. J Genet Genomics 36: 363-370.

14. Rai SN, Rani V, Kojima T, Ogihara Y, Singh KP, et al. (2006) RAPD and ISSR fingerprints as useful genetic markers for analysis of genetic diversity, varietal identification, and phylogenetic relationships in peanut (Arachis hypogaea) cultivars and wild species. Genome 44: 763-772.

15. Banziger M, Long J (2000) The potential for increasing the iron and zinc density of maize through plant breeding. Food Nutr Bull 21: 397-400.

16. Kaman Z, Kara B (2003) Genotypic variations for mineral content at different growth stages in wheat (Triticum aestivum L.). Cereal Res Commun 31: 459-466.

17. Nagaraju J, Kathirvel M, Kumar RR, Siddiq EA, Hasnain SE (2002) Genetic analysis of traditional and evolved Basmati and non-Basmati rice varieties by using fluorescence-based ISSR-PCR and SSR markers. Proc Natl Acad Sci U S A 99: 5836-5841.

18. FAO (2002) A report on: Crops and drops – making the best use of water for agriculture. Food and Agriculture Organization of the United Nations. Rome.

19. Weising K, Atkinson RG, Gardner RC (1995) Genomic fingerprinting by microsatellite-primed PCR: a critical evaluation. PCR Methods Appl 4: 249-255.

20. Möller EM, Bahnweg G, Sandermann H, Geiger HH (1992) A simple and efficient protocol for isolation of high molecular weight DNA from filamentous fungi, fruit bodies, and infected plant tissues. Nucleic Acids Res 20: 6115-6116.

21. Botstein D, White RL, Skolnick M, Davis RW (1980) Construction of a genetic linkage map in man using restriction fragment length polymorphisms. Am J Hum Genet 32: 314-331.

22. Provost, Wilkinson (1999) A new system of comparing PCR primer applied ISSR fingerprinting of potato accessions. Theor Apple Genet 98: 107-117

23. Rohif FJ (2002) NTSYS-pc: Numerical taxonomy and multivariate analysis system (Ed. 2.2), Department of Ecology and evolution, State University of NY, Stony Brook.

24. Saini N, Jain N, Jain S, Jain RK (2004) Assessment of genetic diversity within and among Basmati and non-Basmati rice varieties using AFLP, ISSR and SSR markers. Euphytica 140: 133-46.

25. Singh D, Kumar A, Sirohi A, Kumar R, Yadav R, et al. (2009) Effects of adaptation to irrigate and aerobic agro-eco systems and quality on ISSR and SSR based genetic analysis of Basmati and non-Basmati rice's of Asia: Limitations of molecular tools. In: International Conference on "Current Trends in Biotechnology & Implications in Agriculture" held on 19-21 Feb. 2009 at SVBP University 0f Ag. &Technology, Meerut, Pp. 38-49.

26. Sanger F, Nicklen S, Coulson AR (1977) DNA sequencing with chain-terminating inhibitors. Proc Natl Acad Sci USA 74: 5463-5467.

27. Vikram P, Swamy BPM, Dixit S, Ahmed HU, Sta Cruz, et al (2012) Bulk

Segregant Analysis: "An effective approach for mapping consistent-effect drought grain yield QTLs in rice" Field Crops Res. 134: 185-192.

28. Ram SG, Thiruvengadam V, Vinod KK (2007) Genetic diversity among cultivars, landraces and wild relatives of rice as revealed by microsatellite markers. J Appl Genet 48: 337-345.

29. Upadhyay P, Singh V K, Neeraja C N (2011) Identification of Genotype Specific Alleles and Molecular Diversity Assessment of Popular Rice (Oryza sativa L.) Varieties of India. International Journal of Plant Breeding and Genetics 5: 130-140.

30. Neeraja CN, Vemireddy LR, Malathi S, Siddiq EA (2009) Identification of alternate dwarfing gene sources to widely used Dee-Gee-Woo-Gen allele of sd1 gene by molecular and biochemical assays in rice (Oryza sativa L.). Electronic Journal of Biotechnology

31. Narshimulu G, Jamaloddin M, Vemireddy LR, Anuradha G, Siddiq E (2011) Potentiality of evenly distributed hypervariable microsatellite markers in marker-assisted breeding of rice. Plant Breeding 130: 314-320.

32. Zhang SB, Zhu Z, Zhao L, Zhang YD, Chen T, et al. (2007) Identification of SSR markers closely linked to eui gene in rice. Yi Chuan 29: 365-370.

33. Ma H, Yin Y, Guo ZF, Cheng LJ, Zhang L, et al. (2011) Establishment of DNA fingerprinting of Liaojing series of japonica rice. Middle-East Journal of Scientific research 8: 384-392.

34. Peng ST, Zhuang JY, Yan QC, Zheng KL (2003) SSR markers selection and purity detection of major hybrid rice combinations and their parents in China. Chin J Rice Sci 17: 15.

35. Zhou HF, Xie ZW, Ge S (2003) Microsatellite analysis of genetic diversity and population genetic structure of a wild rice (Oryza rufipogon Griff.) in China. Theor Appl Genet 107: 332-339.

36. Zietkiewicz E, Rafalski A, Labuda D (1994) Genome fingerprinting by simple sequence repeat (SSR)-anchored polymerase chain reaction amplification. Genomics 20: 176-183.

14

AMMI Analysis of Tuber Yield of Potato Genotypes Grown in Bale, Southeastern Ethiopia

Miheretu Fufa*

Oromia Agricultural Research Institute, Sinana Agricultural Research Center; Horticulture and Spice Technology Generation Team, Ethiopia

Abstract

The problem of genotype by environment interactions that often complicates the interpretation of multi-locations trail analysis making the prediction of genotype performance difficult can be eased with the adoption of the Additive Main Effects and Multiplication Interaction (AMMI) model analysis. The AMMI model was used to evaluate tuber yield stability of twelve potato (*Solanumtuberosum L.*) genotypes in randomized complete block design with three replications at Sinana, Shallo and Dinsho during 2009, 2010 and 2011. The objectives were to estimate the nature and magnitude of GEI for tuber yield and to identify stable potato genotypes for general adaptation and unstable genotypes for specific adaptation. Combined analysis of variance showed highly significant difference between the genotypes, locations and GEI. Proportion of variation captured by genotypes, locations and GEI is 13.65, 51.64 and 34.81, respectively indicating more effects of locations as compared to genotypes.

Keywords: AMMI analysis; Tuber yield of potato and stability; Environment interactions; Genotypes of potato.

Introduction

Potato (*SolanumtuberosumL.*), belonging to the family Solanaceae, is an important food and cash crop ranking fourth after maize, wheat and rice annual production in the world [1,2]. It is the world's number one none-grain crop to ensure food security due to its growing demand [3]. It is a high biological value crop that gives an exceptionally high yield with more nutritious content per unit area per unit time than any other major crops. Thus, it can play a remarkable role in human diet as a supplement to other food crops such as wheat and rice [4]. Furthermore, the contribution of potato to the diversification of the cereal mono-cropping in Bale is great.

Despite the importance of potato in the country agriculture, its productivity has shown a decreasing trend even if its production is expanding steadily [5,6]. One of the major factors contributing to reduction in yield of potato is inadequacy of improved cultivars with wide adaptability and stability in tuber yield. Thus, evaluating genotypes across various environments for their stability of performance and range of adaptation is crucial and is an important component of the research activity of the national as well as regional research program.

Evaluating genotypes over diverse environments is universal practice to ensure the stability of performance of the genotypes [7]. Stability in performance is one of the most desirable properties of a genotype to be released as a variety for wide cultivation [8]. However, the activity of identification, selection and recommendation of superior genotypes is complicated and severely limited by genotype × environment interaction that is inevitable in multi-environmental trails [9-13]. The presence of genotype x environment interaction may confound the genotypic performance with environmental effects [14].

Several statistical models and procedures have been developed and exploited for studying the genotype x environment interaction effects, stability of genotypes and their relationships in varietal development process [9,11,15]. A combined analysis of variance (ANOVA) can quantify the interactions, and describe the main effects. However, it is uninformative for explaining genotype x environment interaction. To increase accuracy, additive main effects and multiplicative interaction (AMMI) is the model of first choice when main effects and interaction are both important [13]. It is a powerful tool for effective analysis and interpretation of multi-environment data structure in breeding programs and is useful for understanding genotype x environment interaction [7,9]. Plant breeders frequently apply AMMI model for explaining genotype x environment interaction and analyzing the performance of genotypes and test environments [16,17]. Therefore, this paper assesses genotype x environment interaction and tuber yield stability of potato genotypes under Bale highlands, Southeastern Ethiopia.

Material and Methods

In this experiment, twelve genotypes of potato were evaluated in randomized complete block design with three replications at Sinana, Shallo and Dinsho during 2009, 2010 and 2011. The plot area used was 3m x 3m with the spacing of 30 cm within rows and 75 cm between rows respectively. The sample data were recorded from the two middle rows. All agronomic and cultural practices were followed as per the general recommendation. No attempt was made to control the late blight with fungicide. At physiological maturity, the tubers were harvested from two middle rows and washed with clean tap water to remove soils. The clean tubers were sorted and graded into large, medium and small based on their size. The weight of the tubers per plot (kg) was recorded and their mean was subjected to analysis.

Statistical analysis

Analysis of variance (ANOVA) was carried out on tuber mean on plot basis and pooled over locations and seasons using the Generalized Linear Model (GLM) procedures of the Statistical Analysis System (SAS) version, 9.1.3 [18]. The Additive Main Effects and Multiplicative Interactions (AMMI) statistical model was produced using Irristat

*Corresponding author: Miheretu Fufa, Oromia Agricultural Research Institute, Sinana Agricultural Research Center, Horticulture and Spice Technology Generation Team, PO. Box 208, Bale Robe, Ethiopia E-mail: miheretufufag@gmail.com

software [19] to analyze the yield data and to produce biplot that shows both main and interaction effects for both genotypes and environments.

Furthermore, AMMI's stability value (ASV) was calculated in order to rank genotypes in terms of stability using the formula suggested by Purchase (1997) as shown below: where, SS = Sum of squares; IPCA1 = interaction principal component analysis axis 1and IPCA2 = interaction principal component analysis axis 2 (Figure 2).

Results and Discussion

Combined analysis of variance showed that there was a highly significant difference (p<0.01) among the genotypes for their tuber yield indicating that there is fluctuation of genotypes in their response to the different environments. Because of the highly significant difference existing in tuber yield among the genotypes, the AMMI analysis was used to estimate the highest stable genotypes. The majority of the total variation was accounted for by location (27.60%) while that of genotype is only 9.56%. This variability may be due to the variability of soil and rainfall across locations.

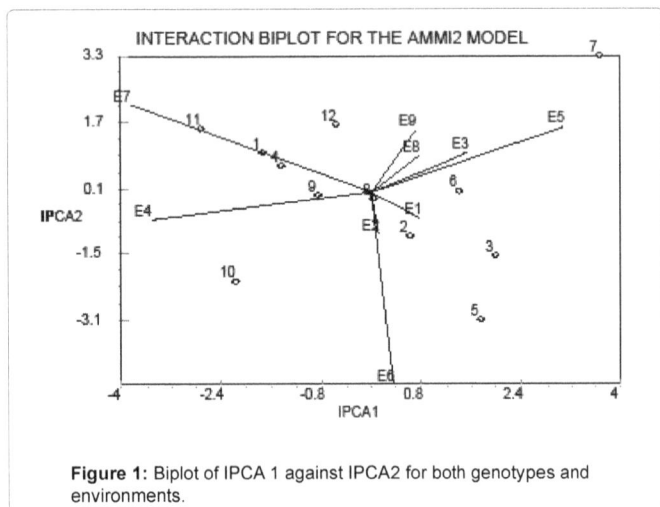

Figure 1: Biplot of IPCA 1 against IPCA2 for both genotypes and environments.

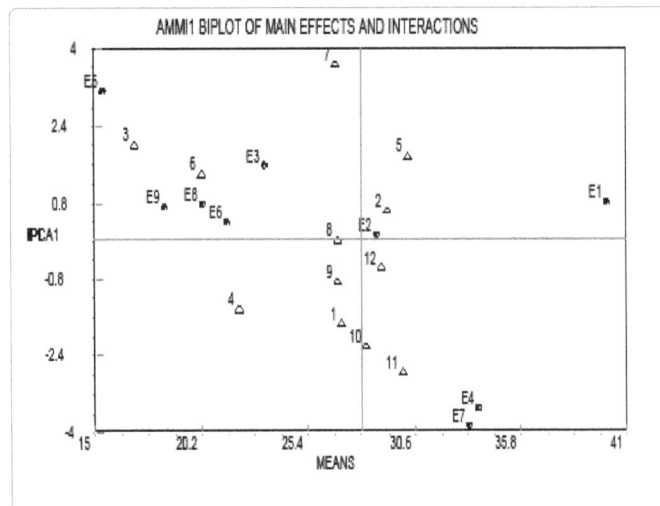

Figure 2: IPCA 1 scores ploted against tuber mean yield of the 12 geno-types evaluated in 9 environments.
E1= Sinana 2009, E2= Shallo 2009, E3= Dinsho 2009,
E4=Sinana 2010, E5= Shallo 2010, E6=Dinsho 2010,
E7=Sinana 2011, E8=Shallo 2011, E9=Dinsho 2011,

The AMMI analysis of variance showed that all the components were highly significant (Table 1). The environment had the greatest effect and accounted for 51.94% of the total sum squares; genotypes accounted for 13.65% and GXEI had accounted for 34.41% which is the next highest contribution. A large sum of squares for environments indicates that the environments were diverse, with large difference among environmental means causing most of the variation in tuber yield. The variation in soil moisture across the different environment was considered as the major underlying causal factor for the GXE interaction. The magnitude of the GEI sum of squares was 2.52 times larger than that for genotypes, indicating that there were substantial differences in genotype response across environments. Results from AMMI analysis (Table 1) also showed that the first principal component axis (PCA 1) of the interaction captured 38.94% of the interaction sum of squares. Similarly, the second principal component axis (PCA 2) explained a further 27.75% of the GEI sum of squares. Furthermore, PCA 1 and PCA 2 had sum squares greater than that of genotypes.

The mean squares for the PCA1 and PCA 2 were significant at p=0.01 and cumulatively contributed to 66.69% of the total GEI. Both the IPCA 1 and IPCA 2 scores revealed that genotype 8 was the most stable genotype. Similarly, the calculated ASV indicated that genotype is the only stable genotype while others showed considerable interaction with the environments (Table 1). So, genotype 8 showed negligible interaction and is found to be the most stable genotype showing broad adaptation across environments. The rest genotypes

Source	DF	SS	MS	F Value	Explained (%)
GEN	11	5198.75229	472.61384	5.72***	9.56
LOC	2	15009.47866	7504.73933	90.81***	27.60
Year	2	3499.53862	1749.76931	21.17 ***	6.44
LY (REP)	18	1600.53208	88.91845	1.07 ns	2.94
GEN*LOC	22	3584.42150	162.92825	1.97 **	6.59
GEN*Year	22	3553.21118	161.50960	1.95**	6.53
LOC*Year	4	1274.04735	318.51184	3.85**	2.34
GEN*LOC*Year	44	5966.98924	135.61339	1.64**	10.97
Error	198	14700.16631	74.24326		
Corrected Total	323	54387.13723			
TREATMENTS	11	1732.92	157.538		13.65
ENVIRONMENT	8	6594.35	824.294		51.94
TREATMENT X SITES	88	4368.21	49.6387		34.41
AMMI COMPONENT 1	18	1701.01	94.5004	2.480**	38.94
AMMI COMPONENT 2	16	1212.20	75.7624	2.812**	27.75
AMMI COMPONENT 3	14	682.203	48.7288	2.522**	15.62
AMMI COMPONENT 4	12	381.963	31.8302	2.280**	8.74
GXE RESIDUAL	28	390.836			
TOTAL	107	12695.5			

*** highly significant; **significant , ns=non-significant at p<0.001; df= degree freedom. SS= Sum of Square, MS= Mean Square, Gen= Genotype and LOC= Location.

Table 1: The analysis of variance and AMMI of tuber yield for 12 potato geno-types tested over three locations.

Genotype	Yield Code	t/ha	AMMI model Rank	IPCA1	IPCA2	ASV	Rank
Hunde	1	27.02	8	-1.724	0.9620	2.603	7
392637-500	4	22.04	10	-1.435	0.6476	2.115	5
90147-15	7	26.80	9	3.669	3.290	6.110	12
90170-37	10	26.33	4	-2.187	-2.148	3.746	10
394640-539	2	29.21	2	0.6317	-1.070	1.390	2
387967-3	5	30.27	1	1.754	-3.079	3.942	9
390012-2	8	26.82	7	0.2655	-0.1682	0.409	1
Jalane	11	30.00	3	-2.732	1.533	4.129	11
Local	3	16.94	12	2.004	-1.538	3.205	8
90147-46	6	20.18	11	1.403	0.8300	2.137	6
Ararsa	9	26.88	6	-0.8574	0.7024	1.393	3
90147-41	12	28.93	5	-0.5522	1.632	1.807	4

ASV= AMMI stability Value, t/ha= tone per hectare
Table 2: Mean tuber yield (t/ha), scores for AMMI and rank of 12 potato genotypes tested for 3 years per location in Southeastern Ethiopia during 2009-2011.

showing considerable interaction with the environments were highly interactive and were highly unstable across environments. According to SVA calculated, the most unstable genotype is 7 followed by 11, 5, 10 and 3. The underlying causes of the interaction observed can therefore be based on both the genetic differences between these genotypes. Except the local check, which gave the lowest tuber yield, most of the unstable genotypes had the best performance in their tuber yield. The AMMI biplot provides a visual expression of the relationships between the first interaction principal component axis (IPCA1) and means of genotypes and environments (Figure 1) with the biplot accounting for up to 78.99% of the total sum of squares. Genotype, Environment and PCA 1 respectively accounted for 13.65%, 51.94% and 13.39% of the total sum of squares. Genotypes 7, 11, 10 and 5 were the highest yielding but most unstable genotypes. Genotype 3 was the low yield unstable genotype.

In conclusion, the AMMI analysis showed the presence of Genotype X Environment interactions among the 12 potato genotypes. Genotype 2 was found to be the most stable high yielding genotype. It is found useful to use other stability parameters in assessing the stability of potato (*Solanumtuberosum L.*) genotypes under the studied environments of South East Ethiopia. Although AMMI was found to be more informative in depicting the adaptive response of the genotypes [20-22], the joint regression analysis also remains a good option (Table 2).

Acknowledgment

My special thanks go Oromia Agricultural Research Institute for funding this experiment. I also express my gratitude to GutaLeggesse, MulugetaGadif and Adam Abdulhamid for data collection and compilation.

References

1. http://www.potato2008.org/en/potato/economy.html

2. Lemaga B, Kakuhenzire R, Gildemacher P, Borus D, Gebremedhin WG, et al. (2009) Current status and opportunities for improving the access to quality potato seed by small farmers in Eastern Africa. Symposium 15th triennial of the symposium of the international society for tropical root crops 2-6.

3. FAO (2008) Workshop on opportunities and challenges for promotion and sustainable production and protection of the potato crop in Vietnam and elsewhere in Asia 25-28.

4. Badoni A, Chauhan, JS (2010) Potato seed production of cultivar Kufri Himalini, In vitro. stem cell 1:1-6.

5. CSA (2003) Ethiopian agricultural sample enumeration, 2001/02 (1994 E.C.). Statistical report on socio-economic characteristics of the population in agricultural...Part I, Chapter 5, Area and Production of Crops and Crop Utilization. Addis Ababa 167.

6. CSA (2008) Agricultural Sample Survey 2007/2008 (2000 E.C.) volume1 report on area and production of crops for private peasant holdings. Addis Ababa 44.

7. Sadeghi SM, Samizadeh H, Amiri E, Ashouri M (2011) Additive main effects and multiplicative interactions (AMMI) analysis of dry leaf yield in tobacco hybrids across environments. Afr J of biotechnology 10: 4358-4364.

8. Singh RK, Chaudhary BD (1977) Biometrical methods in quantitative genetic analysis. Kalyani publishers, New Delhi, India.

9. Asfaw A, Alemayehu F, Gurmu F, Atnaf M (2009) AMMI and SREG GGE biplot analysis for matching varieties onto soybean production environments in Ethiopia. Scientific research and essay 4: 1322-1330.

10. Eberhart SA, Russell WA (1966) Stability parameters for comparing varieties. Crop Sci 6: 36-40.

11. Finlay KW, Wilkinson GN (1963) The analysis of adaptation in a plant breeding program. Aust J Agr Res 14: 742-754.

12. Shafii B, Price WJ (1998) Analysis of genotype-by-environment interaction using the additive main effects and multiplicative interaction model and stability estimates. J agr biol environ stat 3: 335-345.

13. Zobel RW, Wright MS, Gauch HG (1988) Statistical analysis of a yield trial. Agron J 80: 388-393.

14. Thillainathan M, Fernandez GCJ (2002) A novel approach to plant genotypic classification in multi-site evaluation. Hort. science 37: 793-798.

15. Crossa J (1990) Statistical analysis of multi-location trials. Adv agron 44: 55-85.

16. Gauch HG (2006) Statistical analysis of yield trials by AMMI and GGE. Crop Sci 46: 1488-1500.

17. Yan W, Kang MS, Ma B, Wood S, Cornelius PL (2007) GGE biplotvs AMMI analysis of genotype-by-environment data. Crop sci 47: 643-655.

18. SAS (2008) SAS software version 9.2. SAS Institute INC Cary NC USA.

19. SAS Institute Inc. SAS 9.1.3.for windows.exe.

20. IRRI Stat (2003) International rice research institute. Metro Manila, Philippines.

21. Purchase JL (1997) Parametric stability to describe G x E interactions and yield stability in winter wheat. PhD Thesis, department of agronomy, faculty of agri, univ of orange free state, Bloemfontein, South Africa.

22. Hassanpana D, Azimi J (2010) Yield Stability Analysis of Potato Cultivars in Spring Cultivation and after Barley Harvest Cultivation. American-Eurasian J. Agric. & Environ Sci 9: 40-144.

15

Efficiency of *In Vitro* Regeneration is Dependent on the Genotype and Size of Explant in Tef [*Eragrostis tef* (Zucc.) Trotter]

Sonia Plaza-Wüthrich[1], Regula Blösch[1] and Zerihun Tadele[1,2*]

[1]*Institute of Plant Sciences, University of Bern, Altenbergrain 21, 3013 Bern, Switzerland*
[2]*Institute of Biotechnology, Addis Ababa University, P.O. Box 32853, Addis Ababa, Ethiopia*

Abstract

Tef [*Eragrostis tef* (Zucc.) Trotter] is the major cereal crop in the Horn of Africa particularly in Ethiopia where it is staple food for about 50 million people. Its resilience to extreme environmental conditions and high in nutrition makes tef the preferred crop among both farmers and consumers. The efficiency of *in vitro* regeneration plays significant role in the improvement of crops. We investigated the efficiency of regeneration in 18 tef genotypes (15 landraces and three improved varieties) using three sizes of immature embryos (small, intermediate and large) as an explant. *In vitro* regeneration was significantly affected by the genotype and the size of the immature embryo used as a donor. Intermediate-size immature embryos which were 101-350 μm long led to the highest percentage of regeneration. Interestingly, the three improved varieties presented very low regeneration efficiencies whereas the landrace Manyi resulted in consistently superior percentage of *in vitro* regeneration from all three sizes of explants. The findings of this work provide useful insight into the tef germplasm amenable for the regeneration technique which has direct application in techniques such as transformation. It also signifies the importance of using tef landraces instead of improved varieties for *in vitro* regeneration.

Keywords: *Eragrostis tef*; Immature embryo; Plantlet; Regeneration; Embryogenesis; tef

Introduction

Tef [*Eragrostis tef* (Zucc.) Trotter] is a cereal crop extensively cultivated in the Horn of Africa where it is annually cultivated on about 3 million hectares of land in Ethiopia alone [1]. This extensive cultivation of the crop is related to some traits beneficial for farmers and consumers including, i) its tolerance to extreme environmental biotic and abiotic conditions, ii) its gluten-free seeds, hence considered as a healthy food, and iii) high palatability of its straw by livestock. Despite all these useful traits, tef is considered as an orphan crop due to the little scientific research done on the crop. As a result, the crop remains largely unimproved which is associated with poor productivity lodging or displacement of the plant from its upright position is the major cause for inferior yield in tef [2]. Tef has a very tall and weak stem which falls on the ground due to wind and rain. The majority of research on tef improvement has been done at the Ethiopian Institute of Agricultural Research where conventional techniques of selection and hybridization are widely implemented to release 35 improved varieties which are suited to diverse agro-ecological regions [3]. The widely cultivated and popular variety called Quncho was developed by the intra-specific crossing between two improved cultivars [4]. The recently published tef genome [5] will accelerate the breeding program if integrated with improvement methods such as tissue culture and genetic transformation.

Tissue culture or also commonly known as *in vitro* regeneration plays a key role in crop improvement. In addition to its significant contribution in genetic transformation of plants, tissue culture is also useful in developing large-scale clonal propagation of genotypes of interest and producing and propagating disease-free plants [6]. The somaclonal variations induced in the tissue culture are also source of variability in plant breeding [7,8]. The percentage of initial explants converted to plantlets or whole plants referred to the culture efficiency of regeneration. This efficiency is mainly affected by the genotypes and explants. The presence of a strong genetic effect was reported for *Arabidopsis* [9], wheat [10-12] and rice [13]. Considerable differences in regeneration ability were observed among four *Arabidopsis* ecotypes, namely Columbia, Landsberg erecta, Cape Verde Island and Wassilewskija based on the source of explant and composition of the culture medium [9]. In wheat, embryogenic capacity or number of somatic embryos formed from cultured immature embryos was mainly altered by the genotype whereby the best performing cultivars scored 1.4-1.8 plants/explant [10].

The source and size of explant affects the efficiency of regeneration. The study in malting barley called Morex showed that smaller embryos (0.5-1.5 mm) had higher regeneration efficiency than larger embryos (1.6-3.0 mm) [14]. Similarly, in Sudan grass (*Sorghum sudanense* Piper) smaller immature embryos (0.7-1.5 mm) were better than larger ones (1.6-2.5 mm) in the speed and frequency of callus and shoot formation [15].

Diverse *in vitro* regeneration techniques were studied for tef. The explants used for these investigations were seedlings, roots, and leaves [16,17], seeds [18], immature spikelets or panicles [19,20], and immature embryos [21] in which the latter resulted in substantially high percentage of regeneration. However, since earlier study using immature embryo was made on only two tef genotypes, it did not represent the existing tef germplasm with huge variations. Hence, the present study was made to investigate the efficiencies of *in vitro* regeneration in 18 tef genotypes with diverse morphological and agronomic properties [22].

Material and Methods

Plant material

Fifteen selected landraces and three improved varieties of tef were used. The 15 landraces were obtained from the National Plant

*Corresponding author: Zerihun Tadele, Institute of Plant Sciences, University of Bern, Altenbergrain 21, 3013 Bern, Switzerland
E-mail: zerihun.tadele@ips.unibe.ch

Germplasm System (NPGS) of the United States Department of Agriculture [23]. They are: *Ada* (NPGS accession number: 524433), *Addisie* (524434), *Alba* (524435), *Balami* (524436), *Beten* (524437), *Dabbi* (524438), *Enatite* (524439), *Gea Lamie* (524440), *Gommadie* (524441), *Karadebi* (524442), *Manyi* (524443), *Red dabi* (524457), *Rosea* (524444), *Tullu Nasy* (524445) and *Variegata* (524446) while the three improved varieties were *Dukem* (DZ-01-974), *Magna* (DZ-01-196), and *Tsedey* (DZ-Cr-37). Donor plants were grown for three weeks under long-day conditions (16 h light, 8 h dark at 21 ± 1°C) before plantlets were transferred to short day conditions (8 h light, 16 h dark at 20 ± 2°C). The soil used consisted of 5/11 parts of topsoil, 4/11 parts of turf and 2/11 parts of quartz sand. Plants were fertilized once a week with Hauert Plantaktiv 16-6-26 N-P-K (Hauert HBG Dünger Schweiz, Grossaffoltern, Switzerland).

Embryo isolation and *in vitro* regeneration

The procedure for isolating immature embryos from panicles was based on earlier work [21]. Panicles were surface sterilized for 10 minutes with 1% HCl followed by three washings with sterile water. Immature embryos were separated from the sterilized caryopses by squeezing them out through an incision made at its base. Three different sizes of embryos were selected: small and globular (50-100 μm), intermediate (101-350 μm) and large embryos (351-750 μm) (Figure 1). Another distinction between the small and intermediate embryos was the loss of the globular shape in the intermediate ones. Large embryos were extracted from solid endosperm. Thirty immature embryos were plated on 3.5 cm diameter petri dishes containing K99 medium [24] placing the scutellum facing up and incubating in the dark at 25°C ± 2°C. The K99 media contains 90 g/l maltose, 1 g/l glutamine and 2 mg/l of 2,4-D. After two to three weeks in the dark, somatic embryos were transferred to K4NB medium [25] containing 36 g/l maltose, 0.15 g/l glutamine and 0.22 mg/l BAP. The pH of the medium was adjusted to 5.8 where 0.4% phytagel was used as a gelling agent. Plantlets were regenerated under 14-hour photoperiodic conditions for 4 weeks with a sub-culture to fresh medium after 2 weeks. The growth conditions consist of a relative humidity of 50% all day-long and a temperature of 21 ± 2°C during the dark and 25 ± 2°C during the light. The photon luminosity was set to 70 μmol/m² s during the light period. Plantlets with well-developed root systems were transferred to soil and grown under the same conditions than for the donor plants (see above). After three weeks of hardening in long-day conditions, plants were transferred to short-day room for the production of seeds. Five replicates each containing 30 immature embryos were tested for each tef cultivar and size of the explant (Figure 1).

The efficiencies in somatic embryogenesis, regeneration and culture were enumerated as follows:

Embryogenesis efficiency, direct (%)=number of embryos/number of explants × 100

Embryogenesis efficiency, indirect (%)=number of callus/number of explants × 100

Regeneration efficiency, direct (%)=number of plantlets/number of embryos × 100

Regeneration efficiency, indirect (%)=number of plantlets/number of callus × 100

Culture efficiency=number of plantlets/number of explants × 100 [26].

Determination of morphological, phenotypic and yield related traits

Morphological traits including numbers of tillers, panicles and internodes, and lengths of culm, panicle and the second culm internode (starting from the base of the plant) were determined at the flowering time. The length of second culm internode was earlier reported to determine the lodging tolerance [27]. Days to heading or flowering and days to maturity and the grain filling period were also determined. Days to heading was defined as the number of days from sowing until the first flower appeared and days to maturity was the number of days for all the grains of a panicle to mature. The time between flower emergence and grain maturity was defined as grain filling period. At harvesting time, shoot biomass was separated into culms and panicles

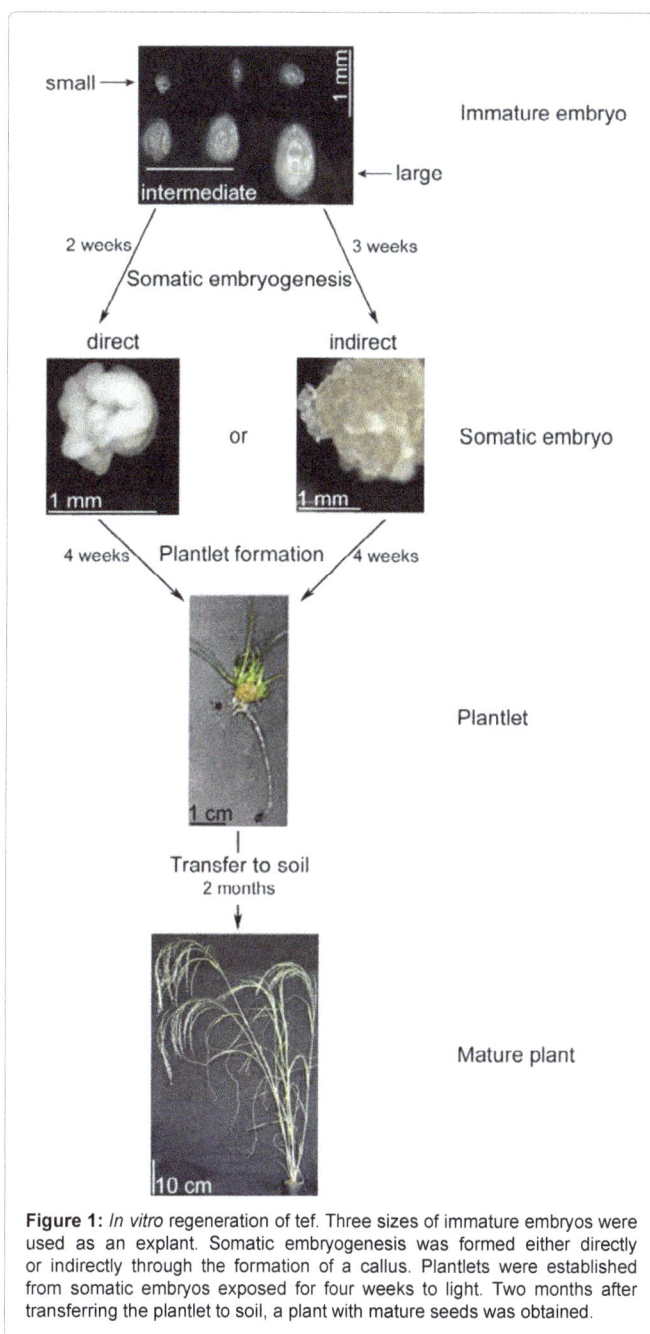

Figure 1: *In vitro* regeneration of tef. Three sizes of immature embryos were used as an explant. Somatic embryogenesis was formed either directly or indirectly through the formation of a callus. Plantlets were established from somatic embryos exposed for four weeks to light. Two months after transferring the plantlet to soil, a plant with mature seeds was obtained.

and dried for 24 hours at 60°C in order to determine the dry weight of all culms and panicles. Harvest index was calculated as the ratio of the grain yield to the shoot biomass.

Statistical analysis

Statistical analysis was performed using SPSS Statistical 17.0 (IBM, Chicago, IL). Non-parametric tests were chosen as it is appropriate for the number of replicates used and the non-homogeneity of the variance in order to compare differences between the treatments (p ≤ 0.05). For K independent samples, Kruskal-Wallis tests were used, whereas for two independent samples, Mann-Whitney U tests were employed. For correlation analysis, the Pearson correlation test was used on the mean values of selected traits among efficiencies of embryogenesis, regeneration, and culture.

Results

In vitro regeneration

Steps and the timeline from excising the three sizes of immature embryos to the two paths of somatic embryogenesis, plantlet formation and to finally grown on soil to full maturity are shown in Figure 1. The efficiency of somatic embryogenesis was determined for the three sizes of immature embryos used as an explant as well as for the two types of embryos formed. While the three groups of immature embryos were small (50-100 μm), intermediate (101-350 μm) and large (351-750 μm). Somatic embryos made from immature embryos can be either directly without passing an intermediate callus forming step or indirectly through callus. While direct embryogenesis took two weeks once the embryos were placed on the appropriate media, the indirect embryogenesis took an additional week. The whole procedure starting from embryo isolation to fully mature plants on soil takes 12-13 weeks or about 3 months.

A high diversity was found in the somatic embryogenesis depending on the size of the explant and the type of the tef ecotype. While intermediate-size immature embryos resulted in high percentage of somatic embryogenesis, only low proportion of small explants formed somatic embryos (Figure 2). Regarding small immature embryos, the efficiency of somatic embryos formed varied from less than 10% in Ada, Enatite, Rosea, and the two improved varieties (Magna and Dukem) to more than 70% in Manyi (Figure 2A). Surprisingly, small-size explants from the landrace Rosea did not produce any somatic embryos. The proportions of embryos formed through direct and indirect somatic embryogenesis were similar except in Gommadie where significantly higher percentage was formed through direct somatic embryogenesis. In the case of intermediate immature embryos, the percentage of somatic embryogenesis ranged from less than 20% for Ada to around 90% for Gommadie and Manyi (Figure 2B). Unlike small-size embryos which formed the direct or indirect somatic embryogenesis in a similar proportion, in the intermediate-size embryos, the indirect embryogenesis was dominant over the direct one. This favor to the indirect embryogenesis was significant in six landraces, namely Alba, Enatite, Gea Lamie, Karadebi, Rosea and Variegataas well as all the three improved varieties. For large immature embryos used as explants, the percentage of somatic embryogenesis varied from less than 20% in Balami, Rosea and Magna to more than 65% in Addisie, Gommadie and Manyi (Figure 2C). Similar to the intermediate-size embryos, the proportion of indirect embryogenesis was significantly higher than the direct ones for large-size explants except in Gea Lamie land race where the difference between direct and indirect embryogenesis was negligible.

The efficiency of regeneration which refers to the proportion of somatic embryos that result in plantlet formation were quantified for the three types of explants (small, intermediate and large) and the two forms of somatic embryos (direct and indirect) (Figure 3). Substantial variability in regeneration capacity was observed among the ecotypes, sources of explants and forms of embryos. Using small explants, four genotypes, namely Ada, Dabi, Rosea and Magna, did not form any plantlet while Alba and Balami did not regenerate from indirectly formed embryos (Figure 3A). However, over 70% of regeneration was obtained for Karadebi and Manyi. Regeneration efficiencies were variable between direct and indirect embryogenesis in each genotype although significantly higher values were obtained for directly formed embryos. The proportion of plantlets formed from intermediate-size

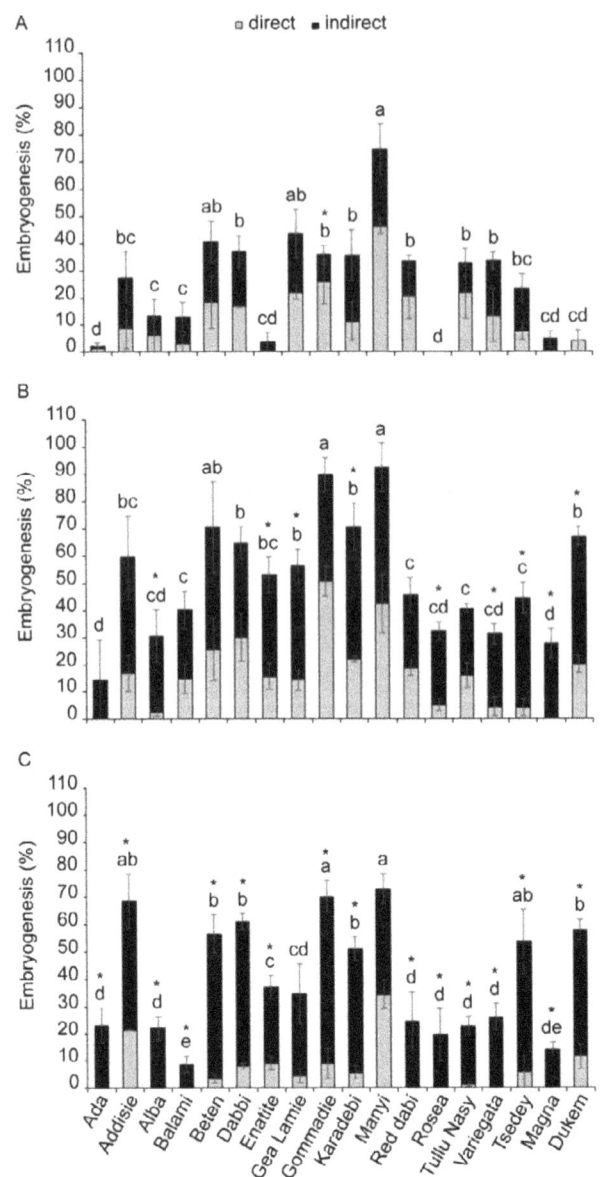

Figure 2: Efficiency of somatic embryogenesis derived from small (A), intermediate (B) and large (C) immature embryos. Values with different letter indicate significant difference (p<0.05) among the genotypes for the total embryogenesis (direct + indirect). An asterisk (*) indicates a significant difference (p<0.05) between direct and indirect embryogenesis.

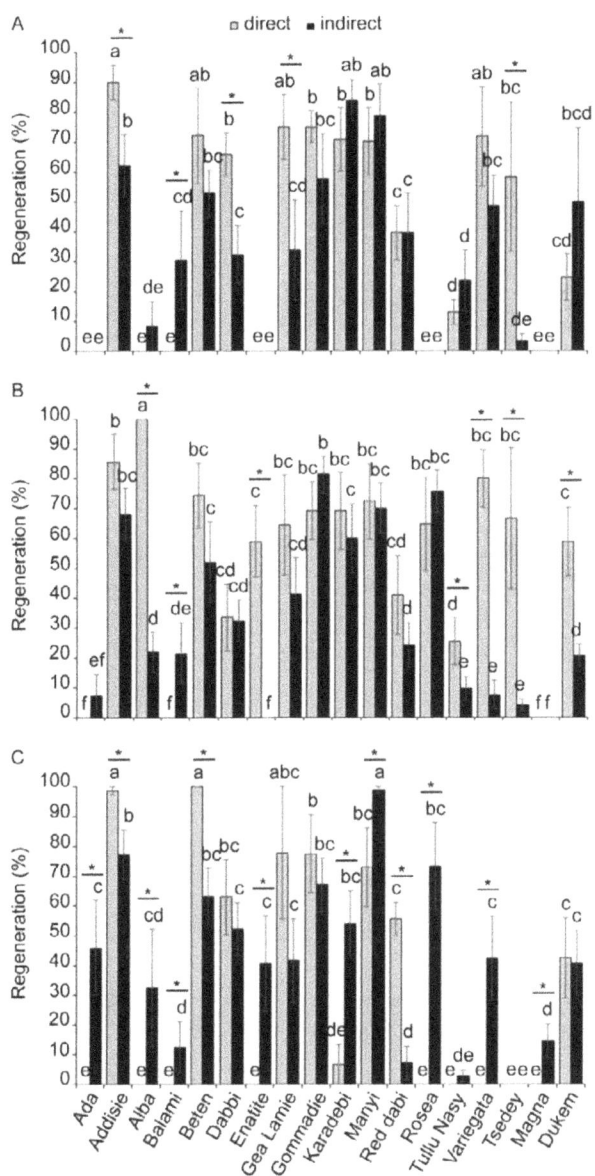

Figure 3: Efficiency of plantlet regeneration derived from somatic embryos which were originally obtained from small (A), intermediate (B) and large (C) immature embryos. Values with different letter represent a significant difference (p<0.05) among the genotypes. An asterisk (*) indicates a significant difference (p<0.05) between plantlets derived from direct or indirect somatic embryogenesis.

immature embryos ranged from less than 10% in Ada, Balami, Enatite and Magna (direct embryogenesis) as well as in Ada, Variegata, Tsedey and Magna (indirect embryogenesis) to more than 70% in Addisie, Beten, Manyi and Variegata (direct embryogenesis), and Gommadie, Manyi and Rosea (indirect embryogenesis) (Figure 3B). Some inconsistencies were observed in some genotypes between the capacity of plantlet formation and somatic embryogenesis. For instance, in Alba cultivar, although the percentage of direct embryogenesis from the intermediate-size embryo was only 2% (Figure 2), the plantlet formation reached 100% for same size of embryos (Figure 3) indicating high capacity of plantlet formation from extremely low rate of somatic embryo development. Regarding the regeneration capacity from large embryos, variations revealed in the other two embryo sizes were also

observed among different genotypes (Figure 3C). Three landraces (Ada and Beten from direct embryogenesis, and Manyi from indirect embryogenesis) gave exceptionally high plantlet formation.

Culture efficiency which refers to proportions of initial immature embryos resulting in plantlets was also investigated. Significant variability was observed among the tef genotypes and the sizes of immature embryos used as an explant (Table 1). The performance of some landraces was high when small starting material was used while in others large explants gave superior results. The large explants from Addise, Gea Lamie and Manyi had extremely high efficiency. Although inconsistencies in performance were observed for the three sizes of embryos, Manyi gave exceptionally high efficiency for all sizes of embryos. Irrespective of the size of immature embryos, all three improved tef varieties performed extremely poor indicating their low application in regeneration and transformation related studies. The main reason for the culture efficiencies above 100% is due to the friable nature of somatic embryos which resulted in the generation of several viable pieces generated from a single somatic embryo.

Correlations among steps of *in vitro* regeneration

Pearson correlation coefficients were used to investigate the relationship among the three parameters of regeneration (embryogenesis-, regeneration- and culture-efficiency) and three sizes of immature embryos used as an explant (small, intermediate and large). Very significant and positive correlations (p<0.01) were observed among the three parameters and three sizes of immature embryos as well as their interaction (Table 2). The only non-significant correlation (p=0.108) was between the embryogenesis from the small explant and the regeneration from the large explant.

Determination of morphological, phenotypic and yield related traits

The existence of substantial variability in diverse morphological

Genotype	Culture efficiency (%)		
Landrace	small	intermediate	Large
Ada	0.0 *d*	1.7 *e*	13.9 *d**
Addisie	40.0 *bc*	69.3 *a*	106.0 *a**
Alba	12.5 *c*	9.3 *cd*	5.5 *e*
Balami	6.6 *d*	7.4 *d*	2.7 *e**
Beten	69.0 *b*	73.4 *a*	66.8 *b*
Dabbi	36.0 *bc*	27.0 *bc*	66.4 *b**
Enatite	0.0 *d**	14.5 *c*	9.3 *e*
Gea Lamie	29.0 *c*	30.8 *b*	115.8 *a**
Gommadie	54.4 *b*	99.3 *a**	54.1 *c*
Karadebi	72.5 *b*	67.7 *ab*	34.7 *d**
Manyi	95.8 *a*	93.9 *a*	93.1 *b*
Red dabi	21.8 *c*	36.5 *b*	5.9 *d**
Rosea	0.0 *d**	36.4 *b*	24.4 *d*
Tullu Nasy	4.7 *d+*	16.0 *c+*	7.9 *d+*
Variegata	28.5 *c*	16.7 *c*	20.7 *d*
Improved			
Dukem	1.8 *d**	11.8 *c*	21.0 *d*
Magna	0.0 *d*	0.0 *e*	3.7 *e*
Tsedey	11.2 *c*	4.8 *d*	0.9 *e*

Table 1: Culture efficiency of 18 tef genotypes regenerated in vitro from small, intermediate and large immature embryos. Means followed by the same letter are not significantly different (P<0.05) for a same column. An asterisk (*) indicates a significant difference (P<0.05) between the sizes of a specific tef genotype or landrace. A plus (+) indicates a significant difference (P<0.05) among the three sizes of explants.

	EB_I	EB_L	EB_T	RG_S	RG_I	RG_L	RG_T	CEF_S	CEF_I	CEF_L	CEF_T
EB_S	.646**	.640**	.763**	.845**	.474*	.392	.484*	.831**	.672**	.643**	.672**
EB_I		.803**	.928**	.762**	.618**	.508*	.573*	.768**	.820**	.625**	.805**
EB_L			.943**	.719**	.487*	.474*	.509*	.681**	.716**	.661**	.744**
EB_T				.811**	.602**	.553*	.609**	.808**	.831**	.725**	.843**
RG_S					.687**	.582*	.698**	.927**	.830**	.669**	.835**
RG_I						.810**	.954**	.703**	.849**	.610**	.866**
RG_L							.934**	.600**	.797**	.710**	.865**
RG_T								.716**	.868**	.682**	.911**
CEF_S									.874**	.651**	.859**
CEF_I										.661**	.943**
CEF_L											.797**

Table 2: Pearson correlation coefficients among three efficiencies.
EB: Embryogenesis; RG: Regeneration; CEF: Culture efficiency
Three sizes of immature embryos S: Small; I: Intermediate; L: Large; T: Total
The * and ** showed statistical significance at the 0.05 and 0.01 probability levels, respectively.

Genotype	NTP		NPP		NIC		CL		PL		SCIL	
							cm		cm		mm	
Landrace	In vitro	seed	In vitro	seed	In vitro	seed	In vitro	seed	In vitro	seed	In vitro	seed
Ada	2.8 c*	1.0d	4.8 b*	1.7d	4.5e	3.6cd	41.0 a	34.7cd	39.0ab	32.1c	8.6 ab	6.4e
Addisie	3.6c	2.0bc	3.9c	2.8abc	5.0 c	4.6bc	37.4 b*	30.6d	37.3 b*	29.2d	6.8e*	8.6cde
Alba	4.0 bc*	0e	5.0b*	1.3e	4.3de	4.3bcd	46.1 a*	37.2cd	47.5 a*	33.7c	10.9 ab	11.2ab
Balami	5.2 b*	0e	4.9bc*	1.0e	5.3 bc*	6.0a	37.5 b*	43.6b	44.9 a	45.1a	9.3b	9.9bc
Beten	7.5a*	1.5bcd	5.7 b*	2.0c	5.7 ab*	4.5bc	39.4 ab	39.5c	34.8 bc	35.7c	7.3e	9.2cd
Dabbi	4.1b*	2.0bc	6.5b	4.0a	4.3 d	4.0c	36.3 b	40.5bc	19.7 e	25.3d	11.3 a	13.2a
Enatite	4.7b*	1.2bcde	5.9 b*	2.3bc	4.6d	5.2abc	38.3 b	34.2cd	38.9 ab	35.4c	8.7c	8.0de
Gea Lamie	6.3 a*	2.0bc	9.0 a*	3.7ab	3.7e	3.0d	31.9 c	35.9cd	19.6 e	21.3d	8.9c*	6.4e
Gommadie	3.5 c*	1.0cd	4.2 c*	1.5e	4.7 cd	4.5bc	36.7 b	36.8cd	30.3 c	26.3d	9.7b	8.6cde
Karadebi	3.7 c*	1.2cd	4.7c*	2.0bc	4.6 d*	4.0cd	35.5 b*	46.7a	25.0 de*	29.4c	9.5b	11.2ab
Manyi	5.1 b*	1.0d	5.6 b*	1.8d	5.9a	5.6ab	37.0 b	40.6bc	36.4 bc	39.8b	6.7e*	9.9bc
Red dabi	5.9 a*	1.6bcd	8.8 a*	3.3abc	4.0e	4.0c	31.7 c	32.0cd	22.5 de	27.4d	8.8c	9.2cd
Rosea	5.5 b*	0.2e	7.9 b*	2.3bc	4.2 e	4.5bc	30.1 c	34.1cd	30.0 cd	34.1c	7.9d*	13.2a
Tullu Nasy	5.0 b*	0.4de	8.0 b*	1.7d	4.2 e	3.7cd	39.9 a*	30.0de	25.9 d	29.7c	10.7 ab*	6.4e
Variegata	3.5 c*	0e	3.4c*	1.5e	4.5 d*	3.7cd	31.7 c	29.1de	24.6 de	22.8d	9.8b	8.6cde
Improved												
Dukem	5.9a	3.0ab	6.7 b*	2.0c	4.2 e	4.0c	33.6 b	33.0cd	40.1 ab	37.8b	8.5c*	9.3bc
Magna	5.7 ab	3.4a	8.1b*	3.8ab	4.1 e	4.6bc	29.2 c*	36.6cd	32.7 c	32.9c	7.8 d	8.4cde
Tsedey	4.9 b*	2.0bc	8.1b*	2.8bc	3.7 e*	4.4bc	28.9 c	29.6e	28.3 d*	32.9c	6.8e	7.5de

Table 3: Selected morphological traits for 18 tef genotypes developed by *in vitro* regeneration. Values followed by the same letter were not significantly different (P<0.05) between plants regenerated by *in vitro* method (this study) and those produced from seeds [22].
NTP: Number of Tillers per Plant; NPP: Number of Panicles per Plant; NIC: Number of Internodes per Culm; CL: Culm Length; PL: Panicle Length; SCIL: Second Culm Internode Length

and phenotypic traits were earlier reported for the same 18 tef ecotypes derived from seed [22,28]. In Table 3, comparisons were made for selected traits between those obtained from the seed [22] and those developed through *in vitro* regeneration (the current study). Tef lines generated through *in vitro* method were more robust than those from seeds for key morphological traits especially in the numbers of tillers and numbers of panicles per plant which have positive impact on the productivity of the crop. While most landraces derived from seeds had a maximum of two tillers per plant, those from the *in vitro* had up to seven tillers per plant. Astonishingly, three landraces (namely, Alba, Balami and Variegata) which did not develop a single tiller when generated from seeds were able to form 3-5 tillers when generated *in vitro*. Number of panicles which is also dependent on the form

of the panicle is normally low for plants with compact panicles (e.g. Gommadie) and high for those with loose panicles (e.g. Gea Lamie). Compared to those developed from seeds, up to 3-fold increase in the number of panicles was obtained for those from *in vitro* regenerated plants. Although substantial variability was obtained among the tef genotypes for the number of internodes per plant, culm length and plant height, differences between those generated from seeds and those from *in vitro* were not obvious for the three traits.

Significant diversity was also observed for phenotypic and yield related traits among the 18 tef genotypes generated from the seed or from *in vitro* method (Table 4). However, the differences between those from seed and those from immature embryos were inconsistent for most of the traits. Amazingly, all the three improved tef varieties

Genotype	DH		DM		GFP		WAC		WAP		GY		HI	
					Days				mg				%	
Landraces	Invitro	seed	Invitro	seed	Invitro	seed	Invitro	seed	Invitro	seed	Invitro	seed	Invitro	Seed
Ada	61.0 e*	90.0a	95.2 d	110.0c	42.8 a*	20.0e	1924.2 a*	292.4de	826.7 a*	159.8ab	251.1 b*	113.7ab	11.9 bc*	28.9a
Addisie	65.7 e	65.5f	94.8 d	90.0ef	31.3 b*	24.5d	744.1 d*	497.8cd	417.8 c*	234.7ab	198.6 bc	168.8a	16.7 b	20.6a
Alba	75.3 cd	78.0cd	120.7 b*	105.0cd	45.3 a*	27.0cde	1229.3 b*	369.5d	497.1 c*	210.3ab	147.7 c	135.8ab	8.5 d*	23.8a
Balami	100.6 b*	86.2a	139.9 a*	108.2c	40.8 a*	22.0e	2137.9 a*	636.5bc	618.2 b*	342.5 a	254.0 b	229.9a	8.1 d*	23.6a
Beten	85.6 c	87.5a	116.5 bc*	108.0c	31.3 b*	20.5e	1863.0 a*	964.6 a	619.3 b*	292.1 a	321.2 b*	131.0a	11.3 c	10.4b
Dabbi	57.1 f	58.0g	86.2 e	87.4fg	29.9 b	29.4cd	538.9 e	309.2de	187.0 d	180.9ab	72.0 e*	38.5bc	7.6 d	8.1bc
Enatite	61.4 e	65.0fg	97.1 d*	118.3a	35.6 a*	53.3a	1340.4 b*	499.6bcd	340.0 c	225.2bc	74.7 e*	147.5a	3.7 e*	20.1ab
Gea Lamie	60.3 e	58.0g	82.4 e	87.5fg	22.3 d*	29.5c	358.6 f*	195.2e	136.2 d	88.3bc	34.9 f	26.9c	4.4 e*	8.7bc
Gommadie	72.3 d	69.5e	100.6 d*	92.5e	28.1 c	23.0e	814.2 c	776.0ab	299.6 c	338.8a	129.5 d	147.7ab	9.8 c*	16.7ab
Karadebi	63.9 e	60.2f	92.7 d	85.2g	27.8 c	25.0d	959.5 b*	550.7bc	317.6 c*	137.3ab	153.9 c*	41.9b	10.6 c*	6.6cd
Manyi	81.2 c	86.8 a	119.4 b*	107.6c	37.5 a*	20.8e	1347.8 b*	630.8bc	308.1 c	359.1a	128.5 d*	290.1a	6.5 d*	31.6a
Red dabi	56.9 f	56.5h	84.2 e	83.5g	28.7 c	27.0d	818.8 c*	450.6cd	335.0 c	228.0ab	99.0 de	96.4b	9.8 c*	16.9ab
Rosea	74.6 cd	70.3e	113.6 bc*	103.6c	39.0 a	33.3bc	1196.4 b*	340.3d	499.3 c*	203.5ab	154.6 c	95.4bc	7.2 d*	17.7ab
Tullu Nasy	47.1 g*	60.0f	70.3 f*	88.8ef	23.6 d	28.8cd	892.7 c*	143.7e	404.7 c*	80.7c	230.9 c*	43.1bc	16.3 b	18.5ab
Variegata	64.3 e*	72.4de	91.6 d	90.4e	27.4 c*	18.0e	850.0 c*	345.0d	192.7 d*	82.3c	79.4 e*	25.5c	6.6 d	5.4d
Improved														
Dukem	114.0 a*	80.0bc	136.7 a*	116.3b	23.5 d*	36.3b	1195.0 b*	660.0ab	1481.7 a*	253.6 a	734.3 a*	79.0b	35.0 a*	8.0c
Magna	84.1 c	83.4ab	108.8 c*	122.8a	24.9 d*	39.4b	680.5 d	665.7ab	423.2 c*	182.7ab	172.6 c*	33.3c	15.4 b*	3.7d
Tsedey	70.1 d*	73.6 de	100.2 d*	106.6d	30.6 c*	33.0c	810.8 c*	317.9de	666.4 ab*	130.9ab	311.2 b*	34.6bc	18.1 ab*	7.6c

Table 4: Phenotypic and yield related traits for 18 tef genotypes regenerated by *in vitro* method. Values followed by the same letter were not significantly different (P<0.05) between plants regenerated by *in vitro* method (this study) and those produced from seeds [22].
DH: Days to Heading; DM: Days to Maturity; GFP: Grain Filling Period; WAC: Weight of all Culms; WAP: Weight of all Panicles; GY: Grain Yield per Plant; HI: Harvest Index.

developed from immature embryos were superior over those from the seed for all traits investigated. This positive effect from tissue cultured plants was revealed on grain yield (up to 8-fold) and harvest index (up to 3-fold) over those from seed. Although improved varieties were mainly selected based on high grain yield and harvest index, the three improved varieties in the current study were inferior to some land races for these two valuable traits.

Discussion

The diversity in regeneration capacity among tef genotypes indicates the existence of a genetic control of this process which differs among diverse tef lines as earlier reported for *Arabidopsis* [9] and rice [29]. This variation among genotypes might be related to the level of endogenous hormones [30] or the effect of exogenous growth regulators on the level of endogenous hormones through influencing their biosynthesis and distribution [31], which subsequently alters the *in vitro* regeneration responses [32-34]. Interestingly, all three improved tef varieties used in the current study had extremely poor regeneration capacity. This was mainly due to the development of tef varieties in the past focused on selecting genotypes with superior grain yield without considering their *in vitro* regeneration efficiency.

The size and/or age of an explant is also responsible for controlling the frequency and speed of regeneration in plants. Although not a single albino plant or plantlet was observed in the current study, up to 50% of shoots of barley derived from large immature embryos were albinos unlike those from small embryos [14]. The absence of albinos in tef plants developed through *in vitro* regeneration increases the acceptability of the technique by researchers as it has positive impact on the viability and productivity of the crop. Large immature embryos of tef tend to initially form a callus before differentiating into somatic embryo. This indirect somatic embryogenesis requires an additional one week to the direct embryogenesis to form somatic embryos. This makes large embryos less favorable to use as an explant. As earlier

reported for Sudan grass and tef, large embryos are more determined to germinate than to produce a callus [15,21]. The strong positive correlation among the three steps of the *in vitro* regeneration suggests the existence of general internal/genetic mechanism which controls the response of a particular genotype to the tissue culture, irrespective of the size of the explant or the step of the *in vitro* regeneration process. This means, a highly-responsive genotype performs better than a low responsive one for all sizes of explant and steps of the tissue culture. In winter wheat, only culture efficiency tended to increase as regeneration capacity was enhanced [35].

The comparison between the 18 tef genotypes developed from the seed [22] and those generated via the tissue culture (the current study) revealed significant differences for major morphological, phenotypic and yield related traits. Plants developed through *in vitro* regeneration were more vigorous and productive than those from seed especially in the number of tillers and panicles.

Since tissue culture or in vitro regeneration could result in useful stable somaclonal variations, thorough investigation of progenies needs to be done as some of the changes could be used in developing tef cultivars with improved traits. Using somaclonal variation, potato cultivar with reduced plant height, and sorghum and rice varieties tolerant to drought were developed [36-38].

In conclusion, based on the regeneration efficiency, the intermediate-size explants are the best. Among the germplasms, the Manyi landrace provides the highest culture efficiency. Since the three improved tef varieties were inferior in the efficiency of regeneration, they are not suggested for use especially with the technique and type of the explant indicated in the current study.

Acknowledgments

Financial and technical support was provided by Syngenta Foundation for Sustainable Agriculture, SystemsX.ch and University of Bern. We would like to thank Christopher Ball and Jasmin Sekulovski for taking care of our plants.

References

1. CSA (2014) Agricultural Sample Survey for 2013/14. Addis Ababa, Ethiopia.

2. Assefa K, Yu JK, Zeid M, Belay G, Tefera H, et al. (2011) Breeding tef [Eragrostis tef (Zucc.) trotter]: conventional and molecular approaches. Plant Breeding 130: 1-9.

3. MoA (2014) Crop Variety Register Issue No. 15. Addis Ababa, Ethiopia.

4. Assefa K, Aliye S, Belay G, Metaferia G, Tefera H, et al. (2011) Quncho: the first popular tef variety in Ethiopia. International Journal of Agricultural Sustainability 9: 25-34.

5. Cannarozzi G, Plaza-Wuthrich S, Esfeld K, Larti S, Wilson YS, et al. (2014) Genome and transcriptome sequencing identifies breeding targets in the orphan crop tef (Eragrostis tef). BMC genomics 15:581.

6. Garcia-Gonzales R, Quiroz K, Carrasco B, Caligari P (2010) Plant tissue culture: Current status, opportunities and challenges. Ciencia E Investigacion Agraria 37: 5-30.

7. Larkin PJ, Scowcroft WR (1981) Somaclonal Variation - a Novel Source of Variability from Cell-Cultures for Plant Improvement. Theoretical and Applied Genetics 60: 197-214.

8. Jain SM (2001) Tissue culture-derived variation in crop improvement. Euphytica 118: 153-166.

9. Candela M, Velazquez I, De la Cruz R, Sendino AM, De la Pena A (2001) Differences in in vitro plant regeneration ability among four Arabidopsis thaliana ecotypes. In Vitro Cellular & Developmental Biology-Plant 37: 638-643.

10. Haliloglu K, Baenziger PS (2005) Screening wheat genotypes for high callus induction and regeneration capability from immature embryo cultures. Journal of Plant Biochemistry and Biotechnology 14: 155-160.

11. Dodig D, Zoric M, Mitic N, Nikolic R, Surlan-Momirovic G (2008) Tissue culture and agronomic traits relationship in wheat. Plant Cell Tissue and Organ Culture 95: 107-114.

12. Maddock SE, Lancaster VA, Risiott R, Franklin J (1983) Plant-Regeneration from Cultured Immature Embryos and Inflorescences of 25 Cultivars of Wheat (Triticum-Aestivum). Journal of Experimental Botany 34: 915-926.

13. Ge XJ, Chu ZH, Lin YJ, Wang SP (2006) A tissue culture system for different germplasms of indica rice. Plant cell reports 25: 392-402.

14. Chang Y, von Zitzewitz J, Hayes PM, Chen THH (2003) High frequency plant regeneration from immature embryos of an elite barley cultivar (Hordeum vulgare L. cv. Morex). Plant cell reports 21: 733-738.

15. Gupta S, Khanna VK, Singh R, Garg GK (2004) Identification of in vitro responsive immature embryo size for plant regeneration in Sudan grass (Sorghum sudanensis Piper). Indian Journal of Biotechnology 3: 124-127.

16. Bekele E, Klock G, Zimmermann U (1995) Somatic embryogenesis and plant regeneration from leaf and root explants and from seeds of Eragrostis tef (Gramineae). Hereditas 123: 183-189.

17. Mekbib F, Mantell SH, BuchananWollaston V (1997) Callus induction and in vitro regeneration of Tef [Eragrostis tef (Zucc.) Trotter] from leaf. Journal of Plant Physiology 151: 368-372.

18. Kebebew A, Gaj MD, Maluszynski M (1998) Somatic embryogenesis and plant regeneration in callus culture of tef, Eragrostis tef (Zucc.) Trotter. Plant cell reports 18: 154-158.

19. Tefera H, Chapman GP (1992) In vitro Normal and Variant Development of Tef (Eragrostis-Tef) Spikelets. Plant Cell Tissue and Organ Culture 31: 233-237.

20. Gugsa L, Sarial AK, Lorz H, Kumlehn J (2006) Gynogenic plant regeneration from unpollinated flower explants of Eragrostis tef (Zuccagni) Trotter. Plant cell reports 25: 1287-1293.

21. Gugsa L, Kumlehn J (2011) Somatic Embryogenesis and Massive Shoot Regeneration from Immature Embry Explants of Tef. Biotechnology Research International: 7.

22. Plaza-Wüthrich S, Cannarozzi G, Tadele Z (2013) Genetic and phenotypic diversity in selected genotypes of tef [Eragrostis tef (Zucc.)] Trotter. African Journal of Agricultural Research 8: 1041-1049.

23. USDA A, National Genetic Resources Program. Germplasm Resources Information Network - (GRIN). [Online Database] National Germplasm Resources Laboratory, Beltsville, Maryland.

24. Deutsch F, Kumlehn J, Ziegenhagen B, Fladung M (2004) Stable haploid poplar callus lines from immature pollen culture. Physiologia Plantarum 120: 613-622.

25. Kumlehn J, Serazetdinova L, Hensel G, Becker D, Loerz H (2006) Genetic transformation of barley (Hordeum vulgare L.) via infection of androgenetic pollen cultures with Agrobacterium tumefasciens. Plant Biotechnology Journal 4: 251-261.

26. Zale JM, Borchardt-Wier H, Kidwell KK, Steber CM (2004) Callus induction and plant regeneration from mature embryos of a diverse set of wheat genotypes. Plant Cell Tissue and Organ Culture 76: 277-281.

27. Hundera F, Nelson LA, Baenziger PS, Bechere E, Tefera H (2000) Association of lodging and some morpho-agronomic traits in tef [Eragrostis tef (Zucc.) Trotter]. Tropical Agriculture 77: 169-173.

28. Ebba T, editor (1975) Tef cultivars. Part II. Dire Dawa, Ethiopia: Addis Ababa University, College of Agriculture.

29. Khalequzzaman M, Haq MD, Hoque E, Aditya TL (2005) Regeneration efficiency and genotypic effect of 15 indica type Bangladeshi rice (Oryza sativa L.) landraces. Plant Tissue Cult 15: 33-42.

30. Hess JR, Carman JG (1998) Embryogenic competence of immature wheat embryos: Genotype, donor plant environment, and endogenous hormone levels. Crop Science 38: 249-253.

31. Kaminek M, Motyka V, Vankova R (1997) Regulation of cytokinin content in plant cells. Physiologia Plantarum 101: 689-700.

32. Valdes AE, Ordas RJ, Fernandez B, Centeno ML (2001) Relationships between hormonal contents and the organogenic response in Pinus pinea cotyledons. Plant Physiology and Biochemistry 39: 377-384.

33. Klems M, Slamova Z, Motyka V, Malbeck J, Travnickova A et al. (2011) Changes in cytokinin levels and metabolism in tobacco (Nicotiana tabacum L.) explants during in vitro shoot organogenesis induced by trans-zeatin and dihydrozeatin. Plant Growth Regulation 65: 427-437.

34. Yasmin S, Mensuali-Sodi A, Perata P, Pucciariello C (2014) Ethylene influences in vitro regeneration frequency in the FR13A rice harbouring the SUB1A gene. Plant Growth Regulation 72: 97-103.

35. Ozgen M, Turet M, Altinok S, Sancak C (1998) Efficient callus induction and plant regeneration from mature embryo culture of winter wheat (Triticum aestivum L.) genotypes. Plant cell reports 18: 331-335.

36. Duncan RR, Waskom RM, Nabors MW (1995) In-Vitro Screening and Field-Evaluation of Tissue-Culture-Regenerated Sorghum (Sorghum-Bicolor (L) Moench) for Soil Stress Tolerance. Euphytica 3: 373-380.

37. Adkins SW, Kunanuvatchaidach R, Godwin ID (1995) Somaclonal viariation in rice-drought tolerance and other agronomic characters. Aust J Bot 43: 201-209.

38. Valkonen JPT, Moritz T, Watanabe KN, Rokka VM (1999) Dwarf (di)haploid pito mutants obtained from a tetraploid potato cultivar (Solanum tuberosum subsp tuberosum) via anther culture are defective in gibberellin biosynthesis. Plant Science 149: 51-57.

16

Analytical Framework for the Assessment of Crop Production Technology

Janusz Gołaszewski[1]*, Dariusz Załuski[1] and Krystyna Żuk-Gołaszewska[2]

[1]*Department of Plant Breeding and Seed Production, University of Warmia and Mazury in Olsztyn, Poland*

[2]*Department of Agrotechnology, University of Warmia and Mazury in Olsztyn, Poland*

Abstract

The objective of the study was to compile a methodical approach for a time-and-cost-efficient test of a crop production technology. The two or three-year procedure consisted of: (i) a survey among farmers on the technological state-of-the-art, (ii) screening the selected agro-technical factors with the use of factorial (FD) and/or fractional factorial design (FFD), (iii) a series of on-farm FD and FFD experiments and estimation of the total and individual contribution of the factors into costs and profitability of production. The test crop for verification of the procedure was pea (*Pisum* spp.). The approach is illustrated by empirical data obtained from the implementation of the procedure for testing of a green pea production technology in north-eastern Poland. The statistical efficiency and economic profitability of agrotechnical factors in the tested technology were given.

Keywords: Smart agriculture; Agricultural experimentation; Survey in agriculture; Factorial and fractional designs; On-farm experiment

Introduction

There is an ongoing process in agriculture to bring the growing demand for higher yields and changing preferences for agricultural products together with improved efficiency of crop production [1,2]. Any modification of the crop production technology involves testing a set of agro-technical factors in a given local habitat and for environmental conditions. It suggests that changes in agro-technical factors, such as machinery, cultivars, fertilizers, pesticides, etc. should be validated with a cost-and-time-efficient empirical testing procedure. There is no prompt methodical approach for screening the interaction between specific agro-technical factors in order to design a new technological process and implement it into agricultural practice. Instead, there is a common practice to conduct single or two-factor field experiments in research stations or to compare the reduced number of technologies, e.g., "traditional" and "new" ones, in a single replicated on-farm comparative field demonstration.

The initial problem in validation of new crop production technology is associated with the selection of key factors responsible for an anticipated level and/or quality of yield and testing them under farm conditions. Useful tools in solving these problems are factorial designs (FD) of type s^k (k factors on s levels) or fractional factorial designs (FFD) of type s^{k-p} (p is a fraction of s^k) Box and Hunter, Załuski and Gołaszewski, Załuski et al. [3-5] and on-farm experiments e.g., On-Farm Trials - Some Biometric Guidelines [6], Blaise et al., Gomez and Gomez, Byerlee, Amir and Knipscheer, Liu et al., Barlow et al. [7-12].

Certain methodological and economical considerations limit the broad application of FDs in agricultural experiments. The higher the number of factors and their levels in an experiment, the higher the number of treatments to be tested and the greater influence of soil variability on the experimental results. At the same time, the cost of the experiment rises.

On-farm experiments include methods, tools and techniques which may be easily applied in the environmental and organizational conditions of a given farm [7,8]. Hence, methodical assumptions of on-farm research should be applied to the production activity of a given farm and consideration given to the self-evident prerequisites of such activity, including the interest of farmers and available farm machinery. In substance, an initial prerequisite for on-farm research

related to plant production technology is verification of a given technology in agricultural practice [9]. In such cases the methodical approach is substantially different than in classic field experimentation. Excluding tested factors, all the remaining factors should be taken at the levels of present farm technology - which means the traditional crop production technology at the farm level.

An indispensable part of the on-farm experimentation methodology is the calculation of technology modification costs. Several authors Amir and Knipscheer; Liu et al.; Barlow et al. [10-12] have discussed some practical issues in improving the economic impact of farming system research with respect to three stages: the ex-ante stage - securing representativeness of the target area, the ongoing stage - using suitable partial budgeting techniques, ex-post stage - an assessment of the wider technological and economic effects together with the identification of enabling and constraining factors in the technological progress.

It is presupposed that the number of treatments and replications of on-farm experiments should be at a minimum. An extreme case is when only two technologies are compared, which results in a relatively low research utility because the main function of such an experiment is a demonstration of new technology.

The objective of this study was to present a methodical approach for testing new crop production technology. The test crop was pea (*Pisum* spp.). The consecutive stages of the procedure assume: 1) detection of the key technology factors on the basis of the results from on-farm survey and FFD field experiments. 2) implementation of the key factors into a series of FD and FDD on-farm experiments 3) calculation of the contribution of the individual agro-technical factors and profitability of the new technology.

*Corresponding author: Janusz Gołaszewski, Department of Plant Breeding and Seed Production, University of Warmia and Mazury in Olsztyn, Plac Łódzki 310-724 Olsztyn, Poland, E-mail: janusz.golaszewski@uwm.edu.pl

Materials and Methods

The results of a questionnaire survey and field experiments according to FFD illustrate an approach in detection of the key factors of technology.

The survey data on pea production technology was collected in the north-eastern part of Poland (2006). It was carried out on 243 farms with a total area 3896 ha and an average yield of 6.20 t ha^{-1} (SE=0.124). Only pea crops produced for consumption and contracted by the local processing industry were considered. This type of production was represented by 75% of farms and accounted for 86% of the surveyed area at the average yield 6.73 t ha^{-1} (SE=0.104). The other surveyed plantings produced fodder pea as dry seeds. The survey questionnaire addressed 27 production factors assigned into four groups: stand quality, agro-technical factors applied before sowing, quality of sowing material, and agro-technical factors associated with sowing and chemical application(s) after sowing (Figure 1). The on-research-station study is represented by data obtained from two-single-replicated factorial experiments of type 3$^{5-1}$ arranged in completely randomized designs. These experiments were carried out in 2003 and 2004 at the Tomaszkowo Experimental Station (53°42'N, 20°26'E). In the experiments, five experimental factors had 3 variants: A - cultivars (Kos, Stig, Set), B - phosphorus and potassium fertilization doses (natural fertility, 70 kg P·ha^{-1} and 100 kg K·ha^{-1}, and 100 kg P·ha^{-1} and 130 kg K·ha^{-1}), C - sowing dates (the earliest possible, 10, and 20 days after the first date, respectively), D - sowing densities (70, 100, 130 kg·ha^{-1}), and E - chemical protection (without protection, seed chemical protection, seed and plant chemical protection).

The factors responsible for a high variability of pea yield were

selected for testing in on-farm experiments (2007). It was planned that a distributed system of on-farm experiments would be built on the basis of the FD of type 2^3 and FFDs of type 2^{3-p} and located at different farms while taking into account the organizational customizability of a given farm. The farms were typified in cooperation with the Warmian Fruit and Vegetable Processing Company, Ltd. in Kwidzyn, Poland. Three factors (A, B, C) were selected for testing in on-farm experiments on the basis of generalized results from the survey and research-station FFD experiments (Table 1):

Fertilization of P and K: "t" - average level of fertilization in farms; "n" – doses of P and K per hectare were increased by 25 kg of P and 20 kg of K, N was fixed - 50 kg.

Sowing date: "t" - the earliest possible, "n" - 10 days after the first date.

Plant protection: "t" - traditional farm approach, "n" full protection (seed dressing, herbicide, fungicide).

At each farm the experiments were replicated in the 2 fields.

Statistical analyses

In the survey analysis, the total variability of yield was fractionated into the contributions of production factors by the use of General Linear Model (GLM). The efficiency of parameter estimation was maximized by using the weighted least squares method with hectares as weights [13]. The sums of Type-III squares in the ANOVA model and those coefficients η^2 (eta square) displaying the relative contribution of a given factor were calculated according to (1) [14].

$$\eta^2 = SS_{Effect} / SS_{Model} \tag{1}$$

where: SS$_{Effect}$ is the sum of squares of a given effect, and SS$_{Model}$ is the

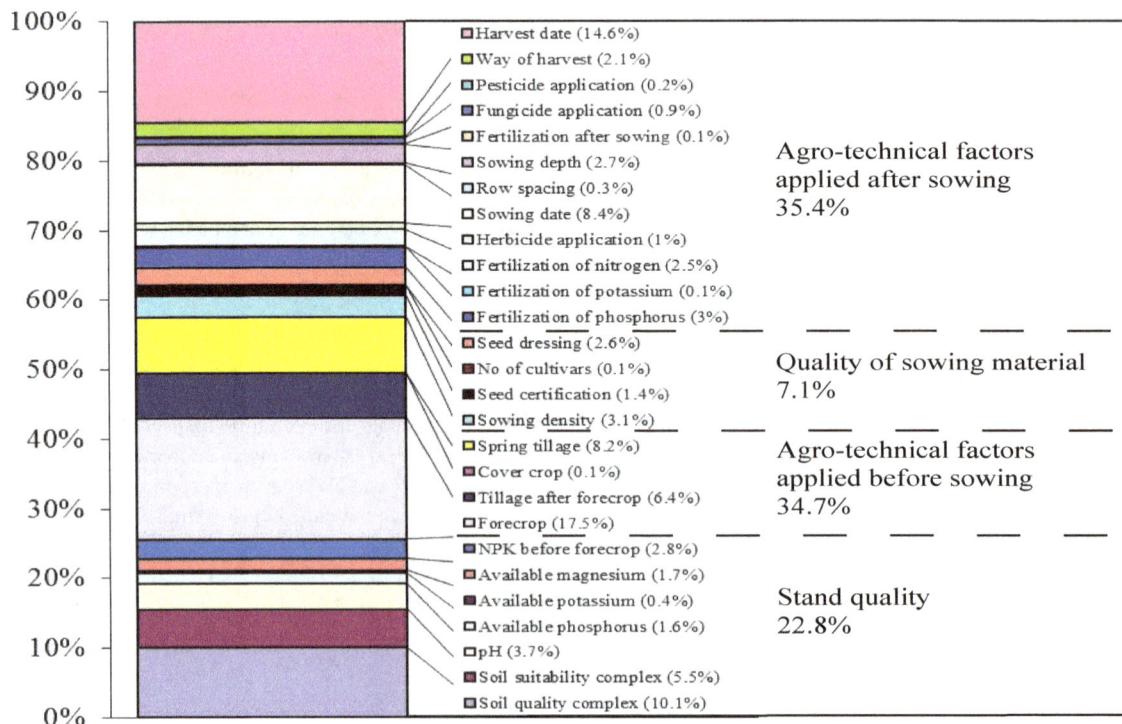

Figure 1: Percentage of contribution of green pea production factors in yield variation.

sum of squares. The statistical analyses were supported by STATISTICA v. 9.1 (StatSoft, Inc.).

Results

Survey

All the studied factors of the green pea production technology contributed 40.1% to the yield variation. The further decomposition of this variation showed that factors associated with stand quality, agricultural activities before sowing, quality of sowing material, and application after sowing contributed in 22.8, 34.7, 7.1, and 35.4 percent, respectively (Figure 1).

The soil gradation and the complex of agricultural suitability had the highest impact on yield variation and its suitability for pea production (stand quality). This was followed by forecrop and tillage after forecrop harvesting (before sowing), seed dressing, sowing density, fertilization of phosphorus, dates of sowing and harvest (seed quality and treatments after sowing).

FFDs

The on-research-station studies on the key agro-technical factors of green pea production are represented by data from the two-year series (E03, E04) of two 3^{5-1} FFD experiments.

On-farm research

The treatment means from experiments fluctuated greatly between the very high yield of 8.30 t ha^{-1} in 1E2 and the very low yield of 1.69 t ha^{-1} in 5E5 (Table 2). In general, yields in experiments E5 and E8 were significantly lower than in experiments with two treatments but the difference between yields of traditional (T) and new (N) technology (technology difference N-T) was nearly at the same level. The only exception was experiment 1E2 where the difference N-T exceeded the others by about twice the amount.

The main and interactional effects estimated for experiments E5 and E8 (Table 3) were the basis for the calculation of the contribution of each factor to the technology. This was then used to assess the

relative costs of modifying the technology. The contribution of factors and their interactions (experiments E5 and E8) to N-T technology differences was measured with estimates of main and interaction effects. Fertilization (A, 33%) had the highest contribution, followed by sowing date (B, 15%) and chemical protection (C, 9%). Among the interactions, fertilization × sowing date × chemical protection (ABC, 28%) and sowing date × chemical protection (BC, 13%) also had significant contributions.

Extra fertilization which involved costs of fertilizers and their application was done at the average cost €9, 82 ha^{-1} (recalculated from 39.29PLN - Polish zloty) (Table 4). Sowing date was a non-cost factor. Decisively, the highest costs were of intensified chemical protection, i.e., fungicides and their application; on average €43.75 ha^{-1}. The total additional costs in new technology accounted for €53.57 ha^{-1}.

The individual contributions of tested factors were estimated using the results from FD and FFD on-farm experiments. In the analysis of added return for each factor in experiment E8, there are noticeable negative values for added return, added profit and marginal ratio for chemical protection (C). The low economic effect stems from the fact that yields in this FD experiment were relatively low and did not compensate for the high costs of fungicides. In addition, as mentioned, interactions BC and ABC had a relatively high share in yield variation, which could have diminished the main effect of C and may indicate the necessity of an analysis of interactions with other factors. The marginal benefit-cost ratio for fertilization was 2.00 and for chemical protection 0.12.

At an average technological difference of 0.83 t ha^{-1}, the average added return and added profit were €134 t^{-1} and €81 t^{-1}, respectively. The marginal rate across the experiments was 2.51, which means that each €1 invested in new technology gave €2.51 in return.

Discussion

This paper presents a methodical procedure for engineering crop production technology in a two-stage process, which consists of detecting key agro-technical factors of a given technology which contributes the most to yield variation and verifying them in the conditions of agricultural practice. The procedure is composed of

Experiment*	Factor-treatment	A-t B-t	A-t B-n	A-n B-t	A-n B-n
xE2 (4 sites) (nnn, ttt)	C-t	ttt	-	-	-
	C-n	-	-	-	nnn
xE5 (1 site) (nnn, nnt, ntn, tnn, ttt)	C-t	ttt	-	-	nnt
	C-n	-	tnn	ntn	nnn
xE8 (1 site) (nnn, nnt, ntn, tnn, ttn, tnt, ntt, ttt)	C-t	ttt	tnt	ntt	nnt
	C-n	ttn	tnn	ntn	nnn

*xE2 – symbol x means the number of farms, E2 means the experiment with 2 treatments. As an example, treatment "tnt" for factors A,B,C means technology where factor A is at the level "traditional" (t), B – "new" (n), and C – "traditional" (t).

Table 1: The scheme of six on-farm experiments with 2-, 5-, and 8-treatments.

Experiment	Mean	Treatments								Technology difference
		nnn	nnt	ntn	ntt	tnn	tnt	ttn	ttt	nnn-ttt
1E2	7.58	8.30	-	-	-	-	-	-	6.85	1.45
2E2	7.00	7.38	-	-	-	-	-	-	6.62	0.77
3E2	5.82	6.14	-	-	-	-	-	-	5.50	0.65
4E2	4.61	4.97	-	-	-	-	-	-	4.24	0.73
5E5	2.09	2.45	2.13	2.15	-	2.03	-	-	1.69	0.76
6E8	2.13	2.50	2.15	1.94	2.44	1.90	2.11	2.09	1.88	0.63
Average	4.87	5.29	2.14	2.04	2.44	1.97	2.11	2.09	4.46	0.83

Table 2: Mean yields and technology difference in on-farm experiments.

Effect	5E5	6E8	Average mean and interaction effects	Adjusted average mean and interaction effects*
A	0.275	0.265	0.270	0.273
B	0.160	0.080	0.120	0.121
C	0.180	-0.035	0.072	0.073
AB		0.058	0.057	0.058
AC		0.038	-0.038	-0.038
BC		0.108	0.108	0.109
ABC	0.145	0.315	0.230	0.233
Sum of average contribution of main and interaction effects into the difference N-T			0.82	0.83

*Average effects from experiments were adjusted to the average difference N-T (0.83)

Table 3: Main and interaction effects in on-farm experiments.

	Rates of return of additional costs on new technology, rates of return of additional costs on new technology per test factor				
Experiment	Technological difference N-T (t ha⁻¹)	Sale price € t⁻¹	Added return € t⁻¹	Added profit € t⁻¹	Marginal benefit-cost ratio
1E2	1.45	160	232	178	4.33
2E2	0.77	160	122	69	2.29
3E2	0.65	165	106	53	1.99
4E2	0.73	165	120	67	2.25
5E5	0.76	163	124	70	2.31
6E8	0.63	163	102	48	1.90
Mean	0.83	163	134	81	2.51

	Rates of return of additional costs on new technology per test factor								
Experiment	Added return			Added profit			Marginal benefit-cost ratio		
	A	B	C	A	B	C	A	B	C
5E5	44.69	26.00	29.25	34.88	26.00	-14.50	4.56	0.00	0.67
6E8	43.06	13.00	-5.69	33.26	13.00	-49.44	4.39	0.00	-0.13
Mean	43.88	19.50	11.78	34.07	19.50	-31.97	4.47	0.00	0.27
Adj. factor*	1.79			3.74			0.53		
Adj. mean	78.46	34.87	21.07	127.50	72.98	-119.64	2.37	0.00	0.14

*adjusting factor results from division of mean added return, added profit and marginal rate by the relevant sum of mean effects for factors A, B, and C, i.e., for added return: 134.39/(43.88+19.50+11.78)=1.79

Table 4: Summary of economic outputs.

three consecutive research methods: a survey of the state of the art of technology and/or on-research-station factorial experiments at the first stage, and a system of on-farm factorial experiments at the second stage. The information provided by these methods is complementary but the crucial stage of this approach is associated with on-farm research.

The use of on-farm field experiments in agricultural experimentation is considered a very useful tool at any stage of research. However, it also underlines the difficulty of testing crop production technology within differentiated farm conditions [2,15-18]. Some of the technologies developed by agricultural scientists in the conditions of experimental stations do not work in practice and some of them may be unique to a specific location of the station and have to be (re)developed from the beginning [19-21].

In this paper, the developed system of various full and fractional factorial designs for testing crop production technology in on-farm experiments was based on the results of a thorough survey study and fractional factorial experiments at a plot scale. A similar compilation of different research methods for an integrated agricultural technology developed system was proposed by Biggs [22], who combined exploratory surveys, diagnosis of farmers' problems, on-farm experiments and development feedback systems. Byerlee et al. [9] distinguished the two main objectives of on-farm research; increasing the body of knowledge and solving specific problems in the farming

systems. These authors argued that location-specific research with a short-term objective of developing improved technologies for a target group of farmers and research conducted over a longer time (to overcome major, widespread constraints in farming systems) are part of an integrated research system in which area-specific research provides the basis for defining longer-term research priorities. Another model for generating an acceptable agricultural technology was developed by Rhoades and Booths [23] and stresses that applied research must begin and end with the farmer.

In our studies the test crop was *Pisum sativum* L. sensu lato cultivated for food use. The study was conducted in the north-eastern part of Poland and covered farms which contracted their production to the local vegetable processing company. The results of the studies from the survey and field plot experiments enabled the identification of key factors responsible for high yields, construction of factorial experimental designs, and eventual implementation of the system of experiments on farms. The system is composed of factorial and fractional designs based on a plan of full factorial design 2^3. Such a systemic approach was advantageous both from the point of view of organization of the research, as well from the point of view of farms where the experiments were located. It was done in an attempt to improve the efficiency of on-farm experiments by implementation of covariates associated with spatial variation of soil properties into statistical analysis. However, the results were not satisfactory. A possible reason for this was the number of on-farm experiments and replications per experiment was too small to prove the effects of spatial covariates because of the small number of degrees of freedom for experimental error.

Field experiments conducted on farms are not common practice in Poland. In general, demonstration experiments involving only two or three treatments without replications or with quasi-replications is rather easily accepted by farmers. However, in the case of most advanced designs requiring some extra organizational activities, and usually extra expenditures, the motivation for experimentation is relatively low. In the current study, only two advanced experiments (E5 and E8) were established on farms. Despite this, the data from the experiments enabled estimation of significant sources and their contribution to yield variation. The farmers' participation in the experiment was crucial for the successful introduction of the proposed experimental system. Some authors report that this is generally advantageous because the farmers' participation in on-farm experiments requires fewer resources and less time than on-research-station or diagnostic survey research [24].

Conclusions

On the basis of the results presented in this paper some methodical conclusions may be drawn.

Survey studies and on-research station multifactorial experiments are effective tools in screening agro-technical factors in the context of crop production technology and detection of key factors responsible for high yield variation. Both the research methods are confirmative and validate each-other.

ANOVA of data from the survey, together with eta-square estimates enables the fractionation of production factors and reveals the structure of their contribution into yield variation.

Single two-treatment experiments with alternative technologies (traditional, new) should be established at as many sites as possible because they are the basis of overall information on the technology gap. In addition, they are the reference point for correction of a single contribution of factors and their interactions to the gap.

The system of FD and FDD on-farm experiments is universal and scalable for a higher number of factors.

The system of on-farm experiments enables flexible organization of experiments, depending not only on the interests of farmers and organizational capacity of farms, but also on the methodical assumptions of research and available funds.

The efficiency of the system is high because the experiments may be analyzed in multiple ways: as a single realization or configured in different sets of experiments (i.e., at the local administration level, for different soils, etc.). It is worth noting that the efficiency of the system and its information provided will be higher, together with the higher number of farms engaged in on-farm experimentation.

The proposed procedure may be adopted by processing companies which contract feedstock from farmers. This is because, for a given feedstock the crop production technology at the contracted farms is common and uniform and any innovative changes in production factors may be quickly and efficiently verified, e.g., when a new production factor should be tested.

Acknowledgements

Funding for this research was provided by the Ministry of Science and Higher Education, Grant No. MSzWiN 529-108-0913. This paper benefited from the discussions and assistance of the representatives of Warmian Fruit and Vegetable Processing Company, Ltd. in Kwidzyn and associated farmers. The authors would like to express special acknowledgements to PhD Student Anna Zaręba and the farmers: Arkadiusz Piskorski (Kwidzyn), Lucjan Kossobudzki (Jazowa), Wiesław Szpura (Królewo), and Leszek Grenda (Stare Pole).

References

1. Clay DE, Shanahan JF (2011) GIS Applications in Agriculture. Volume II: Nutrient Management for Energy Efficiency. CRC Press, Taylor & Francis Group.

2. Gołaszewski J, Załuski D, Żuk-Gołaszewska K, Grzela K (2013) Geostatistical methods as auxiliary tools in field plot experimentation. In: Stafford JV (ed.), Precision Agriculture '13. Wageningen Academic Publishers, pp: 499-506.

3. Box GEP, Hunter JS (1961) The 2^{k-p} Fractional Factorial Designs Part I. Technometrics 3: 311-351.

4. Załuski D, Gołaszewski J (2006) Efficiency of 3^{5-p} fractional factorial designs determined using additional information on the spatial variability of the experimental field. Journal of Agronomy and Crop Science 192: 303-309.

5. Załuski D, Gołaszewski J, Stawiana-Kosiorek A, Zaręba A (2006) Full and fractional design in the practice of field experimentation. Advances in Agricultural Sciences 1: 39-47.

6. On-Farm Trials - Some Biometric Guidelines (1998) Statistical Services Centre, University of Reading, UK, p: 16.

7. Blaise D, Majumdar G, Tekale KU (2005) On-farm evaluation of fertilizer application and conservation tillage on productivity of cotton+pigeon pea strip intercropping on rainfed Vertisols of central India. Soil and Tillage Research 84: 108-117.

8. Gomez AG, Gomez AA (1984) Statistical procedures for agricultural research. John Wiley and Sons Inc.

9. Byerlee D, Harrington L, Winkelmann D (1982) Farming system research: Issues in research strategy and technology design. American Journal of Agricultural Economics 65: 897-904.

10. Amir P, Knipscheer HC (1989) Conducting on-farm animal research: procedures and economic analysis. Winrock International Institute for Agricultural Development, Morrilton, USA. International Development Research Centre, Ottawa, Canada.

11. Liu Y, Swinton SM, Miller NR (2006) Is Site-Specific Yield Response Consistent over Time? Does It Pay? American Journal of Agricultural Economics 88: 471-483.

12. Barlow C, Jayasuriya SK, Price E, Maranan C, Roxas N (1986) Improving the economic impact of farming systems research. Agricultural Systems 22: 109-125.

13. Carroll RJ, Ruppert D (1988) Transformation and Weighting in Regression. Chapman and Hall, New York, USA.

14. Greenacre M, Blasius J (2006) Multiple correspondence analysis and related methods. Chapman & Hall/CRC, Taylor & Francis Group, Boca Raton, London, New York.

15. Farrington J (1988) Farmer participatory research. Editorial introduction. Experimental Agriculture 24: 269-279.

16. Hildebrand PE, Poey F (1985) On-farm agronomic trials in farming systems research and extension. Lynne Rienner Publishers Inc., Boulder, CO, USA, p: 162.

17. Scherr SJ (1991) On-farm research: the challenges of agroforestry. Agroforestry Systems 15: 95-110.

18. Gołaszewski J, Voort MVD, Meyer AA, Baptista F, Balafoutis A, et al. (2014) Comparative analysis of energy efficiency in wheat production in different climate conditions of Europe. Journal of Agricultural Science and Technology B 4: 632-640.

19. Sumberg J, Okali C (1988) Farmers, on-farm research and the development of new technology. Experimental Agriculture 24: 333-342.

20. Matlon P, Cantrell R, King D, Benoit-Cattin M (1984) Coming full circle: Farmers' participation in the development of technology. IDRC-189e, Ottawa, Canada.

21. Stroud A (1993) Conducting on-farm experiments. CIAT Public, No 228.

22. Biggs SD (1980) On-farm research in an integrated agricultural technology developed system: Case study of triticale for the Himalayan hills. Agricultural Administration 7: 133-145.

17

Growth and Yield Response of Cowpea (*Vigna unguiculata* L. Walp.) to Integrated Use of Planting Pattern and Herbicide Mixtures in Wollo, Northern Ethiopia

Getachew Mekonnen[1]*, Sharma JJ[2], Lisanework Negatu[2] and Tamado Tana[2]

[1]*College of Agriculture and Natural Resources, Mizan Tepi University, Mizan Teferi, Ethiopia*
[2]*College of Agriculture and Environmental Sciences, Haramaya University, Dire Dawa, Ethiopia*

Abstract

To assess the integrated effect of planting pattern and low dose herbicide mixtures on weeds and growth, yield attributes and yields of cowpea, and to determine the economic feasibility of different weed management practices in cowpea, a field experiment was conducted at Sirinka Agricultural Research Center experimental sites at Jari and Sirinka in Northern Ethiopia during the 2014 main cropping season. There were 16 treatments comprising the combinations of two planting patterns (60 cm × 10 cm, 45 cm × 15 cm) and eight weed management practices (s-metolachlor 2.0 kg ha^{-1}, s-metolachlor at 1.0 kg ha^{-1}+hand weeding and hoeing 35 weeks after crop emergence (WAE), pendimethalin at 1.0 kg ha^{-1}+hand weeding and hoeing 5 WAE, s-metolachlor at 1.0 kg ha^{-1}+pendimethalin at 1.0 kg ha^{-1}, s-metolachlor at 1.0 kg ha^{-1}+pendimethalin at 0.75 kg ha^{-1}, s-metolachlor at 0.75 kg ha^{-1}+pendimethalin at 1.0 kg ha^{-1}, hand weeding and hoeing 3 WAE and weedy check. The treatments were laid out in factorial combination in a randomized complete block design with three replications. The highest number of pods per plant, number of seeds per pod, and hundred seed weight were obtained from the combination of s-metolachlor at 1.0 kg ha^{-1}+hand weeding 5 WAE along with 60 cm × 10 cm at Sirinka. Higher (3092 kg ha^{-1}) grain yield was recorded at Sirinka than at Jari (2714 kg ha^{-1}). The highest (53460 ETB ha^{-1}) gross benefit was obtained from s-metolachlor at 1.0 kg ha^{-1}+hand weeding and hoeing 5 WAE, followed by pendimethalin at 1.0 kg ha^{-1}+hand weeding and hoeing 5 WAE (46737 ETB ha^{-1}). Therefore, managing the weeds with the application of 1.0 kg ha^{-1} of s- metolachlor+hand weeding and hoeing 5 WAE along with 60 cm × 10 cm proved to be the most feasible practice. Alternate herbicides for the control of X. strumarium infested fields in the study area needs to be explored.

Keywords: Broadleaved and grass weeds; Cowpea; Grain yield; Herbicide mixtures; Integrated management; Pendimethalin; s-metolachlor; *Vigna unguiculata*; Weed

Introduction

Cowpea (*Vigna unguiculata* (L.) Walp.) is one of the most important food grain legumes in the tropics, including Africa, which accounts for 64% of the world production [1]. West Africa represents the largest production zone with modest amounts emanating from the east African countries of Mozambique, Tanzania, Uganda and to some extent Ethiopia [1,2]. In addition to its importance in human food, cowpea is also useful for soil fertilization through symbiotic nitrogen fixation and can be a major animal feed due to the quality of its leaves [3].

It is cultivated around the world primarily for seed, but also as a vegetable (for leafy greens, green pods, fresh shelled green peas, and shelled dried peas), as cover crop and for fodder [4]. In most African countries, cowpea is either grown alone or intercropped with various cereal crops, such as leafy vegetables, maize, millet, sorghum, beans, pigeon peas, bananas and others [5,6]. Since, it is shade tolerant and compatible as an intercrop with cereal crops, it helps to prevent buildup of disease incidence, insect pests and weeds. Its variability of uses, nutritive content and storage qualities have made cowpea an integral part of the farming system in Africa [7].

Cowpea yield loss due to weed interference was described to reach up to 96%, which indicates the importance of weed management in this crop [8]. Chikoye [9] stated that the reduction in yield of cowpea depends on the weed species, weed density and weed dry biomass. Also, Blackshaw [10] stated that cowpea is sensitive to weed competition; for instance, 2 to 100 plants m^{-2} density of *Solanum nigrum* plant, decreased cowpea yield that ranged between 13 and 77%. Fennimore [11] also

reported up to 40% yield loss in cowpea due to the competition with *S. nigrum*. Similarly, Wilson [12] found that for every 100 kg dry weight of weeds, cowpea yield was reduced by about 208 kg ha^{-1}. In Ethiopia, one timely, early weeding at 25 days after emergence resulted in 70% yield increase of common bean and up to 300% increase in cowpea compared to the no-weeding [13].

Different management practices should be employed to reduce yield loss due to weeds. Among those practices, integrated weed management (IWM) involves a combination of cultural, physical, chemical and biological methods for effective and efficient or economical weed control [14]. The principle of IWM should provide the foundation for developing optimum weed control systems and efficient use of improved varieties. Integrating herbicides with cultural methods is an option for better weed management. IWM does not preclude herbicide use, it includes their judicious use along with other agronomic methods that help crops compete with weeds and reduce weed seed production. IWM also involves using an agronomical approach to minimize the

*Corresponding author: Getachew Mekonnen, College of Agriculture and Natural Resources, Mizan Tepi University, PO Box 260, Mizan Tepi, Ethiopia
E-mail: sibuhmekdes@gmail.com

overall impact of weeds and, indeed, maximize the benefits. The use of a single herbicide may result in shift of the weed flora in favour of the species that are not controlled, thus may increase the problem in the future.

Moreover, to manage mixed population of weeds and also to avoid herbicide resistance development by continuous use of a single herbicide, compatible mixtures can be employed to widen the spectrum of weed suppression. Herbicide combinations can give spectacularly good control at doses considerably below those normally applied in a single application. It may be additive or synergistic or prevent rapid detoxification of herbicides and are safer to crops than application of a single herbicide alone. The use of herbicide combinations is not new, but it has not received the attention and input that is necessary to fully understand and implement the practice. Therefore, there is a need for evaluation of a range of herbicides alone and as a tank mixture to have broad spectrum weed management [15].

The present study, therefore, is intended 1) To assess the integrated effect of planting pattern and low dose herbicide mixtures on weeds, nodulation, growth, yield attributes and yields of cowpea, and 2) to determine the economic feasibility of different weed management practices in cowpea.

Materials and Methods

Description of the study area

The experiment was conducted at Jari experimental sites (11°21'N latitude and 39°38'E longitude; 1680 masl. altitude) at Sirinka Agricultural Research Center and Sirinka (11°45'00" N latitude; 39°36'36"E longitude; 1850 masl altitude) in northern Ethiopia during the 2014 main cropping season. The soil of the experimental fields was clay loam and clay, while the pH was 6.98 and 6.94 at Sirinka and Jari, respectively. At Sirinka, the organic carbon was 1.35%, total N was 0.07%, available P 13.7 mg kg^{-1} soil and CEC 56.47 cmolC kg^{-1}, while the respective values at Jari were 1.37%, 0.05%, 11.17 mg kg^{-1} soil and 47.44 cmolC kg^{-1}. The total rainfall received during the crop season was 795.4 and 649.1 mm at Sirinka and Jari with mean maximum and minimum temperatures of 27.0 and 14.2°C, and 30.1°C and 16.0°C, respectively (Figure 1). Soil sample analysis was done at the Sirinka Agricultural Research Center.

Experimental materials

The cowpea variety Asrat (ITS 92KD-279-3), released by Sirinka Agricultural Research Center/Amhara Region Agricultural Research Institute (SRARC/ARARI) in 2001, was used in these experiments. The variety is well adapted to moisture stress areas in the northeast Wollo and similar lowland areas. This variety is suitable for an altitudinal range of 1450-1850 masl and annual rainfall of 660-1025 mm. It is bushy and trailing type I. It attains physiological maturity in 95-100 days [16]. Description of herbicides (s-metolachlor and pendimethalin) used in the experiment has been presented in tabular form hereunder Table 1.

Treatments and experimental design

There were 16 treatment combinations comprising of two planting patterns (60 cm × 10 cm and 45 cm × 15 cm) and eight weed management practices (s-metolachlor at 2.0 kg ha^{-1}, s-metolachlor at 1.0 kg ha^{-1}+hand weeding and hoeing 5 weeks after crop emergence (WAE), pendimethalin at 1.0 kg ha^{-1}+hand weeding and hoeing 5 WAE, s-metolachlor at 1.0 kg ha^{-1}+pendimethalin at 1.0 kg ha^{-1}, s-metolachlor at 1.0 kg ha^{-1}+pendimethalin at 0.75 kg ha^{-1}, s-metolachlor at 0.75 kg ha^{-1}+pendimethalin at 1.0 kg ha^{-1}, hand weeding and hoeing 3 WAE, and weedy check). The treatments were laid out in factorial combination in a randomized complete block design with three replications.

Experimental procedure and management

The experimental field was ploughed to get a fine seedbed using tractor and the plots were leveled manually. The gross plot size was 3.6 m × 2.4 m (8.64 m^2). The pathway between replications and plots were 1 and 0.5 m, respectively. The cowpea variety Asrat was planted on 21 and 22 July 2014 at Jari and Sirinka, respectively. Fertilizer (100 kg DAP; 18 kg N+46 kg P$_2$O$_5$ ha^{-1}) was applied to each plot uniformly at the sowing time. There were 6 and 8 rows per plot under 60 cm and 45 cm row spacing, respectively.

The herbicides were applied as per the treatment in the assigned plots as pre-emergence within one day after planting. Herbicide spray volume with water as carrier was 450 l ha^{-1}. Spraying was done with manually-operated knapsack sprayer (15 l capacity) using flat-fan nozzle. The outermost one row from one side and two rows from another side of in the plots having 60 cm inter row spacing, while two rows from each side of the plots having 45 cm inter row spacing were

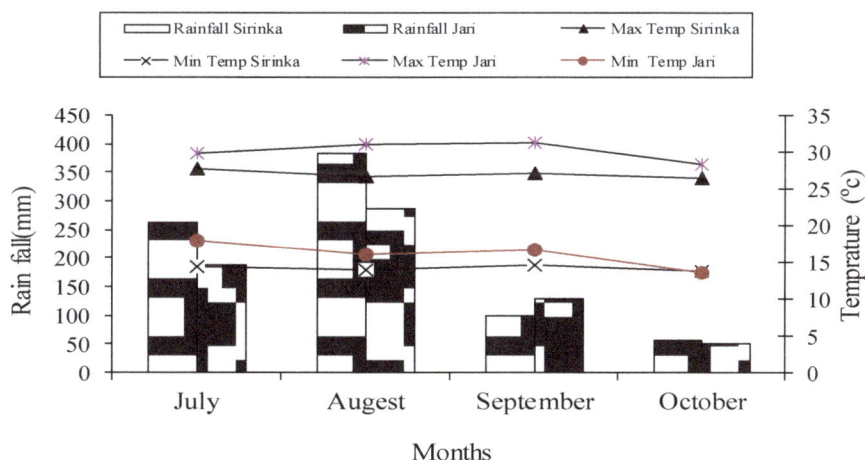

Figure 1: Monthly mean maximum and minimum temperatures (°C) and total rainfall (mm) at Jari and Sirinka in 2014 cropping season.

Common name	Trade name	Chemical name
S-metolachlor	Dual Gold 960EC	[2-chloro-6-ethyl-N-(2-methoxy-1-methylethyl)acet-o-toluidide]
Pendimethalin	Stomp Extra 38.7% CS	[N-(1-ethylpropyl)-2,6-dinitro-3,4-xylidine]

Table 1: Description of herbicides used in the integrated weed management in cowpea experiment at Jari and Sirinka in 2014.

considered as borders. From the end-point of each row, three plants in plots having 10 cm intra row spacing and two plants in 15 cm intra row spacing were considered as borders. Thus the net plot size was 1.8 m × 1.8 m (3.24 m²). All the recommended practices, except the treatments, were followed to raise the crop. The crop was harvested on 29 October and 6 November 2014 at Sirinka and Jari, respectively. The harvested produce was sun-dried for 7-10 days and threshing and winnowing was done subsequently.

Data collection and analyses

Weeds data: Weed aboveground dry biomass (g): For aboveground weed dry biomass, the weeds falling within the quadrate were cut near the soil surface immediately after recording data on weed count and placed into paper bags separately treatment-wise. The samples were sun-dried for 3-4 days and thereafter were placed into an oven at 65°C temperatures till a constant weight and, subsequently, their dry weight was measured. The dry weight was expressed in g m⁻².

Weed Control Efficiency (WCE): It was calculated using the following formula:

$$WCE = \frac{(WDC - WDT)}{WDC} \times 100$$

Where,

WDC=Weed dry weight in weedy check, WDT=Weed dry weight in a particular treatment.

Crop data:

Plant height (cm): It was taken with a ruler from 10 randomly taken and pre tagged plants in each net plot area from the base to the apex of the main stem at physiological maturity.

Number of pods per plant: It was taken from the total pods of the above tagged plants at harvest.

Number of per pod: The total number of seeds from the above pods was taken and counted to average the number of seeds pod⁻¹.

Hundred seed weight (g): Out of seeds from the above pods, 100 seeds were counted and their weight was recorded at 10.5% moisture content for hundred seed weight.

Aboveground biomass (g): This parameter was determined by harvesting ten plants in each plot at physiological maturity and their dried aboveground biomass was recorded. Treatment-wise per plant dry weight of straw was multiplied by the number of plants in respective treatments. This was considered as the aboveground dry biomass weight.

Grain yield (kg ha⁻¹): The grain yield was measured after threshing the sun-dried plants harvested from each net plot and the yield was adjusted at 10.5% seed moisture content. The grain weight obtained in ten plants was added to the final yield.

Harvest index (%): This parameter was calculated by dividing the grain yield by the aboveground biomass yield and multiplied by 100.

Data analyses: Data on weed density, weed dry biomass, growth, yield attributes and yield were subjected to analysis of variance (ANOVA) using GenStat 15.0 computer software [17]. Fisher's protected least significant difference (LSD) test at p≤0.05 was used to separate differences among treatment means [18]. As the F-test of the error variances for most parameters of the two sites was homogeneous, combined analysis of data was used.

Partial budget analysis

The concepts used in the partial budget analysis were the mean grain yield of each treatment in both locations, the field price of cowpea (sale price (ETB 15 kg⁻¹) minus the costs of harvesting, threshing and winnowing (ETB 165/100 kg) bagging (ETB 4.0 per 100 kg) and transportation (ETB 5 per 100 kg), the gross field benefit (GFB) per hectare (the product of field price and the mean yield for each treatment), the field price of s-metolachlor ETB 417 kg⁻¹, cost of pendimethalin ETB 620 kg⁻¹ (the herbicide cost plus the cost of transportation from the point of sale to the farm), the total costs that varied (TCV) included the sum of field cost of herbicide and its application (spraying ETB 99 ha⁻¹). The net benefit (NB) was calculated as the difference between the GFB and the TCV. All costs and benefits were calculated on hectare basis in Ethiopian Birr (ETB). Actual yield was adjusted downward by 10% to reflect the difference between the experimental yield and the yield farmers could expect from the same treatment. It was assumed that there was optimum plant population density, timely labor availability and better management (e.g., weed control, better security) under experimental conditions [19,20].

Results and Discussion

Weed parameters

Weed dry biomass: The minimum (30.4 g m⁻²) weed dry weight recorded at Sirinka from s-metolachlor at 1.0 kg ha⁻¹+hand weeding and hoeing 5 WAE was statistically at par with the application of s-metolachlor at 2.0 kg ha⁻¹, pendimethalin at 1.0 kg ha⁻¹+hand weeding and hoeing 5 WAE, s-metolachlor at 1.0 kg ha⁻¹+pendimethalin at 1.0 kg ha⁻¹ and hand weeding and hoeing at 3 WAE at Sirinka and s-metolachlor at 1.0 kg ha⁻¹+hand weeding and hoeing 5 WAE at Jari. Hand weeding and herbicide significantly encouraged vigorous cowpea growth.

In line with the current research result, Ahmad [21] reported that pre-emergence application of pendimethalin at 1.25 and 1.50 kg ha⁻¹+hand weeding were equally and even much more effective in reducing dry weight of weeds than other treatments. The better weed suppression due to herbicide mixtures may be due to effective suppression of both types of weeds. Also, the low weed density observed in herbicides treated plots could be attributed to effective weed control of the herbicides and their ability to manage weeds beyond the critical period of cowpea growth. Also, the adequate weed cover by cowpea vine led to smothering effect of the weeds judging from the low weed population and low weed dry weight, which invariably led to increase in weed smothering efficiency [22]. They also found lower weed dry matter and higher weed control efficiency with herbicides+hand weeding than other treatments included in their experiment.

Sharma [23] also concluded that dry weight of weeds was significantly reduced in herbicide-treated plots of common bean. In pigeon pea, effective weed control has been reported with integrated use of pendimethalin and hand weeding [24]. However, lower performance of intra-group herbicides might be due to lower doses than their recommended doses, which needs to be investigated at recommended doses of individual herbicides in mixture [25].

The location and weed management practices interaction further showed that the maximum (472.4 g m^{-2}) weed dry weight obtained in weedy check at Jari was significantly higher than all the other interactions (Table 3). These results are consistent with the findings of Arif and Marwat [26,27] who reported more weed dry biomass in weedy check than pre-emergence herbicides (s-metolachlor and pendimethalin) application in canola (*Brassica napus* L.) for weed management.

It was also recognized that there was no significant difference in weed dry weight obtained in weedy check at Sirinka with s-metolachlor at 2.0 kg ha^{-1}, s-metolachlor at 1.0 kg ha^{-1}+pendimethalin at 1.0 kg ha^{-1}, s-metolachlor at 1.0 kg ha^{-1}+pendimethalin at 0.75 kg ha^{-1} and hand weeding at 3 WAE at Jari, and s-metolachlor at 1.0 kg ha^{-1}+pendimethalin at 0.75 kg ha^{-1} and s-metolachlor at 0.75 kg ha^{-1}+pendimethalin at 1.0 kg ha^{-1} at Sirinka. The moderate increase in weed dry weight could be attributed to frequent reoccurrence and persistent characteristics of weeds. Furthermore, like weed density, the weed dry biomass was also lower at Sirinka than at Jari. Herbicide molecules tend to bind with soil clay and organic matter particles, and thus become unavailable for weed killing purposes.

The lower weed dry matter accumulation may be attributed to lower weed density at Sirinka than at Jari (Table 2). The lowest (29.8 g m^{-2}) weed dry weight was recorded with 60 cm × 10 cm planting pattern when treated with s-metolachlor at 1.0 kg ha^{-1}+hand weeding and hoeing 5 WAE, which was statistically at parity with pendimethalin at 1.0 kg ha^{-1}+hand weeding at 5 WAE, s-metolachlor at 1.0 kg ha^{-1}+pendimethalin at 1.0 kg ha^{-1} under both the planting patterns. Jafari [28] stated that pre-emergence herbicides reduced the weed density and dry weight significantly as compared to weedy check in common bean. Similarly, Masoumeh [29] found application of pendimethalin 0.5 kg ha^{-1}+hand weeding 30 days after sowing though was comparable with other treatments, but gave lower weed dry weight after weed-free check in soybean.

The highest (327.5 g m^{-2}) weed dry weight was found in weedy check under 45 cm × 15 cm planting pattern, followed by 60 cm × 10 cm planting pattern and both these interactions resulted in higher increase in weed dry weight than all the other planting pattern and weed management practices interactions (Table 2). A high weed density recorded in the weedy plots invariably resulted in high weed dry weight that could be attributed to low ground cover of cowpea vines.

This could be attributed to faster and better canopy cover of the crop under narrow spacing resulting in better suppression of weeds than in wide spacing. Reduction in weed dry biomass due to narrow rows has been reported by Adigun and Joseph [30,31]. This current result is consistent with the findings of other others in lentil [21,32,33].

Weed control efficiency: The data on weed control efficiency indicated that all the treatments in general gave more than 53% weed control efficiency over the weedy check. The maximum (91.6%) weed control efficiency was observed in s-metolachlor at 1.0 kg ha^{-1}+hand weeding and hoeing at 5 WAE under 60 cm × 10 cm planting pattern, which was statistically at par with the interaction of same weed management practice and 45 cm × 15 cm planting pattern as well as the interaction of combined application of metolachlor and pendimethalin each at 1.0 kg ha^{-1} under 60 cm × 10 cm planting pattern (Table 3).

The current finding is in agreement with the investigation of Shinde [34] who reported that integration of pendimethalin with hand weeding 40 days after sowing is known to provide high weed control efficiency in pigeon pea. Priya [35] also found the lowest weed dry matter and higher weed control efficiency with herbicides+hand weeding in soybean. A similar trend was also reported by Jafari [28] in common bean, where pre-emergence herbicides application gave high weed control efficiency by reducing the weed density and dry weight significantly as compared to the weedy check. Sylvestre [36] also reported that unweeded check showed lower weed control efficiency than the rest of pre-emergence herbicide treatments in soybean, while the lowest weed control efficiency was obtained in s-metolachlor at 1.0 kg ha^{-1}+pendimethalin at 0.75 kg ha^{-1} and s-metolachlor at 0.75 kg ha^{-1}+pendimethalin at 1.0 kg ha^{-1} in planting pattern of 60 cm × 10 cm, respectively (Table 3).

A similar trend was also reported by Jafari [28] in common bean, where pre-emergence herbicides gave high weed control efficiency by reducing the weed density and dry weight significantly as compared to weedy check.

Initially, the weed flora may be suppressed because the toxic effect of herbicide normally appears immediately after application when their concentration in soil is highest. Later on, microorganisms take part in degradation process and herbicide concentration and its toxic

	Location (L)		Planting Pattern (P)	
	Jari	Sirinka	S1	S2
Weed management practices (W)				
S-metolachlor at 2.0 kg ha^{-1}	100.6^{d-g}	72.1^{e-h}	88.0^{def}	84.7^{def}
S-metolachlor at 1.0 kg ha^{-1}+hand weeding and hoeing 5 WAE	50.4^{gh}	30.4^{h}	29.8^{g}	50.9^{fg}
Pendimethalin at 1.0 kg ha^{-1}+hand weeding and hoeing 5 WAE	100.7^{d-g}	53.8^{fgh}	81.8^{d-g}	72.7^{efg}
S-metolachlor at 1.0 kg ha^{-1}+pendimethalin at 1.0 kg ha^{-1}	115.2^{cde}	77.2^{e-h}	62.0^{fg}	130.4^{bcd}
S-metolachlor at 1.0 kg ha^{-1}+pendimethalin at 0.75 kg ha^{-1}	162.9^{bc}	107.8^{def}	152.6^{bc}	118.0^{cde}
S-metolachlor at 0.75 kg ha^{-1}+pendimethalin at 1.0 kg ha^{-1}	202.0^{b}	110.0^{cde}	183.9^{b}	128.1^{cd}
Hand weeding and hoeing 3 WAE	109.3^{cde}	73.6^{e-h}	91.8^{def}	91.0^{def}
Weedy check	472.4^{a}	136.9^{cd}	281.8^{a}	327.5^{a}
LSD (5%) L *x* W/ P *x* W			54.31	
CV (%)			38.1	

Means followed by the same letters are not significantly different at 5% level of significance, LSD=least significant difference, CV=Coefficient of variation, DAE=days after crop emergence, S1=60 cm × 10 cm; S2=45 cm × 15 cm

Table 2: Interaction effect of location with weed management practices and planting pattern with weed management practices on total weed dry biomass (g m^{-2}) at harvest in 2014 cropping season.

	Planting Pattern (P)	
	S1	S2
Weed management practices (W)		
S-metolachlor at 2.0 kg ha⁻¹	75.9ᶜᵈ	80.7ᵇᶜ
S-metolachlor at 1.0 kg ha⁻¹+hand weeding and hoeing 5 WAE	91.6ᵃ	87.1ᵃᵇ
Pendimethalin at 1.0 kg ha⁻¹+hand weeding and hoeing 5 WAE	74.9ᶜᵈ	81.3ᵇᶜ
S-metolachlor at 1.0 kg ha⁻¹+pendimethalin at 1.0 kg ha⁻¹	82.2ᵃᵇᶜ	70.0ᵈ
S-metolachlor at 1.0 kg ha⁻¹+pendimethalin at 0.75 kg ha⁻¹	53.3ᵉ	74.3ᶜᵈ
S-metolachlor at 0.75 kg ha⁻¹+pendimethalin at 1.0 kg ha⁻¹	54.2ᵉ	73.1ᶜᵈ
Hand weeding and hoeing at 3 WAE	74.5ᶜᵈ	78.6ᵇᶜᵈ
Weedy check	0.0ᶠ	0.0ᶠ
LSD (5%) (P x W)	9.6	
CV (%)	12.7	

Means followed by the same letters are not significantly different from each other at 5% level of significance, LSD=least significant difference, CV=Coefficient of variation, DAE=days after crop emergence, S1=60 cm × 10 cm; S2=45 cm × 15 cm.

Table 3: Interaction effect of planting pattern with weed management practices on weed control efficiency (%) in cowpea at harvest in 2014 cropping season.

effect decreases [37]. This could be one of the reasons for lower weed control efficiency with the application of s-metolachlor at 2.0 kg ha⁻¹ than s-metolachlor at 1.0 kg ha⁻¹+hand weeding and hoeing 5 WAE. S-metolachlor dissipation may be due to photo-degradation losses occurring since its application [38]. Chauhan [39] found decrease in bioavailability of s-metolachlor with the increase in days after sowing and it was 45% of the original amount applied 33 days after sowing. However, the lower dose of s-metolachlor initially suppressed the weed competition, which was further enhanced by integrating hand weeding at 5 WAE that kept the crop weed free during critical periods of 5 WAE, which offered prolonged and efficient weed control. Mondal and Warade [40,41] also observed similar results in onion.

Crop parameters

Growth parameters

Plant height: Application of herbicides alone or in combination as well as hand weeding resulted in significantly taller (86.0 cm to 98.0 cm) plant height than in the weedy check (Table 4). The current results are also in agreement with findings of Jafari [28] who stated that pre-emergence herbicides increased plant height in common bean significantly as compared to the weedy check. Similarly, plant height was also remarkably increased in wheat by all weed management methods compared to the weedy check [42].

Yield components, yield and harvest index:

Number of pods per plant: The highest (31.8 plant⁻¹) number of pods was recorded from s-metolachlor at 1.0 kg ha⁻¹+hand weeding and hoeing 5 WAE at Sirinka, which was significantly higher than that was obtained with different management practices at Jari and s-metolachlor at 1.0 kg ha⁻¹+pendimethalin at 0.75 kg ha⁻¹ and hand weeding at 3 WAE at Sirinka (Table 5). The current result is in agreement with this findings of Priya and Abouziena [35,43] who reported the highest number of pods per plant with single herbicide and two hand weeding was at par with herbicide supplemented with hand weeding in peanut (*Arachis hypogaea* L.) and soybean, respectively. This can be ascribed to the fact that the effective management of weeds led to the favourable environment for growth and photosynthetic activity of the crop resulting in improvement in the number of per pods. Ayaz [44] stated that the number of pods produced per plant or maintained to final harvest depends on a number of environmental and management practices. Mirshekari [45] also showed that the presence of weeds is a prominent factor in reducing the number of pods in cowpea plant.

Further, Dadari [46] reported that competition between weeds and crop starts right from germination of the crop up to harvest affecting both growth and yield parameters adversely.

The weedy check plots had the lowest number of pods per plant at both locations. At Sirinka, all the weed management practices had significantly higher number of pods per plant than the weedy check. However, it did not differ significantly with the combined application of s-metolachlor and pendimethalin and hand weeding and hoeing 5 WAE at Jari. In line with this result, Paudel [47] revealed that the average number of pods per plant was affected by different treatments of pre-emergence herbicides against weeds in cowpea and the treatments showed a significant difference from the uncontrolled plots. This result is in agreement with that of Jafari [28] who stated that pre-emergence herbicides increased the number of pods per plant significantly as compared to the weedy check in common bean. It was also found that, under all weed management practices at Sirinka, the number of pods per plant was significantly higher than at Jari.

The results obtained in this experiment also agree with the findings of Mousavi [48] who reported that the effect of s-metolachlor on cowpea pods per plant was significant. Also, Sylvestre [36] has documented earlier the role of yield contributing factors that enhanced yield on account of herbicidal control of weeds.

Number of seeds per pod: Application of s-metolachlor at 1.0 kg ha⁻¹ supplemented with hand weeding and hoeing at 5 WAE resulted in significantly higher (14.3 pod⁻¹) number of seeds than the other weed management practices except the application of pendimethalin at 1.0 kg ha⁻¹ supplemented with hand weeding and hoeing at 5 WAE (Table 6). Further, the latter treatment had no significant difference from the combined application of s-metolachlor and pendimethalin each at 1.0 kg ha⁻¹. The results also revealed that the weedy check plots produced significantly lower number of seeds per pod than the other treatments. The lower weed density and dry weight might have contributed to the significant increase in number of seeds per pod over the weedy check as suggested by Prakash [49] that number of seeds per pods in fieldpea increased with decrease in weeds density. This result agrees with the findings of Tenaw and Sharma [50,51] who reported that the number of seeds per pod was significantly reduced with the increased weed infestation and significantly increased with the weed-free period in common bean. In agreement with this observation, Jafari [28] also stated that pre-emergence herbicides increased the number of seeds per pod significantly as compared to the weedy check. Similarly, Muhammad [52] recorded a maximum number of seed per pod in

	Plant height (cm)
Weed management practices	
S-metolachlor at 2.0 kg ha[-1]	87.7[b]
S-metolachlor at 1.0 kg ha[-1]+hand weeding and hoeing 5 WAE	96.9[a]
Pendimethalin at 1.0 kg ha[-1]+hand weeding and hoeing 5 WAE	87.2[b]
S-metolachlor at 1.0 kg ha[-1]+pendimethalin at 1.0 kg ha[-1]	98.0[a]
S-metolachlor at 1.0 kg ha[-1]+pendimethalin at 0.75 kg ha[-1]	86.0[b]
S-metolachlor at 0.75 kg ha[-1]+pendimethalin at 1.0 kg ha[-1]	87.6[b]
Hand weeding and hoeing 3 WAE	88.2[b]
Weedy check	71.1[c]
LSD (5%)	8.4
CV (%)	11.7

CV=Coefficient of variation, DAE=Days after crop emergence, LSD=Least significant difference, Means followed by the same letters are not significantly different from each other at 5% level of significance.
Table 4: Main effect of weed management practices on plant height of cowpea in 2014 cropping season.

	Location (L)	
	Jari	Sirinka
Weed management practices (W)		
S-metolachlor at 2.0 kg ha[-1]	17.8[ef]	28.3[ab]
S-metolachlor at 1.0 kg ha[-1]+hand weeding and hoeing 5 WAE	22.9[cd]	31.8[a]
Pendimethalin at 1.0 kg ha[-1]+hand weeding and hoeing 5 WAE	21.3[de]	30.3[ab]
S-metolachlor at 1.0 kg ha[-1]+pendimethalin at 1.0 kg ha[-1]	16.7[e-g]	27.9[ab]
S-metolachlor at 1.0 kg ha[-1]+pendimethalin at 0.75 kg ha[-1]	15.7[fg]	26.1[b-d]
S-metolachlor at 0.75 kg ha[-1]+pendimethalin at 1.0 kg ha[-1]	16.2[fg]	27.3[a-c]
Hand weeding and hoeing 3 WAE	16.8[e-g]	26.0[b-d]
Weedy check	12.5[g]	12.2[g]
LSD (5%) (L x W)	4.9	
CV (%)	19.4	

CV=coefficient of variation, DAE=days after crop emergence, LSD=least significant difference, Means followed by the same letters are not significantly different from each other at 5% level of significance.
Table 5: Interaction effect of location with weed management practices on number of pods per plant in cowpea in 2014 cropping season.

Factors	Number of seeds pod[-1]	Hundred seed weight (g)
Location:		
Jari	11.94[a]	12.4[b]
Sirinka	11.16[b]	12.7[a]
LSD (5%)	0.73	0.20
Weed management practices:		
S-metolachlor at 2.0 kg ha[-1]	11.5[c]	12.8[bc]
S-metolachlor at 1.0 kg ha[-1]+hand weeding and hoeing 5 WAE	14.4[a]	14.0[a]
Pendimethalin at 1.0 kg ha[-1]+hand weeding and hoeing 5 WAE	13.1[ab]	13.1[b]
S-metolachlor at 1.0 kg ha[-1]+pendimethalin at 1.0 kg ha[-1]	11.7[bc]	12.6[cd]
S-metolachlor at 1.0 kg ha[-1]+pendimethalin at 0.75 kg ha[-1]	11.0[c]	12.3[d]
S-metolachlor at 0.75 kg ha[-1]+pendimethalin at 1.0 kg ha[-1]	10.9[c]	12.5[cd]
Hand weeding and hoeing 3 WAE	10.9[c]	12.5[cd]
Weedy check	8.9[d]	10.6[e]
LSD (5%)	1.5	0.4
CV (%)	15.5	3.9

CV=Coefficient of variation, DAE=Days after crop emergence, LSD=Least significant difference, Means followed by the same letters are not significantly different from each other at 5% level of significance.
Table 6: Main effect of location and weed management practices on number of seeds per pod and hundred seed weight of cowpea in 2014 cropping season.

fieldpea with application of s-metolachlor, while the minimum number of seed per pod was obtained in the weedy check plots. Also, the size of pods increased with application of s-metolachlor and hence resulted in maximum number of seeds per pod and vice versa.

Hundred seed weight: Hundred seed weight at Sirinka was 2.4% higher than the hundred seed weight at Jari. The highest (14.0 g) 100

seeds weight was recorded with the application of s-metolachlor at 1.0 kg ha[-1]+hand weeding and hoeing 5 WAE, which was significantly higher (6.4 - 24.3%) than the other treatments (Table 6). The application of the pre-emergence herbicides s-metolachlor and pendimethalin had been found to increase 100 seed weight of canola plant [26,27,53]. It was also found that pendimethalin at 1.0 kg ha[-1] supplemented with hand weeding and hoeing 5 WAE had no significant difference from s-metolachlor at 2.0 kg ha[-1]. On the other hand, the weedy check plots had significantly the lowest 100 seed weight of all the other treatments. This result is in line with Mohammadi [54] who found that the increased duration of weed interference in chickpea is associated with reduced dry matter to seed production, which results in yield reduction, in particular hundred seed weight per plant. Similarly, Sana [55] recorded the lowest 100 seed weight from the untreated weedy plots of chickpea. These results are in agreement with those of Yadav, Singh and Mohammadi [25,56,57] who found that the increased duration of weed interference in chickpea is associated with reduced dry matter to seed production which results in yield reduction in particular, hundred seed weight plant[-1].

Aboveground dry biomass yield: The highest (10157 kg ha[-1]) total dry biomass yield was obtained from s-metolachlor 2.0 kg ha[-1] treated plots, which were not significantly different from dry aboveground biomass yield obtained with s-metolachlor at 1.0 kg ha[-1]+hand weeding and hoeing 5 WAE, hand weeding and hoeing 21 days after crop emergence, pendimethalin at 1.0 kg ha[-1]+hand weeding and hoeing 5 WAE, and s-metolachlor at 1.0 kg ha[-1]+pendimethalin at 1.0 kg ha[-1] weed management practices (Table 7). Mizan [58] also reported that the increased dry matter weight of the crop was highly governed by the length of weed-free period. However, high production of total dry matter might not necessarily be of great value when the grain comprises a part of the plant. Aboveground dry biomass showed a significant variation across locations where significantly higher biomass was recorded at Sirinka than at Jari.

Grain yield: The grain yield (3092 kg ha[-1]) obtained at Sirinka was significantly higher by 13.9% than that at Jari. Among different weed management practices, application of s-metolachlor at 1.0 kg ha[-1]+hand weeding and hoeing 5 WAE gave significantly higher (3960 kg ha[-1]) grain yield than the other treatments. This was followed by the application of pendimethalin at 1.0 kg ha[-1] superimposed with hand weeding and hoeing at 5 WAE, which was also significantly higher than all the remaining treatments. Arif and Marwat [26,27] also reported significantly higher grain yield of canola with pre-emergence s-metolachlor and pendimethalin application.

Reduced crop-weed competition due to effective weed management with various treatments resulted in better growth, development and photosynthetic activity of the crop. Thus, the higher yield in these treatments might be attributed to the better weed management, which made better utilization of the resources, like nutrients, solar radiation, water and space by the crop that produced higher grain yield than the untreated control. In line with this, Rao and Begum [59,60] also reported higher yield due to effective management of weeds in early stage, which reduced weed growth and increased the growth and yield of black gram. Suppression of weed competition was further enhanced by integrating pre-emergence herbicides with hand weeding at 5 WAE that kept the crop weed-free during critical period, which offered prolonged and efficient weed control.

Grafton [61] opined better translocation of photosynthates under lesser competition among plants and this could be one of the reasons for obtaining higher yields. Townley [62] stated that good weed management is critical to obtain higher yield from fieldpea. Askew [63] reported that managing weeds and lesser competition within the plant community could result in utilization of the available resources efficiently, which, in turn, is reflected in higher grain yield. Morad [64] observed that yield of broad bean increased in plots treated with pre-emergence herbicides due higher pods per plant, seed number per pod and hundred seed weight.

Hand weeding and hoeing 3 WAE also proved significantly better than the mixture of s-metolachlor at 1.0 kg ha[-1]+pendimethalin at 0.75 kg ha[-1], and s-metolachlor at 0.75 kg ha[-1]+pendimethalin at 1.0 kg ha[-1]. This may be attributed to lower dry matter accumulation by weeds and decrease in their population, which, in turn, increased the yield attributes and ultimately increased the grain yield [65].

On the other hand, significantly lower yield was obtained in the weedy check than in the other treatments. Weed management practices significantly encouraged vigorous cowpea growth with minimal weed competition with cowpea. Thus, the cowpea grain yield obtained in the weedy check was 40.2 to 64.0% lower than the grain yield from other weed management practices as a result of intense weed competition (Table 7). This yield loss depends upon the density and species of weeds, duration of infestation and completing ability of the crop plants with weeds. These results are in line with the findings of Ahmad, Stork, Sangakkora and Tanveer [21,33,65,66] who reported 20 to 50% losses in grain yield if the weed management practices are not properly followed.

Similarly, Mohamed [32] reported that pre-emergence herbicides provided excellent suppression of weeds and the yield was significantly increased over weedy check. Weeds can severely affect the performance of the cowpea bean and the yield loss was 60 to 66% due to weed interference [67]. Prakash [49] found that long season crop-weed competition reduced the fieldpea yield by 44.6 to 55.6%. Similarly, several authors reported that weedy check plots gave the lowest yield in chickpea [25,57,68]. Blackshaw [10] stated that the weeds reduce more than 75% of yield in cowpea crop. However, the contradictory reports on the extent of yield losses due to weeds might be due to the variation in environmental conditions, soil types, the crop varietal characters and the extent of weed interferences at the locations.

Harvest index: Significantly higher (31.9%) harvest index was recorded at Sirinka than at Jari. The result indicated that there was significant variation on harvest index among the weed management treatments evaluated at both locations. S-metolachlor at 1.0 kg ha[-1]+hand weeding and hoeing 5 WAE gave the highest (40.0%) harvest index of all treatments, while the lowest (20.5%) harvest index was recorded from weedy check plots (Table 7). Increase in shoot weight with increasing weed interference might have increased the vegetative growth duration and decreased root/shoot ratio resulting in reduced harvest index. Soltani [69] reported that the harvest index of cowpea increases with increasing seed production. This result is in line with that obtained by Mousavi [48] who reported that the application of s-metolachlor herbicide application on cowpea increased harvest index.

Partial budget analysis

The result of the partial budget analysis and the data used for the partial budget analysis is given in a tabular form Table 8. The partial budget analysis was performed as described by CIMMYT (1988) where the variable costs that vary included the cost of inputs (herbicide) as well as the cost involved in their application.

However, for ease of calculation in place of field price of the crop, the cost incurred for harvesting, threshing, winnowing, packing and transportation was added to the variable input cost. The yield difference per hectare recorded from the different treatments accounted for the variation observed in value of gross benefit in both locations. The partial budget analysis indicated that the highest (ETB 53460 ha[-1]) gross benefit was obtained from s-metolachlor at 1.0 kg ha[-1]+ hand

Factors	Grain yield (kg ha⁻¹)	Above ground biomass yield (kg ha⁻¹)	Harvest index (%)
Location			
Jari	2714[b]	8873[b]	30.8[b]
Sirinka	3092[a]	9683[a]	31.9[a]
LSD (5%)	116.7	350.1	1.2
Weed management practices			
S-metolachlor at 2.0 kg ha⁻¹	3169[c]	10157[a]	31.9[c]
S-metolachlor at 1.0 kg ha⁻¹+hand weeding and hoeing 5 WAE	3960[a]	10099[a]	40.0[a]
Pendimethalin at 1.0 kg ha⁻¹+hand weeding and hoeing 5 WAE	3462[b]	9655[ab]	36.4[b]
S-metolachlor at 1.0 kg ha⁻¹+pendimethalin at 1.0 kg ha⁻¹	3039[c]	9473[ab]	32.5[c]
S-metolachlor at 1.0 kg ha⁻¹+pendimethalin at 0.75 kg ha⁻¹	2683[d]	8678[c]	31.3[c]
S-metolachlor at 0.75 kg ha⁻¹+pendimethalin at 1.0 kg ha⁻¹	2383[e]	9052[bc]	26.8[d]
Hand weeding and hoeing 3 WAE	3106[c]	10063[a]	31.4[c]
Weedy check	1424[f]	7043[d]	20.5[e]
LSD (5%)	233.5	700.2	2.5
CV (%)	9.9	9.2	9.8

CV=coefficient of variation, DAE=days after crop emergence, LSD=least significant difference, Means followed by the same letters in a column are not significantly different from each other at 5% level of significance.

Table 7: Main effect of location and weed management practices on grain yield, aboveground dry biomass, and harvest index of cowpea in 2014 cropping season.

	Average yield (kg ha⁻¹)	Adjusted yield (kg ha⁻¹) 10% down	Total variable cost (ETB ha⁻¹)	Gross return (ETB ha⁻¹)	Net return (ETB ha⁻¹)
Weed management practices					
S-metolachlor at 2.0 kg ha⁻¹	3169	2852	5687	42782	37095
S-metolachlor at 1.0 kg ha⁻¹+hand weeding and hoeing 5 WAE	3960	3564	6984	53460	46476
Pendimethalin at 1.0 kg ha⁻¹+hand weeding and hoeing 5 WAE	3462	3116	6440	46737	40297
S-metolachlor at 1.0 kg ha⁻¹+pendimethalin at 1.0 kg ha⁻¹	3039	2735	5695	41027	35332
S-metolachlor at 1.0 kg ha⁻¹+pendimethalin at 0.75 kg ha⁻¹	2683	2415	5006	36221	31215
S-metolachlor at 0.75 kg ha⁻¹+pendimethalin at 1.0 kg ha⁻¹	2383	2145	4606	32171	27564
Hand weeding and hoeing 3 WAE	3106	2795	6144	41931	35787
Weedy check	1424	1282	2136	19224	17088

Cost of hand weeding and hoeing 2 WAE 45 persons, 5 WAE 16 persons @ETB 33 person⁻¹, ETB=USD 0.0498

Table 8: Results of partial budget analysis of weed management practices in cowpea at Sirinka and Jari in 2014 cropping season.

weeding and hoeing 5 WAE, followed by gross benefit (ETB 46737 ha⁻¹) obtained from treatment with pendimethalin at 1.0 kg ha⁻¹+hand weeding and hoeing 5 WAE, while the lowest price was recorded from the weedy check plots. Singh [70] also reported a high economic return with butachlor+one hand weeding in rice, while Gupta [71] observed that the use of butachlor took equivalent to 186 hrs, while two-hand weeding took 604 hrs ha⁻¹ in rice.

In agreement with the present result, most studies showed that applying herbicide or herbicide plus manual weeding was more economical than manual or hand weeding alone [72]. The result of this experiment indicated that the use of herbicide though reduced the cost of production, the poor management of weeds resulted in significantly low yield compared to the combinations of s-metolachlor at 1.0 kg ha⁻¹+hand weeding and hoeing at 35 days after emergence and pendimethalin at 1.0 kg ha⁻¹+hand weeding and hoeing at 35 days after emergence. Therefore, managing weeds with the application of s-metolachlor at 1.0 kg ha⁻¹+hand weeding and hoeing 5 WAE proved to be the most profitable practice in weed management for sustainable cowpea production in northen Ethiopia and elsewhere cowpea is cultivated.

Summary and Conclusions

The weed density and dry weight at harvest was considerably

due to treatments with the application of both s-metolachlor and pendimethalin each at 1.0 kg ha⁻¹ supplemented with hand weeding and hoeing at 5 WAE at Sirinka, and s-metolachlor at 1.0 kg ha⁻¹+hand weeding and hoeing 5 WAE at Jari. However, the performance of both planting patterns was similar under s-metolachlor and pendimethalin each at 1.0 kg ha⁻¹ supplemented with hand weeding and hoeing at 5 WAE. The highest weed control efficiency was obtained with the combination of s-metolachlor at 1.0 kg ha⁻¹+hand weeding and hoeing 5 WAE, and 60 cm × 10 cm spacing was statistically at parity with s-metolachlor at 1.0 kg ha⁻¹+pendimethalin at 1.0 kg ha⁻¹, and s-metolachlor at 1.0 kg ha⁻¹+hand weeding and hoeing 5 WAE, respectively, under 60 cm × 10 cm and 45 cm × 15 cm plant spacing.

Plant height was significantly influenced by weed management practices. Number of pods per plant was significantly influence by location and weed management practices, whereby at Sirinka all weed management practices resulted in significantly higher pods per plant than the weedy check. On the other hand, at Jari weedy check had no significant difference from treatment with herbicide mixtures and hand weeding. Location and weed management practices significantly affected the number of grains per pod and 100 seed weight. Application of low dose herbicides supplemented with hand weeding resulted in higher number of seeds per pod and hundred seed weight than that of the control plots. Weedy check had significantly lower seed weight and number of seeds per pod than the other treatments.

Aboveground dry biomass weight and grain yield and harvest index were significantly higher at Sirinka than at Jari. Application of s-metolachlor at 1.0 kg ha-1+hand weeding and hoeing 5 WAE gave significantly higher grain yield and harvest index than other treatments, while it did not significantly differ from treatment with pendimethalin at 1.0 kg ha^{-1} supplemented with hand weeding for the aboveground dry biomass yield.

From the result of the study, it can be concluded that managing the weeds with the application of s- metolachlor at 1.0 kg ha^{-1}+hand weeding and hoeing at 5 WAE along with 60 cm × 10 cm planting pattern proved to be the most profitable practice. Based upon availability, alternatively pendimethalin could also be used in supplement with hand weeding at 5 WAE. Further, to prevent the weed shift, these two herbicides (s-metolachlor and pendimethalin) should be used as herbicide rotation. In future, there is a need to explore the effectiveness of various combinations of these two herbicides for cost effective and broad spectrum weed control in cowpea production.

Acknowledgements

We are grateful to the Ministry of Education, Federal Democratic Republic of Ethiopia, for the financial support, and Sirinka Agricultural Research Center for providing research facilities and Mizan Tepi University for permitting the corresponding author for PhD study at Haramaya University.

References

1. Timko MP, Singh BB (2008) Cowpea, a multifunctional legume. In: Moore PH, Ming R, (Eds.) Genomics of tropical crop plants, Springer, New York, pp: 227-257.

2. Pottorff M, Ehlers JD, Fatokun C, Philip A, Close RTJ (2012) Leaf morphology in Cowpea [Vigna unguiculata (L.) Walp]: QTL analysis, physical mapping and identifying a candidate gene using synteny with model legume species. BMC Genomics 13: 234.

3. Diouf D (2011) Recent advances in cowpea [Vigna unguiculata (L.) Walp.] "omics" research for genetic improvement. African Journal of Biotechnology 10: 2803-2810.

4. Andargie M, Remy P, Gowda B, Muluvi G, Timko M (2011) Construction of a SSR-based genetic map and identification of QTL for domestication traits using recombinant inbred lines from a cross between wild and cultivated cowpea [Vigna unguiculata (L.) Walp.]. Molecular Breeding 28: 413-420.

5. Bittenbender HC, Barrett RP, Indire- Lavusa BM (1984) Beans and cowpea as leaf vegetables and grain legumes. In: Occasional Monograph Series, No. 1, Bean/Cowpea, Michigan State University, East Lansing, MI, USA, pp: 1-21.

6. Singh BB, Mohan RDR, Dashie LK, Jackai LEN (1997) Origin, taxonomy and morphology of cowpea [Vigna unguiculata (L.)Walp.]. In: Co-publication of International Institute of Tropical Agriculture (IITA) and Japan International Research Center for Agricultural Sciences (JIRCAS) IITA, Ibadan, Nigeria, pp: 1-12.

7. Eaglesham AR, Ayanaba JA, Rama VR, Eskew DL (1982) Mineral nitrogen effects on cowpea and soybean crops in a Nigeria soil I. Development, nodulation, acetylene reduction and grain yield. Journal of Plant and Soil 68: 171-181.

8. Amador-Ramirez MD, Wilson RG, Martin AR (2001) Weed control and dry bean (Phaseolus vulgaris) response to in-row cultivation, rotary hoeing and herbicides. Weed Technology 15: 429-436.

9. Chikoye D, Weise SF, Swanton CJ (1995) Influence of common ragweed (Ambrosia artemisiifolia) time of emergence and density on white bean (Phaseolus vulgaris L.). Weed Science 43: 375-380.

10. Blackshaw RE (1991) Hariy Nightshade (Solanum sarrachoides) interference in dry bean (Phaseolus vulgaris L.). Weed Science 39: 48-53.

11. Fennimore SA, Mitch LW, Radsevich SR (1994) Interference among bean (Phaseolus vulgaris), barnyard grass (Echinochloa crusgalli) and black nightshade (Solanom nigrum). Weed Science 32: 336-342.

12. Wilson JR, Wicks GA, Fenter CR (1980) Weed control in field bean (Phaseolus vulgaris) in western Nebraska. Weed Science 28: 295-299.

13. Rezene F, Kedir N (2006) Review of weed research in highland and lowland pulses. In: Abraham Tadesse (ed.) Increasing Crop Production through Improved Plant Protection – Volume I. Addis Ababa, Ethiopia, pp: 133-167.

14. Swanton CJ, Weise SF (1991) Integrated weed management: Rational and approach. Weed Technology 5: 657-663.

15. Dixit A, Singh VP (2008) Efficacy of ready mix application of carfentrazone plus isoproturon (affinity) to control weed in wheat (Triticum aestivum L.). Indian Journal of Agricultural Science 78: 495-497.

16. MoARD (Ministry of Agriculture and Rural Development) (2009) Animal and Plant Health Regulatory Directorate. Crop Variety Register, Addis Ababa, Ethiopia, p: 213.

17. Payne RW, Murray DA, Harding SA, Baird DB, Soutar DM (2009) GenStat for Windows Introduction. VSN International, Hemel, Hempstead, p: 204.

18. Gomez KA, Gomez AA (1984) Statistical Procedures for Agricultural Research. 2nd edn. John Willey and Sons, p: 680.

19. CIMMYT (International Maize and Wheat Improvement Center) (1988) From Agronomic Data to Farmer Recommendations: An Economics Training Manual, Completely revised edition. Mexico, p: 79.

20. Moro BM, Nuhu IR, Toshiyuki W (2008) Determining optimum rates of mineral fertilizers for economic rice grain yields under the "Sawah" System in Ghana. West African Journal of Applied Ecology 12: 1-12.

21. Ahmad S, Abid SA, Cheema ZA, Tanveer A (1996) Study of various chemical weed control practices in lentil (Lens culinaris L. Medic.). Journal of Agricultural Research 34: 127-134.

22. Sunday O, Udensi E (2013) Evaluation of Pre-Emergence Herbicides for Weed Control in Cowpea [Vigna unguiculata (L.) Walp.] in a Forest-Savanna Transition Zone. American Journal of Experimental Agriculture 3: 767-779.

23. Sharma V, Thakur DR, Sharma JJ (1998) Effect of metolachlor and its combination with atrazine on weed control in rajmash (Phaseolus vulgaris L.). Indian Journal of Agronomy 43: 677-680.

24. Tomar J, Singh HB, Vivek HB, Tripathi SS (2004) Integrated weed management in intercropping of mungbean (Vigna radiata) and cowpea fodder (Vigna unguiculata) with pigeonpea (Cajanus cajan) under western condition. Indian Journal of Weed Science 36: 133-134.

25. Yadav SK, Singh SP, Bhan VM (2010) Weed Control in chickpea. Tropical Pest Management 29: 297- 298.

26. Arif M, Ihsanulla H, Khan S (2001) Effect of various weed control methods on the performance of canola. Sarhad Journal of Agriculture 17: 321-324.

27. Marwat KB, Saeed M, Gul B, Hussain Z, Khan NI (2005) Efficacy of different pre- and post-emergence herbicides for weed management in canola in higher altitude. Pakistan Journal of Weed Sciences Research, 11: 165-170.

28. Jafari R, Rezai S, Shakarami J (2013) Evaluating effects of some herbicides on weeds in field bean (Phaseolus vulgaris L.). International Research Journal of Applied and Basic Science 6: 1150-1152.

29. Masoumeh Y, Das TK, Sharma AR (2013) Effect of tillage and tank mix herbicide application on weed management in soybean (Glycine max). Indian Journal of Agronomy 58: 372-378.

30. Adigun JA, Lagoke STO, Kumar V, Erinle ID (1994) Effect of intra-row spacing, nitrogen level and period of weed interference on growth and yield of transplanted tomato (Lycopersicom esculentum Mill.) in the Nigerian savanna. Samaru Journal of Agricultural Research 11: 31-42.

31. Joseph A, Osipitan AO, Segun TL, Raphael OA, Stephen OA (2014) Growth and Yield Performance of Cowpea (Vigna Unguiculata (L.) Walp) as Influenced by Row Spacing and Period of Weed Interference in southwest Nigeria. Journal of Agricultural Science 6: 188-198.

32. Mohamed ES, Nourali AH, Mohammad GE, Mohamed MI, Saxena MC (1997) Weeds and weed management in irrigated lentil in northern Sudan. Weed Research 37: 211-218.

33. Stork PR (1998) Bio-efficacy and leaching of controlled-release formulations of triazine herbicides. Weed Research 38: 433-441.

34. Shinde SH, Pawar VS, Suryawanshi GB, Ahire NR, Surve US (2003) Integrated weed management studies in pigeonpea+pearl millet intercropping (2:2) system. Indian Journal of Weed Science 35: 90-92.

35. Priya G, Thomas G, Rajkannan B, Jayakumar R (2009) Efficacy of weed control practices in soybean crop production. Indian Journal of Weed Science 41: 58-64.

36. Sylvestre H, Kalyana MKN, Shankaralingappa CR, Devendra MTS, Ramachandra C (2013) Effect of pre- and post-emergence herbicides on weed dynamics, growth and yield of soybean (Glycine max L.). Advances in Applied Science Research 4: 72-75.

37. Radivojevic L, Santric L, Stankovic-Kalezic R, Janjic V (2004) Herbicides and soil microorganisms. Pesticides 32: 475-478.

38. Mathew R, Khan SU (1996) Photodegradation of metolachlor in water in presence of soil mineral and organic constituents. Journal of Agriculture and Food Chemistry 44: 3996-4000.

39. Chauhan BS, Gill GS (2009) Application of timing affects s-metolachlor bioavailability in soil. Indian Journal of Weed Science 41: 54-57.

40. Mondal DC, Hossain A, Duray B (2005) Chemical Weed control in onion (Allium cepa L.) under lateritic belt of West Bengal. Indian Journal of Weed Science 37: 281-282.

41. Warade AD, Gonge VS, Jog-Dande ND, Ingole PG, Karunakar AP (2007) Integrated weed management in onion. Indian Journal of Weed Science 2: 205-208.

42. Pradhan AC, Pravir C (2010) Quality Wheat Seed Production through Integrated Weed Management. Indian Journal of Weed Science 42: 159-162.

43. Abouziena HF, Abd El Wahed MSA, Eldabaa MAT, El-Desoki ER (2013) Effect of Sowing Date and Reduced Herbicides Rate with Additives on Peanut (Arachis hypogaea L.) Productivity and Associated Weeds. Journal of Applied Sciences Research 9: 2176-2187.

44. Ayaz S, McNeil DL, McKenzie BA, Hill GD (2001) Density and sowing depth effects on yield components of grain legumes. Proceeding of Agronomy Society, New Zealand, 29: 9-15.

45. Mirshekari B (2008) Time interaction effect of (Amaranthus retroflexus L.) weed on yield of cowpea [Vigna unguiculata (L.) Walp.]. Knowledge of Modern Agriculture Journal 4: 71-81.

46. Dadari SA (2003) Evaluation of herbicides in cowpea or cotton mixture in Northern Guinea Savannah. Journal of Sustainable Agriculture and Environment 5: 153-159.

47. Paudel L, Bishnoi UR, Kegode GO, Cebert E (2008) Influence of Timing of Herbicide Application on Winter Canola Performance. World Journal of Agricultural Science 17: 908-913.

48. Mousavi M (2009) Performance evaluation of herbicides for weed control imazethapyr red beans and pinto beans. Master's thesis and combat weeds, University of Shushtar Branch.

49. Prakash V, Pandey AK, Singh RB, Mani VP (2000) Integrated weed management in garden pea under mid hills of northwest Himalayas. Indian Journal of Weed Science 32: 7-11.

50. Tenaw W, Beyenesh Z, Waga M (1997) Effect of variety, seed rate and weeding frequencies on weed infestation and grain yield of haricot bean. In: Proceeding of the 2th and 3th Annual Conference of the Ethiopian Weed Science Society, Addis Ababa, Ethiopia, p: 61.

51. Sharma GD, Sharma JJ, Sood S (2004) Evaluation of alachlor, metolachlor and pendimethalin for weed control in rajmash (Phaseolus vulgaris L.) in cold desert of northwestern Himalayas. Indian Journal of Weed Science 36: 287-289.

52. Muhammad S, Abdur R, Noor UA, Fazaliwahid IJ, Imtiaz A, et al. (2012) Effect of herbicides and row spacing on the growth and yield of pea. Pakistan Journal of Weed Science Research 18: 1-13.

53. Khan IA, Hassan G, Ihsanulla H (2003) Efficacy of pre-emergence herbicides on the yield and yield components of canola. Asian Journal of Plant Science 2: 251-253.

54. Mohammadi GH, Javanshir RA, Rahimzadeh KF, Mohammadi A, Zehtab SS (2004) Effect of weed interference on growth of shoots and roots and harvest index of chickpea. Iranian Journal of Crop Sciences 6: 181-191.

55. Sana UC, Javed I, Muzzammil H (2011) Weed management in chickpea grown under rice based cropping system of Punjab. Crop and Environment 2: 28-31.

56. Singh RP, Singh UP (1998) Effect of weed management practices on yield and economics of crops under upland rice (Oryza sativa)-based cropping system. Indian Journal of Agronomy 43: 213-218.

57. Mohammadi G, Javanshir A, Khooie FR, Mohammadi SA, Salmasi SZ (2005) Critical period of weed interference in chickpea. Weed Research 45: 57-63.

58. Mizan A, Sharma JJ, Gebremedhin W (2009) Estimation of Critical Period of Weed-Crop Competition and Yield Loss in Sesame (Sesamum indicum L.). Ethiopian Journal of Weed Management 3: 39-53.

59. Rao AS, Murthy KVR (2004) Effect of sequential application of herbicides on nutrient uptake by rice fallow black gram. The Andhra Agriculture Journal 50: 360-362.

60. Begum G, Rao AS (2006) Efficacy of herbicides on weeds and relay crop of black gram. Indian Journal of Weed Science 38: 143-147.

61. Grafton KF, Schneiter AA, Nagle BJ (1988) Row spacing, plant population and genotype x row spacing interaction effects on yield and yield components of dry bean. Agronomy Journal 80: 631-634.

62. Townley SL, Wright AT (1994) Fieldpea cultivars and weed response to crop seed rate in western Canada. Canadian Journal of Plant Science 74: 387-393.

63. Askew SD, Wilcut JW, Cranmer JR (2002) Cotton (Gossypium hirsutum L.) and weed response to flumioxazin applied pre-plant and post-emergence directed. Weed Technology 16: 184-190.

64. Morad, S. 2013. Effect of cultivation time and weeds control on weeds and some characteristics of broad bean (Vicia faba L.). Advanced Agricultural Biology 1: 51-55.

65. Sangakkora UR (1999) Effect of weeds on yield and seed quality of two tropical grain legumes. Tropical Science 39: 227-232.

66. Tanveer A, Ali A (2003) Weeds and their control. Published by Higher Education Commission, Islamabad-Pakistan, p: 162.

67. Arnold RN, Murray MW, Gregory EJ, Smeal D (1996) Weed control in pinto beans with imzethapyr alone or in combination with other herbicides. New Mexico State University.

68. Aslam M, Ahmed HK, Ahmad E, Ullah H, Khan MA, et al. (2007) Effect of sowing methods and weed control techniques on yield and yield components of chickpea. Pakistan Journal of Weed Science Research 13: 49-61.

69. Soltani N, Bowiey S, Sikkema P (2005) Responses of Blach and Cranberry beans (Phaseolus vulgaris L.) to post-emergence herbicides. Crop Protection 24: 15-21.

70. Singh G, Chaunhan RS (1978) Weed management in upland paddy. Indian Journal of Weed Science 10: 83-86.

71. Gupta PC, O'Toole JC (1986) Upland rice: global perspective. The International Rice Research Institute, Los Banos, Philippines p: 372.

72. Ismaila U, Kolo MGM, Gbanguba UA (2011) Efficacy and Profitability of Some Weed Control Practices in Upland Rice (Oryza sativa L.) at Badeggi, Nigeria. American Journal of Experimental Agriculture 1: 174-186.

18

Pre-harvest Microbial Contamination of Tomato and Pepper Plants: Understanding the Pre-harvest Contamination Pathways of Mature Tomato and Bell Pepper Plants Using Bacterial Pathogen Surrogates

Seelavarn Ganeshan[1] and Hudaa Neetoo[2]*

[1]*Mauritius Sugar Industry Research Institute, Réduit, Moka, Mauritius*
[2]*Faculty of Agriculture, University of Mauritius, Réduit, Moka, Mauritius*

Abstract

Tomatoes and bell peppers have been previously incriminated in outbreaks of foodborne illnesses due to contamination by human pathogens such as *E. coli* O157:H7 and *Listeria monocytogenes* in the field. The objectives of the present study were to investigate (i) the potential entry of *E. coli* (EC) and *L. innocua* (LI) from soil to various non-edible and edible parts of the tomato and pepper plants, and (ii) the ability of EC and LI to survive in the plant environment (soil, rhizosphere and phyllosphere). Mature tomato and bell pepper plants cultivated in a greenhouse were soil-inoculated with a bacterial suspension (ca. 10^8 cfu/ml) of EC or LI. Tomatoes and peppers were also artificially contaminated on the surface with 1 ml of an overnight culture of EC and LI (ca. 10^9 cfu/ml). Samples of vegetables as well as non-edible parts (soil, roots, stem, foliage) were subjected to microbiological analyses by plating on Eosin Methylene Blue Agar and Listeria Identification Agar to recover EC and LI respectively. Although these bacteria were recovered at population densities of 3.0-3.6, 1.8-2.2 and <0.7 log cfu/g in the bulk soil, roots and foliage respectively, we were unable to recover these bacteria from the edible tomato and pepper fruits. When tomatoes and peppers were spot-inoculated on the surface with EC or LI, the vegetables analyzed were shown to harbor viable bacterial cells for up to 48 h after inoculation. Overall, the potential for systemic uptake and translocation of human pathogens from soil to the edible plant parts was found to be negligible in tomato and pepper plants. However, overhead (spray or sprinkler) irrigation with contaminated water could create opportunities for the deposition and subsequent persistence of human pathogens on the edible surface of vegetables even after harvest. These findings therefore underscore the need for adoption of Good Agricultural Practices (GAPs) by growers and Good Manufacturing Practices (GMPs) by post-harvest handlers of fresh produce.

Keywords: Contamination; *Escherichia coli*; *Listeria innocua*; Tomato; Pepper; *Ralstonia solanacearum*; *Pseudomonas fluorescens*

Introduction

Fresh vegetables contain rich sources of many nutrients and provide numerous health benefits, so nutritionists and health professionals highly recommend increasing consumption of these important foods [1]. Tomatoes and peppers represent some of the vegetables that are most commonly consumed in the raw state. However, these vegetables have also been the source of recent outbreaks of foodborne illnesses in developed countries, which have caused sickness, hospitalizations, and deaths of consumers, as well as serious adverse economic impact on growers and processors [2]. Since 1990, up to 15 outbreaks of salmonellosis have been linked to the consumption of fresh tomato fruits in developed countries such as the United States [3]. Trace-back investigations of outbreaks linked to tomatoes have concluded that the fruits were generally contaminated in the field [4]. Suggested sources ranged from animals in nearby pastures or wetlands to water used for irrigation or pesticide applications [4]. Orozco et al. [5] detected *Salmonella* in 1.8% of tomatoes grown hydroponically in a greenhouse prior to an extreme weather event during which time floodwaters entered several of the houses. Immediately after the floodwaters had disappeared, the contamination rate increased to 9.4% [5]. Bell peppers also represent a major world commodity by virtue of their high content in vitamin A and C as well as the presence of the compound responsible for the irritation ("hotness") called capsaicin [6]. The production of hot and sweet peppers for vegetable uses has increased by more than 21% since 1994 [6]. Peppers are commonly used fresh in condiments, sauces, salads, meats and vegetable dishes [6]. Unfortunately, peppers form increasingly recognized vehicles for transmission of foodborne pathogens [7]. A study conducted on the prevalence of *Salmonella* in peppers showed that 10 out of a total of 27 samples from a pepper production system tested positive for *Salmonella* and were identified as either *Salmonella enterica* serovar Typhimurium (91% of 54 cases) or *Salmonella enterica* serovar Enteritidis (9% of cases) [6].

Given the high frequency of microbial contamination of raw tomatoes and peppers, there has been a concern regarding the potential for human pathogens to become internalized within plant tissue [8]. In the current study, tomato and pepper plants, belonging to the family of *Solanaceae*, were used as model host systems to study their susceptibility to uptake and persistence of bacterial human pathogens. *E. coli* O157:H7 is one of the most common zoonotic enteric pathogens associated with vegetables given its widespread presence in animal manure used in produce cultivation [9]. *Listeria monocytogenes* on the other hand, is a common geophilic (soil-borne) bacterium and is ubiquitous in vegetation [10]. In addition, the role of plant commensal bacteria such as plant pathogen *Ralstonia solanacearum* and plant beneficial bacteria *Pseudomonas fluorescens* in enhancing or hindering internalization of human pathogens in vegetables is of equal interest.

***Corresponding author:** Hudaa Neetoo, Faculty of Agriculture, University of Mauritius, Réduit, Moka, Mauritius, E-mail: s.neetoo@uom.ac.mu

Indeed, previous research has suggested that bacterial plant pathogens can enhance infiltration or internalization of human pathogens in the roots, leaves and fruits of food crops. Moreover, *P. fluorescens* represents one of the most abundant soil resident species that usually confer several benefits to the plants. It is thus hypothesized that the presence of phyto-pathogenic species such as *R. solanacearum* might enhance the uptake of human pathogens in plants due to ability of *R. solanacearum* to produce plant lesions and wounds which may act as sites of co-infection by human pathogens. Plant pathogens may also have the ability to depress the defense mechanisms of plants, thus enhancing colonization and persistence of human pathogens. On the other hand, it is hypothesized that non-pathogenic *P. fluorescens* will discourage uptake or internalization of human pathogens since literature has shown that it acts as an excellent plant competitor against non-resident human pathogenic bacteria.

The objectives of the present study were therefore to: (i) investigate the potential uptake, infiltration or internalization of bacterial human pathogens from soil into the edible parts of tomato and bell pepper plants, (ii) investigate the influence of plant pathogen and plant beneficial bacteria on the uptake or internalization of human pathogens and (iii) investigate the survivability of human pathogens in the soil, rhizosphere and phyllosphere of tomato plants.

Materials and Methods

Assessing the potential for systemic uptake of *E. coli* and *L. innocua* in tomato and pepper plants

Soil sterilization: The oven was preheated to 82-88°C (180°-190°F). Ten kg of soil was spread evenly in a large pan to a maximum depth of 10 cm. The pan was sprayed with water to moisten slightly and then covered tightly with aluminum foil. At the center of the covered baking pan, a thermometer probe was inserted into the soil and the pan placed into the oven. Once the soil temperature reached 82-88°C, the temperature was maintained for 60 minutes following which the pan was removed from the oven and allowed to completely cool. Once cooled, soil was transferred to clean gunny bags. Given the limited capacity of the oven, multiple cycles were run to sterilize several batches of soil.

Plant preparation: Tomato (*Solanum lycopersicum* var. St Pierre) and bell pepper (*Capsicum annum* var. Nikita) seeds were used. Briefly, seeds were disinfected with 70% ethyl alcohol (EtOH) for 3 min, rinsed in sterile water, and soaked in Javel commercial bleach (0.525% sodium hypochlorite) for 15 min. Seeds were then rinsed in sterile water three times (5 min each rinse). Subsequently, they were sowed in steam-sterilized soil contained in Styrofoam plug trays and grown in a Biosafety

Level 1 (BSL-1) greenhouse located at the Mauritius Sugar Industry and Research Institute, Reduit. Plants were watered on a daily basis with sterile water. Seedlings were transplanted at 2 weeks of age to potting bags containing steam-sterilized soil (⊠1 kg) placed in plastic saucers to serve as a water reservoir for indirect irrigation. The pH and water activity of the soil were regularly monitored with a pH meter (Mettler Toledo) and a water activity meter (Novasina) respectively. Over the period of October 2013 to December 2014 chamber temperatures ranged from 21 to 32°C (daytime) and 12 to 23°C (nighttime) and the relative humidity varied between 65 to 81%. The saucer was refilled with ca. 50 ml sterile water daily. Additionally, the soil was supplemented with 'Terreau' or peat (Stender) as per the manufacturer's instructions to maintain plant growth, to speed up harvest time and increase yields.

Experiment Design: Two plant types (tomato and pepper) were investigated in this part of the study. The plants were given one of 7 treatments (Sterile water, EC, EC + R, EC + P, LI, LI +R, LI + P) where EC, LI, P and R stand for *Escherichia coli, Listeria innocua, Pseudomonas fluorescens* and *Ralstonia solanacearum* respectively. Each inoculation treatment was carried out in duplicates. The plants were grown in two separate batches. A total of 56 plants (7 treatments × 2 plant types × 2 plants per treatment × 2 batches) were considered. The different treatments given to the plants are summarized in the (Table 1).

Soil Inoculation:

Bacterial cultures: *E. coli* ATCC 25922 strain was provided by the Food Technolog Laboratory of the Ministry of Agro-Industry and Food Security of Mauritius. The strain was plated onto Eosin Methylene Blue medium (HiMedia) and incubated for 24 h at 37°C for confirmatory identification of *E. coli*. *Pseudomonas fluorescens* ATCC 13525 (Microbiologics Ltd) and was revived on Pseudomonas CFC medium. Colonies that were straw coloured with a greenish tinge were presumed to be *P. fluorescens* and confirmed by oxidase and catalase tests. *Listeria innocua* ATCC 33090 (Microbiologics Ltd) and was revived on Polymyxin Acriflavin Lithium-Chloride Ceftazidime Aesculin Mannitol (PALCAM) medium (HiMedia). Olive green colonies with dark sunken centers and black haloes were confirmed to be *L. innocua*. *L. innocua* hydrolyzes aesculin to form aesculetin and dextrose. Aesculetin reacts with ammonium ferric citrate and forms a brown-black complex seen as a black halo around colonies. An environmental isolate of *Ralstonia solanacearum* was generously provided by Dr S. Ganeshan, from the Mauritius Sugar Industry Research Institute. The isolate was obtained from the ooze of a tomato plant suffering from bacterial wilt disease. The isolate was plated onto triphenyl tetrazolium chloride (TTC) medium (Sigma) and incubated overnight at 27°C. Strains were stored at -80°C in glycerol stocks.

TREATMENTS	DETAILS OF INOCULATION OF POTTED VEGETABLE PLANTS
Water	Addition of 100 ml of sterile water to the potted vegetable
E	Inoculation of each potted vegetable type with 200 ml of diluted suspension of overnight culture of *E. coli* with a cell density of ca. 10^8 cfu/ml; twice a week
EC+P	Inoculation of each potted vegetable type with 200 ml of diluted suspension of overnight culture of *E. coli* with cell density of ca. 0^8 cfu/ml & 200 ml of diluted suspension of overnight culture of *P. fluorescens* with cell density of ca. 10^7 cfu/ml on alternate days; twice a week
EC+R	Inoculation of each potted vegetable type with 200 ml of diluted suspension of overnight culture of *E. coli* with cell density of ca. 10^8 cfu/ml & 200 ml of overnight culture of *R. solanacearum* with cell density of ca. 10^7 cfu/ml on alternate days; twice a week
LI	Inoculation of each potted vegetable type with 200 ml of diluted suspension of overnight culture of *L. innocua* with cell density of ca. 10^8 cfu/ml; twice a week
LI + P	Inoculation of each potted vegetable type with 200 ml of diluted suspension of overnight culture of *L. innocua* with cell density of ca. 10^8 cfu/ml & 200 ml of diluted suspension of overnight culture of *P. fluorescens* with cell density of ca. 10^7 cfu/ml on alternate days; twice a week
LI + R	Inoculation of each potted vegetable type with 200 ml of diluted suspension of overnight culture of *L. innocua* with cell density of ca. 10^8 cfu/ml & 200 ml of diluted suspension of overnight culture of *R. solanacearum* with cell density of ca. 10^7 cfu/ml on alternate days; twice a week

Table 1: Inoculation treatments of plants.

Inoculum preparation: The cells of the four cultures were adapted to grow on Plate Count Agar supplemented with 100 μg/ml of nalidixic acid (Sigma) (PCA-N) to select for Nalidixic-acid (NA) resistant strains of *E. coli, L. innocua, P. fluorescens* and *R. solanacearum*. NA-resistant mutant strains were subsequently transferred on fresh Plate Count Agar supplemented with 100 μg/ml of NA and plates incubated overnight at 35°C to yield solid cultures. Stock cultures of NA resistant strains of *E. coli, L. innocua, R. solanacearum* and *P. fluorescens* were also stored in TSB-N broth containing 25% glycerol (Sigma) at −18°C. To prepare liquid cultures, a single colony of each NA-resistant strain was transferred to 200 ml of tryptic soy broth (TSB-N) and placed on an orbital shaker at 35°C for 18 h.

Soil inoculation of plants: On the day of inoculation of the plants, 100 ml of each culture was mixed with 900 ml of sterile water (10-fold dilution of an overnight culture) to serve as the inoculum for the plants. The concentration of each culture was determined by serial dilution and plating on PCA-N. In addition, the population density of *E. coli* and *L. innocua* recovered from the soil immediately after inoculation was also determined. Various treatments were given to the tomato and pepper plants upon fruit set (Table 1). Plants serving as negative controls were treated with sterile water. Tomato plants were staked and strung to bamboo sticks to ensure upright growth. All plants were watered once or twice daily as needed.

Microbiological analysis of vegetables at harvest: Vegetables reaching commercial maturity were harvested by plucking tomato and pepper fruits. Tomatoes and peppers were blended with 0.1% Buffered Peptone Water at a 1:4 ratio. Vegetable samples were blended with 0.1% Buffered Peptone Water at a 1:4 ratio. Vegetables were macerated for 10 minutes into a slurry. The slurry and its serial dilutions were then plated onto Eosin Methylene Blue agar or PALCAM agar supplemented with 100 μg/ml of Nalidixic acid and plates incubated at 44 or 35°C respectively for 48 h. In addition, primary samples suspected to be contaminated with *E. coli* or *L. innocua* were subjected to primary enrichment in Lauryl Tryptose broth (LTB) and Half-Fraser broth respectively and incubated at 44 and 35°C for 24 h. Broths were supplemented with NA to a final concentration of 100 ug/ml. Aliquots of LTB and Half-Fraser Broth were then transferred for secondary enrichment into EC and Fraser broths supplemented with NA, and incubated at 44°C and 35°C for 24 h respectively. A loopful of secondary enrichment broth was then streaked onto EMB-N or PALCAM-N and plates incubated at 44 or 35°C respectively for 24 h. Colonies with characteristic green metallic sheen on EMB-N or olive green colonies with a surrounding black halo on PALCAM-N were presumed to be Nalidixic-acid resistant *E. coli* or *L. innocua* respectively.

Assessing the translocation potential of *E. coli* and *L. innocua* into different sections of the tomato plant

This experiment was conducted to investigate the translocation potential of soil-inoculated *E. coli* and *L. innocua* into different parts of the tomato plants (*S. lycopersicum* cv. St Pierre). Mature tomato plants (past fruit set) were soil-inoculated with 200 ml of a 10-fold dilution of a late-log phase culture of NA-resistant *E. coli* or *L. innocua*. The population density of the suspension was ca. 8 log cfu/ml. After 24 h, the plants were cut into 3 sections: the roots, stems and foliage.

Assessing the persistence of *E. coli* and *L. innocua* in rhizosphere soil

Soil microcosms were set up consisting of a polypropylene tray containing 2 kg (dry wt) of soil mixed with live roots of an un-inoculated tomato plant. Initial water activity of the soil-roots mix was ca. 0.3. The microcosm was inoculated with 200 ml of a suspension of NA-resistant *E. coli* or *L. innocua* having a cell density of ca. 10^8 cfu/ml resulting in a theoretical final population density of ca. 10^7 cfu/g of soil. The inoculum was homogeneously stirred into the soil-roots mix and the microcosm covered with aluminum foil. Microcosms were incubated in the dark at 25°C for 7 days with daily addition of 100 ml of sterile water. Soil was collected daily and subjected to microbiological, water activity and pH analyses. In order to determine the population density of bacteria present in the microcosms at daily intervals, about 25 g of soil was taken and mixed with 225 ml of 0.1% buffered peptone water in a sterile stomacher bag. This soil suspension was ten-fold serially diluted in 0.1% buffered peptone water and plated on EMB-N and PALCAM-N agar. Plates were subsequently incubated for up to 48 h at 44°C and 35°C respectively. Soil water activity and pH were determined using a dew point water-activity meter (Novasina) and a pH meter (Mettler-Toledo) respectively.

Assessing the survivability of *E. coli* and *L. innocua* on the surface of tomato and pepper fruits

Tomato and pepper plants were cultivated as described previously. At fruit set, a spot inoculation method was used to artificially contaminate the tomatoes and peppers since it allows the deposition of a known amount of cells onto the surfaces, regardless of weight/size. A total of 54 tomatoes and 30 peppers were used for the spot-inoculation study. Mature red ripe tomato and pepper fruits were spot-inoculated with 1000 ul of late-log phase cultures of Nalidixic-acid resistant *L. innocua* or *E. coli* on the pericarp and calyx using an appropriate micropipettor. In addition, tomatoes and peppers were also spotted with sterile water as a negative control. Tomatoes and peppers were aseptically harvested after 24 h and 48 h by plucking the fruits together with the stem or peduncle. After aseptically removing the peduncle and calyx, each fruit was then placed in an individual sterile Whirl-Pak filter bag containing 40 ml of 0.1% BPW. To recover bacteria from the surface of fruits, each tomato or pepper fruit was gently hand-massaged for 2 min, and then the rinsate was diluted 10-fold in 0.1% Buffered Peptone Water, and 0.1-ml aliquots of the appropriate dilutions were spread-plated onto EMB-N or PALCAM-N. Plates were incubated and enumerated after 24 h as described previously.

Results and Discussion

Translocation of *E. coli* and *L. innocua* in tomato and pepper plants

In this part of the study, *E. coli* ATCC 25922 and *L. innocua* ATCC 33090, non-pathogenic surrogate microorganisms were used in lieu of the enteric pathogens *Salmonella* or *E. coli* O157:H7 and the ubiquitous soil-borne pathogen *L. monocytogenes* respectively, to avoid introduction of pathogenic agents in the BSL-1 greenhouse. Other authors including Ingham et al. [11] and Wood et al. [12] have also resorted to non-pathogenic surrogates to circumvent this limitation. Examples of surrogates that have been used *in planta* studies include *E. coli* Shiga toxin-negative *E. coli* O157:H7 [9,13], *Listeria innocua* [14], and avirulent *Salmonella* [9]. In using these surrogates, the assumption has been made that they would respond similarly as the pathogenic agent.

(Table 2) summarizes the results obtained for the soil-inoculation experiment of tomato plants. The population density of *E. coli* and *L. innocua* recovered from all tomato fruits was below the limit of detection of the plating methodology (<1.7 log cfu/g) and the bacteria were not

Bacterial Human Pathogen Surrogates (BHPS)	Inoculum Level of BHPS (log cfu/ml)	Plant Commensal Bacteria (PCB)	Inoculum Level of PCB (log cfu/ml)	BHPS Population in fruits (log cfu/g)	# Presumptive Positive Samples/ Total Samples
------	0	------	0	< 1.7	0/22
EC	8	------	0	< 1.7	**2/34**
EC	8	RS	7	< 1.7	0/24
EC	8	PF	7	< 1.7	**1/35**
LI	8	------	7	< 1.7	0/18
LI	8	RS	7	< 1.7	0/15
LI	8	PF	7	< 1.7	0/17

Table 2: Internalization rate of *E. coli* (EC) and *L. innocua* (LI) in tomato fruits via soil.

Bacterial Human Pathogen Surrogates (BHPS)	Inoculum Level of BHPS (log cfu/ml)	Plant Commensal Bacteria (PCB)	Inoculum Level of PCB (log cfu/ml)	BHPS Population in fruits (log cfu/g)	# Presumptive Positive Samples/ Total Samples
------	0	------	0	< 2.2	0/9
EC	8	------	0	< 2.2	0/17
EC	8	RS	7	< 2.2	0/12
EC	8	PF	7	< 2.2	1/18
LI	8	------	7	< 2.2	0/11
LI	8	RS	7	< 2.2	0/12
LI	8	PF	7	< 2.2	0/20

Table 3: Internalization rate of *E. coli* (EC) and *L. innocua* (LI) in pepper fruits via soil.

detected after enrichment and streaking in most of the tomato samples tested except for three samples highlighted in bold. These suspect *E. coli* isolates originating from three tomato samples yielded negative results upon biochemical identification, thus confirming their absence. (Table 3) indicates that similar to tomato fruits, *E. coli* and *L. innocua* were also undetectable (<2.2 log cfu/g) by plating in pepper fruits following artificial contamination of the soil. In other words, our study failed to demonstrate the translocation of these bacteria from soil to fruits despite optimizing the cultivation conditions to promote uptake of the inoculated bacteria in the plant. In our study, *E. coli* ATCC 25922 and *L. innocua* ATCC 33090 were used as non-pathogenic surrogates to mimic *Salmonella* spp. or *E. coli* O157:H7 and *L. monocytogenes* respectively. Similar to our findings, other authors have also reported the inability to detect *Salmonella* in tomatoes that have been artificially contaminated with the microorganisms via soil [15,16]. Contrary to our findings however, Zheng et al. [17] has shown that *Salmonella* was capable of internalizing in tomato plants through the roots provided there are favorable conditions for this to occur. Zheng et al. [17] also indicated that uptake of *Salmonella* through the roots of *S. lycopersicum* Micro-Tom grown in sandy loam soil led to the contamination of developing tomato fruits. The authors further noted that fruit contamination rate was much higher with *Salmonella* introduction through flowers (70.4%) than through the rhizosphere (5.5%). Hence, the phenomenon of *Salmonella enterica* internalizing tomato plants through the root system remains a largely controversial issue.

Tables 2 and 3 also compared the translocation potential of *E. coli* and *L. innocua* in the presence of plant pathogen *Ralstonia solanacearum* and plant beneficial bacterium *Pseudomonas fluorescens*. *R. solanacearum* is a soil-borne pathogen that infects the roots of plants including tomatoes and peppers leading to bacterial wilt disease. Good Agricultural Practices (GAP) guidelines urge growers not to harvest fruits from diseased plants infected by plant pathogens in fear that the plant's compromised immune system would make them more susceptible to human pathogens such as *S. enterica*, *E. coli* O157:H7 or even *L. monocytogenes* [18]. Indeed, *R. solanacearum* when added to soil has the ability to infect the plant through natural openings

or through wounds in the roots [18], thus potentially increasing the chances for ingress of human pathogens. In this study, the influence of *R. solanacearum*, a plant pathogen, on the uptake of pathogen surrogates in food crop plants was thus of interest. (Tables 3 and 4) indicate that systemic uptake of *E. coli* and *L. innocua* from roots to fruits did not occur in the presence of either plant pathogen *R. solanacearum* or plant beneficial bacterium *P. fluorescens*. Contrary to our findings,

Population density (log cfu/g) of *E. coli* on the surface of tomatoes			
Sample ID	Day 0	Day 1	Day 2
Sample 1	7.8	3.6	< 0.7 (-)
Sample 2	7.7	3.1	< 0.7 (-)
Sample 3	8.4	3.7	< 0.7 (-)
Sample 4	8.6	4.3	< 0.7 (+)
Sample 5	7.2	4.4	< 0.7 (+)
Sample 6	8.8	3.7	< 0.7 (-)
Sample 7	8.2	3.0	< 0.7 (-)
Sample 8	7.7	3.0	< 0.7 (-)
Sample 9	8.3	4.1	< 0.7 (-)
Mean	8.1 ± 0.48	3.6 ± 0.51	< 0.7 (2/9)

Table 4a: Survival of *E. coli* spot-inoculated on tomatoes.

Population density (log cfu/g) of *L. innocua* on the surface of tomatoes			
Sample ID	Day 0	Day 1	Day 2
Sample 1	7.4	< 0.7	< 0.7 (-)
Sample 2	7.7	< 0.7	< 0.7 (-)
Sample 3	8.3	< 0.7	< 0.7 (-)
Sample 4	7.6	< 0.7	< 0.7 (-)
Sample 5	7.2	< 0.7	< 0.7 (-)
Sample 6	8.1	4.96	< 0.7 (-)
Sample 7	7.4	< 0.7	< 0.7 (+)
Sample 8	8.8	< 0.7	< 0.7 (-)
Sample 9	8.2	< 0.7	< 0.7 (-)
Mean	8.2 ± 0.49	1.2 ± 0.00	< 0.7 (1/9)

Table 4b: Survival of *L. innocua* spot-inoculated on tomatoes.

Pollard et al. [19] demonstrated that *R. solanacearum* could enhance *S. enterica* survival and its transportation throughout the internal tissues of tomato plants, causing an increase in *S. enterica* populations on plants [19]. This is because phytopathogenic bacteria, such as the wilt pathogen *R. solanacearum*, have the ability to digest pit membranes, having pores of about 0.3 um [20], allowing water to move freely from the stem into a petiole [21-30]. Indeed, certain laboratory models have demonstrated internalization of wilt pathogen *Ralstonia solanacearum* by tomato roots and then movement up the xylem of the plant [4]. Overall, findings of the current work indicate that the presence of a prototypic plant pathogen exemplified by *R. solanacearum* and a typical beneficial plant bacterium such as *P. fluorescens* did not have any effect on the susceptibility of tomato and pepper plants to uptake of bacterial human pathogens. According to Van der Schoot [22], certain cultivars of *Solanaceae* may possess a type of resistance against wilt pathogens rendering their pit membranes resistant to digestion. Resistance to infection by the plant pathogen or resistance to colonization by the plant beneficial bacteria could have explained the inability to detect any of the plant commensal bacteria or pathogenic surrogates.

Translocation of *E. coli* and *L. innocua* to different sections of the tomato plant

The localization and population density of *E. coli* and *L. innocua* in different parts of the tomato plant is depicted in (Figure 1). Our study indicated that *E. coli* and *L. innocua* were recovered from bulk soil and roots at population densities of 3.0-3.6 log cfu/g and 1.8-2.2 log cfu/g respectively 24 hrs post-inoculation. However they were undetectable (<0.7 log cfu/g) in the main stem and foliage (fruits, flowers, stemlets, petiole and leaves) of the tomato plant. Jablasone et al. [23] similarly applied water contaminated with *Salmonella* directly onto the soil of pots containing tomato plants (*S. lycopersicum* cv. Cherry Gold) and also could not recover *Salmonella* from the stems or fruits of the tomato plant although populations in the soil ranged from 2.3 to 3.7 log cfu/g. In addition, another study found no evidence of *Salmonella enterica* serovar Montevideo on the stems, leaves, or fruit of tomato plants (*S. lycopersicum* L. cv. Trust) when soil-inoculated with contaminated water [4]. This is very similar to our data where plants artificially soil-contaminated with *E. coli* and *L. innocua* did not show evidence of translocation of the bacteria to the aerial parts of the plant. However, presence of *E. coli* and *L. innocua* in the bulk soil as well as in the roots was observed as indicated in Figure 1. Contrary to our findings where we observed a relatively lower population of these bacteria on roots (1.8-2.2. log cfu/g) than in the bulk soil (3.0-3.6 log cfu/g), Semenov et al. [24] found the densities of *S.* Typhimurium and *E. coli* O157:H7 in bulk soil and rhizosphere (roots) to be similar following addition of manure to soil. Similarly, Habteselassie et al. [25] found comparable numbers of *E. coli* cells in bulk and rhizosphere soil when manure was added to pots in which lettuce was being grown. Overall, presence of human pathogens in the bulk soil and on the rhizoplane may not necessarily guarantee entry of the bacteria through the roots to the aerial parts of the plant such as the leaves, flowers or fruits. Contrary to our observation, Hintz et al. [16] reported that repeated application of *Salmonella enterica* serovar Newport to the root zone via irrigation water has the potential to contaminate various tissues of the tomato plant *Solanum lycopersicum* cv. Solar Fire. Likewise, Zheng et al. [17] demonstrated that of 22 tomato plants grown with *Salmonella*-infested soil, 22% (4 out of 18) contained endophytically colonized *Salmonella* based on direct plating or enrichment procedures, including two stem samples (11.1%), one leaf sample (5.5%), and one fruit sample (5.5%). *S. enterica* serovar Saintpaul was also isolated from a single positive leaf

sample and *S.* Newport was found on the surface and within the single positive tomato sample (5.5%).

Survival of *E. coli* and *L. innocua* in soil mixed with live roots

Plant roots are known to modify their immediate habitat by changing the soil porosity and clustering properties [26] and such physical alterations are likely to impact the microbial community near those roots (i.e., the rhizosphere community). Live roots release root exudates that have the potential to significantly affect the microbial population including the fate of pathogens in the rhizosphere of food crops [27]. These exudates serve as nutrient sources for the bacteria in the vicinity of the roots and could therefore promote the extended survival of pathogens in soil. Taking this into consideration, we thus designed a microcosm consisting of a mix of autoclaved soil and live roots, since previous research has shown that *E. coli* O157:H7 survived longer in rhizosphere soil compared to free soil [28].

The survival curves of *E. coli* and *L. innocua* in the soil-root mix is shown in Figure 2. Both bacterial species exhibited a slow decline from an initial population of 5.2-5.3 to <0.7 log cfu/g but persisted for up to 96 hours in the soil-roots mix. The death curve of *E. coli* had a

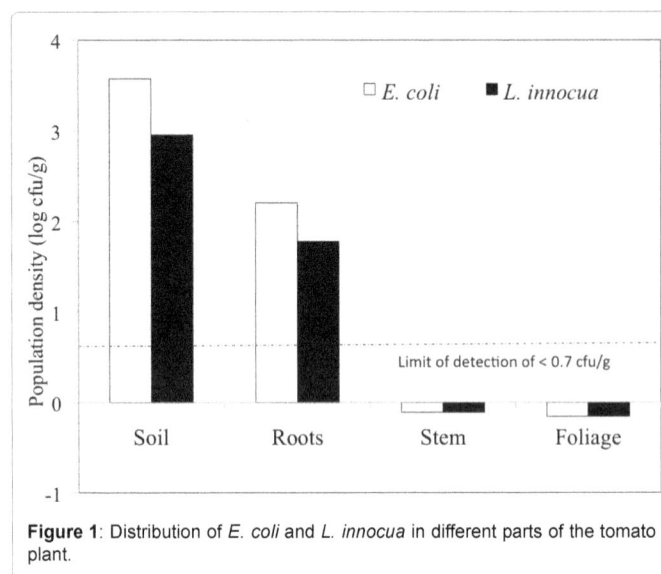

Figure 1: Distribution of *E. coli* and *L. innocua* in different parts of the tomato plant.

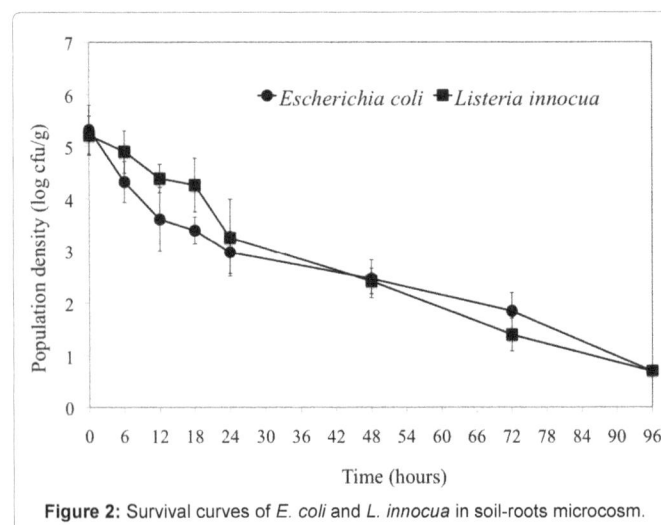

Figure 2: Survival curves of *E. coli* and *L. innocua* in soil-roots microcosm.

characteristic concave curvature with slightly higher death rate in the first 12 hrs. Islam et al. [29] indicated that survival curves generally exhibit a concave curvature with initial decreases that are log-linear. It is to be noted that conditions of the microcosm were particularly optimized to promote survival of the inocula in the soil-roots mix by protecting against dessication. This was achieved by daily watering with sterile water (soil a_w ~ 0.97-0.99) and shielding against UV radiation by covering with foil. Indeed, a critical factor influencing bacterial persistence in the soil is the moisture availability.

Literature has shown the variable persistence of different microorganisms in different agricultural niches [30]. Bell et al. [31] and Micallef et al. [32] indicated that *Salmonella* can also persist in the tomato-growing environment including the soil. Bernstein et al. [33] reported that *S.* Newport is capable of persisting in potting medium for 4.7 to 10 weeks. Even among *Samonella* serovars, there were considerable differences in their persistence; *S.* Newport and *S.* Javiana appeared to persist in sandy loam soil more efficiently than other serovars, including *S.* Montevideo, *S.* Saintpaul, and *S.* Typhimurium. In addition to *Salmonella*, enteric bacteria such as *E. coli* O157:H7 as well as other fecal microorganisms have been demonstrated to survive for extended periods in soils. Reported survival times of *E. coli* O157:H7, *E. coli* O26, *Salmonella*, *Listeria*, *Campylobacter* and *Cryptosporidium* in soil are up to 6 months, 3 years, 2 years, 20 days and 3 months respectively [34]. Indeed, there is considerable evidence to support the fact that pathogens can survive for widely varying periods of time in the soil and even on produce [33]. The relatively short survival times of *E. coli* and *L. innocua* noted in our study (≤ 4 days) could be due the high air temperatures (27-31°C) of the greenhouse during the experiment. Indeed lower survival rates were noted by Fremaux et al. [35] at higher air and soil temperatures. Semenov et al. [24] also reported that the survival of *S.* Typhimurium and *E. coli* O157:H7 declined with increasing mean soil temperature.

Survival of *E. coli* and *L. innocua* on the surface of tomato and pepper fruits

(Tables 4a and 4b) show the population density of *E. coli* and *L. innocua* recovered after 24 and 48 h from tomatoes that have been spot-inoculated with the bacteria. Our results show that tomatoes surface-contaminated with *E. coli* still harbored the bacteria after 24 h at varying density of 3.0-4.4 log cfu/g. However, after 48 h, *E. coli* was below the limit of detection of the plating methodology (<0.7 log cfu/g). Nevertheless, *E. coli* was still detected on the samples after enrichment and streaking in 2 out of 9 samples. *L. innocua* lost their viability quicker, dropping from an initial of 8.2 log cfu/g to a mean density of 1.2 log cfu/g after 24 h. After 48 h, *L. innocua* was detected in only 1 out of 9 samples.

Peppers were surface-inoculated with *E. coli* or *L. innocua* at a mean population density of 7.3 to 7.8 log cfu/g respectively. The population declined to 3.5-4.2 log cfu/g after 24 h; after 48 h the bacteria were undetectable by plating although *E. coli* was detected after enrichment in 6 out 14 samples (Table 5a). *L. innocua* on the other hand was undetectable in all samples tested after 48h (Table 5b). Taken together, our findings highlight the differential survival of *E. coli*, a zoonotic bacterium of an intestinal origin, and *L. innocua*, an environmental bacterium that predominantly resides in soil, on the surface of fruits. The relatively poor colonizing abilities of these bacteria as epiphytes could partly be attributed to the waxy cuticle and regular topography (smoothness) of the tomato and pepper exocarp. Guo et al. [4] also mentioned that bacteria can more readily colonize and penetrate fruit

Population density (log cfu/g) of *E. coli* on the surface of peppers			
	Day 0	Day 1	Day 2
Sample 1	8.5	4.8	< 0.7 (-)
Sample 2	8.2	4.7	< 0.7 (+)
Sample 3	7.9	4.6	< 0.7 (+)
Sample 4	8.0	4.6	< 0.7 (+)
Sample 5	7.4	3.9	< 0.7 (-)
Sample 6	7.6	4.2	< 0.7 (-)
Sample 7	8.4	4.0	< 0.7 (-)
Sample 8	7.2	4.1	< 0.7 (+)
Sample 9	8.1	4.5	< 0.7 (-)
Sample 10	7.4	4.4	< 0.7 (+)
Sample 11	7.6	< 0.7	< 0.7 (-)
Sample 12	7.8	2.8	< 0.7 (-)
Sample 13	8.0	4.2	< 0.7 (+)
Sample 14	7.5	3.5	< 0.7 (-)
Overall	7.8 ± 0.40	4.2 ± 0.54	< 0.7 (6/14)

Table 5a: Survival of *E. coli* spot-inoculated on peppers.

Population density (log cfu/g) of *L. innocua* on the surface of peppers			
	Day 0	Day 1	Day 2
Sample 1	7.9	4.1	< 0.7 (-)
Sample 2	6.8	3.0	< 0.7 (-)
Sample 3	7.4	3.0	< 0.7 (-)
Sample 4	7.2	4.9	< 0.7 (-)
Sample 5	6.9	3.6	< 0.7 (-)
Sample 6	6.7	5.2	< 0.7 (-)
Sample 7	7.6	2.0	< 0.7 (-)
Sample 8	7.5	3.0	< 0.7 (-)
Sample 9	6.9	2.0	< 0.7 (-)
Sample 10	7.9	< 0.7	< 0.7 (-)
Sample 11	7.6	< 0.7	< 0.7 (-)
Sample 12	6.8	< 0.7	< 0.7 (-)
Sample 13	7.9	3.2	< 0.7 (-)
Sample 14	7.2	2.8	< 0.7 (-)
Sample 15	6.9	4.2	< 0.7 (-)
Sample 16	7.1	4.8	< 0.7 (-)
Overall	7.3 ± 0.42	3.5 ± 1.10	< 0.7 (0/16)

Table 5b: Survival of *L. innocua* spot-inoculated on peppers.

tissue in the early stages of fruit development prior to deposition of the waxy materials. Erickson et al. [13] noted that *E. coli* O157:H7 cells had a greater propensity to attach to coarse, porous, or injured surfaces than uninjured smooth surfaces of green peppers. The smooth and topographically uniform surface of peppers is thought to be devoid of any microenvironments that can afford protection to the deposited inoculum. Hence, it is not surprising to observe a rapid decline in the bacterial population from an initial 7.8 log cfu/g to 4.2 and <0.7 log cfu/g after 24 h and 48 h respectively.

Vegetables can be indirectly contaminated in the field when the soil in which they are cultivated becomes contaminated for e.g., during drip-irrigation with contaminated water. In addition to drip-irrigation, vegetables can also be directly contaminated during overhead irrigation with contaminated water through splash dispersal of the bacteria onto the fruit surface [36]. Wei et al. [37] previously demonstrated the survival and growth of *Salmonella* deposited as an aqueous cell suspension on natural openings of the tomato fruit such as the stem scar. Contrary to Wei et al. [37], we noted that the inoculum deposited on the surface did not grow; instead the population declined rapidly to below detectable levels after 48 h post-inoculation. Wei et al. [37] mentioned that survival of the bacteria was most likely dependent on the inoculum size; when small populations of *S.* Montevideo of 2.8-3.9 log cfu/ml were placed on the smooth periderm of tomato fruits, none could be detected after overnight storage. However, when the concentration of inoculum was increased to 9.5 log cfu/ml, the bacterium could be detected up to three days later. In our experiment, a volume of 1 ml of the overnight culture having a cell density of ca. 9 log cfu/ml was aliquoted on the fruit resulting in the deposition of ca. 10^9 cells on the fruit. In spite of the high inoculum, the population rapidly declined to 3-4 log cfu/g and to undetectable levels after 24 and 48 h respectively. It has also been mentioned elsewhere that better survival of the inocula was observed when the bacterial cells were suspended in a buffer as compared with distilled water. In our experiment, a 10-fold dilution of the culture was effected in distilled water rather than buffer. The use of plain water over buffer could have contributed to the poor viability of the culture. Finally, the disparity between Wei et al. [37] and our results could be due to the different bacterial species used in our inoculum.

Within the plant production systems, two very different environments are encountered, the rhizosphere (below-ground bacterial habitat) and phyllosphere (above-ground surfaces of a plant as a habitat for microorganisms). This pioneering study examined how introduction of bacterial human pathogens in the rhizosphere and phyllosphere of commercially important food crops affected their microbial safety. Our findings revealed that artificial introduction of *E. coli* and *L. innocua* in the rhizosphere of tomato and bell pepper plants did not result in translocation of the bacteria into the fruits 24 h post-inoculation although a relatively high surviving population was noted in the bulk soil and in the roots. Moreover, the presence of plant pathogen *Ralstonia solanacearum* and plant beneficial bacteria *Pseudomonas fluorescens* did not influence the systemic uptake of human pathogenic bacteria from the soil to the aerial parts of the plants. However, when *E. coli* and *L. innocua* were deposited onto the surface of tomato and pepper fruits, they remained viable for up to 48 h. Hence, a preventative approach to minimizing the risks of pre-harvest contamination of tomatoes and peppers is through avoiding contact between mature fruits and environmental sources of human pathogens such as overhead or sprinkler irrigation water.

References

1. Garrow JS, Ralph A, James WPT (2000) Human Nutrition and Dietetics, Elsevier, Amsterdam.

2. Beuchat LR1 (2002) Ecological factors influencing survival and growth of human pathogens on raw fruits and vegetables. Microbes Infect 4: 413-423.

3. Erickson MC, Webb CC, Diaz-Perez JC, Phatak SC, Silvoy JJ, et al. (2010) Surface and internalized Escherichia coli O157:H7 into field-grown spinach and lettuce treated with spray-contaminated irrigation water. J Food Prot 73: 1023–1029.

4. Guo X1, Chen J, Brackett RE, Beuchat LR (2001) Survival of salmonellae on and in tomato plants from the time of inoculation at flowering and early stages of fruit development through fruit ripening. Appl Environ Microbiol 67: 4760-4764.

5. Orozco L1, Rico-Romero L, Escartín EF (2008) Microbiological profile of greenhouses in a farm producing hydroponic tomatoes. J Food Prot 71: 60-65.

6. Stine SW1, Song I, Choi CY, Gerba CP (2005) Effect of relative humidity on preharvest survival of bacterial and viral pathogens on the surface of cantaloupe, lettuce, and bell peppers. J Food Prot 68: 1352-1358.

7. Burke G (2008) Mexican peppers posed problem long before outbreak. Assoc. Press. Yahoo News.

8. Beuchat LR (1996) Pathogenic microorganisms associated with fresh produce. Journal of Food Protection 59: 204-216.

9. Cooley MB1, Miller WG, Mandrell RE (2003) Colonization of Arabidopsis thaliana with Salmonella enterica and enterohemorrhagic Escherichia coli O157:H7 and competition by Enterobacter asburiae. Appl Environ Microbiol 69: 4915-4926.

10. Critzer FJ1, Doyle MP (2010) Microbial ecology of foodborne pathogens associated with produce. Curr Opin Biotechnol 21: 125-130.

11. Ingham SC, Losinski JA, Andrews MP, Breuer JE, Breuer JR, et al. (2004) Escherichia coli contamination of vegetables grown in soils fertilized with non-composted bovine manure: Garden-scale studies. Applied and Environmental Microbiology 70: 6420-6427.

12. Wood JD, Bezanson GS, Gordon RJ, Jamieson R (2010) Population dynamics of E. coli inoculated by irrigation into the phyllosphere of spinach grown under commercial production conditions. International Journal of Food Microbiology 143: 198-204.

13. Erickson MC1, Webb CC, Diaz-Perez JC, Phatak SC, Silvoy JJ, et al. (2010) Infrequent internalization of Escherichia coli O157:H7 into field-grown leafy greens. J Food Prot 73: 500-506.

14. Girardin H, Morris CE, Albagnac C, Dreux N, Glaux C, et al. (2005) Behaviour of the pathogen surrogates Listeria innocua and Clostridium sporogenes during production of parsley in fields fertilized with contaminated amendments. FEMS Microbiology and Ecology 54: 287-295.

15. Guo X, van Iersel MW, Chen J (2002) Evidence of association of salmonellae with tomato plants grown hydroponically in inoculated nutrient solution. Applied and Environmental Microbiology 68: 3639-3643.

16. Hintz LD, Boyer RR, Ponder MA, Williams RC, Rideout SL (2010) Recovery of Salmonella enterica Newport introduced through irrigation water from tomato (Lycopersicum esculentum) fruit, roots, stems, and leaves. Hort Science 45: 675-67.

17. Zheng J1, Allard S, Reynolds S, Millner P, Arce G, et al. (2013) Colonization and internalization of Salmonella enterica in tomato plants. Appl Environ Microbiol 79: 2494-2502.

18. Upreti R, Thomas P (2015) Root-associated bacterial endophytes from Ralstonia solanacearum resistant and susceptible tomato cultivars and their pathogen antagonistic effects. Front Microbiol 6: 255.

19. Pollard S, Barak J, Boyer R, Reiter M, Gu G (2014) Potential interactions between Salmonella enterica and Ralstonia solanacearum in tomato plants. Journal of Food Protection 77: 320-324

20. Goodman RN, Kira´ly Z, Zaitlin M (1967) the biochemistry and physiology of infectious plant disease. Van Nostrand, Princeton, NJ.

21. Prior PH, Beramis M, Chillet M, Schmit J (1990) Preliminary studies for tomato bacterial wilt (Pseudomonas solanacearum E.F.Sm.) resistance mechanisms. Symbiosis 9: 393-400.

22. Van der Schoot C, van Bel AJE (1989) Architecture of the intermodal xylem of tomato (Solanum lycopersicum) with reference to longitudinal and lateral transfer. Amer J Bot 76: 487-503.

23. Jablasone J, Brovko LY, Griffiths MW (2004) a research note: the potential for transfer of Salmonella from irrigation water to tomatoes. J Sci Food Agric 84: 287-289.

24. Semenov AV1, van Bruggen AH, van Overbeek L, Termorshuizen AJ, Semenov AM (2007) Influence of temperature fluctuations on Escherichia coli O157:H7 and Salmonella enterica serovar Typhimurium in cow manure. FEMS Microbiol Ecol 60: 419-428.

25. Habteselassie MY1, Bischoff M, Applegate B, Reuhs B, Turco RF (2010) Understanding the role of agricultural practices in the potential colonization and contamination by Escherichia coli in the rhizospheres of fresh produce. J Food Prot 73: 2001-2009.

26. Feeney DS1, Crawford JW, Daniell T, Hallett PD, Nunan N, et al. (2006) Three-dimensional microorganization of the soil-root-microbe system. Microb Ecol 52: 151-158.

27. Haichar FZ1, Marol C, Berge O, Rangel-Castro JI, Prosser JI, et al. (2008) Plant host habitat and root exudates shape soil bacterial community structure. ISME J 2: 1221-1230.

28. Ibekwe AM1, Watt PM, Shouse PJ, Grieve CM (2004) Fate of Escherichia coli O157:H7 in irrigation water on soils and plants as validated by culture method and real-time PCR. Can J Microbiol 50: 1007-1014.

29. Islam M1, Morgan J, Doyle MP, Phatak SC, Millner P, et al. (2004) Fate of Salmonella enterica serovar Typhimurium on carrots and radishes grown in fields treated with contaminated manure composts or irrigation water. Appl Environ Microbiol 70: 2497-2502.

30. Barak JD1, Liang AS (2008) Role of soil, crop debris, and a plant pathogen in Salmonella enterica contamination of tomato plants. PLoS One 3: e1657.

31. Bell RL, Cao G, Meng J, Allard MW, Keys C (2012) Salmonella Newport contamination of produce: ecological, genetic, and epidemiological aspects. In: AS Monte, PED Santos (eds), Salmonella: classification, genetics and disease outbreaks. Nova Science Publishers, Inc., Hauppauge, NY.

32. Micallef SA, Rosenberg RE, Goldstein A, George L, Kleinfelter MS, et al (2012) Occurrence and antibiotic resistance of multiple Salmonella serotypes recovered from water, sediment and soil on mid-Atlantic tomato farms. Environ Res 114: 31-39.

33. Bernstein N, Sela S, Neder-Lavon S (2007) Effect of irrigation regimes on persistence of Salmonella enterica serovar Newport in small experimental pots designed for plant cultivation. Irrig Sci 26: 1-8.

34. Nicholson FA1, Groves SJ, Chambers BJ (2005) Pathogen survival during livestock manure storage and following land application. Bioresour Technol 96: 135-143.

35. Fremaux B1, Delignette-Muller ML, Prigent-Combaret C, Gleizal A, Vernozy-Rozand C (2007) Growth and survival of non-O157:H7 Shiga-toxin-producing Escherichia coli in cow manure. J Appl Microbiol 102: 89-99.

36. Jablasone J, Warriner K, Griffiths M (2005) Interactions of Escherichia coli O157:H7, Salmonella Typhimurium, and Listeria monocytogenes plants cultivated in a gnotobiotic system. International Journal Food Microbiology 99: 7-18.

37. Wei CI, Huang JM, Lin WF, Tamplin ML, Bartz JA (1995) Growth and survival of Salmonella Montevideo on tomatoes and disinfection with chlorinated water. J Food Prot 8: 829-836.

19

Long Term of Cattle Manure Amendments and Its Impact on Triticale (X. *Triticosecale Wittmack*) Production and Soil Quality

Enrique Salazar Sosa[1], Hector I Trejo Escareno[3]*, Jesus Luna Anguiano[2], Miguel A. Gallegos Robles[3], Enrique Salazar Melendez[2], Jose Dimas Lopez Martinez[3] and Orona Castillo Ignacio[3]

[1]Technology Institute of Torreon, Mexico
[2]Agricultural Sciences and Forestry, Mexico
[3]College of Agriculture and Animal Husbandry of Durango University of Durango State (FAZ-UJED), Ejido Venecia Municipal of Gomez Palacio, Durango. Km 28 Gomez Palacio-Tlahualilo, Mexico

Abstract

Organic amendment is a good alternative to improve soil fertility to maintain or increase crop forage and grain production. After several times of organic applications (crop cycles), it is important to follow soil physical and chemical parameters to avoid soil pollution such as salinity and nitrate. The main objective of this study was to maintain good triticale forage production and soil quality after seven years in plots where two factors were studied: cow manure amendments; 0, 40, 80, 120 and 160 t ha^{-1} and one chemical level with 150-100-00 kg ha^{-1} of Nitrogen, Phosphorus and potassium, respectively just to compare manure amendments. After this, to decrease soil salinity and high levels of nitrate, triticale forage was planted in the same plots using two varieties without manure and chemical fertilizer application. Triticale variables measured were green forage and ential (Ph) and Nitrates (NO$_3$). Results indicated that triticale forage production was high in all plots were cow manure was applied after three years than the control and chemical fertilizer level, also, the chemical soil parameters such as, salinity and nitrate decrease to adequate levels of: 4 mmhos cm^{-1} in salinity and less than 20 ppm of nitrates. Triticale forage production was better in all plots with cow manure application with more than 25 mg ha^{-1}, that's the triticale average production in this region and more than 100% of the control and chemical fertilizer plots. According to these results, cow manure amendments is a good alternative to get high triticale forage production and maintain a good soil quality.

Keywords: Manure; Soil pollution; Nitrate; Production of the crop and salinity

Introduction

Mexico has a production of 61 million tons of manure considering only the feedlot cattle and partial barn, where the main basin of this important residue are the Laguna Region, the Juarez Valley and other estimated areas of northern and northeastern Chihuahua [1]. Dairy cattle manure and other organic fertilizers used in agriculture have the potential to be a cost-effective source of nutrients for crops. Land application determines an increase in fertility as well as improved physical properties [2].

The organic waste is accumulated in the places where this waste is generated or applied commonly to score some agricultural land, which can cause degradation of the quality of soil and groundwater [3]. The amount of manure nutrients available in the soil for plants, is perhaps one of the most common questions without exact answer because of the many physical, chemical and biological factors involved in the process of decomposition of organic materials [4]. One of the most precise ways is by evaluating the decomposition of manure in the field directly [5].

The use of this fertilizer is unquestionably beneficial, but there are difficulties to predict its effect in every situation due to the great variability of the material covered and the differences created by the previous management [6]. In this case a classification of the type of material is necessary to predict their nutrient to crops [7] this prediction is important because it may cause environmental pollution when applying excessive doses, either excessive loss from gas losses N (denitrification processes and ammonia volatilization loss as NO$_3$ leaching [8,9].

Producers who use this fertilizer, use it indiscriminately applying a high dose of 200-250 tons per hectare per year, making it necessary to carry out an analysis of the salt balance and soil quality without forgetting that the key is to avoid inappropriate use of a resource to protect the quality of soil and water [10,11].

The fodder produced in the spring-summer cycle, is not enough to feed the livestock during the winter season, this caused the producers to look for alternatives to help them supplement the food supply, both in quantity and quality this season [12].

The Comarca Lagunera is the most important dairy region of Mexico and Latin America. A good alternative that has been used to replace alfalfa as a protein source is growing forage triticale [13]. This is a winter crop, which has been included in the diets of dairy cattle, since this crop brings a high potential for biomass production (10.59 mg ha^{-1} of dry matter), with adequate nutritional value (16.76% PC), it is also very tolerant to adverse environmental factors.

Materials and Methods

Geographic location

The Laguna Region is located in the north-central part of the Mexican Republic [14]. This is between the meridians 102° and 104°

*Corresponding author: Hector I. Trejo Escareno, College of Agriculture and Animal Husbandry of Durango University of Durango State (FAZ-UJED), Ejido Venecia Municipal of Gomez Palacio, Durango. Km 28 Gomez Palacio-Tlahualilo, Mexico, E-mail: fazujed@yahoo.com.mx

47' 22' west longitude and 24°22' and parallel 26°23' North latitude. The average height above sea level is 1,139 m. It consists of a hilly expanse and other flat where agricultural and urban areas, comprising an area of 4, 788, 750've^{-1} are located [11]. According to Koeppen classification modified by Cervantes. The climate is dry desert or warm steppe with summer rains and cool winters. The average annual temperature is 21°C, with an average annual evaporation of about 2,396 mm [15]. The rainfall is 258 mm [15]. In the region the predominant clay soils are heavy duty, medium sandy loam [15]. Soil type that was used is clay type, which interferes with ground mineralization as reported by Vazquez et al. [15]. This research was conducted in the experimental agricultural field of the Faculty of Agriculture and Animal Husbandry-UJED, which is located at km 28 of the Gomez Palacio-Tlahualilo, Durango road, to nearby Venice ejido, municipality of Gomez Palacio, Durango.

Soil characteristics and manure

Three random samples of soil and manure were collected to be analyzed at the laboratory of FAZ- UJED [14], to determine the conditions in which the soil and manure were before the establishment of the experiment.

Establishment of experiment

The experiment was conducted in the fall-winter cycle of the years 2008, 2009 and 2010, after planting corn every year since 1978. Initially, since this year the experiment was established with doses of bovine manure of 0, 40, 80, 120 and 160 Mg ha^{-1} with characteristics shown in Table 1 and further treatment with chemical fertilizer 150-150-0 kg ha^{-1} of nitrogen, phosphorus and potassium, respectively. These treatments were distributed in field under a randomized block design arrangement in strips, where each group contained three replicates in an experimental unit of eight meters wide by eight meters long with. Statistical analysis was performed with the Statistical Analysis System software package Ver. 9.

Soil depth cm	pH	C.E. dS m^{-1}	M.O.%	NO_3 mg kg^{-1}	P mg kg^{-1}	K mg kg^{-1}	N-NH_4
0-15	8.14	1.36	1.93	14	7.5	1360.0	9.8
15-30	8.25	1.33	1.58	7	6.5	892.5	12.95
30-60	8.20	1.20	1.24	3	11.0	572.5	13.65
60-90	8.24	3.16	0.89	4	3.5	410.0	14.35
90-120	8.14	3.93	0.27	2	3.5	202.5	12.95

Table 1: Soil Chemical Characteristics Before the caw manure Application. C.A.E.-FAZ-UJED 1998.

After 2004 manure doses were reduced by 50% because it was detected through the soil analysis that the salinity levels increased higher than 4 cm mmhos cm^{-1} (Table 2). This action was not enough to reduce the salinity to permissible values lower than the 4 mmhos cm^{-1}, so it was decided to plant a crop with total coverage of soil surface and to consume the maximum amount of salts with good yields and high quality of forage mainly with protein content similar to alfalfa but, with less water consumption to make it attractive crop forage for the dairy protein producer in the region. The triticale forage crop is an excellent fodder for milk production livestock in this region, where there are more than 500,000 heads of cattle for this purpose making it the most important dairy region of Latin America. Triticale sowing was carried out since 2008 after planting corn in the summer season, but with absolutely no manure applied in each plot, where the organic fertilizer was applied since 1978. This is how the way to not only have a high protein forage and performance, but also have a forage crop that absorbs as much salts and improves soil quality as well as to take advantage of all the residual nitrogen accumulated in it was sought.

Harvest

Harvesting took place 110 days after planting, when the crop was in boot stage and about 10% flowering, this was at 1 m^2 per experimental unit to determine its performance. The variables evaluated in the soil were: organic matter (OM) with Walkley and Black method [16], Electrical Conductivity extract with resistivity, pH with pH meter extract, and nitrates (NO_3) with colorimetric [16]. The variables evaluated on the ground allowed us to determine what the best treatments were. Forage yield and plant height also were measured to evaluate treatments of manure applied.

Results and Discussion

Production of green forage

The production of green forage was statistical different for treatments of manure in the three years of evaluation (Table 3 and Figure 1), with higher production in the treatment of 40 Mg ha^{-1} with 55.2% more than in the control for the year 2008. This results being statistically equal with each other treatment, except the control which obtained lower production. In 2009 29.7% more forage yield was obtained in the treatment of 20 Mg ha^{-1} of bovine manure applied and resulting statistically equal to 40, 80 Mg ha^{-1} and the chemical fertilizer treatment. In 2010 the output was 74.3% more in the treatment of 60 Mg ha^{-1} of bovine manure which showed statistically equal to treatment

Sample number	Prof. cm	% N total	PX	K%	Ca%	Na%	Mn ppm	Fe ppm	Zn ppm	Cu ppm	Bp ppm
1	0-15	1.51	0.356	3.27	3.38	0.97	560	10960	200	49	390
2	15-30	1.39	0.388	3.32	3.47	1.02	620	12300	198	45	450
3	30-45	1.3	0.344	3.4	3.41	1.07	600	11250	206	53	410
4	45-60	1.27	0.358	3.3	3.31	0.98	590	11200	198	47	400

Table 2: Manure Chemical Characteristics C.A.E.-FAZ-UJED 1998.

Contents	FD	Years		
		2008	2009	2010
		-------------Pv>F-------------		
R (Replications)	2	0.0076	0.1415	0.5951
FA (Triticale Varieties)	1	0.0698	0.0864	0.1100
FB (Manure Levels)	5	0.0480**	0.0512*	0.0007*
DMS (Manure Levels)		9.65	4.95	15.45

**Statistically significant at the 5 % and *Statistically highly significant at 1%

Table 3: Means for corn yield per year of study.

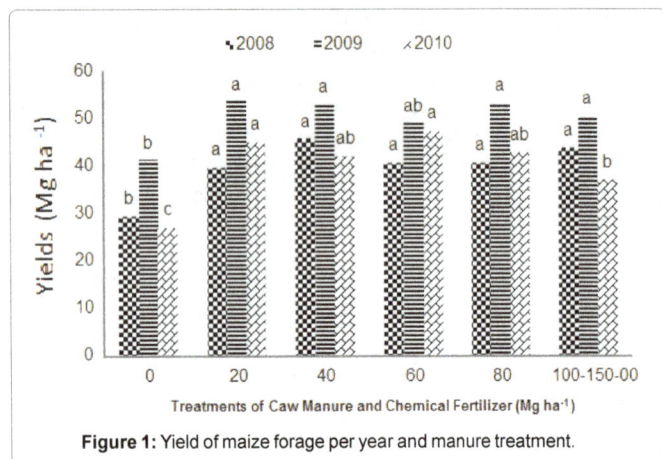

Figure 1: Yield of maize forage per year and manure treatment.

of 20 Mg ha^{-1} of bovine manure. Also, the treatments of 40 and 80 Mg ha^{-1} of manure applied were statistically equal with 57.86% from the control which showed a lower yield, chemical fertilizer was higher than the control but less than all cattle manure treatments reaching 36.8% more yield forage compared with the control. These results clearly indicated that the manure increased forage yields after several times that was applied in the soil, for this reason it is a good alternative for farmers to save money in chemical fertilizer and to get good yields in this region, similar results were found by Salazar et al. [17-19].

Soil and manure characteristics

Table 1 shows the soil test before starting the original experiment in 1998. It shows that the ground has normal characteristics of the soils in the region with a low content of organic matter (OM), alkaline pH and low salt Electric Conductivity (EC) less than 4 mmhos cm^{-1} and very low levels of nitrate (NO$_3$), ammonium (NH$_4$), phosphorus (P$_2$O$_5$), potassium (K$^+$), that had very low fertility to 120 cm depth. Regarding the manure (Table 1 and 2) and after analyzing its features in a pile of accumulation at different depths up to 50 cm, a total range of nitrogen was found from 1.27 to 1.51 with wide variations in other nutrients as; P$_2$O$_5$, K$^+$, Ca^{2+}, Mg^{2+}, Na$^+$, Mn^{2+}, Fe^{2+}, Zn^{2+}, Cu$^+$ and Bo; which allows for a quantitative support to calculate the quantity of nutriment applied when different amounts are added to the soil. This does not mean that these values when multiplied by the dry residue of the plant gives the amount of nutrient available to it, because other nitrogen transformation must be considered such as mineralization, also how much is immobilized, or volatilized and leached so extensive care should be taken to determine the amount of nutrient available for biodegradation of manure after application in soil.

Table 4 shows the chemical characteristics of the soil after application of manure for 9 years in the different treatments of manure

at two depths: 0 cm -15 cm and 15 cm - 30 cm, where increased EC, MO, NO$_3$, K$^+$, Ca^{2+} and Na were observed mainly in treatments from 40 to 160 t ha^{-1} of manure applied. The average EC levels were higher than four mmhos cm^{-1} permissible in a farm field and the contents of OM with more than 5% macronutrient levels over 150 PPM and so with other micronutrients prompting that doses to decline in the following years to 50%. Not finding an acceptable decrease in the concentration of these parameters on the floor it was decided to plant in winter triticale with complete, high soil extraction but without applying manure for three years after planting corn in summer.

Figures 2, 3, 4 and 5 show the average concentration of pH, OM, NO$_3$ and CE for the treatment of manure after planting triticale because with these parameters where found high and very variable concentrations in years prior to 2010. They can be seen as the OM, CE and NO$_3$ decreased to acceptable levels considerably lower than the 4 mmhos cm^{-1} in EC demonstrating that culturing with full coverage and decreased the

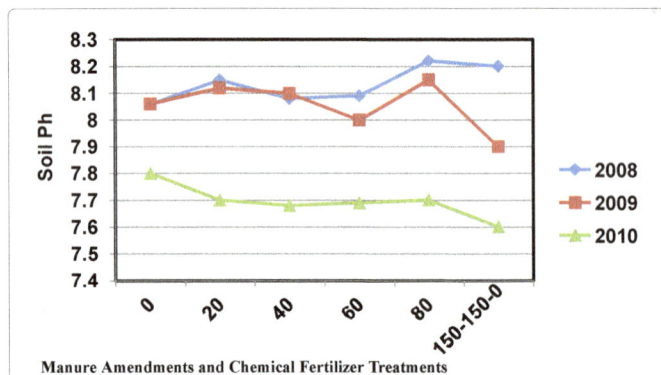

Figure 2: Average distribution of pH in Soil per Treatment of Cow Manure and Fertilizer amendments on Triticale Fertilizer from 2010 to 2012. FAZ-UJED.

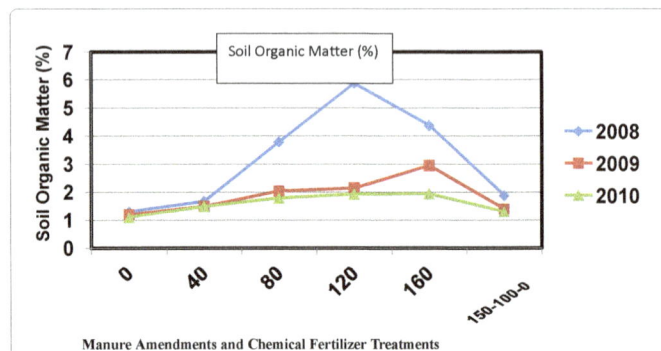

Figure 3: Average distribution of Organic Matter in Soil per Treatment of Caw Manure and Chemical Amendments on Triticale from 2010 to 2012. FAZ-UJED.

Manure Treatments	pH		CE mm cm^{-1}		OM%		NO$_3$ Mg l^{-1}		K$^+$ Mg l^{-1}		Ca^{++} Mg l^{-1}		Na$^+$ Mg l^{-1}	
	*	**	*	**	*	**	*	**	*	**	*	**	*	**
0 t ha l^{-1}	7.32	7.4	1.91	1.47	1.35	1.21	18	17	14	115	9.2	8	13.2	6.5
40 t ha l^{-1}	7.11	7.23	3.77	3.08	1.44	1.72	89	76	14.7	33	23.2	16.2	17.2	14.4
60 t ha l^{-1}	6.93	7.14	6.2	3.26	5.52	2.07	136	87	15.4	11.3	32.2	14.5	35.4	15.5
120 t ha l^{-1}	6.99	6.99	6.22	6.42	5.52	5.92	70	83	97	15.4	39.4	27.6	36.4	35.9
160 t ha l^{-1}	6.8	6.92	6.28	5.2	6.62	2.42	168	102	17.8	20.3	22.4	15.6	32.5	22.9
150-100-0	6.24	6.42	1.74	1.62	1.93	1.51	3	21	15	1.6	11.6	10.4	17.2	8.6

*Soil depth 0 cm-15 cm; **Soil depth 15 cm-30 cm

Table 4: Soil Chemical Characteristics After the cow manure Application. C.A.E.-FAZ-UJED 2007.

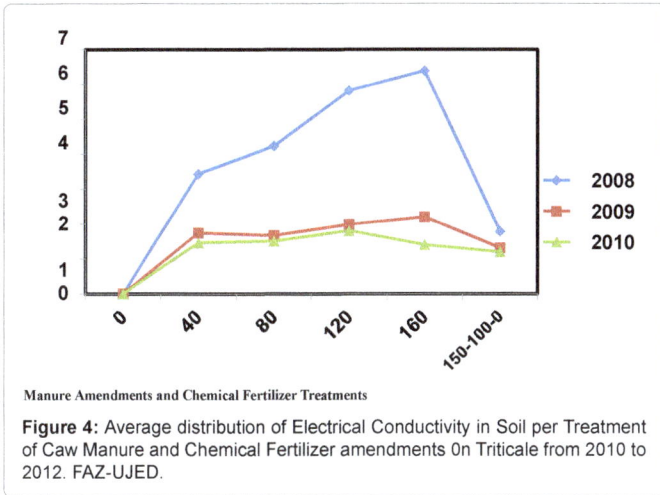

Figure 4: Average distribution of Electrical Conductivity in Soil per Treatment of Caw Manure and Chemical Fertilizer amendments 0n Triticale from 2010 to 2012. FAZ-UJED.

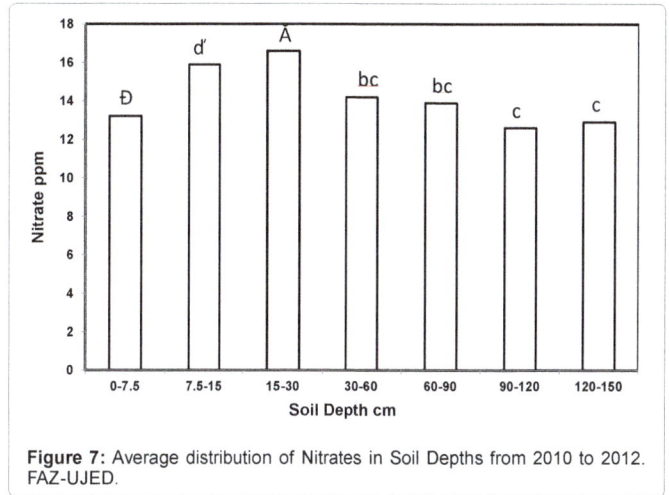

Figure 5: Average distribution of of Nitrates in l Soil per Treatment of Caw Manure and and Chemical Fertilizer Amendments on Triticale from FAZ-UJED.

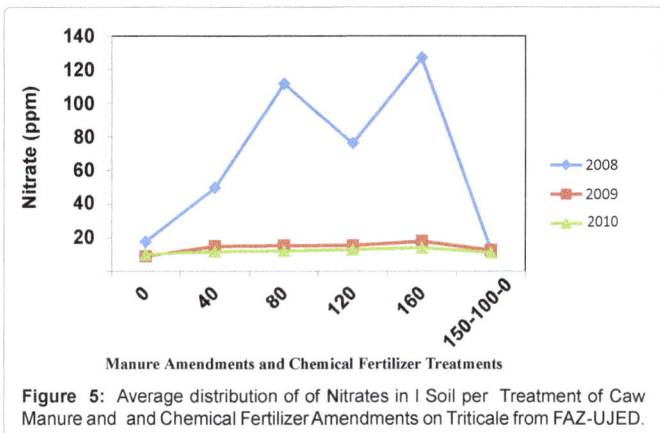

Figure 7: Average distribution of Nitrates in Soil Depths from 2010 to 2012. FAZ-UJED.

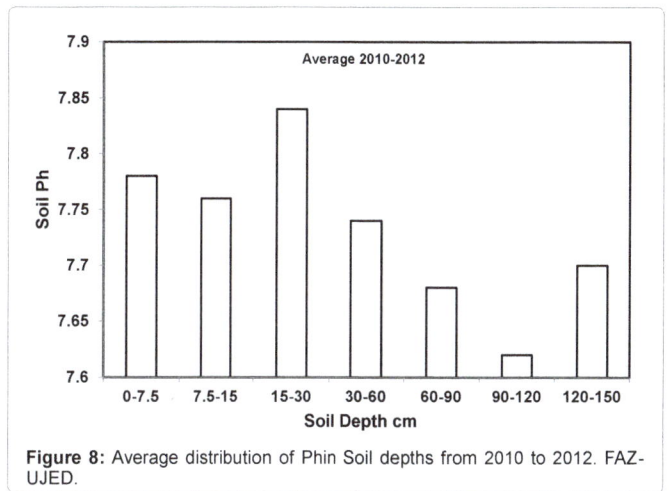

Figure 8: Average distribution of Phin Soil depths from 2010 to 2012. FAZ-UJED.

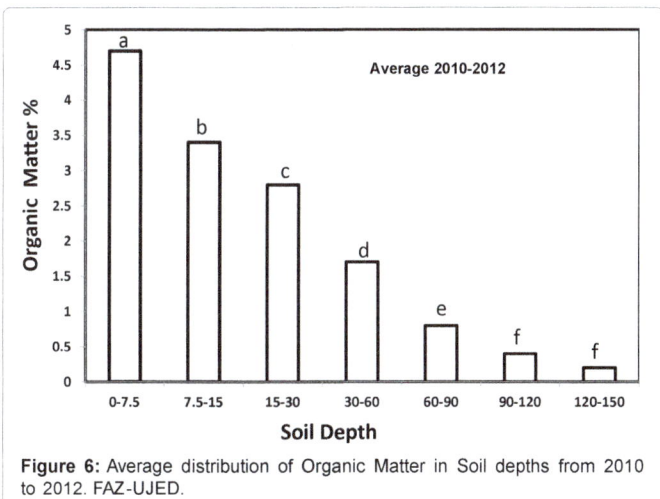

Figure 9: Average distribution of Electrical Conductivity in Soil Depths from 2010 to 2012. FAZ- UJED.

dose applied to the soil manure allowed preserve the quality of soil and maintain acceptable fertility to plant any crop.

Additionally and for detecting the concentration especially OM, CE and nitrates throughout the soil profile up to 1.50 cm deep, Figures 6, 7, 8 and 9 show these observed results. As the EC and Nitrates are highly soluble show high concentrations after the 30 cm depth, away from the area of maximum absorption due to the high concentrations

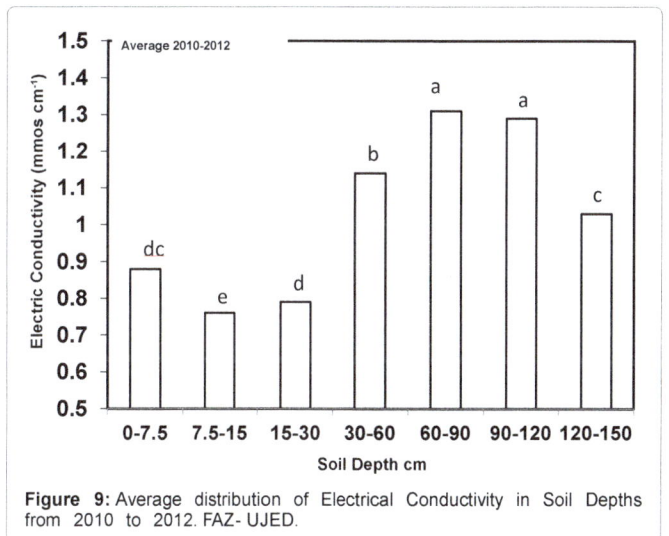

Figure 6: Average distribution of Organic Matter in Soil depths from 2010 to 2012. FAZ-UJED.

found in the first 30 cm in the first 9 years of the experiment which, consequently, ample care should have especially nitrates, as these to be found in high concentrations after the first 30 cm because they are a potential contamination of the underground aquifer in this and other regions [20].

Discussion

According to the results found, cow manure amendments are important to improve soil crop forage production. However, special care is necessary to maintain good soil quality with respect to salinity and high concentration of nitrates. Triticale forage crops with all soil surfers cover is a good option to take up high concentration of salt and nitrate after several years. In this study, three years after seven applications of cow manure consequently, the soil salinity and nitrate were decreased to lower levels than the maximum permissible of 4 mmhos cm^{-1} in salinity and less than 20 ppm of nitrates. Triticale forage yields also were higher than 100% in all plots were caw manure was applied than the control and chemical fertilizer treatments. That means that soil must be monitored over time when organic amendments are used to avoid soil pollution due to high mineralization of

organic amendment that induce to more salinity and nitrate concentration in soil mainly. Additionally, soil nitrate and organic matter were higher in the first centimeters of depth (15) due to more mineralization than in the others soil depths decreasing at 150 cm. And because high concentrations of the solubility of nitrate and salt after 30 centimeters were found, soil monitoring of these parameters are important to avoid possible aquifer pollution in the future. Finally, caw manure amendments is a good option for farmers of this and other regions to improve soil quality and crop forage production.

References

1. Cortes JJM (2007) Residual use of manure on wheat in the Yaqui Valley of Sonora. 19th international week of face-UJED agronomy, November, p: 209.

2. Alexander M (1978) Introduction to soil microbiology. Soil science 125: 331.

3. Allison FE (1995) The enigma of soil nitrogen of soil nitrogen balance sheets. Advances in agronomy 7: 213-250.

4. Lozano RA (2002) The magazine Marsh producer. Agricultural Laguna. Bimonthly publication of the Agricultural Cooperative Society of the Laguna Region, SCL.

5. Flores JP, Flores RM, Palomo MB, Corral D (2003) Evaluation of biosolids with feed crops in the Valley of Juarez, Chihuahua. Validation project. INIFAP, Campo Experimental Valle de Juarez, Municipal Water and Sanitation Board of Cd. Juarez, Chihuahua Produce Foundation. Final Project Report.

6. Jimenez LM, Larreal N (2004) Walnut effects of bovine manure on chemical properties of degraded ultisol marchites area, colon state of Zulia. Journal of the Faculty of Agronomy, University of Zulia, p: 21.

7. Cadahia CF (1998) Horticultural and ornamental crops. 2nd edn, Spain, pp: 46-47.

8. Aguirre LO (1987) Climate Guide to Laguna Region- SARHCIAN-INIA. Matamoros, Coahuila, Mexico, p: 174.

9. Infoagro (2002) Organic Fertilization.

10. Powers WL, Wallingford GW, Murphy LS, Whitney DA, Manguez HL, et al. (1974) Guidelines for Applying beef feedlot manure to fields. Publication C-502. Kansas State University, Cooperative Extension Service. Manhattan, KA.

11. Márguez FJP (2003) Management and biodegradation of biosolids applied to agriculture. In: Payne WJ (1973). Reduction of nitrogenous oxides by microorganisms. Bacteriul Rev 37: 409- 452.

12. Rowell DM, Prescott CE, Preston CM (2001) Decomposition and nitrogen mineralization from biosolids and other organic materials: With initial relationship chemistry. J Environ Qual 30: 1401-1410.

13. Vazquez VC, Salazar SE, Leos RJA, Fortis HM, Zuniga TR, et al. (2007) Impact of the application of cattle manure on soil quality and production of tomato (*Lycopersicon esculentum* Mill) organic fertilizers, pp: 63-64.

14. SAGARPA (2002) Statistical Year book of Agricultural Production. SAGARPA. Laguna Region, Lerdo de Tejada, Durango, Mexico.

15. Vazquez VC, Salazar SE, Figueroa VR, Fortis HM (2001) Effect of padding and cattle manure in modifying some soil characteristics of the Laguna Region. 13th International Week of Agriculture, FAZ-UJED. Gomez Palacio, Durango, pp: 178-182.

16. Paul EA (1989) Soil Microbiology and Biochemistry. In: Ed Academic Press, San Diego, California, pp: 115-130.

17. Salazar SE, Leos RJA, Fortis HM, Vazquez VC (2002) HM Recovery and Nitrogen uptake by wheat and sorghum in stubble and no-tillage system. Agrociencia 36: 433-440.

18. Salazar SE, Vazquez VC, Leos RJ, Fortis HM, Montemayor TJ (2003) Mineralization cattle manure and its impact on soil quality and production of tomato (*Lycopersicon esculentum* Mill) under subsurface irrigation. Journal of Python.

19. Salazar SE, Vazquez VC, Rivera O (2002) Management and biodegradation of bovine manure in the Laguna Region, Proceedings of the 15th International Week of Agriculture, Faculty of Agriculture and Animal Husbandry of the Juarez University of Durango State.

20. Salazar SE, Vazquez VC, El Trejo (2002) El of organic fertilizers and plasticultura. UJED-SMCS nitrogen cycle.

Occupational Health Hazard in *Abelmoschus esculentus (Bhindi)* Picking and Mitigating Measures

Gandhi S[1], Mehta M[2] and Dahiya R[3]

[1]Scientist Department of FRM, India
[2]College of Home Science, CCS Haryana Agricultural University, HISAR, Haryana, India
[3]DES Home Science Krishi Vigyan Kendras, Fatehabad, Haryana, India

Abstract

Vegetable *A. esculentus (Bhindi)* picking is labour-intensive work that requires painstaking physical effort, patience and perseverance. Women usually use their hands resulting in physical and mental fatigue, hardship, exploitation and pain. The present study was carried out to study the occupational health hazard in vegetable *A. esculentus (Bhindi)* picking and mitigating measures. The experimental study was carried out on 10 farm women labourers in Majra village of Fatehabad district in *A. esculentus (Bhindi)* farms, studying use of four types of gloves as existing protective measures used by the respondents and then introduced 3 appropriate technologies viz. protective gloves, capron and vegetable picking bags as mitigating measures. Women workers faced severe health hazards in picking *A. esculentus (Bhindi)* (lady finger) in terms of cuts and wounds in hands, hardness of skin, itching, blisters and abrasions. Various mitigating measures used by the respondents during *A. esculentus (Bhindi)* picking were homemade mittens of cloth materials (40%), surgical gloves (30%), cotton (20%) and woolen gloves (10%). Performance evaluation of existing mitigation measures was studied in terms of durability, safety and comfort in terms of sweating, it was highlighted that mittens made of denim cloth were preferred most (Rank I) followed by cotton gloves (Rank II), woolen gloves (Rank III) and surgical gloves (Rank IV). Use of improved technology in *A. esculentus (Bhindi)* picking not only reduced the health hazards, but also increased the output by 40 percent, thereby, increasing the efficiency of the worker.

Keywords: Vegetable picking; Health hazard; A. esculentus (Bhindi); lady finger; Agricultural operations

Introduction

Most of the works performed by farm women are tedious, tiring as well as time-consuming. These tasks are performed manually or by traditional tools. Workers in agricultural operations for both crop and animal production typically use repetitive motions in awkward positions and which can cause musculoskeletal injuries. (Kirkhorn [1]) Ergonomic risk factors are found in jobs requiring repetitive, forceful, or prolonged exertions of the hands; frequent or heavy lifting, pushing, pulling, or carrying of heavy objects; and prolonged awkward postures. Women are extensively involved in various farm operations like transplanting, weeding, harvesting, processing, marketing and selling of food grains, fruits and vegetables etc. These tasks not only demand considerable time and energy but also are sources of drudgery. Drudgery is generally conceived as physical and mental strain, agony, monotony and hardship experienced by farm women while performing these farm operations. The drudgery prone condition leads to various health and mechanical hazards which creates physical exhaustion fatigue and low productivity. Vegetable A. esculentus (Bhindi) picking is labor-intensive work that requires painstaking physical effort, patience and perseverance. Women usually use their hands resulting in physical and mental fatigue, hardship, exploitation and pain.

Health hazard in A. esculentus (Bhindi) picking

Meyers et al. [2] stated that occupational musculoskeletal disorders (MSDs) might affect muscles, tendons, joints, nerves and related soft tissues anywhere in the body. The lower back and upper extremities, including the neck and shoulders, are the most common sites. Because repeated risk factor exposure of the same muscle, tendon, or region may result in injury and inflammation to the affected area, names such as cumulative trauma disorder, repetitive motion injury, repetition strain injury, and occupational overuse syndrome have been applied to these disorders. Women workers faced severe health hazards in picking *A. esculentus (Bhindi)* (lady finger) in terms of cuts and wounds in hands, hardness of skin, blisters and abrasions. Moreover, skin allergies due to chemical sprays were commonly an acute problem to 30% of the *A. esculentus (Bhindi)*-pickers. They were using their own devised methods for protecting themselves against these hazards. Bhattacharya and Chakarbarti [3] reported high prevalence of musculoskeletal disorder among tea leaf pluckers. Shoulders, back, neck and fingers were the most affected organs. Musculoskeletal disorders were mostly related to the work habit *i.e.* awkward posture, repetitiveness and duration. Hence, urgent need was felt to design a plucking device to lower down the possibilities of MSDs among workers. Park and associates [4] found that farmers have reported having daily LBP for a week (31%), which is significantly greater than the general working population (18.5%). Kaur and Sharma [5] studied that a survey was conducted by taking 200 farm women of Punjab State. The results showed regarding the level of work related body disorders in agriculture by women included pain in many parts of body followed by numbness or stiffness. Some farm women also felt itching and swelling in hands while working in the fields and some felt burning in abdomen and chest especially during spraying of pesticides in the fields due to inhalation.

The reasons of pain or stiffness may be due to the poor body postures while performing certain farm operations and lack of awareness regarding the right body postures. Sometimes, they did not even take rest in between which is essential to make our body stress free.

*Corresponding author: Mehta M, CCS Haryana Agricultural University, HISAR, Haryana, India, E-mail: mm1964@rediffmail.com

Methodology

The experimental study was carried out on 10 farm women labourers in *A. esculentus (Bhindi)* farms of Majra village of Fatehabad district during Front Line Demonstration. use of 4 types of gloves as existing protective measures used by the respondents were studied and 3 appropriate technologies viz. protective gloves, capron/Protective gloves and vegetable picking bags were introduced as mitigating measures.

Capron/protective gloves are made of PVC material on the one side and fusion of cotton or hosiery on the other side. The blending of cotton and hosiery avoids sweating and skin allergies. These gloves protect hands from injury, blisters, cuts, hardness of skin and abrasions. Being a user-friendly technology, these gloves not only protect from health hazards but increase work efficiency of workers.

Protective mask are made of cotton fabric and these cover neck, head and the face of the user. It can be made up to a designed length, depending upon type of use. It avoids dust, husk and straw to penetrate round the neck area. A muslin mask around nose protects from insecticide/pesticide hazards during picking of *A. esculentus (Bhindi)*. The protective mask protects from a harsh environment and direct sun rays.

Pick bag is used to collect vegetables. It is made of cotton cloth and is designed according to anthropometric measurements of women. Shaped pockets are made in front of waist so that picking can be made comfortable as it reduces the hand movements up and down. Cushion belts on shoulders make it easy to carry vegetable loads. Five-six kg *A. esculentus (Bhindi)* can be picked in one loading.

Results

Existing mitigating measures used by the respondents

Various mitigating measures used by the respondents during *A. esculentus (Bhindi)* picking were mitten and gloves of different materials (Table 1). Mittens made of denim, and using cloth material with cello tape covering on fingers were mainly used as protective measure against cuts, wounds and itching by 40% of the respondents, followed by surgical gloves (30%), cotton gloves (20%) and woolen gloves (10%). In addition to gloves, others measures were also adopted like wearing suit with full sleeves shirt (70%), using full sleeves gents' shirt over the dress (40 %), and covering face with head cloth (20%) to protect themselves from itching (Figure 1).

Performance evaluation of existing methods of *A. esculentus (Bhindi)* picking

Performance evaluation of existing methods of *A. esculentus (Bhindi)* picking was studied in terms of durability, safety and comfort in terms of sweating. Among the existing methods of *A. esculentus*

Mitigating measures	Frequency	Percentage
Gloves		
Woollen gloves	1	10
Mittens-Denim, cloth with tape covering	4	40
Surgical gloves	3	30
Cotton gloves	2	20
Others measures		
Wearing suit with full sleeves shirt	7	70
Using full sleeves gents' shirt over the suit	4	40
Covering face with head cloth	6	60

Table 1: Existing mitigating measures used by the respondents (n=10)

(Bhindi) picking mittens made of Denim cloth were preferred most (Rank I) with average mean score of 1.9, cotton gloves (Rank II) scored almost same mean score (mean score of 1.8). Woolen gloves (Rank III) scored 1.2 and surgical gloves (Rank IV) scored minimum (1.0) on all these parameters (Table 2). Although they were using their own devised methods for protecting themselves against these hazards but still they experienced drudgery and had lesser output efficiency (Figure 2).

To minimize these health hazards and enhancing work efficiency of women labourers in *A. esculentus (Bhindi)*-picking, 10 farm women workers were provided protective garments, viz protective gloves, capron and vegetable picking bags i.e. all these three technologies designed/tested by AICRP, FRM (Figure 3).

Reduction in health hazards using improved technology in *A. esculentus (Bhindi)* picking

With the use improved technology in *A. esculentus (Bhindi)* plucking there was reduction in health hazards. Table 3 depict that the utmost reduction (80%) was in incidence of cuts and wounds in hands, hardness of skin and irritation in hands. Blisters and abrasions during *A. esculentus (Bhindi)* picking were lessen by 70%. The headache was lowered in 20 % workers and other hazards namely suffocation, skin allergies and infection due to spray were reduced by 10% each. However there was no effect of improved technology on chest pain and soar eyes. For this purpose, 75 farm women were selected from five different villages of Ludhiana district. Bal et al. [6] introduced tools viz maize sheller, improved sickle and ring cutter among farm workers and reduction in drudgery was assessed. The parameters for

Figure 1: Protective measures used in bhindi picking

Figure 2: Use of protective clothing's in *bhindi* picking

Figure 3: Performance evaluation of *bhindi* picking with protective clothing's

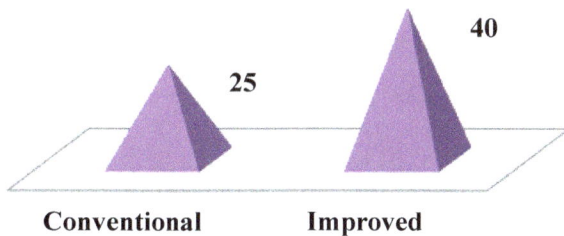

Amount of bhindi picked (kg/day)

Figure in parentheses indicate percentage
Figure 4: Amount of bhindi picked (kg/day)

Characteristics	Durability	Comfort in terms of sweating	Safety	Average mean score	Rank
Woollen gloves	1.2	1.2	1.3	1.2	III
Denim, cloth with tape covering	2.1	1.7	1.9	1.9	I
Surgical gloves	1.0	1.0	1.0	1.0	IV
Cotton gloves	1.4	2.0	2.0	1.8	II

Table 2: Mean score of performance level of existing methods of *bhindi* picking (n=10)

Health hazards	Conventional	Improved	Reduction in health hazards
Cuts and wounds in hands	8 (80)	-	8 (80)
hardness of skin	9 (90)	1 (10)	8 (80)
blisters and abrasions	7 (70)	-	7 (70)
Irritation in hands	9 (90)	1 (10)	8 (80)
headache	4 (40)	2 (20)	2 (20)
Suffocation	3 (30)	2 (20)	1 (10)
Chest pain	2 (20)	2 (20)	0
skin allergies	3 (30)	2 (40)	1 (10)
Infection due to spray	3 (30)	2 (30)	1 (10)
Soar eyes	2 (20)	2 (20)	0

Table 3: Reduction in health hazards using conventional and improved technology in *bhindi* picking (n=10)

assessment of drudgery experience were drudgery scores, Ovako Work Assessment System, Cardiac Strain Index and Angle of Deviation. The results showed significant reduction in these parameters when improved tools were used as compared to traditional tools. Tiwari and Gite [7] conducted an experiment to study the influence of four work-rest schedules on physical workload during power tiller operation and reported that that the work-rest schedules did influence the physiological and postural workload as evidenced by the differences in working heart rate and postural discomfort. It was concluded that

to avoid excessive postural discomfort the minimum duration of rest pauses should be of 15 min. The duration of the lunch break should be more than 45 min.

Comparative output given by the respondents in conventional and improved methods of *A. esculentus (Bhindi)* picking

Use of improved technology in *A. esculentus (Bhindi)* picking not only reduced the health hazards, but also increased the output thereby increasing the efficiency of the worker. The output increased by 40 percent with the use of these protective technologies (Figure 4). Kaur et al. [8] conducted ergonomic evaluation of vegetable plucking activity with traditional (ordinary knife) and improved tool (ring cutter). Ergonomic assessment of both the method showed that by using ring cutter physiological and muscular stress of workers in terms of heart rate, energy expenditure rate, physiological cost of work and grip fatigue were reduced as compared to traditional method. Thus, new tool *i.e.* ring cutter was found to be beneficial to improve work efficiency of farm women.

Turner [9] reported that instituting micro pausing might reduce discomfort and pain by reducing muscle and nerve tension. Micro pausing to prevent fatigue is more effective than resting than to recover from it. Micro pauses may be passive where the employee rests and active where the employee undertakes a range of stretching type exercises.

Conclusion

Women workers faced severe health hazards in picking A. esculentus (Bhindi) (lady finger) in terms of cuts and wounds in hands, hardness of skin, itching, blisters and abrasions. In order to overcome these they were using various mitigating measures like homemade mittens of cloth materials (40%), surgical gloves (30%), cotton (20%) and woolen gloves (10%). Performance evaluation of existing mitigation measures was studied in terms of durability, safety and comfort in terms of sweating, it was highlighted that mittens made of denim cloth were most preferred followed by cotton gloves, woolen gloves and surgical gloves . Use of improved technology in *A. esculentus (Bhindi)* picking not only reduced the health hazards, but also increased the output by 40 percent, thereby, increasing the efficiency of the worker.

References

1. Kirkhorn SR, Earle-Richardson G, Banks RJ (2010) Ergonomic Risks and Musculoskeletal Disorders in Production Agriculture: Recommendations for Effective Research to Practice. Journal of Agromedicine 15: 281-299.

2. Meyers J, Bloomberg L, Faucett J, Janowitz I, Miles JA (1995) Using ergonomics in the prevention of musculoskeletal cumulative trauma injuries in agriculture: Learning from the mistakes of others. Journal of Agromedicine 2: 11-24.

3. Bhattacaharya N, Chakarbarti D (2010) Occupational analysis of tea leaf plucking task in tea gardens of Assam. Proceedings of HWWE 2010 Chennai, 83.

4. Park H, Sprince NL, Whitten PS, Burmeister LF, Zwerling C (2001) Risk factors for back pain among male farmers: Analysis of Iowa farm family health and hazard surveillance study. Am J Indu Med 40: 646-654.

5. Kaur H, Sharma S (2009) Work related body disorders and health hazards faced by farm women of Punjab. Paper presented in International Ergonomic Conference 2009, Humanizing Work and Work Environment in University of Calcutta India17- 19.

6. Bal SK, Shivani S, Harpinder H (2013) Assessment of Drudgery Experience of Rural Women while Performing Different Farm Operations. Advance Research Journal of Social Science 4: 68-71.

7. Tiwari PS, Gite LP (2006) Evaluation of work-rest schedules during operation of a rotary power tiller. International Journal of Industrial Ergonomics 36: 203-210.

8. Kaur H, Sharma S (2010) Ergonomic evaluation of vegetable plucking with traditional and modern tool. Proceedings of HWWE, 2010 Chennai, 76.

9. Turner WED (2004) Prevention of Work Related Musculoskeletal Disorders (WMSD) An Evidence Based Approach.

21

Pesticide Use Practices and Perceptions of Vegetable Farmers in the Cocoa Belts of the Ashanti and Western Regions of Ghana

Victor Afari-Sefa[1]*, Elvis Asare-Bediako[2], Lawrence Kenyon[3] and John A. Micah[4]

[1]AVRDC - The World Vegetable Center, Eastern and Southern Africa, P. O. Box 10, Duluti Arusha, Tanzania
[2]University of Cape Coast, School of Agriculture, Department of Crop Science, Private Mail Bag Cape Coast, Ghana
[3]AVRDC- The World Vegetable Center, Headquarters P. O. Box 42 Shanhua Tainan 74199, Taiwan
[4]University of Cape Coast, School of Agriculture, Department of Agricultural Economics and Extension Private Mail Bag Cape Coast, Ghana

Abstract

Chemical pesticides are commonly used in the management of pests and diseases in vegetable production in Ghana. However, there is increasing concern about the adverse effects this use has on public health and the environment. A study was conducted to assess how much farmers' know about the safe handling and use of pesticides, and what they perceive to be the hazards around their use. In-depth field surveys was undertaken with 437 sampled vegetable producers and complimented with focus group discussions and field observation. The results revealed that knapsack sprayers were the most widely used equipment for spraying pesticides (92.4%), followed by hand-held applicators (4.5%) whereas only 3.1% used motorised sprayers. Only 15.6% of the respondents fully protect themselves during spraying operations; others either wore partial protective clothing (38%) or did not wear any protective clothing at all (46.4%), there by coming into direct contact with pesticides. Over 80 % of the respondents re-entered their farms within 3 days of pesticide application; harvest their produce within 7 days, without observing safe harvest interval protocols. The study also revealed that the farmers were aware of and had experienced pesticide hazards such as headache, dizziness, body weakness, and itching. Three per cent of the farmers also mentioned burning sensation, catarrh, stomach pain, unconsciousness, itching of eyes and body pains as side effects from pesticides application. Females and illiterates were found to be more vulnerable to these hazards than their male and literate counterparts. The study findings show that most farmers dispose of empty pesticide containers (59.8%) and wash water from sprayers (79.2%) by throwing or disposing them on their farms. The study concludes that farmers are misapplying pesticides by disregarding the potential harmful effects of pesticides on human health and the environment.

Keywords: Pest management; Pest control; Crop protection practices; Agricultural knowledge and information systems; Farmers' perceptions; Hazard; Pesticide policy

Introduction

Vegetables are the most important ingredients of human diets for the maintenance of good health and prevention of diseases. Cultivation of vegetables is an excellent source of employment for both rural and urban dwellers as it takes place in many rural areas through truck farming and in the outskirts of towns and cities in the form of market and backyard gardening to supply fresh produce to urban markets [1]. It thus plays an important socio-economic role as well as in diversifying diets for improved nutrition [2]. Ghana benefits from considerable foreign exchange through the export of vegetables such as okra and chillies to European countries including Belgium, Britain, Germany, Italy, and Switzerland [3,4]. Chilli exports for instance have ranged between 26,000 and 41,000 MT per annum over the past 5 years with corresponding foreign exchange from US$18.2 to US$28.7 million [4].

As vegetables are generally susceptible to a wide range of pests and diseases, these are major constraints to vegetable production in Ghana and require intensive effort in their management [2,5]. The increased demand for food, particularly to feed the growing urban population in Ghana, has necessitated an expansion and intensification of agriculture and horticulture and a concomitant increase in the use of synthetic pesticides for food production [6], particularly for the production of high-value cash crops and vegetables [1,7]. However, these pesticides are often applied indiscriminately and inappropriately, resulting in adverse environmental and health effects, and negative effects on other economic activities such as fisheries and tourism [2,8,9]. In the cocoa belts of Ghana it is likely that pesticides approved for controlling diseases and pests on cocoa are used instead on vegetable crops or vegetables are contaminated with these pesticides when intercropped with cocoa.

The World Health Organization (WHO) reports that 20% of pesticide use in the world is concentrated in developing countries posing a danger to human health and the environment [10]. Families residing in agricultural areas were found to have elevated levels of pesticides in their bodies [11]. These were greater in homes located closer to fields [12]. Problems experienced by farmers during and after the application of pesticides have been well documented in Ghana [1,13]. A survey carried out by the Northern Presbyterian Agricultural Services [14] on 183 farmers in 14 villages in the Upper East region of Ghana found that more than a quarter had recently suffered from directly inhaling chemicals and one fifth from spillage of chemicals on the body. A study on the analysis of pesticide contamination on farmers in Ghana also found the presence of organo chlorine pesticide residues, including dichlorodiphenyltrichloroethane (DDT), in the breast milk and blood of vegetable farmers [13].

---

*Corresponding author: Victor Afari-Sefa, AVRDC - The World Vegetable Centre, Eastern and Southern Africa, P. O. Box 10, Duluti Arusha, Tanzania
E-mail: victor.afari-sefa@worldveg.org

The Ghanaian public and government are becoming aware of the increasing and excessive use of chemical pesticides by vegetable farmers and that if agricultural production is to be safe and sustainable then this trend should be reversed [15]. While the better educated or more informed populace in Ghana are increasingly concerned about the long-term adverse effects of pesticides on the environment and the health of the country's resources, little scientific research has been done to properly characterize, quantify or address the issue. This study was aimed at better understanding how vegetable farmers in the cocoa belts of Ashanti and Western Regions of Ghana use pesticides, and theirknowledge and perceptions of appropriate pesticide use and the associated risks involved. These are important pre-requisites for the development of more appropriate and sustainable pest management options and tools, and better pesticide policies or guidelines.

Materials and Methods

Population and sampling approach

The study population was vegetable farmers in the cocoa belts of the Ashanti and Western regions of Ghana. Reconnaissance surveys in Ashanti and Western regions allowed familiarization with the study are after which multi-stage sampling was used for the in-depth one-on-one interviews. Through interaction with the Vegetable Growers' Association and the regional directorate of the Ministry of Food and Agriculture, information such as which were the most important vegetable growing areas in each region, was obtained to guide the sampling plan. Offinso North, Atwima Nwabiagya, and Amansie East districts in the Ashanti Region, and Sefwi Bekwai, Bibiani-Anwiaso-Bekwai, and Prestea-Huni Valley districts in the Western Region, were then selected for the survey. These districts were selected purposely from each region to reflect the importance and scale of vegetable production, diversity of vegetables grown, and technology levels in vegetable production in the study area. The sample size of respondent farmers was proportionate to the total number of vegetable farmers in each region and district. A total of 437 farmers were interviewed.

Up to five communities from each district were selected for the study based on criteria including poverty and population density thresholds, access to pesticides and produce markets and other institutions, natural resources integrity, and farming systems.

Instrumentation and data collection

The study used participatory tools and techniques for data collection, including in-depth interviews with vegetable producers, focus group discussion (FGD) with farmer groups, use of observational checklists on the selected communities and in farmers' fields, and key informant interviews with officials from the Plant Protection Regulatory Services Division of the Ministry of Food and Agriculture, including pesticide inspectors. Secondary data on the list of registered pesticides was obtained from the Environmental Protection Agency (EPA). Four survey instruments were developed, and their content validated for use with the specific respondent types. Ten enumerators from the area were trained in the administration of the data collection instruments. Structured interview schedule with both open ended- and closed ended- questions was prepared and pre-tested to determine the ability of enumerators to administer it. The questions were written in English and administered in the corresponding lingua franca (Akan) of each community. The one-on-one survey instrument comprised of two categories of questions based on (i) household socio-economic and farm characteristics of respondent farmers (i.e., age, sex, educational background work experience, size of farm under vegetable cultivation) and (ii) pesticide use practices and management (i.e., sources and types of pesticide acquisition, time and frequency of pesticide application, the use of protective clothing, knowledge of pesticide hazards, re-entry and pre-harvest intervals, disposal of pesticide containers etc.).

Information from the structured interviews were complemented with informal focus group discussions comprising of 6 to 12 respondents per group with the help of a facilitator. Two to three farmer groups were purposively selected from each of the four communities from each district with help of agricultural extension agents. At the group meetings information was gathered on vegetable cropping systems, type of vegetables grown, farmers' perception of pesticide use, constraints in the use of pesticides, and poverty indices among others.

Farmers' perceptions on pesticides and health risks were assessed with the farmer groups through a modified ranking game described by Warburton et al. [16] and modified by Ntow et al. [2]. All participants were individually shown empty containers and/or labels of pesticides commonly available in the study area. The containers and/or labels were shown one by one, and the names of the pesticides were read out to ensure that each participant knew what it was. They were asked which ones they recognised (but not necessarily used); the unfamiliar pesticides were removed and noted in the questionnaire. From the familiar containers and/or labels the respondents were asked which ones were thought to be generally effective in controlling popular identified pests (fungi, insects, weeds, and nematodes) hence dividing the containers and/or labels into two piles: 'effective' and 'ineffective'. The 'ineffective' pile was removed and the 'effective' pesticides were ranked according to degree of effectiveness in destroying/controlling specific identified pest and perceived hazardous side effect on humans resulting from application on crops. Once the pesticides were ranked as per the stated criteria, respondents were asked to describe how effective each pesticide was on a likert scale of 1-5 (Table 1). Respondents were finally asked their reasons for ranking a particular pesticide as the most effective. In situations where empty containers and/or labels of the target pesticideswere not available for recall purposes, farmers were asked to list the pesticides they use and rank them accordingly. Ranking of pesticides in terms of their hazard levels followed a similar pattern. Farmers were asked to select and rank empty containers and/or labels of pesticides thought to be hazardous in terms of their combined perceived side effects (i.e., contact with undiluted concentrate, ingestion of undiluted concentrate, inhaling-in chemical during spraying) on human health through application. Responses were again ranked on a likert scale of 1-5 (Table 1) with reasons assigned for ranking a particular pesticide as most hazardous.

Effectiveness	
1	Very effective: 75-100% of identified pest killed
2	Effective: 50-75% of identified pest killed
3	Small effect only: <50% of identified pest killed
4	No effect
5	Makes the effects of the identified pest problem worse
Hazard	
1	Extremely hazardous: likelihood of hospitalisation Or long-term illness
2	Moderately hazardous: likelihood of more than 2 days sick and need to see a doctor
3	Slightly hazardous: likelihood of dizziness orvomiting or blurred vision or skin sores
4	Least hazardous: likelihood of some dizziness, tiredness, or headache
5	No effects

Table 1: Description of ranking levels used for pesticide ranking game.

The relationship between farmers' perception of pesticide hazard and pesticides' perceived effectiveness against pests was assessed by the $\chi 2$ test proposed by Ntow et al. [2].

Using a checklist, observations were made on the selected communities for socio-economic indices such as road access to the community, housing types, water delivery system, marketing facilities, and education infrastructure and facilities. Observations were also made on selected farmers' fields for vegetable cropping system, pests and disease incidence and severity, pesticide use and management, pesticide types, pesticide storage system, type of irrigation facilities, type of land preparation methods etc. Collected data was analyzed using descriptive statistics (means, frequency distributions, and percentages), inferential statistics (chi-squared test), partial budgeting and cost analysis. Microsoft 'EXCEL' and IBM 'Statistical Package for the Social Sciences (SPSS)'Software Package Version 16 were used to process the elicited data.

Results and Discussion

Household and farm characteristics of respondent farmers

Most of the respondent household heads were male (75.9%), within 40-49 years age bracket (35.7%) as indicated in (Figure 1) and had achieved a Middle School or Junior High School certificate (42.6%) (not reported). This is an interesting result since farmers within 40-49 years age bracket are a very component of the active labour force are mostly literate. Thus adoption of improved farming methods and/or technology for this age group can be better enhanced. This makes the prospects of an improved vegetable production very positive. However, 22.2% of the household heads are illiterates. This is of critical concern for the growth of the vegetable production industry since abuse of pesticide by vegetable farmers has been partly attributable to their high illiteracy levels [17]. The size of farms of the majority of the respondents ranged from less than (0.4 ha) up to 4 ha and is in conformity with the observation that, the majority of respondents in the study area are smallholder farmers. It is indeed reported that agriculture is predominantly on a smallholder basis in Ghana [18]. Vegetable

farmers with less than 10 years experience in farming constitute 40.3% but the number of farmers decreased with increased number of years in farming (Figure 1). This indicates that the number of people engaging in vegetable production is increasing, perhaps because of increasing demand for vegetables for the increasing urban population.

Pesticide acquisition and storage

Majority of the farmers (90.8%) obtain their pesticides from local agrochemical input dealers (Figure 2). This is not surprising as the majority of the respondent base are unable to distinguish between different pest and disease pathogens and control measures such as insecticides and fungicides and rely on information and advice provided by local agro-input dealers for the decision making. In the course of the focus group discussion however, farmers did classify and confirm that some pesticides were deemed effective on some vegetables and not on others Consequently, farmers perception of the effectiveness of chemical pesticides based on advice provided by agro-input dealers in managing pests and diseases in vegetables might be a major contributing factor to their excessive use and hence their abuse or misuse, as has already been earlier reported.

Continuous use of the same pesticide against a particular pest can lead to the development of resistance by the pest against the pesticide, thereby rendering the pesticide ineffective [19]. Unfortunately, this is the currently being practised by most of the vegetable farmers (77.4%) in the surveyed communities and perhaps several other parts of Ghana. Field observations showed for example that, in Sefwi Bekwai, a large cultivated cabbage farm had been abandoned by a respondent prior to crop maturity. This observation is partly attributed to ineffectiveness of various pesticides applied on the field to control the diamond back moth that is the major pest of crucifers. The concerned respondent confirmed that he applied four different types of insecticides on the cabbage plants with the perception that diverse combination of chemicals might be more effective than a single type. After probing further, it was later discovered, that the 4 different insecticides that were applied had the same active ingredient and similar concentrations, albeit with different trade names, a fact that is hardly understood by the farmer.

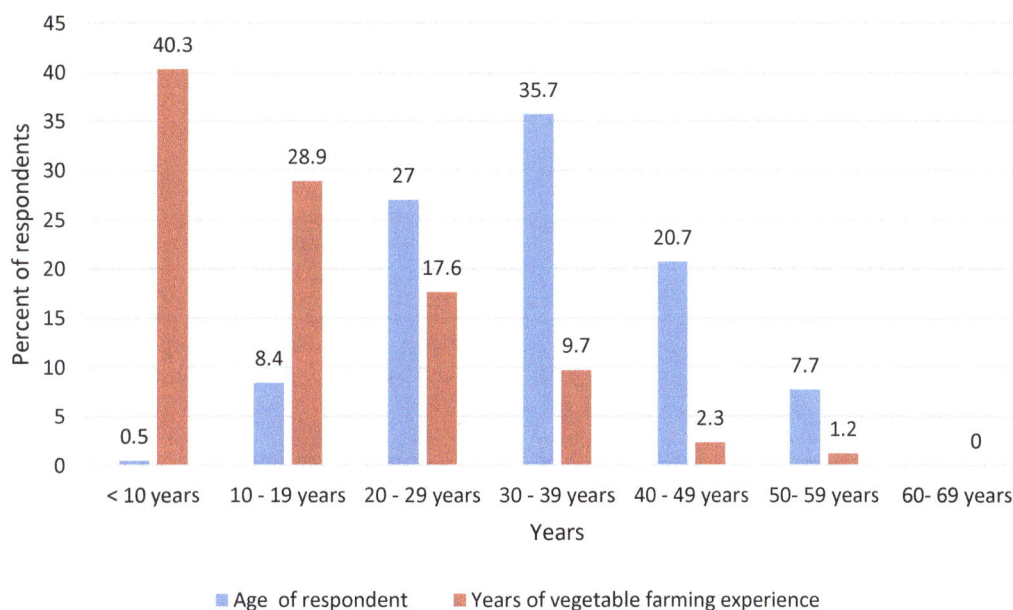

Figure 1: Household and farm characteristics of respondent farmers.

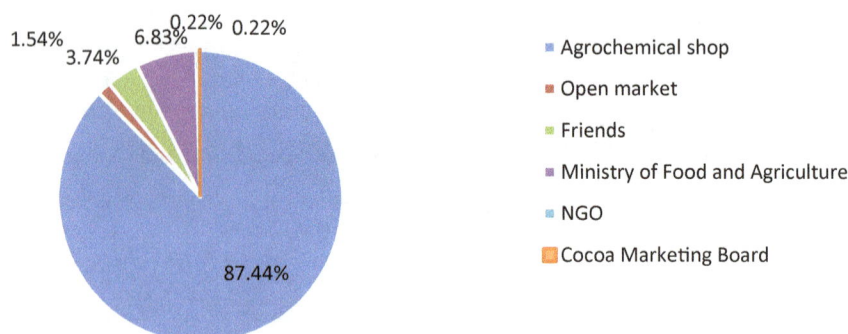

Figure 2: Source of agrochemicals by respondents.

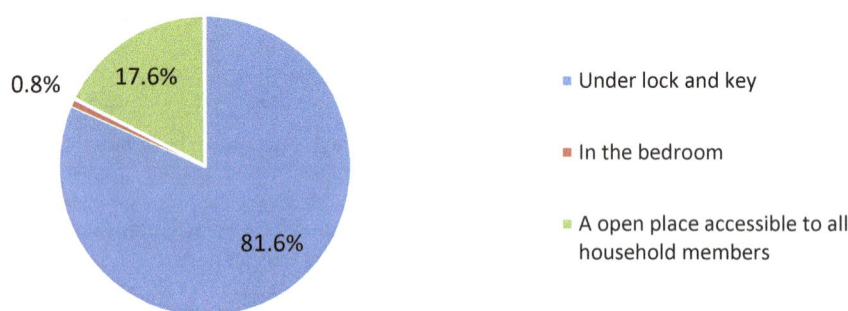

Figure 3: Mode of pesticide storage after harvest.

Results from (Figure 3) show that, even though majority of the respondents (81.6%) appear to store pesticides at a safe place that is under lock and key after procurement, a large number of them (17.6%) store them in their bedrooms, thereby exposing them to toxicity through direct inhalation of the pesticides. [6] Ntow for example confirms that storing pesticides in open accessible places such as bedrooms may lead to acute and/or chronic exposures, with adverse health consequences [20]. It has also been reported that in late 2010, 15 farmers died from suspected pesticide poisoning in Upper East region of Ghana and most of these deaths resulted from poor storage of pesticides, which seeped into food stocks [14].

Pesticide use by vegetable farmers

Over half of the farmers (56.8%) had received training on safe handling and application of pesticides, while 43.2% had received no training (Table 2A). The observed results are consistent with the findings of Fianko et al. [21]. For the Densu River basin of Ghana. The relatively large cohort of respondents with no technical knowledge in pesticide use can be a major source of worry given the absence of farmer training has been found to further increase the heightened danger of pesticide misuse and abuse in vegetable production [22]. The misuse of pesticide by the farmers can also endanger their health and that of consumers as well as the environment [8]. It appears that even those who claim to have received some form of official training seemed to be still misusing and abusing pesticides in their vegetable fields. For instance, results from (Table 2C) indicate that most farmers apply the pesticide either at the sight of a pest and/or disease (52.2%) or according to crop calendar (45%), corroborating the report of Amoako et al. [23].

As noted in (Table 2E), pesticide mixture preparation for spraying

were done mainly by shaking knapsack sprayers as represented by 47.8% of respondents or using a stick (43.8%). Some respondents (6.2%) however confirmed mixing pesticides with their bare hands. This is obviously unacceptable and disturbing since the farmers concerned will be directly exposed to the hazardous of the pesticides. As reported by Amoako et al. [23], most farmers mix two or more pesticides together without considering their compatibility or active ingredients but rather rely on the perceived efficacy based on their trade names. Mixing of pesticides was encouraged by the farmers' desire to have rapid knockdown of pests or the economics of managing both pests and diseases at a single spraying operation. This idea is however, questionable [24] at least as practised [2], because the combinations used could be indiscriminate and incompatible resulting in ineffectiveness of the pesticides to manage the pests and diseases. This findings are also consistent with that of Biney [25] who attributed the increase in incidences of insect pest infestation of tomato in Ghana to the practice of using indiscriminate combinations of pesticides, particularly of insecticides.

With respect to pesticide application procedures, the knapsack was the most popular spraying equipment used (88%), though a few farmers did use motorised sprayers/mist blowers (3%) and hand held applicators (4.3%) as confirmed in Table 3B. Lack of capital was the main reason for farmers' inability to buy required equipments such as motorised sprayer, hence their intended use of knapsack sprayers. Over 17% of respondents were found not to own a sprayer at all. In the course of the focus group discussions, some farmers without access to a knapsack sprayer reported using a brush, broom or leaves tied together to splash pesticides from a bucket as their means of spraying. Consequently, such practices expose users to the harmful effects of pesticides, especially as most of the farmers do not wear protective clothing when spraying.

Variable	Total respondents	
	N	%
A. Have you ever received training on safe handling and application of pesticides?		
Yes	248	56.8
No	187	43.2
Total	435	100
B. Source of farmers' knowledge onpesticide application rates		
Pesticide dealer	54	20.1
Fellow farmer	48	17.8
Agricultural extension agents	149	55.4
Media (Radio, TV, Newspaper)	22	8.2
Pesticide label	6	2.2
COCOBOD	5	1.9
NGOs	6	2.2
School	2	0.7
Government	1	0.4
Total	269	100
C. Timing of pesticide application		
When first symptoms of pests /diseases are observed	229	52.2
Based on severity of pest infestation / disease infection	30	6.9
Based on crop calendar / date of transplanting	197	45
Based on advice from Agricultural Extension Agents	1	0.2
D. Do you mix different kinds of pesticides?		
Yes	186	42.6
No	251	57.4
Total	437	100
E. How do you mix pesticides?		
With bare hands	25	6.2
With stick	178	43.8
Shaking the sprayer	194	47.8
Wear hand gloves and protective eye goggles	9	2.2
Total	406	100
F. What direction do you spray?		
With the wind	283	76.1
Against the wind	6	1.6
Perpendicular to the wind	33	8.9
Don't consider the wind	50	13.4
Total	372	100
G. Do you eat or smoke while spraying?		
Yes	8	1.8
No	429	98.2
H. Sign to indicate to people field has being sprayed with pesticide		
Sign board	27	6.6
Red flag	24	5.9
Empty pesticides bottle	31	7.6
None	327	80
Total	409	100
I. Farmer re-entry interval (days)		
Same day	23	5.7
1-3	321	79.65
4-7	56	13.9
8-14	3	0.7
> 14	0	0
Total	403	100
J. Pre-harvest interval (in days)		
Same day	6	1.4
1-3	38	9.1
4-7	319	76.3
8-14	48	11.5
> 14	7	1.7
Total	418	100.0

Table 2: Pesticide application by vegetable farmers.

Variable	Total respondents	
	Frequency	Percentage
A. Disposal of empty pesticide containers		
Incineration	52	13.1
Burying	100	25.3
Throw away on farm	237	59.8
Throw away in town or village	7	1.8
Total	396	100.0
B. Type of sprayer used		
Knapsack sprayer	388	92.4
Motorized sprayer/Mist blower	13	3.1
Hand held applicator	19	4.5
Total	420	100.0
C. Own a sprayer		
Yes	350	82.7
No	73	17.3
Total	423	100.0
D. Do you wash sprayer after use?		
Yes	335	76.7
No	102	23.3
Total	437	100
E. Disposal of waste water		
On the field	346	79.2
In nearby stream	7	1.6
On floor within premises of own household	14	3.2
In a septic tank within compound	2	0.4
At nearby bush	24	5.4
Bury waste in a hole	1	0.2
Spray over the waste (garbage heap)	1	0.2
In a dug pit	1	0.2
Total	396	100.0

Table 3: Disposal of pesticides containers and waste water from sprayers.

Most farmers own and utilize aknapsack sprayer, yet the use of this type of sprayer in itself presents some danger to the user. According to Ntow et al. [2] it is prone to leakage, especially as the spray equipment ages. Matthews etal. [9] have identified causes of leakage from the knapsack and have emphasised the need to provide better-quality equipment at an affordable cost that will be more durable in a hot and humid tropical environment such as sub-Saharan Africa.

Most farmers are adopting safer pesticide application practices such as spraying against the dwind irection, not eating or smoking during spraying so as to prevent respective potential dermal and oral contamination with pesticides. However, majority of the respondents do not display warning signs after spraying so as to prevent public or any member of the household from entering a sprayed field. This is not surprising because majority of the farmers even re-enter a sprayed field within 24 hours. This could be a major reason why pesticide poison is common among most smallholder farmers in Ghana.

The study further revealed that most vegetable farmers harvest their produce within 7 days after spraying pesticides with some harvesting their produce on the same day after spraying, thereby endangering the lives of consumers. Amoako et al. [23] also reported that majority of cabbage farmers in the Ejisu-Juaben Municipality of the Ashanti Region of Ghana continue spraying pesticides during produce harvesting, hence no waiting period is observed, thereby exposing consumers to high pesticide residue levels. Residues of Chlopyrifos (Dursban), lindane, endosulfan, Karate and DDT have been detected beyond maximum permitted residue levels in samples of lettuce from major markets from Kumasi, Accra and Tamale, Amoah et al. [23] Darko and Akoto [26] also assessed contamination levels and health risk hazards of organo phosphorus pesticides residues in tomatoes and eggplant and showed that health risks are associated with levels of pesticides exceeding the recommended doses for these vegetables. Death cases resulting from consumption of pesticide contaminated vegetables have already been reported in some parts of the country. In early December 2010, the then Upper East Regional Minister, Mark Woyongo, announced that 12 farmers had died after eating food contaminated with pesticides, and that a further 63 had been treated and discharged from hospital [27]. Personal communication with some consumers indicates that they are very wary of consuming vegetables such as cabbage and okra which they believe are contaminated with pesticides.

Disposal of pesticides containers and waste water by the vegetable farmers

The commonest way of disposing of empty pesticide containers (59.8%) and waste water from the sprayer (79.2 %) among the respondent farmers was by throwing or discharging them on the field (Tables 3A and 3E), as also confirmed by Ntow et al. [2]. During field observation, empty pesticide containers were found loitered in some farms (Plates 1 and 2), and in some farms the containers were found close to water bodies. This can potentially pollute the water bodies which are sources of livelihood for human communities and support varied animal and plant life as also reported by Ntow et al. [2]. The authors further asserts that accumulation of such agrochemical pollutants in the tissues of non-target fauna and flora, which ultimately accumulate in the food chain, may restrict the consumption of valuable food resources such as fish. It is therefore not surprising that organo chlorine pesticide residues (DDE) was detected in tilapia fish and water samples from Lake Bosumtwi [26,28], and in fish samples in four Lagoons in Ghana [29]. The health risk associated with pesticide contamination of fish from the Densu River Basin in Ghana have also been reported by Fianko et al. [21].

It was also revealed during the group discussion that some farmers use the empty pesticide containers for storing food items such as salt and palm oil, and as containers for kerosene. This practice appears to be common among farming communities in Ghana. NPAS [14] has also reported a widespread re-use of containers for storing food or water for humans or livestock.

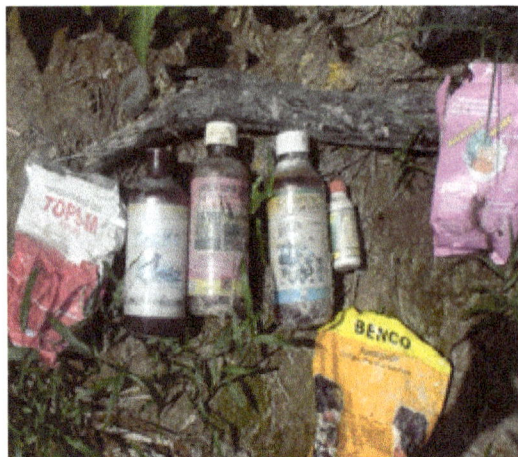

Plate 1: Insecticide containers found in a cabbage farm at Sefwi Bekwai in the tomato field at Akumadan in the Western Region of Ghana.

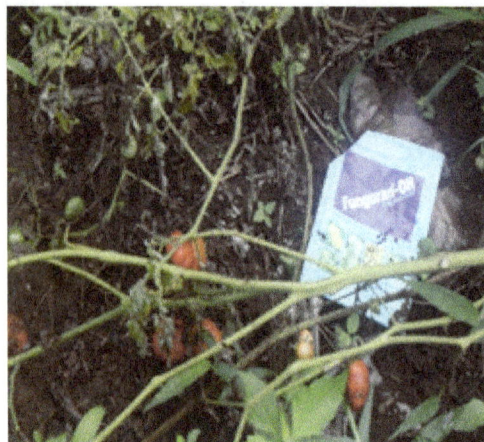

Plate 2: Fungicide container found in Ashanti Region of Ghana.

Farmers' knowledge of pesticides hazards

Results from Table 4E show that, the most common pesticide poisoning side effects mentioned by the farmers based on multiple responses were itching (64.3%), headache (26.1%), weakness (24.3%) and dizziness (11.7%). Some farmers also mentioned burning sensation, catarrh, stomach pain, unconsciousness, itching of eyes and body pains as hazards associated with use of pesticides. The findings of this study therefore corroborates that of Ntow et al. [2] and NPAS [14] who concluded that the most common symptoms of pesticide poisoning among Ghanaian farmers include skin irritations, headaches, general body weakness, difficulty in breathing and dizziness.

Only 15.6% of the respondents protect themselves fully during their spraying operations; others either wear partial protective clothing (38%) or do not wear any protective clothing at all (46.4%) and come into direct contact with the pesticides (Table 4B). This may partly explain why majority of the respondents said that the pesticide solutions come into contact mainly with their hands (53.8%), back (48.1%), feet (33.2 %), and face (3.1%) resulting in poisoning. A survey of pesticides use among farmers in the 3 Northern regions of Ghana also revealed that most farmers fail to use any protective equipment while virtually no farmers use all the recommended equipment [14]. It is reported that Ghanaian farmers who use chemical pesticides to control insects and diseases on their crops are potentially exposed to pesticides through the skin, on the eyes or through inhalation or ingestion, with key risks being death, cancer, birth defects and damage to the nervous system [14]. Apart from not wearing protective clothing, about 25% of the vegetable farmers do not change their clothing immediately after spraying operation, leading to their long exposure to the pesticides, with consequent poisoning.

Relationships between pesticide side effects and gender, education and age of the farmers

The susceptibility of male and female farmers to the hazards of pesticides differed among the farmers interviewed (Table 5). For common symptoms of pesticide poisoning such as headache, dizziness, vomiting and stomach pains, females were found to be more vulnerable than males. However, male farmers were equally as susceptible as their female counterparts with respect to itching which is the most common symptom reported by both genders. This perhaps explains why pesticide application is usually carried out by males.

Variable	Total respondents	
	Frequency	Percentage
A. Likert scale ranking of pesticide hazards on a scale 1-10 (1 is lowest risk and 10 is highest risk)		
1	19	5
2	35	9.3
3	29	7.7
4	65	17.2
5	47	12.4
6	56	14.8
7	20	5.3
8	70	18.5
9	4	1.1
10	33	8.7
Total	378	100
B. Type of protective cover used		
No protective cover	193	46.4
Partial protective cover	158	38.0
Full protective cover	65	15.6
Total	416	100
C. Part of human body where does pesticide solution comes into contact?		
Hand	235	53.8
Feet	145	33.2
Back	210	48.1
Face	14	3.1
Lower abdomen, waist and thighs	4	0.8
Any part	1	0.2
D. Do you change clothes right after spraying?		
Yes	310	74.9
No	104	25.1
Total	414	100
E. Common symptoms associated with frequent pesticide poisoning among farmers		
Headache	114	26.1
Dizziness	51	11.7
Vomiting	6	1.4
Weakness	106	24.3
Itching	281	64.3
Stomach pain	7	1.6
Unconsciousness	7	1.6
Burning sensation	14	3.1
Catarrh	11	2.5
Itching of eyes	4	0.8
Body pains	1	0.8
Nausea	1	0.2

Table 4: Farmers' knowledge of pesticide application side effects.

Generally, farmers with no formal or non-formal education were more likely to be affected by the pesticides as they complained of more poisoning symptoms than those who have received formal education (Table 6). This findings partially agrees with Asante and Ntow [2] who reported that workers exposed to pesticides are often illiterates, and lack training, equipment, and the necessary safety information. There appears to be no clear association between pesticide poisoning symptoms and the farmers of the various age groups. However, only young farmers within the age group of 10-19 years did experience body pains (Table 7). This indicates the vulnerability of young farmers to pesticide poisoning. This agrees with Ntow et al. [2] who reported that generally, possible poisoning cases are reported more among the young than the aged farmers. Caution should however be exercised not to overemphasize our present study findings, as only few farmers (2%) were within this age category. Ntow et al. [2] however, observed that

the percentage of farmers reporting about body weakness and itching/irritation increased from a younger group to a relatively much older aged group whereas headache/dizziness were reported more among the younger group. It was also observed that farmers with all age distribution reported more of itching than the other symptoms.

Types of pesticides applied in vegetable production

A total of 43 pesticides were found in use for vegetable farming in the Ashanti and Western regions of Ghana. The pesticides consisted of 7 fungicides, 9 herbicides and 30 insecticides (Table 8). It is important to note that one systemic insecticide, Carbofuran is used by most farmers both as an insecticide and nematicide as they perceive and have also found it effective in the short-run. The class of pesticides commonly used by vegetable farmers in the surveyed area was insecticide (61.7%), followed by fungicide (32.7) and herbicides (5.5%). On the contrary, in a similar work conducted in Ghana by Ntow et al. [2] herbicide was found to be the most commonly used pesticide (44%), followed by insecticide (33%) and fungicide (23%). This supports the respondents' general perception that pesticides were chosen mainly for the control of pest organisms and hardly for disease pathogens. In addition, farmers desire to satisfy consumers taste (preference for unblemished, cosmetically perfect produce with extended shelf and storage life) and to produce high yields could account for the high proportions of fungicides and insecticides used, as reported by Thomas [30]. The classification of these pesticides by the category of pests they control, active ingredient, chemical group and the World Health Organization (WHO) Hazard category is presented in Table 8 as well. Three out of 7 fungicides used by the farmers are not registered by the Ghana Environmental Protection Agency. Even those that were registered are strictly meant for cocoa and /or coffee, and belong to WHO Hazard Category III, designated as highly hazardous. This is quite serious and requires immediate action to prevent harmful toxic public health hazards and environment pollution. Generally, most farmers use insecticides, fungicides and herbicides which are not cleared for use on target vegetable crops or they are applying pesticides which are not registered for vegetable production. The findings of the presents study thus supports the report of Amoako et al. [23], who confirmed that certain banned chemicals (i.e., Lindane, Endosulfans and DDT) and those not recommended for vegetables (i.e., *Akatemaster* which contains bifenthrin, *Confidor* which contains Imidacloprid and thiamethoxam and Cocostar (contains bifenthrin and pirimiphosmethyl) are being used for cabbage production by farmers in the Ejisu-Juaben Municipality of the Ashanti Region of

Symptoms of pesticide poisoning	Female (n=105)		Male (n=330)	
	F	%	F	%
Headache	35	33.3	79	23.9
Dizziness	25	23.8	26	7.9
Vomiting	3	2.9	3	0.9
Weakness	34	32.4	72	21.8
Itching	65	61.9	215	65.2
Stomach pain	1	1.00	6	1.8
Unconsciousness	3	2.9	4	1.2
Burns	0	0.0	14	4.2
Body pains	2	1.9	2	0.6
Catarrh	2	1.9	8	2.4
Eye irritation	0	0.0	4	1.2
Nausea	0	0.0	1	0.3
Sneezing	0	0.0	1	0.3

Table 5: Relationship between pesticide poisoning and gender.

Type of poisoning effect	Educational background of farmer					
	No Formal Education (f=109)	Non- Formal (f=8)	Primary School (f=67)	Middle or Junior Secondary High School (f=184)	Junior Secondary High School(f=77)	Tertiary Education (f=7)
Headache	25.7	50.0	19.4	29.3	16.9	28.6
Dizziness	9.2	50.0	9.0	12.5	10.4	57.1
Vomiting	1.8	25.0	3.0	0.5	1.3	28.6
Weakness	24.8	12.5	26.9	26.6	10.4	28.6
Itching	71.6	50.0	67.2	66.8	36.4	57.1
Stomach pain	1.8	25	3.0	1.1	1.3	100.0
Unconsciousness	2.8	12.5	0.0	1.6	0.0	0.0
Burns	5.5	0.0	0.0	2.3	2.6	0.0
Body pains	0.0	50.0	3.0	1.1	0.0	0.0
Catarrh	0.0	12.5	0.0	4.3	0.0	0.0
Eye irritation	0.0	0.0	0.0	1.1	2.6	0.0
Nausea	0.0	0.0	0.0	0.5	0.0	0.0
Sneezing	0.9	0.0	0.0	0.0	0.0	0.0

Table 6: Relationship between pesticide poisoning and educational background of the farmers.

Hazard	Age distribution of farmer						
	10-19 (f=2)	20-29 (f=36)	30-39 (f=116)	40-49 (f=153)	50-59 (f=89)	60-69 (f=33)	Total (n=429)
headache	0.0	19.4	25.9	27.5	30.3	15.2	25.9
dizziness	0.0	16.7	11.2	9.2	14.6	12.1	11.7
vomiting	0.0	0.0	0.9	1.3	2.2	3.0	1.4
weakness	0.0	30.6	23.3	22.2	29.2	18.2	24.2
itching	50.0	77.8	59.5	66.0	65.2	60.6	64.6
stomach pain	0.0	0.0	2.5	2.6	0.0	0.0	1.6
unconsciousness	0.0	5.6	0.9	0.0	2.2	6.1	1.6
Burns	0.0	2.8	2.5	3.3	3.4	3.0	3.0
Body pains	100.0	0.0	0.9	1.3		3.0	1.4
Catarrh	0.0	0.0	32.5	3.3	1.1	0.0	2.1
Eye irritation	0.0	2.8	0.0	1.3	1.1	0.0	0.9
Nausea	0.0	0.0	0.0	0.0	0.0	3.0	0.2
Sneezing	0.0	0.0	0.0	0.0	0.0	3.0	0.2

Table 7: Relationship between pesticide poisoning and age of the farmers.

	Common Name	Active Ingredient	Registration status	Crops	Hazard Class
Fungicide (32.7%)	Champion	Copper Hydroxide	FRE	Cocoa and Coffee	III
	Funguran-OH	Copper Hydroxide	FRE	Cocoa	III
	Cobox	Copper Oxychloride	UNREG		
	Kocide	Cupric Hydroxide	FRE	Cocoa	III
	Dithane	Mancozeb	UNREG		
	Ridomil Gold	Metalaxyl-M + Cuprous oxide	FRE	Cocoa	III
	Topsin	Thiophanate methyl	UNREG		
Herbicide (5.5%)	Condemn		UNREG		
	Agrazine	Atrazine	PCL	Various Crops	II
	Adwumapa	Glyphosate	FRE	Vegetables and cereals	III
	Adwumawura	Glyphosate	FRE	Vegetables and cereals	III
	Roundup	Glyphosate	FRE	Various Crops	III
	Sunphosate	Glyphosate	PCL	Various Crops	III
	Weed-out	Glyphosate	FRE	Various Crops	III
	Gramoquat	Paraquat dichloride	FRE	Various Crops	II
	Gramozone	Paraquat dichloride	UNREG		

	Poison		UNREG		
	Buffalo	Acetamiprid	FRE	Vegetable and Fruits	III
	cocoprid	Acetamiprid	FRE	Cocoa	II
	Golan	Acetamiprid	FRE	Vegetables and Fruits	III
	Phostoxin	Aluminium Phosphide	FRE	stored grains	Ib
	Akate Master	Bifenthrin	FRE	Cocoa	II
	Multifos	Chlorpyrifos	UNREG		
	Conpyrifos	Chlorpyrifos ethyl	FRE	Vegetables and cereals	II
	dursban	Chlorpyrifos ethyl	FRE	Various Crops	II
	Sunpyriphos	Chlorpyrifos ethyl	UNREG		
	Termicot	Chlorpyrifos ethyl	PCL	Various Crops	II
	Polythrine C	Cypermethrin	FRE	Vegetables	II
	DDT	DDT	BANNED		
Insecticide (61.7%)	Akatesuro	Diazinon	PCL	Cocoa	II
	Attack	Emamectin benzoate	UNREG		
	Control	Emamectin benzoate	FRE	Vegetables	II
	Confidor	Imidacloprid	FRE	Cocoa	II
	Consider	Imidacloprid	FRE	Vegetables	II
	bossmate	Lambda Cyhalothrin	UNREG		
	Clear	Lambda Cyhalothrin	FRE	Vegetables	III
	Kombat	Lambda Cyhalothrin	UNREG		
	Karate	Lambda Cyhalothrin	FRE	Vegetables	II
	Lambda	Lambda Cyhalothrin	FRE	Vegetables	II
	K-optimal	Lambda Cyhalothrin	FRE	Vegetables	II
		+ Acetamiprid			
	Bypel	Perisrapae Granulosis Virus	PCL	Cabbage	IV
		+ Bacillus thuringiensis			
	Actellic	Pirimiphos-methyl	FRE	Various Crops	III
	Super Agro Blaster	Pyrethrum	FRE	Stored produce	II
	Actara	Thiamethoxam	FRE	Banana	III
Insecticide/ Nematicide (0.1%)	Furadan	Carbofuran	FRE	Various Crops	II

Note: PCL: Provisional Registered List; FRE: Fully Registered List; UNREG: Unregistered

Table 8: Types of pesticides and active ingredients used by respondents.

Ghana. This suggests that such farmers are misusing such pesticides thereby affecting the quality and safety of vegetables for consumption. It is suggested that vegetable farmers should sensitized and be trained in integrated crop and pest management practices so as to minimize use of pesticide and also desist from using unregistered and unapproved pesticides. One official from Plant Protection and Regulatory Services Division of the Ministry of Agriculture however claims that there is more to this as "the farmers have been adequately informed about the consequences of pesticide abuse but they are just refusing to heed to the advice". During the farmers' interview and the focus group discussion, this assertion was partially confirmed was clear that most of them know that some pesticides are not meant for application on vegetables yet they applied them because they claimed they were effective, disregarding their potential risk to themselves and the environment. For instance *Akatemaster* (Bifenthrin) which is fully registered for cocoa pests only are commonly used by vegetable farmers because they claimed it was very effective.

Conclusions

It is clear from the results of the study that the majority of the vegetable farmers obtain pesticides from agrochemical shops mainly for pest agent control, mix different chemicals together and apply the pesticides without wearing protective clothing. Knapsack was the most popular spraying equipment used though a few farmers did use motorised sprayers/mist blowers and hand held applicators. However most farmers enter fields treated with pesticides within 7 days without observing re-entry waiting period and also harvest produce within 7 days after spraying pesticide without observing the recommended waiting period. It was further shown that most farmers dispose of empty pesticides containers by throwing them on the field. The study also revealed that farmers were mostly aware of the hazardous risks which they are exposed to from pesticide application but show a negative attitude to taking the necessary precautionary measures. The most common consequences of pesticide exposure mentioned by the farmers were itching, headache, weakness, and dizziness. Some farmers also mentioned burning sensation, catarrh, stomach pain, unconsciousness, itching of eyes and body pains. Females and illiterates were found to be more likely to be affected than their males and literate counterparts. It can therefore be concluded that the farmers are misapplying the pesticides by disregarding the dangers they cause to human health and the environment as a result of which re-enforcement awareness creation and behavioural change communication are required to change their attitudes in addition to the need for training in integrated crop and pest management practices to minimize pesticide application for safer vegetable production.

Acknowledgements

The financial support of the Integrated Agricultural Systems of the humid tropics program (Humid tropics), a CGIAR Research Program through AVRDC – The World Vegetable Center for this research study is gratefully acknowledged.

References

1. Owusu-Boateng G, Amuzu KK (2013) A survey of some critical issues in vegetable crops farming along River Oyansia in Opeibea and Dzorwulu, Accra-Ghana. Global Advanced Research Journal of Physical and Applied Sciences 2: 024-031.

2. Ntow WJ, Gijzen HJ, Kelderman P, Drechsel P (2006) Farmer perceptions and pesticide use practices in vegetable production in Ghana. Pest Manag Sci 62: 356-365.

3. Gyau A, Spiller A (2007) The role of organizational culture in modelling buyer seller relationships in the fresh fruit and vegetable trade between Ghana and Europe, African Journal of Business Management1 : 218-229.

4. Armah M (2010) Investment opportunity in Ghana: Chili pepper. Millenium Development Authority, Accra, Ghana.

5. Dinham B (2003) Growing vegetables in developing countries for local urban populations and export markets: problems confronting small-scale producers. Pest Management Science 59: 575-582.

6. Amoah P, Drechsel P, Abaidoo RC, Ntow WJ (2006) 'Pesticide and pathogen contamination of vegetables in Ghana's urban markets. Archives of Environmental Contamination and Toxicology 50: 1-6.

7. Gerken A, Suglo JV, Braun M (2001) Pesticide use and policies in Ghana. MoFA/PPRSD, ICP Project, Pesticide Policy Project/GTZ, Accra, 185.

8. Ajayi OC (2005) Biological capital, user costs and the productivity of insecticides in cotton farming systems in sub-Saharan Africa. International Journal of Agricultural Sustainability 3: 154-166.

9. Matthews GA (2008) Attitudes and behaviors regarding use of crop protection products- A survey of more than 8500 smallholders in 26 countries. Crop Protection 27: 834-846.

10. Hurtig AK, San Sebastián M, Soto A, Shingre A, Zambrano D, et al. (2003) Pesticide use among farmers in the Amazon basin of Ecuador. Arch Environ Health 58: 223-228.

11. McCauley L, Beltran M, Phillips J, Lasarev M, Sticker D (2001) The Oregon migrant farm workers community: an evolving model for participatory research. Environ Health Perspect 109: 449-455.

12. Quandt SA, Arcury TA, Rao P, Snively BM, Camann DE, et al. (2004). Agricultural and residential pesticides in wipe samples from farmworker family residences in North Carolina and Virginia. Environ Health Perspect 112: 382-387.

13. Ntow WJ (2008) The use and fate of pesticides in vegetable-based agro ecosystems in Ghana. PhD Thesis. Wageningen University, The Netherlands

14. Ghana's pesticide crisis: The need for further Government action (2012) Northern Presbyterian Agricultural Services (NPAS) Northern Presbyterian Agricultural Services and Partners. Ghana 50.

15. Okorley EL, Zinnah MM, Bampoe EA (2002) Promoting participatory technology development approach in integrated crop protection among tomato farmers in Anyima in the Kintapo district of Brong Ahafo region, Ghana. AIAEE 2002 Proceedings of the 18th Annual Conference Durban, South Africa 337-343.

16. Warburton H, Palis FG, Pingali PL (1995) Farmer perceptions, knowledge and pesticide usepractices. In: Impact of pesticides on farmer health and the rice environment, Natural Resource Management and Policy 7: 59-95.

17. Asante KA, Ntow WJ (2009) Status of Environmental Contamination in Ghana, the Perspective of a Research Scientist. Council for Scientific and Industrial Research - Water Research Institute (CSIR WRI), P.O. Box AH 38, Achimota-Accra, Ghana.

18. Agriculture in Ghana. Facts and Figures (2011) Ministry of Food and Agriculture, (MoFA) Statistical, Research and Information Division, Accra.

19. University of Californis IPM guidelines (2009) Floriculture and ornamental nurseries: Managing pesticide resistance.

20. Ngowi AVF, Mbise TJ, Ijani ASM, London L, Ajayi OC (2007) Pesticides use by smallholder farmers in vegetable production in Northern Tanzania. Crop Prot 26: 1617–1624.

21. Fianko JR, Donkor A, Lowor ST, Yeboah PO, Glover ET et al. (2011) 'Health risk associated with pesticide contamination of fish from the Densu River Basin in Ghana'. Journal of Environmental Protection 2: 115-123.

22. Abang AF, Kouame CM, Abang M, Hannah R, Fotso AK (2013) Vegetable growers perception of pesticide use practices, cost, and health effects in the tropical region of Cameroon. International Journal of Agronomy and Plant Production 4: 873-883.

23. Amoako PK, Kumah P, Appiah F (2012) Pesticides usage in Cabbage (Brassica oleracea) Cultivation in the Ejisu-Juaben Municipality of the Ashanti Region of Ghana. International Journal of Research in Chemistry and Environment 2: 26-31.

24. Medina CP (1987) Pest Control Practices and pesticide perceptions of Vegetable farmers in Loo Valley, Benguet, Philippines. In: Management of Pests and Pesticides: Farmers' Perceptions and Practices, ed. by Tait J and Napompeth B, Westview Press, London 150-157.

25. Biney PM (2001) Pesticide use pattern and insecticide residue levels in Tomato (Lycopersicum esculentum) in some selected production systems in Ghana. Mphil Thesis, University ofGhana, Legon, Ghana 127.

26. Darko G , Akoto O (2008) 'Dietary intake of organo phosphorus pesticide residues through vegetables from Kumasi, Ghana', Food and Chemical Toxicology 46: 3703-3706.

27. Anon (2010)'Ghana: Twelve die after poisonous food', 4 December 2010, www. afriquejet.com

28. Afful S, Awudza JAM, Osae S, Twumasi S K (2013) Persistent organochlorine compounds in the water and sediment samples from the Lake Bosomtwe in Ghana. American Chemical Science Journal 3: 434-448.

29. Essumang D, Togoh GK, Chokky L (2009) Pesticide residues in the water and fish samples from lagoons in Ghana. Bulletin of the Chemical Society of Ethiopia 23: 19-27.

30. Thomas MR (2003) Pesticide usage in some vegetable crops in Great Britain: real on-farm applications. Pest Management Science 59: 591-596.

22

Inoculated Traps, an Innovative and Sustainable Method to Control Banana Weevil *Cosmopolites sordidus* in Banana and Plantain Fields

Aby N[1]*, Badou J[2], Traoré S[1], Kobénan K[1], Kéhé M[1], Thiémélé DEF[1], Gnonhouri G[1] and Koné D[2]

[1]*Laboratory of Entomology and Plant Pathology, National Centre of Agronomic Research, Research Station of Bimbresso; 01 BP 1536 Abidjan 01*

[2]*Laboratory of plant physiology, University of Félix Houphouët Boigny Abidjan, 01 BP 852 Abidjan 01*

Abstract

The banana weevil *Cosmopolites sordidus* (Germar) is the most important insect pest of banana and plantain (Musa spp.). The larvae by its trophic activities bore into the corm, present roots are destroyed and the emergence of new roots is delayed, reducing nutrient uptake and weakening the stability of the plant. Attacks in newly planted banana stands can lead to crop failure. In established fields, weevil damage can result in reduced bunch weights, mat die-out and shortened stand life. Damage and yields losses tend to increase with time. Banana weevil has been implicated as a primary factor contributing to the decline and disappearance of East African highland cooking banana from its traditional growing areas in central Uganda. Control measures are essentially based on the use of chemicals. Adult resistance, environmental pollution, most toxicity to famers also to consumers, and its effect on non target insect were negative side effects of chemicals. Alternative sustainable methods (use of parasitoïds, predators, mass trapping with or without pheromon) were experimented, but results were few interest. Biological control using entomopathogenic fungi within genera *Metarhizium* and *Beauveria bassiana* were performed. Although promise results are shown in preliminary studies, efficient delivery systems still need to be developed. Research protocols for the development of a microbial control program of banana weevil using *Metarhizium* has focused in Côte d'Ivoire since 2008. It based on inoculated disk on stump traps (culombia trap), pheromon traps, pseudostem traps including its gregariousness to disseminate inoculum among weevils for sanitation of established field and biorationals. So these methods appear to be the promise ones significantly least fastidious and economic.

Keywords: *C. sordidus*; Entomopathogenic fungus; Mass trapping; Inoculated trap; Sordidin pheromone; Organic substrate

Introduction

The banana weevil *Cosmopolites sordidus* (Germar) is the most important insect pest of bananas and plantains (Musa spp.). Yields losses up to 100% in cases of severe infestation. The control is based on farming practices including the use of healthy planting material, systematic adult trapping, the removal of crop residues and especially the use of chemicals. Chemicals resistance [1], the elimination of useful non-target fauna and the accumulation of pesticide residues are the unmanageable consequences of chemical control. The use of resistant or tolerant plant [2] is an attractive alternative. However entomopathogenic fungi is more interesting because of its environmental aspects. Isolates belonging two genera *Beauveria* and *Metarhizium* were experienced in the banana producing countries. Mortality rates in the laboratory reach 100% [1].

In Côte d' Ivoire, 11 high pathogenic isolates of *Metarhizium* from mycosed weevils were collected in banana plantations. Pathogenicities tests performed in laboratory showed TL50 ranging from 6 to 10 days and TL90 6 to 40 days [3]. In order to confirm the performances and create optimal application conditions, an isolate was therefore tested in different planting methods of application in the traps commonly used by farmers particularly to control the banana weevil. This paper reviews the research activities since 2008 starting with the bait of weevils naturally infected by entomopathogenic fungi by using pseudo stem traps. Then, the efficiencies of Columbian traps (disc on stump trap) on which inoculum was laid in plate (plate traps) or on washers plywood on corm (plywood traps). These activities aimed to reach a practical, efficient, less costly, less time consuming and less labor intensive method that is currently acute. The document is therefore not an exhaustive review of results obtained but its presents series of methods conducted in Côte d'Ivoire exploiting local isolates of *Metarhizium* sp. for the field control of banana weevil *Cosmopolites sordidus* .

Autochthonous strains of entomopathogenic fungi Research

A survey of 17 banana plantations reaching 6500 ha was conducted from April to August 2008, during which 3.171 pseudo term traps (Figure 1) were baited in 294 plots. 11 isolates of *Metarhizium* sp. were isolated from mycosed banana weevils naturally infected.

Baiting and weevils removing: Traps were visited twise, 6th and

Figure 1: Pseudo stem trap used to collect infected weevils.

*Corresponding author: Aby N'goran, CNRA, de Research Station of Bimbresso, BP 1536 Abidjan 01, Côte d'Ivoire, E-mail: aby_ngoran@yahoo.fr

14th days after baiting. The adults weevils were counted in the 1st visit and left in the traps to ensure natural infection by soilborn fungi. Then, they were recounted and collected to the laboratory in transparent plastic containers (17 × 13 × 6 cm) containing pieces of fresh corm in the 2nd visit. Dead weevils were removed daily.

Isolation of fungi from mycosed weevils: Isolation of fungi from mycosed weevils was made following [4] method. Mycosed weevils were first disinfected in 1% sodium hypochlorite for 30 s and rinsed 3 min for 3 times with sterile distilled water. They are then wiped with sterile blotting paper. Disinfected weevils were placed on filter paper moistened with distilled water in Petri dishes and incubated in the dark room at ambient temperature until abundant mycelial appearance (Figure 2). Weevils were then singly immersed in 5% citric acid for 10 min and rinsed 3 times with distilled sterile water and then drained with filter paper. The mycelial fluff was then seeded in sterile Petri dishes on PDA medium. The Petri dishes were subsequently incubated in a dark room at 28 ± 2°C for 3 days. Mycelial colonies obtained were replanted on PDA medium for purification. An isolate is defined by colonies obtained on a dead weevil. The colony color of each isolate was noted. Isolate is then identified using method of Barry and Barnett (1972) key.

Inoculum production on rice substrate: After pathogenicity test in laboratory, the isolate of *Metarhizium* BME2 has been selected for field trials [5].The inoculum is prepared in 700 ml jars. After sporulation on PDA medium in Petri dishes for 15 days, 15 to 20 g of this culture was mixed with sterile parboiled rice filling 3/4 of the jar and homogenized by shaking. The jars are then incubated in a dark room at room temperature. They are shaken daily to prevent caking rice grains by the fungus (Figure 3a). The conidial concentration per gram of rice was count for jar 15 days later (Figure 3b). The average reached 12.10^8 spores/g.

Research on fields methods for inoculum implementation

The objective was to find field application method of the entomopathogenic fungus that could optimize its pathogenicity. Two types of traps were tested, plates trap and plywood trap.

The plate trap: The technique was to promote weevils attraction by kaïromone from the corm. The inoculum was placed in a plate on the corm sectioned horizontally (Figure 4). This plate prevents contact between inoculum and banana sap. Then inoculum was covered with

Figure 3: Sporulation of *Metarhizium* sp. grown on rice substrate (A) Vegetative growth. (B)Complete sporulation.

Figure 4: Plate trap with inoculum of *Metarhizium* [5]. The black arrows indicate the enter space for weevils.

another plate, leaving an entrance space of weevil adjusted by two sticks. The second plate protects inoculum for rainwater and irrigation. The device is finally protected by a gale weight piece of pseudostem about 10 cm length (Figure 4). Weevils access to the inoculum were performed by a piece of pheromone « COSMOPLUS » in the middle of the inoculum.

The plywood discks trap: This type of trap is similar in processing with plate trap. But no hole is dug in the corm to drop the disk. Plywood disk (15 cm diameter and 5 mm thick) were made on the flattened corm (Figure 5) surface. The inoculum was spread on washers plywood with 50 to 60 g per trap. The average concentration of conidia was 12.10^8 spores/g.

Mortality rate and LT50 generated by both methods plates trap and plywood trap were respectively 50%, 36 days and 90%, 6 days [6].

In addition the implementation of plate trap is tedious due to the notch made in the stump. Also, the notch serves as a shelter for weevil whose access to the inoculum rate becomes lower with time, whereas most borers on plywood traps become infected (Figure 6).

Plate and plywood traps deficiencies and shift work

Limit of both traps: Trapping made with plates or plywood involves:

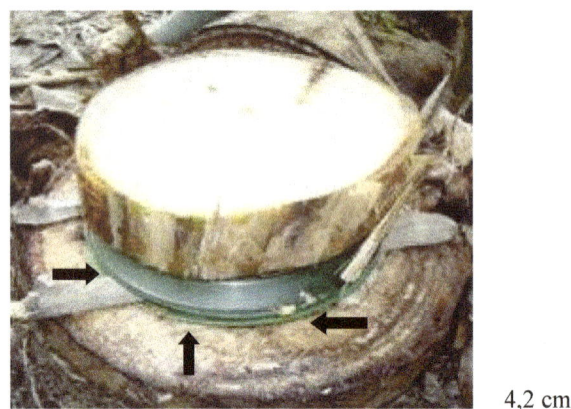
Figure 2: Weevil mycosed by *Metarhizium* sp.

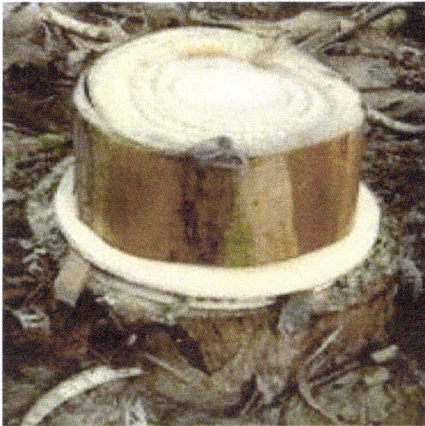

Figure 5: Plywood trap containing the inoculum isolate BME2 of *Metarhizium* [5].

Figure 6: Death Weevil in leaf sheaths following infection by *Metarhizium* sp. applied on plywood trap.

- limiting nutrient reserves of sucker provided by harvested plant;

- Increase pseudo stem residues in plots, nutrient and reproduction niches for weevils;

- Trapping efficiency is related to the corm freshness and to the number of traps / ha required (30 to 40 traps). The both plates trap and plywood trap involve only one trap for banana plant. Thus,

- The traps positions are also conditioned, and the homogeneity distribution of traps in the plot is also reduced.

Innovative and sustainable methods

1 Pseudo stem traps inoculated: Pseudo term trap inoculated aims to limit the sources of reproduction, hiding food and adult constituted by residues of stems in the plots after trap were maid presented above (Plywood trap and plate traps). The inoculum is placed on rectangles plywood (Figure 7) a biodegradable material for avoiding direct contact with soil microorganisms. Attracting adult on inoculum is further enhanced with a piece of pheromone « COSMOPLUS » covered with another piece of plywood on which a slice of pseudo term is deposited while promoting access to inoculum using chopsticks (Figures 8 and 9). The whole is covered by a Polyethylene film. The polyethylene film

protects trap for irrigation and rain water, extends the denaturation of pseudo stem-trunk covering the inoculum and the sporulation of the fungus. More the polyethylene film retains the weevil on the inoculum (Figure 10) by creating microclimate for chimiotropisme and chemotactisme conditions (humidity, darkness ...).

The mortality rate of weevils collected in these traps inoculated

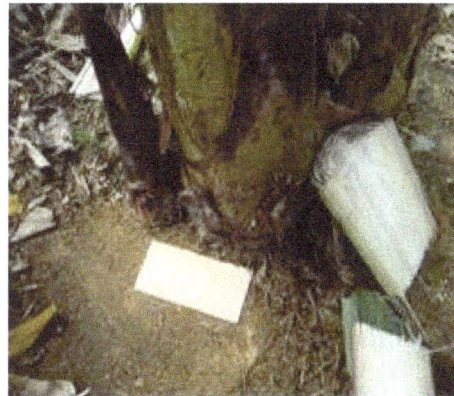

Figure 7: First stage of inoculated pseudo stem trap baiting.

Figure 8: 2nd stage of inoculated pseudo stem trap baiting.

Figure 9: 3rd stage of inoculated pseudo stem trap baiting.

exceeds 50%, otherwise 3 to 4 traps can be made with a pseudo stem freshly harvested bananas. In addition, the trapping provides greater freedom in the placement, increasing the number of traps and the homogeneity desired of their distribution on the area. The return of the fungus in its natural habitat is also facilitated by rapid re-colonization of the soil through the biodegradable material (plywood).

Pheromone traps inoculated for fallows sanitation before replanting: Adult weevils are absorbed in fallow by pheromone traps

Figure 10: Weevil on the inoculum into inoculated pseudo stem trap.

(pheromone + soapy water). These traps are visited weekly, and labor absorbing. This constraint could be lifted by applying the inoculum of *Metarhizum* grown on rice substrate (Figure 11). Twenty (20) pheromone traps inoculated were made per ha. The lower bowl is filled with sterilized soil. The inoculum is reviewed monthly [5].

Production of inoculum on organic substrates: In order to widespread use of the fungus, the foodstuff rice is expected to be substituted by an organic substrate as agricultural residues: cocoa waste, coffee waste, rice bran and sawdust (Figure 12). The both coffee and cocoa wastes residues were already used for organic fertilization in banana plantation.

Field application of inoculum produced on organic substrate

The inoculum produced on the organic substrate is applied in established stands throughout pseudo term traps (Figure 13a). Fungus was then incorporated into the soil (Figure 13b) enhancing its persistence in the plot.

For fallow sanitation before replanting, the fungus produced on coffee or cocoa waste substrates was associated to pheromone (Figure 14) us described above.

Conclusion

The methods implemented will contribute to the control of banana

Figure 11: Pheromone trap inoculated with BME2 an isolate of *Metarhizium* sp.

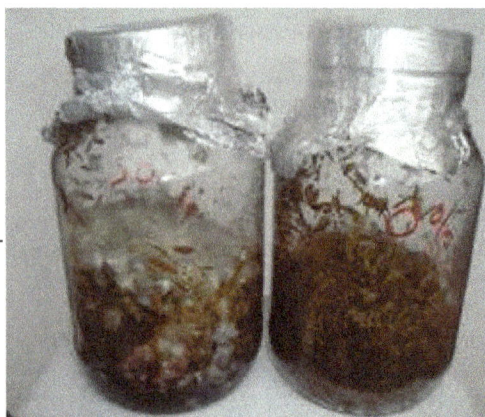

Figure 12: The isolate BME2 of *Metarhizium* sp. grown on organics substrates: (A) Coffee waste. (B)Rice bran.

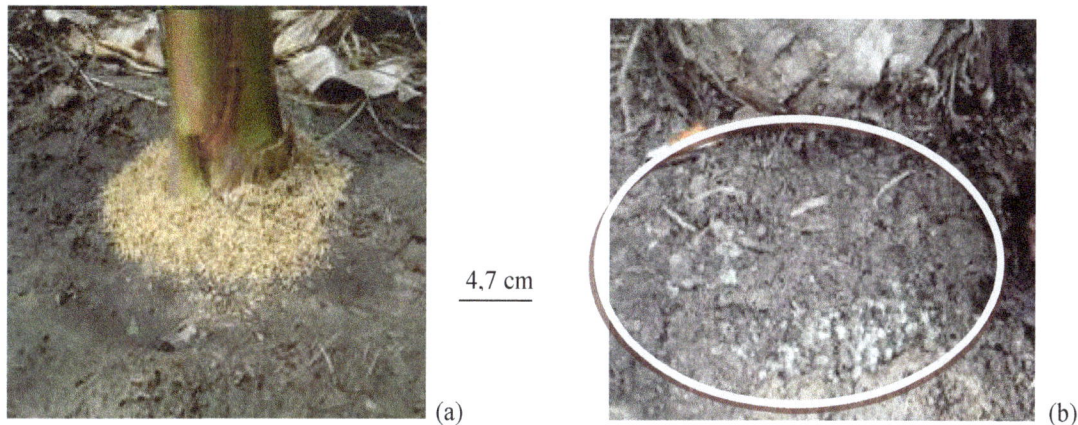

4,7 cm

(a)   (b)

Figure 13: Methods to apply inoculum of *Metarhizium* sp. grown on organic substrates.
(A) Preventive applying of inoculum arround banana plant. (B) Inoculum incorporated in the soil after the disappearance of pseudo stem trap.

GX 0,5   GX 4

Figure 14: Banana weevil catched in Pheromone trap containing *Metarhizium* sp. grown on coffea waste (fallow sanitation).

weevil, integrating prohylactic methods whereas chemical could be practiced on biological warning.

Acknowledgements

All these methods are being tested in banana plantations in Côte d'Ivoire. The project was financed by FIRCA (Interprofessional Fund for Agricultural Research and Council). We are grateful to plantations managers.

References

1. Gold CS, Jorge PE, Karamura EB, Rukazambuga ND (2003) Biology and integrated pest management for the banana weevil Cosmopolites sordidus (Germar). IPM Rev 6: 79-155.

2. Kiggundu A, Gold CS, Labauschaagne MT, Vuytlsteke D, Louw S (2003) Level of host plant resistance to banana weevil, Cosmopolites sordidus (Germar) (Coleoptera: Curculionidae) in: Uganda Musa germplasm. Euphytica 133: 267-277.

3. Aby N, Kobenan K, Kéhé M, Gnonhouri P, Koné D et al. (2010) Metarhizium anisopliae: parasite of banana weevil Cosmopolites sordidus in banana Ivorian. Journal of Animal & Plant Sciences 7: 729-741.

4. Zimmermann G (1998) Suggestions for a standardised method for reisolation of entomopathogenic fungi from soil using the bait method. IOBC/WPRS. Bulletin, Insect Pathogens and Insect Parasitic Nematodes 21: 289.

5. Aby N, Kobénan K, Traoré S, Koné D (2011) Metarhizium anisopliae as potential biological agent of banana weevil Cosmopolites sordidus in Côte d'Ivoire International Journal of Biological and Chemical Sciences IJBCS. IJBCS 5: 2454-2463.

6. Aby N, Kobénan K, Traoré S (2013) IPM against the banana weevil Cosmopolites sordidus in banana culture dessert in Ivory Coast, stage 1 of the project report; convention FIRCA/CNRA 22.

23

Promoter Analysis of a Powdery Mildew Induced *Vitis vinifera* NAC Transcription Factor Gene

Solomon Abera Gebrie*

Ethiopian Institute of Agricultural Research, Holoeta Research Center, Holeta, Ethiopia

Abstract

Grapevine is an economically important crop throughout the world. However commonly cultivated species *Vitis vinifera* is susceptible to powdery mildew fungi. The objective of this study was to analyze the promoter of VvNAC36 transcription factor gene, which was induced by powdery mildew infection, as it was shown by previous microarray results. Analysis of VvNAC36 transcription factor promoter allows identifying the important cis-elements, which are responsible for gene regulation during infection. Knowing that the responsible cis-element of the gene is crucial for further study, the role of this gene is in powdery mildew infected grapevine. The study was performed using circumspect methodology. To confirm the presence of transgene in transgenic plants the PCR and glufosinate total herbicide application was used. The transplanted transgenic seedlings were infected with *Golovinomyces orontii*. To detect the GUS reporter gene expression, histochemical GUS staining and spectrophotometric assays were applied in each deletion variant. Finally the promoter sequence was analyzed by the PLACE (a database for PLAntCis-acting Elements) program for identifying cis-elements. The results of the study showed significant difference in mock treated and induced expression in transgenic plants transformed with VvNAC36 promoter segments of different lengths, fused to GUS reporter gene. In basal expression level, the deletion variants 1178 bp and 257 bp differed significantly. Powdery mildew infection significantly increased the expression in all promoter variants, except the two shortest fragments. Based on the promoter motif analysis, it can be concluded, that the GT1-box (GAAAAA) supposed to have crucial role in regulating of VvNAC36 transcription factor gene during powdery mildew infection.

Keywords: VvNAC36 transcription factor; Powdery mildew; Gene expression; Cis-elements

Introduction

Grapevine (*Vitis vinifera*) is one of the oldest plants in the world (65 million years ago) [1]. It is an important fruit crop to produce wine, jam, juice, jelly, raisins, vinegar, grape seed extracts and oil. The wine sector is estimated to contribute more than 51.8 billion dollars to the US economy [2]. However the most grapevine growing country uses *Vitis vinifera* varieties, which are liable for best quality, but are vulnerable to powdery mildew fungus. Powdery mildew (PM) fungi are widespread and ever present among crop plants [3]. The reduction of yield due to PM each year probably exceeds the reduced yield, caused by any other plant disease. These pathogens rarely kill their hosts, but due to the utilization of their nutrients it lead to, a decrease in photosynthesis, increase of respiration and transpiration, impair me of growth and finally the plant produces 20-40% diminished yield [4].

Powdery mildew is controlled by systemic fungicides application. Beside that these chemical treatments lead to develop fungicide resistant mutants, which are superadded to the environmental pollution too. This implies that chemical treatment is not the best conceivable controlling method. Therefore elucidating the molecular level of plant-pathogen, interaction would approximate the background of resistance.

A large part of plant genes code transcription factors. In *Arabidopsis* ~2500 genes code transcription factors [5]. Transcription factors are the members of multigene families. The NAC transcription factor gene family is well studied. NAC transcription factor (no apical meristem growth gene, *Arabidopsis* activation factor gene, cup-shaped cotyledon response gene) genes are present only in plants [1,6]. The N-terminal region of the protein is divided into five conserved region and is used in DNA binding and dimerization. The C-terminal region has variable sequence and is involved in transcription activation or repression.

In addition, common feature of NAC protein C-terminal regions is characterized by the occurrence of simple amino acid repeats such as serine, threonine, proline and glutamine, or acidic residues [7]. The regulation of NAC transcription factors depend on the environmental condition. Approximately 20 to 25% of NAC transcription factor genes are expressed in response to one type of abiotic or biotic stress [5]. NAC transcription factor has excessive role in plant metabolic activity during any change to transcribe different genes, which are responsible for that influential factor.

In *Arabidopsis* and rice NAC, genes have been reported to regulate programmed cell death in xylem, induction of flowering, leaf senescence, xylary fiber development root and shoot apical meristem development. Additionally, NAC transcription factors have important role in abiotic and biotic stress responses. ATAF1 was found to negatively regulate the defense reactions against necrotrophic fungal pathogens [1]. Over expression of OsNAC gene increased drought resistance in rice [8,9]. Most NAC gene in their promoter region conserved with hormonal and fungal elicitor responsive motifs [10,11]. Fourteen Chitin-responsive transcription factor genes as part of the plant defense reaction were significantly up regulated after chitin treatment in *Arabidopsis* [12]. HBOH13 and HBOH14 transcription factor genes were induced upon infection with the powdery mildew during early stages of infection in rubber tree (*Hevea brasiliensis*) through mRNA Differential Display [13].

*****Corresponding author:** Solomon Abera Gebrie, Ethiopian Institute of Agricultural Research, Holeta, Ethiopia, E-mail: solomonabera19@gmail.com

VvNAC transcription facto genes have different roles in grape vine like tissue development, biotic and abiotic response. For instance VvNAC18, VvNAC29, VvNAC37, VvNAC65 and VvNAC70 were expressed at high levels at inflorescence; VvNAC26 plays key roles in response to abiotic stress; in biotic stress response 7 VvNAC increased expression after virus-infection and 10 were up-regulated after Bois noir infection [1]. In addition 7 VvNAC transcription factor genes have been up regulated by powdery mildew elicitor at different levels in the given hour post infection (phi) [14]. Among these 7 transcription factor genes VvNAC36 has been significantly up regulated by powdery mildew infection.

There are numerous studies in relation to genome-wide identification researched NAC domain genes in plants, of which the whole genome sequences were known. The genomes of *Arabidopsis* thaliana (117), *Populustrichocarpa* (120), *Glycine max* (152), *Oryza sativa* (151), grapevine (79) and banana (167) contain numerous NAC genes respectively [5,15]. The frequency of transcription factor indicates the genome multiplication. For example *V. vinifera* genome, which has 79 NAC genes, changes more gradually than other plant genomes.

Materials and Methods

In the experiment the putative full length NAC promoter (3900 bp), four deletion variants (2935, 2456, 1178, 257 bp) and a promoter-less construct were tested for induction by powdery mildew with three replication. These promoter fragments were fused to GUS reporter gene and each of them was transferred into wild type *Arabidopsis* thaliana by Gateway binary vector pGWB633 DNA. The selection marker was glufosinate total herbicide resistance 'bar' gene in the transgenic plants. For the evaluation the following methodology was investigated.

Transgenic line selection with glufosinate

The third generation (T3) independent lines transgenic *Arabidopsis* thaliana seeds were sown on autoclaved soil (Compo Sana). The two week old seedlings were sprayed by glufosinate containing total herbicide and due to the glufosinate resistance gene (bar gene), the positive transformants survived the application. Subsequently one week after application of herbicide, seedlings show clear selection, the positive transformants grow well, the plants, which did not contain the transgene, yellowed and died. Based on the Mendelian law of segregation the lines that contained 75% of surviving and 25% dying plants were selected for further analysis. Finally the surviving seedlings of these lines were transplanted into pots. The experiment had three replications of each type of transgenic plants.

Confirmation of transgene presence using PCR

In addition to glufosinate application the integration of transgene was confirmed by PCR method. The DNA was extracted with a quick DNA preparation protocol [16]. Three transgenic plant leaves were placed into microcentrifuge tube and after adding 400 μl of extraction buffer (200 mM Tris-Cl (pH 7.5), 250 mM NaCl, 25 mM EDTA, 0.5% SDS) were ground with a micropestle (Figures 1 and 2).

The samples were centrifuged at 13000 rpm for 5 min. Subsequently 300 μl of the supernatant was transferred into a clean tube, 300 μl of 2-propanol was added and mixed by shaking. The mixture was centrifuged at maximum speed for 5 min and the supernatant was discarded carefully. The pellet was rinsed with 70% ethanol, drained, dried then dissolved in 100 μl of distilled water. 2.0 μl was used for the PCR reaction from the isolated DNA. For PCR reaction 1 μl WestTeam

Figure 1: Application of glufosinate to select transgenic lines (red arrows indicate the dying negative and green arrows indicate the surviving positive transgenic plants).

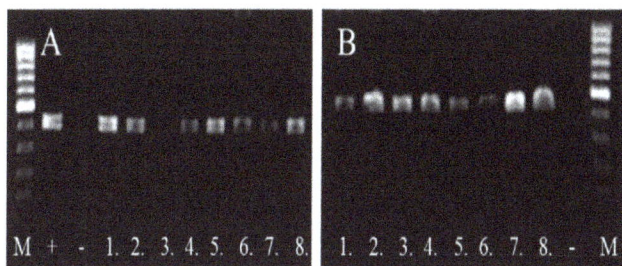

Figure 2: Confirmation of transgenes by PCR in transgenic line (A) GUS (357 bp) and (B) BAR gene (396 bp).

DNA polymerase buffer (10 ×), 1 μl dNTP (2 mM), 0.5 μl forward primer (10 μM), 0.5 μl reverse primer (10 μM), 0.25 μl WestTeam DNA polymerase enzyme (1 U/μl), 0.2 μl $MgCl_2$ (20 mM) and 6.05 μl distilled sterile water was mixed. Both transgenes, the bar selection marker and NAC promoter: GUS fusion were tested by PCR applying the following primers: 5'AAACCCACGTCATGCCAGTT-3' (BAR gene forward) and 5'AAGCACGGTCAACTTCCGTA-3' (BAR gene reverse) and 5'CCTAAAGCTTGGACAAACAGGAG3' (NAC forward primer), 5'GCCCAACCTTTCGGTATAAAGAC-3' (NAC reverse primer). Fragments were amplified using an initial denaturation step at 94°C for 2 min was followed by 30 cycles of 94°C for 35 s, 60°C for 30 s and 72°C for 45 s; finally followed by a final extension at 72°C for 15 min.

Powdery mildew infection

Three weeks after transplantation mock treatment and powdery mildew infection was applied on transgenic plants. For infection *Golovinomyces orontii* infected tobacco leaves were collected, the spores were washed off in 100 ml distilled water. This solution was sprayed on transgenic plants. For mock treatment healthy tobacco leaves were collected and washed in 100 ml distilled water. The infected and the mock treated transgenic *Arabidopsis* thaliana plants were maintained in a growth chamber at 22°C to 24°C in a 16 h light cycle to allow systemic infection.

GUS staining and microscopy observation

Fifteen days post infection the leaves were harvested for histochemical GUS assay. The infected leaves were placed into microcentrifuge tube and sufficient assay solution was added to cover them. To 1 ml of assay solution 10 μl of 30% hydrogen peroxide (H_2O_2) and 10 μl X-Gluc (50 mg/ml) solution was pipetted to them. The leaf samples were incubated overnight at 37°C. Finally 70% ethanol was added to the samples for the removal of chlorophyll from the tissues. The pictures were taken of these stained leaves. To confirm the GUS expression at the infected area microscopic pictures were taken from each deletion variant. To stain the fungus the chitin coloring cotton blue stain was used.

Spectrophotometric assay

Infected leaves were collected for spectrophotometric assay and placed immediately into liquid nitrogen. The frozen plant leaves were ground in GUS extraction buffer (50 mM Na-phosphatebuffer pH: 7.0, 10 mM β-mercaptoethanol and 0.1% triton-X 100; 1 μl extraction buffer/1 mg plant material). The homogenate was centrifuged in a microcentrifuge at 12,000 rpm for 10 min at 4°C. One mM p-nitrophenylβ-D-glucuronide (PNPG Sigma) substrate (final concentration) was added to the supernatant and it was incubated overnight at 37°C. The reaction was stopped by freezing the samples at -20°C. The GUS enzyme activity was measured at absorbance of 405 nm using Nano Drop 1000 spectrophotometer. The promoter-less samples were used as blank for both mock and infected samples. The absorbance of p-nitrophenol was measured according to Gilmartin and Bowler [17] and applied for estimating GUS expression.

Cis-element search of VvNAC promoter

The sequenced 3900 bp long VvNAC36 transcription factor promoter was analyzed by database of PLAntCis-acting Elements (PLACE). The inducible and repressor regulatory cis-elements were identified in the sequence at the website of http://www.dna.affrc.go.jp [18].

Statistical analysis

To analyze the quantitative spectrophotometric data the powerful and comprehensive data analyzing R-commander software was applied [19].

Results

Selection of transgenic *arabidopsis* plants

The selections of herbicide resistant seedlings after glufosinate spraying were identified. Surviving green planlets were further analyzed by PCR to prove the integration of both the bar and NAC promoter sequences.

Promoter sequence analysis

The results of the GUS activity was driven by the longest VvNAC promoter, four deletion, 2935 bp (removed 965 bp from full length), 2456 bp (removed 1444 bp from full length), 1178 bp (removed 2722 bp from full length) and 257 bp (removed 3643 bp from full length) and a promoter-less construct on the effect of mock treatment and powdery mildew infection are shown in Figure 3.

Promoter deletion activity analysis (only in mock)

The spectrophotometrically measured GUS reporter gene expression of 2935, 2456 and 1178 bp promoters did not differ significantly in the mock treatments. However the full length promoter 3900 bp showed significantly lower expression than the above three deletion variants (Figure 3). Likewise the plant carrying the 257 bp

promoter, the GUS expression showed 22 times lower expression than deletion variant 1178 bp. Beside the spectrophotometric data the histochemical GUS staining assay indicates the same expression levels; significant differences were observed in full length and 257 bp promoter carrying plants compared to the other deletion variants (Figure 4). These data show that the 1178 bp to 2935 bp long NAC promoters activate GUS expression even without PM infection (Table 1).

Promoter deletion activity analysis after PM infection

On the effect of powdery mildew infection the GUS expression increased in all five variants (Table 2). The highest level was achieved in VvNAC36 promoter deletion variants 2935 bp and 2456 bp, the difference between these two ones was not significant. In contrast the plant carrying 3900 bp (full length VvNAC36 promoter), 1178 bp and 257 bp VvNAC36 promoter displayed significantly lower expression compared to the plants carrying the 2935 bp and 2456 bp long promoter sequence (Figures 3 and 4).

The longest (3900 bp promoter showed a 15 fold change in the response to infection. The 2456 bp deletion variant changed 2.3 times 2935 bp and.1178 bp deletion variants which increased 1.3 times the expression of GUS due to the pathogen. Besides, the difference between gene regulations of mock and infected treatment is the variation which was observed within each respective pathogen which induces the expression of deletion variants too. Comparing the 1178 bp and 257 bp deletion variant to 2456 bp promoter a 3 fold (1178 bp) and 27 fold (approaching zero expression, 257 bp) reduce was observed in the regulation of GUS respectively. However the induced regulation of 3900 bp and 1178 bp deletion variants showed almost the same GUS expression.

The pictures of GUS staining assay confirm the results of spectrophotometry, where the reporter gene expression level was influenced by the deletion variants (Figure 4). The infected area of 3900 bp, 2935 bp, and 2456 bp promoter containing leaves showed high GUS expression in the microscopic photos (Figure 5). In contrast the infected leaves of plants containing 1178 bp VvNAC36 promoter did not show high increase of GUS expression in the infected area. Unlike the VvNAC36 257 promoter length that showed a lack of GUS expression in the infected area of leaves.

Putative cis-elements responsive to inducer or repressor agent

Using the plant Cis-acting elements, plant promoter analysis program at http://wwwdna.affrc.go.jp several cis-elements were identified in VvNAC36 transcription factor promoter, which response to inducer or repressor (silencer) agent. Table 3 details the identified cis-elements with the respective inducer or repressor agent and location on strands (+/-).

	Df	Sum Sq	Mean Sq	F value	Pr (>F)
Deletion variant	5	0.7549	0.15098	3.359	0.0397*
Residuals	12	0.5393	0.04495		

S^2=0.173

Table 1: ANOVA summary of PNP absorbance data of mock treated plants.

	Df	Sum Sq	Mean Sq	F value	Pr (>F)
Deletion variant	5	1.8972	0.3794	14.21	0.000109***
Residuals	12	0.3204	0.0267		

S^2=0.133

Table 2: ANOVA summary of PNP absorbance data of infected plants.

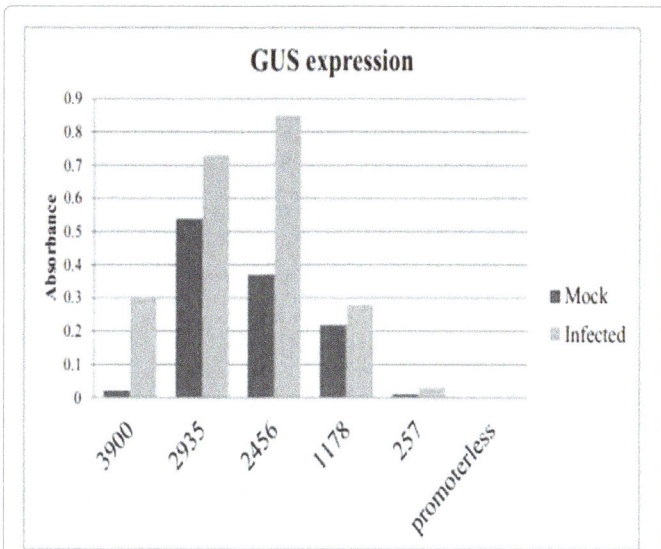

Figure 3: β-glucuronidase expressions measured by PNPG-based spectrophotometric assay of infected and mock treated plants.

Figure 4: GUS staining assays of infected and mock treated plants (*VvNAC* promoter and deletion variants).

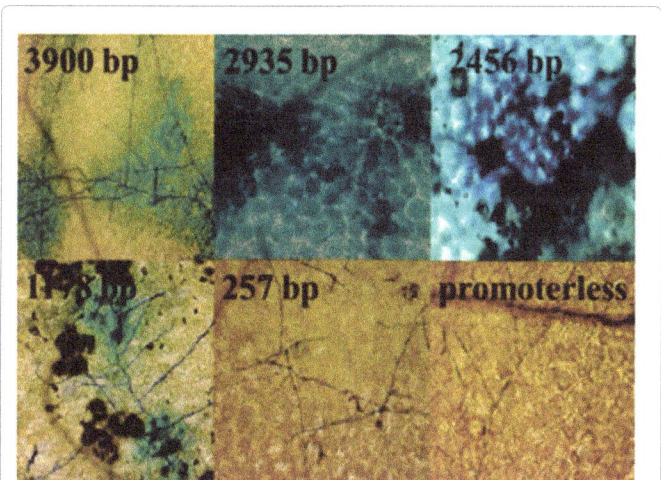

Figure 5: Microscopic picture of the infected leaves by the binding of cotton blue the fungi chitin.

As the PLACE identified the VvNAC36 promoter it contains several inducible and some repressor responsive motifs. Among the inducible cis-elements, pathogen elicitor, salicylic acid (SA), cytokinin and dehydration or dark stress motifs were the most conserved elements in the promoter region. The GT-1 box (GAAAAA) and the W- box core motifs (TGAC) were conserved mostly among the pathogen elicitor induced cis-elements. The PLACE identified sugar and light-repressive cis-element especially in the upstream region of the full length promoter. Two sugar repressive cis-elements were identified TATCCAY at 3381(-), 3401(-), and TTATCC at 1506 (-), 3383 (-), 2553 (+) and 2553 (-). Light repressive cis-element TGTATATAT was detected at 2989 (+) and 513 (-).

Discussion

Vitis vinfera NAC36 transcription factor gene was up regulated by powdery mildew pathogen. Based on the pathogen induced expression level of the longest (full length) promoter::GUS fusion in comparison to basal expression, it indicates that VvNAC36 transcription factor is a powdery mildew dependent gene and may be part of defense system. The basal level expression of GUS reporter gene in the full length fusion containing plants was significantly low (p-value=0.0256); the infection induced GUS reporter gene expression, showed a 15 fold increase. However when it was compared to the highly expressed 2456 bp deletion variant, it showed significantly 3 times less expression.

This result reveals that the full length promoter has some negative regulatory motifs in its sequence. Interestingly based on PLACE plant promoter analysis program, this region contains conserved TATCCAY and TTATCC sugar repressive cis-elements (Figure 6B). Mostly the plants have more sugars in their leaves due to photosynthesis metabolic activity. Therefore if the sugar repressive cis-element is conserved in this powdery mildew inducible promoter upstream region, the deletion of these repressive cis-elements causes a 3 fold up regulation of GUS reporter gene in 2456 bp deletion. Furthermore two sites of the full length promoter region contain light repressive cis-elements and these may contribute too, in the reduction of GUS regulation of full length promoter containing plants. Further investigations are necessary to experimentally confirm these assumptions.

In addition, the second longest promoter (2935 bp) which contains sugar repressive TACGTA cis-elements at 2553 (+) and at 2553 (-) position and showed a little bit lower expression than the 2456 bp promoter fusion in infected treatment, but the difference was not significant. This motif may cause reduction of GUS reporter gene expression by blocking the elicitor responsive inducer. However the induction of basal expression by this promoter was higher than in 2456 bp fusion. This indicates that the repressor site may not influence the general transcription factor protein induced basal expression. In mock treated plant, the full length promoter::GUS reporter gene fusion showed 27 times less GUS expression than the deletion variant 2935 bp probably due to the presence of these sugar and light repressive cis-elements.

Therefore this result revealed that the light and sugar repressive binding cis-elements cause a significantly low inducement of GUS reporter gene by full length VvNAC36 promoter in both mock and infected treatment. However in this promoter::GUS, fusion showed 15 times higher up regulation in response to infection compared to mock treatment, which confirms that VvNAC36 is a powdery mildew inducible gene.

The basal expression of the reporter genes in plants carrying 1178

Cis-element	Inducer	Location of cis -element on both Strand
TATTCT	Blue, white or UV-A light	2168(-)
ACGTG	Dehydration stress or dark.	1732 (+) and 2781 (+)
TACGTA	Sugar-repression	2553 (+) and 2553 (-)
ACGT	Dehydration stress or dark-.	2553 (-), 1732 (+),2554 (+), 2781 (+), 2884(+), 1732 (-), 2554 (-), 2781(-) and 2884(-)
TATCCAT	Sugar starvation	3381(-) and 3401(-)
AAACAAA	Anaerobically	569(+), 3365(+), 2128 (-), 2440 (-), and 3429(-)
NGATT	Cytokinin	2173 (+), 1231 (+), 2675 (+), 2877 (+), 553 (+), 815 (+), 1687 (+), 1709 (+), 1863 (+) , 2217 (+), 116 (+), 1494 (+), 1598 (+),1721 (+), 1982 (+), 2063 (+), 2679 (+),3020 (+), 3079 (+), 3182 (+), 3581 (+), 3890 (+), 127 (-), 395 (-), 465 (-), 528 (-), 712 (-), 1128 (-),1384 (-), 2038 (-), 2460 (-), 2580 (-), 2616 (-), 2656 (-), 3114 (-), 3446 (-), 3461 (-)3745 (-)
TGACG	Auxin and/or salicylic acid	270 (+),1735 (+), 2779 (+) and 2603 (-)
ACCWWCC	Elicitor treatment or UV-B irradiation	2525 (+)
TTGACC	Pathogen elicitor	3235 (+) , 99 (-), 959 (-) and 1698 (-)
AWTTCAAA	Ethylene	2903 (+), 3410 (+),3518 (+), 493 (-), 1008 (-) and 2509 (-)
GCCGCC	Jasmonate or ethylene	1293 (-)
GAAAAA	Pathogen-or salt	442 (+), 1254 (+),1557 (+), 2308 (+),1539 (-), 1842 (-), 1887 (-), 2281 (-), 2682 (-), 3542 (-), 3584 (-),3673 (-),and 3715 (-)
CTCTT	Mycorrhizal	5 (+), 11 (+), 1399 (+), 3359 (+),354 (-) , 446 (-), 734 (-) and 1081 (-)
TGTATATAT	Light-repressed	2989 (+) and 513 (-)
TTATCC	Sugar-repressive	1506 (-), 3383 (-)
AATACTAAT	Sucrose	853 (+)
AACGTG	Jasmonate (JA)	1731 (+)
TATCCAY	Sugar-repression	3381 (-), and 3401 (-)
TTTGACY	Elicitor-induced	3234 (+) ,1312 (-) and 1698 (-)
TTGAC	Salicylic acid (SA)-induced	485 (+) , 754 (+), 1573 (+), 2101 (+), 2786 (+), 3235 (+), 3609 (+), 100 (-), 960 (-) 1313 (-),1699 (-), and 2604 (-)
CTGACY	Elicitor-induced	2449 (-) and 3808 (-)

Table 3: Identified *cis* elements of *VvNAC36* promoter region responsible in interaction with inducer or repressor. Green shading indicates repressor responsive motifs and yellow highlighting indicates pathogen-related motifs.

VvNAC36 promoter was significantly 22 times fold than the 257 bp fragment containing plants. Evaluation of GUS activity in mock treated plants carrying 1178 bp and 257 bp deletion constructs indicates that reduction of the 1178 bp length VvNAC36 promoter to 257 bp results significantly decrease in expression. This showed that positive regulatory cis-element is apparently located between 1178 to 257 bp in the promoter region, and this is responsible for basal expression. The powdery mildew induced GUS reporter gene expression significantly differ in 2456 bp and in 1178 bp VvNAC36 promoter construct meaning that elicitor responsive motifs may be found in 2456 to 1178 bp region. Two putative regulatory motifs, pathogen elicitor responsive cis-acting elements were identified in the promoter region of the VvNAC36 gene. These cis-elements, W-boxes (TGAC) and GT1-boxes (GAAAAA), are considered being pathogen elicitor responsive in plants. However, based on the detailed evaluation the promoter region of 2456 bp was highly significant from 1178 bp and 257 bp VvNAC36 promoter deletion variant respectively (Figure 6A).

In the VvNAC36 2456 bp promoter region fifteen W-boxes and eight GT1-boxes were detected on both strands (Figure 6A). Seven W-boxes out of the total fifteen boxes and seven GT1-boxes of the total eight boxes were dropped out from VvNAC36 1178 bp promoter sequence. The 1178 bp VvNAC36 deletion variant promoter::GUS reporter gene fusion showed significantly low expression when compared to the VvNAC36 2456 bp promoter construct due to the deletion of seven W-boxes and seven GT1-boxes. In addition the total eight GT1-boxes and fourteen W-boxes were absent in VvNAC36 257 promoter regions, the single W-box is not able to regulate neither the basal and induced expression of 257 bp promoter::GUS fusion. The 257 bpVvNAC36 deletion variant promoter::GUS reporter gene fusion

showed a large decrease in expression due to the totally lost GT1-boxes and only a single W-box presence.

In this study the VvNAC36 transcription factor promoter was analyzed to identify the motif(s), which is/are responsible for up regulation of the gene in powdery mildew infection. In infected treatment among six transgenic lines including full length promoter and promoter less construction, the promoter length of 1178 bp and 257 bp showed significantly low GUS gene expression. In the mock treatment the reporter gene expression was significantly low only in the plant transformed with the 257 bp deletion variant. The results suggest that putative motifs found between 1178 bp to 2456 bp deletion variant are critical for powdery mildew elicitor induction and is between 1178 bp to 257 bp deletion variant. The core motif found in the sequence is required for basal expression regulation of the VvNAC36 transcription factor gene.

The detected responsible motifs found in 2456 bp and 1178 bp deletion variants were putative motifs depending on powdery mildew elicitor. A total fifteen core W-boxes (TGAC) and eight GT1-boxes (GAAAAA) were identified in the promoter of 2456 bp length while there were only eight W-boxes (TGAC) and one GT1- box in the 1178 bp promoter region. These pathogen responsive motifs have outstanding role in up regulation in the gene by pathogen elicitor. For example, in soybean suspension-culture cells the expression of SCaM-4 gene, which contains GT1-boxes in its promoter, was dramatically induced within 1 h after treatment with *Pseudomonas syringaepvglycinea P. syringaepv*, while the application of exogenous hydrogen peroxide, salicylic acid, Jasmonic acid did not increase the transcription of this gene. However mutations in the GT1-boxes significantly reduced the expression of this gene [20]. W-boxes are the major elicitor responsive motifs in most pathogen related protein coding genes [21].

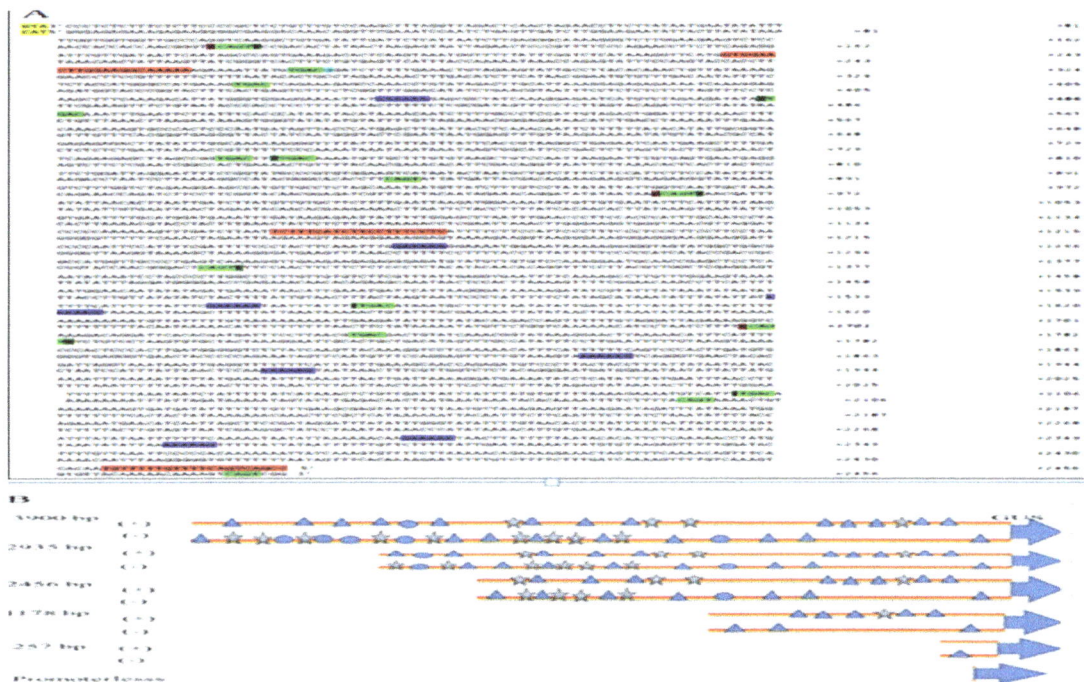

Figure 6: Site of W and GT1 boxes on NAC promoter (A) Original sequence of 2456 bp NAC promoter; W and GT1 boxes, 2456 bp, 1178 bp, 257 bp primer sequences are indicated (B) Schematic diagram of location of responsive *cis*-elements of *VvNAC* deletion variants and GUS reporter gene fusions: ▲ W-box core motif (TGAC), ★ Elicitor responsive GT1-boxes (GAAAAA) and ● Sugar repressive *cis*-element.

The GT1 and W-boxes motifs in *Vitis vinifera* NAC36 transcription factor promoter play important role for expression regulation in powdery mildew infection especially GT1-boxes, which occurs only once in promoter length 1178 and are totally absent in the 257 deletion variant, involve in the responsible transcription factor binding site. Total 8 W-boxes, only one GT1-box was detected in the 1178 deletion variant. Therefore based on the significant expression drop between 2456 bp and 1178 bp promoters the conclusion can be drawn that, the GT1 box has an important regulation role in the PM induced gene expression of VvNAC36 transcription factor gene. The powdery mildew regulated VvNAC36 transcription factor gene may have positive role in defense system against powdery mildew infection. As the previous microarray study indicated, the high expression of VvNAC36 transcription factor gene was only regulated by powdery mildew and not by exogenous salicylic acid treatment [14]. This study identified the responsible motif for regulation of this gene, the GT-1 box, which is regulated by elicitor not by salicylic acid.

Therefore it would be worthwhile examining, how the mutation in GT1 box of VvNAC36 promoter influences the expression on the effect of powdery mildew infection and to study the functional role of VvNAC36 transcription factor gene using complementation test.

Conflicts of Interest

The authors declare that they have no conflicts of interest.

Acknowledgements

First I would like to give my great acknowledgement to GOD, who never aparts me so far and forever in each my activity for this work too. Secondly I would like to thank Zsofia Toth for her great assist in both laboratory experiments and in editing this work. Many thank for her general consultation to Dr. Erzsebet Kiss too. I appreciate the support of FAO and Hungarian government.

References

1. Wang N, Zheng Y, Xin H, Fang L, Li S (2012) Comprehensive analysis of NAC domain transcription factor gene family in Vitis vinifera. Plant Cell Reports 32: 61-75.

2. Broome J, Warner C, Keith D (2008) Agro-environmental partnerships facilitate sustainable wine-grape production and assessment. California Agriculture 62: 133-141.

3. USEPA (2005) Use of genetic toxicology information for risk assessment. USEPA-US.

4. Agrios GN (2005) Plant Pathology. Elsevier Academic Press, Amsterdam, p: 952.

5. Puranik S, Sahu PP, Srivastava PS, Prasad M (2012) NAC proteins: regulation and role in stress tolerance. Plant Science 17: 369-381.

6. Kikuchi K, Ueguchi TM, Yoshida KT (2001) Molecular analysis of the NAC gene family in rice. Molecular and General Genetics 262: 1047-1051.

7. Olsen AN, Ernst HA, Lo LL, Skriver S (2005) NAC transcription factors: structurally distinct, functionally diverse. Trends in Plant Science 10: 79-87.

8. Carviel JL, Al-Daoud F, Neumann M, Mohammad A, Provart NJ, et al. (2009) Forward and reverse genetics to identify genes involved in the age-related resistance response in Arabidopsis thaliana. Molecular Plant Pathology 10: 621-634.

9. Mitsuda N, Seki M, Shinozaki K, Ohme-Takagi M (2005) The NAC Transcription Factors NST1 and NST2 of Arabidopsis Regulate Secondary Wall Thickenings and Are Required for Anther Dehiscence the Plant Cell. American Society of Plant Biologists 17: 2993-3006.

10. Jensen MK, Kjaersgaard T, Petersen K, Skriver K (2010) Time-specific regulators of hormonal signaling in Arabidopsis. Plant Signaling and Behavior 5: 907-910.

11. Fujita M, Yasunari F, Yoshiter N, Takahashi F, NarusakaY, et al. (2006) Crosstalk between abiotic and biotic stress responses: a current view from the points of convergence in the stress signaling networks. Current Opinion in Plant Biology 9: 436-442.

12. Libault M, Wan J, Czechowski T, Udvardi M, Stacey G (2007) Identification of 118 arabidopsis transcription factor and 30 ubiquitin-ligase genes responding to chitin, a plant-defense elicitor. Mol Plant-Microbe Interact 20: 900-911.

13. Li X, Zhenghong B, Rong D, Peng L, Qiguang H, et al. (2016) Identification of Powdery Mildew Responsive Genes in Hevea brasiliensis through mRNA Differential Display. International Journal of Molecular Sciences 17: 181.

14. Fung RWM, Martin G, Csaba F, Laszlo GK, Yan H, et al. (2008) Powdery mildew induces defense-oriented reprogramming of the transcriptome in a susceptible but not in a resistant grapevine. Plant Physiol 146: 236-249.

15. Cenci A, Guignon V, Roux N, Rouard M (2014) Genomic analysis of NAC transcription factors in banana (Musa acuminata) and definition of NAC orthologous groups for monocots and dicots. Plant Molecular Biology 85: 63-80.

16. Glazebrook J (2002) Arabidopsis a Laboratory Manual. Cold Spring Harbor Laboratory Press, New York.

17. Gilmartin PM, Bowler C (2002) Molecular Plant Biology. Oxford University Press, London, UK.

18. Higo K, Ugawa Y, Iwamoto M, Korenaga T (1999) Plant cis-acting regulatory DNA elements (PLACE) database. Nucleic Acids Research 27: 297-300.

19. Fox J, Weisberg S (2011) An R and S-plus companion to Applied Regression. Sage, India.

20. Park HC, Kim ML, Kang YH, Jeon JM, Yoo JH, et al. (2004) Pathogen- and NaCl-induced expression of the SCaM-4 promoter is mediated in part by a GT-1 box that interacts with a GT-1-like transcription factor. Plant physiology 135: 2150-2161.

21. Eulgem T (2005) Regulation of the Arabidopsis defense transcriptome. Trends Plant Science 10: 71-78.

24

Optimization of EMS Mutagenesis on Petunia for TILLING

Jiang P*, Chen Y and Wilde HD[1]

Horticulture Department, University of Georgia, Athens, Georgia, USA

Abstract

The development of a chemically-mutagenized population of Petunia hybrida could enable the identification of novel alleles for crop improvement. Conditions were determined for mutagenizing petunia with ethyl methanesulfonate (EMS), while minimizing deleterious effects on viability and fertility. A mutagenized population of the doubled haploid *P. hybrida* line 'Mitchell Diploid' was developed as a resource for TILLING. Three mutagenesis parameters were investigated in this study: the imbibition of seeds prior to EMS treatment, the EMS concentration, and EMS exposure time. Based on these results, the mutagenesis of 2000 petunia seeds for TILLING was conducted with a 12 hour imbibition followed by exposure to 0.1% EMS for 12 hours.

Keywords: Mutation breeding; Ethyl methanesulfonate; TILLING; Petunia

Introduction

Mutagenesis has been used to introduce genetic variation in ornamental plants for several decades [1,2]. More than 560 ornamental varieties from 41 plant species have been officially released from mutation breeding programs [3]. Generally, plants with novel traits were identified phenotypically from large, mutagenized populations. More recently, DNA screening techniques such as TILLING [4] have been developed that allow mutagenized populations to be analyzed genetically, before trait expression. The combination of mutagenesis and DNA screening has enabled the identification of novel alleles in model plants [5] and horticultural species [6].

Petunia hybrida Vilm is an interspecific hybrid that is both a top-selling ornamental [7] and a genetic model [8] Mutagenesis of this herbaceous ornamental can be a useful tool for crop improvement and basic research. A potential application of mutagenesis is in the development of petunias with resistance to pathogens that can infect plants during commercial propagation, including powdery mildew [9] and potyviruses [10,11]. Resistance to powdery mildew and potyviruses can be conferred by mutations in specific genes, which have been targets for TILLING in other crop species [12,13]. The TILLING of petunia for mutant alleles in light signaling and anthocyanin metabolism genes was recently reported [14]. There are published protocols for the mutagenesis of petunia with ethyl methanesulphonate (EMS), but they vary in terms of concentration (0.1% to 0.5%), exposure time (10 to 24 hours), and seed pre-treatment [15-17]. In this paper, we describe the optimization of mutation parameters, the development of an EMS-mutagenized population that could facilitate TILLING.

Materials and Methods

Mutagenesis treatments

Mutagenesis experiments were conducted with seeds of the doubled haploid cultivar *Petunia hybrida* 'Mitchell Diploid'. Three parameters were examined factorially: pre-treatment imbibition (2 treatments), EMS concentration (4 levels), and EMS exposure time (3 intervals). For each of the 24 treatments, there were 3 replicates of 24 seeds. Prior to exposure to EMS (Fluka, USA), seeds were either not treated or were imbibed in 500 µl of sterile deionized water for 12 h in the dark with mild shaking (45 rpm) at room temperature. EMS concentrations (0, 0.1, 0.2 and 0.3% v/v) were tested by adding an appropriate volume of freshly-prepared 0.5% EMS to the imbibed seeds. The seeds were

then incubated in the dark at room temperature for 6 h, 12 h or 24 h with mild shaking (45 rpm). After EMS treatment, the seeds were rinsed 10 times with 1 ml of sterile deionized water and sown in 72 well trays containing soil (Fafard 3B potting mix). Individual seedling were transplanted to pots (4×4×6 inch) and maintained in 16 hour daylight greenhouse (latitude 33.95N and longitude -83.378W).

Mutagenesis data and statistical analysis

For the EMS treatments, a two-factor/one block experimental design was used with imbibition time as a block. Plants were evaluated 10 days after sowing for seed germination and 30 days after sowing for flower number. Analysis of variance (ANOVA) of the obtained data and regression analysis were conducted using Sigma Plot software.

Evaluation of the genotoxic effect

Genome damage was indirectly evaluated by the effect of the mutagen treatments on developmental and physiological parameters in time course analyses from 1 to 30 days after sowing (DAS) and at 15 and 30 DAS for the following characteristics: seed germination (%), height (mm), flower number and anther quality. Data were individually scored and statistically analyzed by ANOVA and regression analyses.

Results and Discussion

Optimization of EMS mutagenesis

Procedures for EMS mutagenesis of *P. hybrida* were optimized to balance high mutation frequency with plant viability and fertility. Three mutagenesis parameters were investigated: the imbibition of seeds prior to EMS treatment, the EMS concentration, and EMS exposure time. The results of combinations of these factors on germination frequency and flower number are shown in Figures 1 and 2, respectively, with summary statistics in Tables 1 and 2. The effect of seed imbibition

---

*Corresponding author: Jiang P, Horticulture Department, University of Georgia, Athens, Georgia, USA 30602, E-mail: pjiang@uga.edu

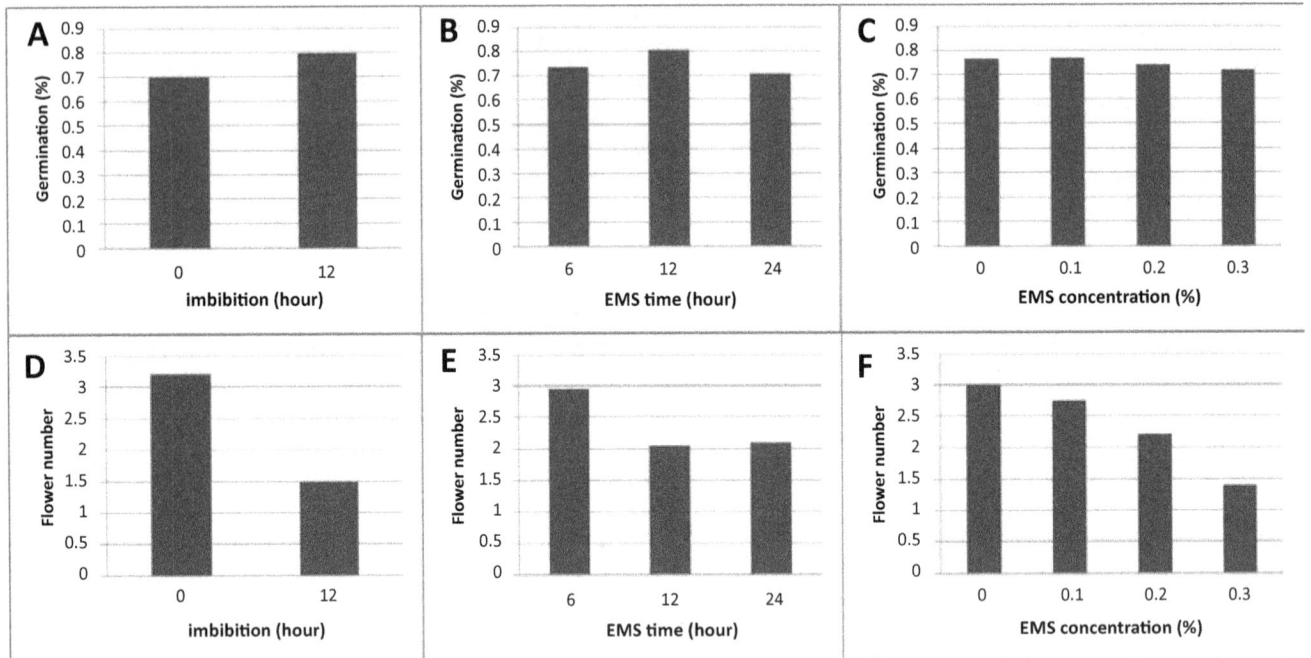

Figure 1. A-C. Germination rates are shown among different imbibition time, EMS treatment time and EMS concentration on 10 days after sowing. D-F. Flower numbers are shown among different imbibition time, EMS treatment time and EMS concentration on 30 days after sowing.

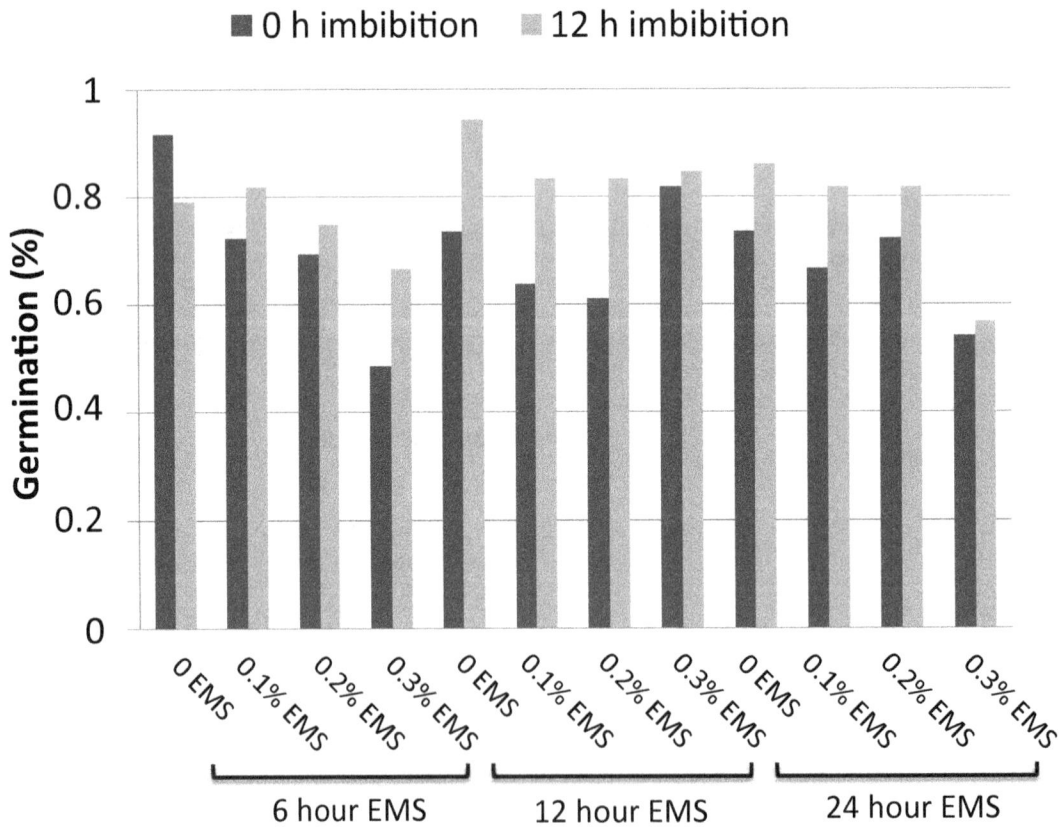

Figure 2: Germination rates are shown among each treatment on 10 days after sowing.

prior to mutagenesis was significant for both germination and flower number. EMS concentration and exposure time had significant effects on flower number, but not on germination. After a 12 h exposure to EMS, the average number of fruit per plant decreased from 2.7 capsules in controls to 0.6, 0.2, and 0 capsules in treatments with 0.1%, 0.2%, and 0.3% EMS, respectively. Based on these results, the mutagenesis of 2000 petunia seeds was conducted with a 12 hour imbibition followed by exposure to 0.1% EMS for 12 hours. After flowering, the M1 plants were self-fertilized manually to produce the M2 generation. More than 1000 M2 plants were generated from M1 seeds.

Overall characteristics of *P. hybrida* mutagenized populations

The petunia seed germination and flower number were significantly affected by the mutagen dosage, treatment time and imbibition time of EMS. Fifteen days after germination, plant height was investigated. The plant heights are negatively affected by increased mutagen dosage of EMS (Figure 3). The plant height of 12 hour EMS treatment among different dosages presented linear correlation coefficients, 99.8% for 12 hour EMS treatment at 99% of probability. Thirty days after germination, anther qualities on each plant were investigated. The anther quality has been classified as four ranks. The quality decrease when the mutagen dosage increase and the longer exposure time (Figure 4).

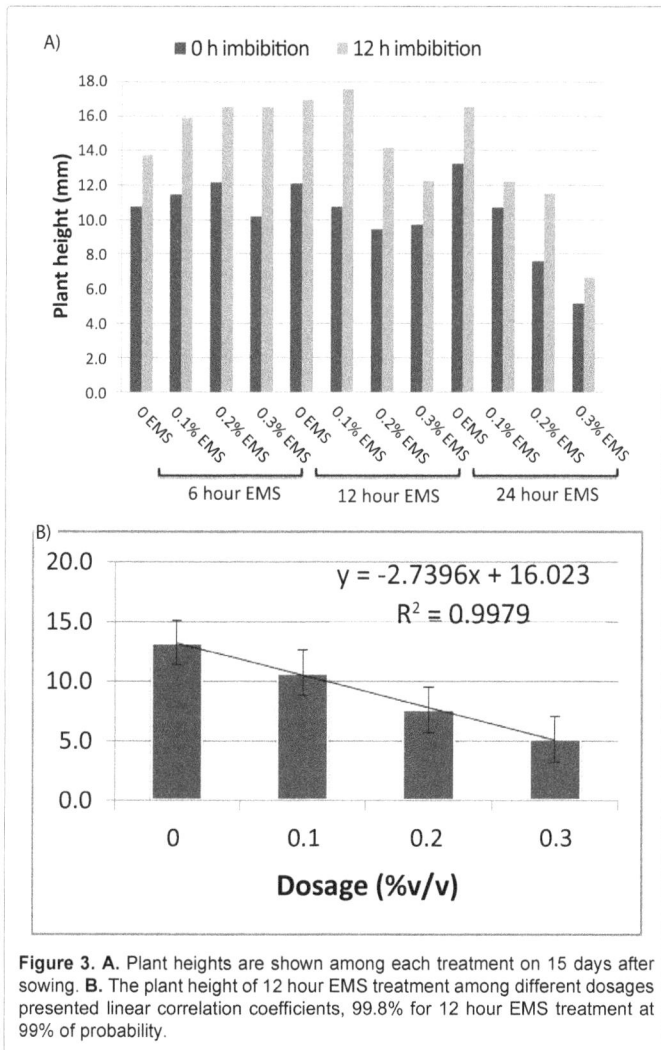

Figure 3. A. Plant heights are shown among each treatment on 15 days after sowing. B. The plant height of 12 hour EMS treatment among different dosages presented linear correlation coefficients, 99.8% for 12 hour EMS treatment at 99% of probability.

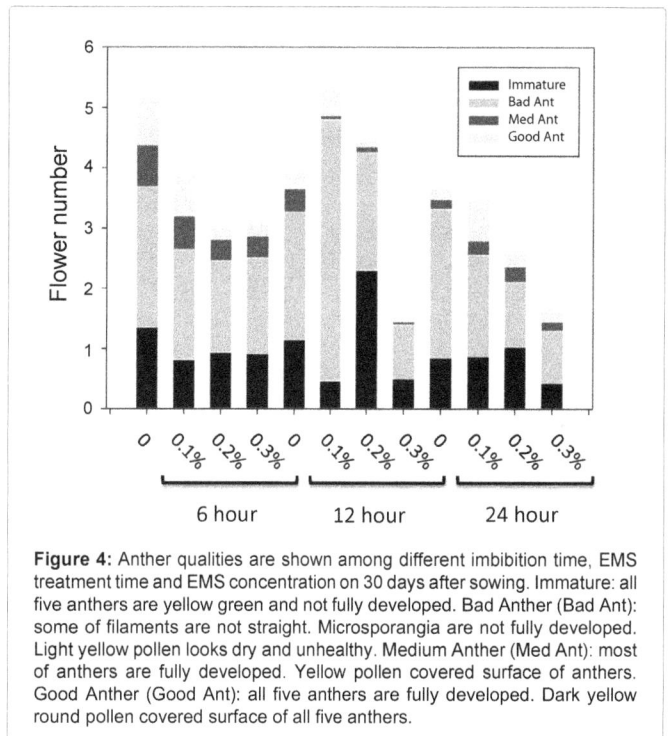

Figure 4: Anther qualities are shown among different imbibition time, EMS treatment time and EMS concentration on 30 days after sowing. Immature: all five anthers are yellow green and not fully developed. Bad Anther (Bad Ant): some of filaments are not straight. Microsporangia are not fully developed. Light yellow pollen looks dry and unhealthy. Medium Anther (Med Ant): most of anthers are fully developed. Yellow pollen covered surface of anthers. Good Anther (Good Ant): all five anthers are fully developed. Dark yellow round pollen covered surface of all five anthers.

Source	DF	SS	MS	F
Imbibition (IMB)	1	0.066544	0.066544	12.73 **
EMS exposure (EXP)	2	0.019351	0.009675	1.85 ns
EMS concentration (CONC)	3	0.003558	0.001186	0.23 ns
IMB x EXP	2	0.012402	0.006201	1.19 ns
IMB x CONC	3	0.064162	0.021387	4.09 *
EXP x CONC	6	0.120158	0.020026	3.83 *
Error	6	0.031369	0.005228	
Total	23	0.317544		

Table 1: Analysis of variance of effect of mutagenesis parameters on seed germination (day 10). *P = 0.05; **P = 0.01; ns = not significant.

Source	DF	SS	MS	F
Imbibition (IMB)	1	18.6345	18.6345	46.38 **
EMS exposure (EXP)	2	4.2732	2.1366	5.32 *
EMS concentration (CONC)	3	8.1507	2.7169	6.76 *
IMB×EXP	2	0.0175	0.0088	0.02 ns
IMB×CONC	3	2.0912	0.3485	0.87 ns
EXP×CONC	6	2.3546	0.7849	1.95 ns
Error	6	2.4107	0.4018	
Total	23	37.9324		

Table 2: Analysis of variance of effect of mutagenesis parameters on flower number (day 30). *P=0.05; **P=0.01; ns =not significant.

The conditions for EMS mutagenesis of *Petunia hybrida* 'Mitchell Diploid' were optimized in preparation for the development of a mutagenized population. Protocols for petunia mutagenesis have been published previously [15-17], but there is little consensus in procedures. This may be related to differences in genotypes. The optimized parameters described in the present work most closely resemble those of Berenschot et al. [15], who used a 14-hour seed imbibition followed by 24 hours of exposure to 0.1% EMS. In their study, EMS concentration had only minor effects on viability after

15 days, but seed capsule production was reduced by 25% with 0.25% EMS [15]. In our hands, treatment of seeds with 0.2% EMS reduced the number of capsules by 79.8%. The application of our EMS mutagenesis protocol to 2000 petunia seeds and manual fertilization of the resulting plants produced an M2 population suitable for TILLING [18].

In order to increase the likelihood of inducing distinct types of genetic variation in *P. hybrida*, a protocol was developed for mutating petunia with EMS while minimizing deleterious effects on viability and fertility. A mutagenized population was produced that is suitable for screening by genetic techniques such as TILLING [19-21].

Acknowledgement

We would like to thank Dr. David Clark (University of Florida) for providing *Petunia hybrida* 'Mitchell Diploid' seeds used in this research.

References

1. Schum A, (2003) Mutation breeding in ornamentals: an efficient breeding method? Acta Hort. 612:47-60.

2. Jain MS, (2006) Mutation-assisted breeding for improving ornamental plants. Acta Hort. 714:85-98.

3. Ahloowalia BS, Maluszynski M, Nichterlein K (2004) Global impact of mutation-derived varieties. Eupytica 135: 187-204.

4. McCallum CM, Comai L, Greene EA, Henikoff S (2000) Targeting induced local lesions IN genomes (TILLING) for plant functional genomics. Plant Physiol 123: 439-442.

5. Dalmais M, Antelme S, Ho-Yue-Kuang S, Wang Y, Darracq O, et al. (2013) A TILLING Platform for Functional Genomics in Brachypodium distachyon. PLoS One 8: 65503.

6. Wilde HD, Chen Y, Jiang P, Bhattacharya A, (2012) Targeted mutation breeding of horticultural plants. Emir. J. Food. Agric. 24: 31-41.

7. USDA-NASS, (2009). Census of Horticultural Specialties.

8. Gerats T, Vandenbussche M (2005) A model system for comparative research: Petunia. Trends Plant Sci 10: 251-256.

9. Kiss L, Jankovics T, Kovács GM, Daughtrey ML, (2008) Oidium longipes, and a new powdery mildew fungus on petunia in the USA: A potential threat to ornamental and vegetable solanaceous crops. Plant Disease 92: 818-825.

10. Lesemann DE, (1996) Viruses recently detected in vegetatively propagated Petunia. Acta Hort.

11. Feldhoff A, Wetzel T, Peters D, Kellner R, Krczal G (1998) Characterization of petunia flower mottle virus (PetFMV), a new potyvirus infecting Petunia x hybrida. Arch Virol 143: 475-488.

12. Gottwald S, Bauer P, Komatsuda T, Lundqvist U, Stein N (2009) TILLING in the two-rowed barley cultivar 'Barke' reveals preferred sites of functional diversity in the gene HvHox1. BMC Res Notes 2: 258.

13. Piron F, Nicolaï M, Minoïa S, Piednoir E, Moretti A, et al. (2010) An induced mutation in tomato eIF4E leads to immunity to two potyviruses. PLoS One 5: e11313.

14. Berenschot AS, Quecini V (2013) A reverse genetics approach identifies novel mutants in light responses and anthocyanin metabolism in petunia. Physiol Mol Plant Path. 1-13

15. Berenschot AS, Zucchi MI, Tulmann-Neto A, Quecini V (2008) Mutagenesis in Petunia x hybrida Vilm and isolation of a novel morphological mutant. Braz. J. Plant Physiol. 20: 95-103

16. Kashikar SG, Khalatkar AS, (1981) Breeding for flower color in Petunia hybrida Hort. Acta Hort. 111:35-40.

17. Napoli CA, Ruehle J, (1996) New mutations affecting meristem growth and potential in Petunia hybrida Vilm. Journal of Heredity 87: 371-377.

18. Jiang P, Chen Y, Wilde HD, (2013) Induction of variation in the petunia Mlo gene for resistance to powdery mildew (abstract). HortScience 48: S290.

19. Kulkarni RN, Sreevalli Y, Baskaran K, Kumar S, (2001) The mechanism and inheritance of intraflower self-pollination in self-pollinating variant strains of periwinkle. Plant breeding 120: 247-250.

20. Iwata F, Ohashi Y, Ishisaki I, Picco LM, Ushiki T (2013) Development of nanomanipulator using a high-speed atomic force microscope coupled with a haptic device. Ultra microscopy 133: 88-94.

21. Pavan S, Jacobsen E, Visser RG, Bai Y (2010) Loss of susceptibility as a novel breeding strategy for durable and broad-spectrum resistance. Mol Breed 25: 1-12.

25

Screening of Different Rice Genotypes against (*Pyricularia grisea*) Sacc. in Natural Epidemic Condition at Seedling Stage in Chitwan, Nepal

Khanal Sabin[1]*, Subedi Bijay[1], Bhandari Amrit[1], Giri Dilli Raman[1], Shrestha Bhuwan[1], Neupane Priyanka[2], Shrestha Sundar Man[2] and Gaire Shankar Prasad[2]

[1]*Institute of Agriculture and Animal Science, Tribhuwan Univeristy, Chitwan, Nepal*
[2]*Department of Plant Pathology, Faculty of Agriculture, Agriculture and Forestry University, Nepal*

Abstract

Numerous research has already establish blast as the continuous and devastating threat to rice production in Nepal and on the contrary Nepalese farmers do not have efficient knowledge and understanding about the complexity of disease for the management of the blast epidemic development. The most effective physical tool seems to be provision of resistant genotypes obtained against screening of different rice genotypes: effective management practices against the complexity of blast pathogen. Experiments were conducted for screening 50 rice genotypes under natural epidemic condition against seedling blast (*Pyricularia grisea*) in Randomized complete block design at Chitwan. Rice grains were sown on July 6, 2015 at field and disease scoring was done on 21, 24, 27 and 30 DAS; Scoring was done based on the standard scale of 0-9 developed by IRRI. Based on the result Taichung-176 and Sankharika showed the highest percentage of incidence and severity of disease. Sabitri, however, was found to be most resistant among genotypes with the lowest percentage of incidence and severity during observation.

Keywords: Rice blast; *Pyricularia grisea*; Sabitri; Blast susceptible

Introduction

Blast is caused by *Pyricularia grisea*. It occurs in nearly all rice growing areas of the world. It is considered the most serious disease in both temperate and tropical rainfed enviroments. With increasing nitrogenous fertilizer and higher plant density, blast is known to be devastating [1]. Blast was first recorded in china in 1637. The causal organism was named *Pyricularia oryzae* by Cavara in italy in 1891 and was renamed by Rossman 1990 to *Pyricularia grisea* [2].

Rice is truly a crop of global importance. Almost half the world's population, particularly in east and south east asia, depends on rice as the major source of nutritional calories [3]. Every year it is estimated that rice blast destroy food more than enough to eat for 60 million people and 50% of the rice yield is lost in the field by the occurrence of blast [4] Rice is the most prestigious food crop of Nepal. It is grown in a diverse environment ranging from tropical plains to foot of the mountain and higher elevation (3050 masl) in Chhumchure, Jumla. Nepal is considered as one of the origin center of rice. It is one of the most important cereal crops in Nepal. Rice is grown in 1440 thousand ha and the productivity is 2.56 t/ha. It contributes nearly 20 per cent to the agricultural gross domestic product. Nepal has released fifty five rice varieties with full package of growing practices in the last 40 years. The coverage by improved varieties is 85 percent of the total rice cultivated land. Popularly cultivated improved varieties are Radha-4, Radha-12, Masuli, Sabitri, CH-45, Bindeswori in terai, Khumal-4, Khumal-11, Taichung-176, chaining in mid-hills and Chandanaath-3 in high hills (NARC 2014). Radha-12, sabitri, janaki possess higher level of resistance [5]. Seedlings of high yielding masuli were affected in late june in saradanagar, Rampur, kiranganj, mangalpur and ratanagar area of the chitwan district [6]. Radha-12 had 7 fold less neck blast than masuli whereas other genotypes showed less neck blast than masuli [5].

Rice blast genetic analysis confirmed gene for gene interaction that control cultivar specificity in fungal plant interactions. Nuclear and mitochondrial genomes molecular analyses suggest that *M. grisea* pathogen remain in nature as different types of genetically distinct asexually reproducing population [7]. An understanding of the molecular mechanism that govern host specificity should aid in the development of new strategies for control of rice blast [8]. Mechanism controlling host species specificity differ in basic compatibility factor that allows pathogen to infect particular species. PWL2 host species specificity gene has properties analogous to classical avirulence genes, which function to prevent infection of certain cultivars of particular host species. The PWL2 gene encodes a glycine-rich, hydrophilic protein with a putative secretion signal sequence [8]. Blast, caused by *Pyricularia grisea* Sacc has been a continuous threat to rice production in Nepal [9,10]. Blast epidemics result in a complete loss of seedlings in the seedbed [6,11-16]

Varying tools have been used as a blast management toolkit such as knowledge tools, communication, physical and policy tools. Each tool is rationalized in terms of having an effect either on initial inoculum or disease (x_0), the epidemic infection rate(R) or duration of epidemic (D). Certain tools like biological control agents and confirmatory serological tools are still unknown to blast control. Nepalese farmers do not have efficient knowledge and understanding about the use of fungicide and the effect of nutrient (nitrogen, phosphorus, potassium), and non-nutrient (silicon) amendments on blast epidemic development. Water management to reduce stress on plants at blast susceptible stages [17] are still in the dark to the Nepalese common farmers. Cultural practices seem ineffective due to no clear cut fallow period between any two rice seasons making blast pathosystem a continuous pathosystem and given the dispersal pattern of conidia initial inoculum will always be available for matching alloinfection. Hence, the most effective physical tool seems to be provision of resistant genotype. Seed possessing resistant genes to blast have been the basis for plant protection for centuries [13].

*Corresponding author: Sabin Khanal, Institute of Agriculture and Animal Science, Tribhuwan University, Nepal, E-mail: savvy.khanal33@gmail.com

Materials and Methods

Experimental setup

Field experiment was set up in Agronomy farm of IAAS, Rampur, Chitwan. The experiment was conducted in single factorial RCBD design with 3 replication. Each plot was 5 mX1 m, in each replication 50 rows was made to sow the 50 different genotypes. Seed was sown randomly in such a way that, genotypes was not repeated in line in the replications. Seeding was done in 2nd week of July.

Genotypes used as treatment:	
1) NR 10676-B-1-3-3-3	26) Madhya dhan -845
2) NR 10490-89-3-2-1	27) Sona Mansuli
3) NR 11105-B-B-27	28) Radha-22
4) NR 11052-B-B-B-B_6	29) Sankharika*
5) 08FAN10	30) Radha-11
6) NR10769-4-2-2	31) IR 87751-20-4-4-2
7) NR 10676-B-5-3	32) NR 11111-B-B-23
8) NR 11011-B-B-B-B-3	33) Manjushree-2
9) NR 11011-B-B-B-B-2	34) Taichung-176
10) Sugandha-2	35) Khumal-4
11) NR 11050-B-B-B-1	36) Kalo Masino
12) NR 11037-B-B-B-B-5	37) Radha-4
13) NR 11022-2-2-3-3-1	38) Sawa Mansuli
14) NR 11092-B-B-B-12	39) Ramdhan
15) NR 11042-B-B-B-1-1	40) Sabitri**
16) NR 11082-B-B-B-5-3	41) Bindheshwari
17) NR 11016-B-5-2-3-3-2	42) Sukkah-3
18) NR 11011-B-B-B-B-6	43) Savasab-1
19) NR 11139-B-B-B-21	44) Jethi Mansuli
20) NR 11050-B-B-B-B-2	45) Basmati Seto
21) NR 11115-B-B-31-3	46) Pusa Basmati
22) NR 11130-B-B-B-19	47) Sarju
23) NR 11105-B-B-16-2	48) Hardinath
24) NR 11111-B-B-23-2	49) Kanchi Mansuli
25) NR 11109-B-B-12-3-2	50) Makwanpure
Note: *= Susceptible Check	
**= Resistant Check	

Observation

Disease assessment

Disease incidence: Appearance of first symptoms of disease among all the plants germinated will be recorded. Here, total no. of plants in a row and Number of plants showing the symptoms will be recorded (Figures 1 and 2).

Percent disease incidence will be calculated by using the formula:

$$\% \text{ DI} = \frac{\text{Number of infected plants}}{\text{total number of plants germinated}} \times 100\%$$

Disease scoring: Disease scoring will be done according to standard scoring scale developed by International Rice Research Institute (IRRI) using a scale of 0-9.

- Small brown specks of pin point size.

- Small roundish to slightly elongated, necrotic gravy spots, about 1-2 mm in diameter, with a distinct brown margin, lesions are mostly found on the lower leaves.

- Lesion type is the same as in 2, but significant number of lesion are on the upper leaves.

- Typical susceptible blast lesions, 3 mm or longer, infecting less than 4% of the leaf area.

- Typical susceptible blast lesions, 3 mm or longer, infecting less than 4-10% of the leaf area.

- Typical susceptible blast lesions, 3 mm or longer, infecting less than 11-25% of the leaf area.

- Typical susceptible blast lesions, 3 mm or longer, infecting less than 26-50% of the leaf area.

- Typical susceptible blast lesions, 3 mm or longer, infecting less than 51-75% of the leaf area, many leaves dead.

- Typical susceptible blast lesions, 3 mm or longer, infecting more than 75% of the leaf area

Disease intensity/index: Disease severity will be scored on the basis of standard scoring scale developed by International Rice Research institute (IRRI). 5 plants from each row showing the symptoms will be selected at random for observation and scored at a scale of 0-9 and average will be taken.

Disease severity will be calculated as:

$$\text{Disease severity}\% = \frac{\text{sum of all numerical rating}}{\text{no. of plants observed} \times 9} \times 100$$

The plants will be scored after 21 days of sowing and 3 days interval henceforth up to 30 days of sowing, giving 4 readings.

AUDPC:

$$\text{AUDPC} = \sum_{i}^{n-1}\left[\left(\frac{y_i + y_{i+1}}{2}\right)(t_{i+1} - t_i)\right]$$

Where, y_i: initial infection percentage (disease score)

Y_{i+1}: progressive infection percentage

$T_{i+1}-t_i$: time interval between the readings

Area under disease progressive curve

Total AUDPC value lied in the range of 17.78-210%. AUDPC 1&2 were calculated based on the disease severity percentage and calculated using formula as presented in the materials and methods above. Lowest total AUDPC was observed on Sabitri whereas highest was observed on Taichung-176 followed by pusa basmati, NR 11111-B-B-23-2, Sankharika and NR 10490-89-3-2-1. Based on the Total AUDPC value rice genotypes were listed on the five categories from resistant to highly susceptible which are shown in the Tables 1 and 2.

Discussion

Disease incidence was observed at 21, 24, 27 and 30 DAS. At 21 DAS, lowest disease incidence was observed in Kanchi Masuli followed by sabitri whereas highest disease incidence was observed in NR 10676-B-1-3-3-3-2 followed by sankharika and Pusa basmati. Similarly, during 24 DAS, 27 DAS and 30 DAS lowest disease incidence was observed in Sabitri whereas highest disease incidence at 24 DAS NR 11050-B-B-B-B-2 followed by NR 11050-B-B-B-B-1, NR 11016-B-5-2-3-3-2; at 27 DAS, NR 11050-B-B-B-B-2 followed by NR 11050-B-B-B-B-2 and NR 11016-B-5-2-3-3-2 were found to be highest and during 30 DAS NR 11050-B-B-B-B-2 followed by NR 11050-B-B-B-1 and NR 11105-

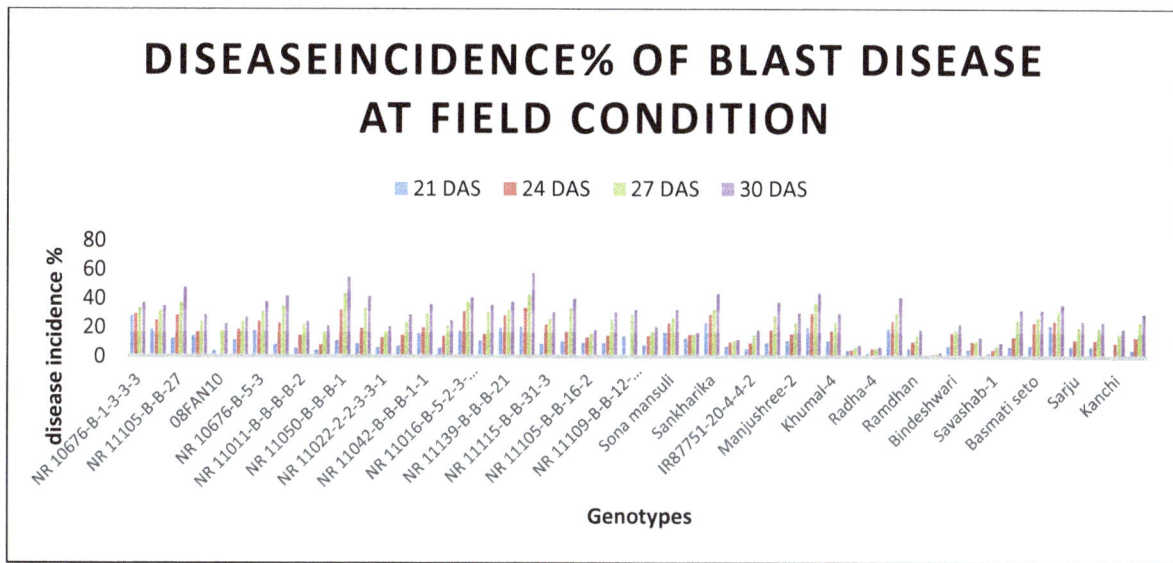

Figure 1: Disease incidence was observed at 21, 24, 27 and 30 DAS. At 21 DAS, lowest disease incidence was observed in Kanchi Masuli followed by sabitri whereas highest disease incidence was observed in NR 10676-B-1-3-3-3-2 followed by sankharika and Pusa basmati. Similarly, during 24 DAS, 27 DAS and 30 DAS lowest disease incidence was observed in Sabitri whereas highest disease incidence at 24 DAS NR 11050-B-B-B-B-2 followed by NR 11050-B-B-B-B-1, NR 11016-B-5-2-3-3-2; at 27 DAS, NR 11050-B-B-B-B-2 followed by NR 11050-B-B-B-B-2 and NR 11016-B-5-2-3-3-2 were found to be highest and during 30 DAS NR 11050-B-B-B-B-2 followed by NR 11050-B-B-B-1 and NR 11105-B-B-27 were found highest.

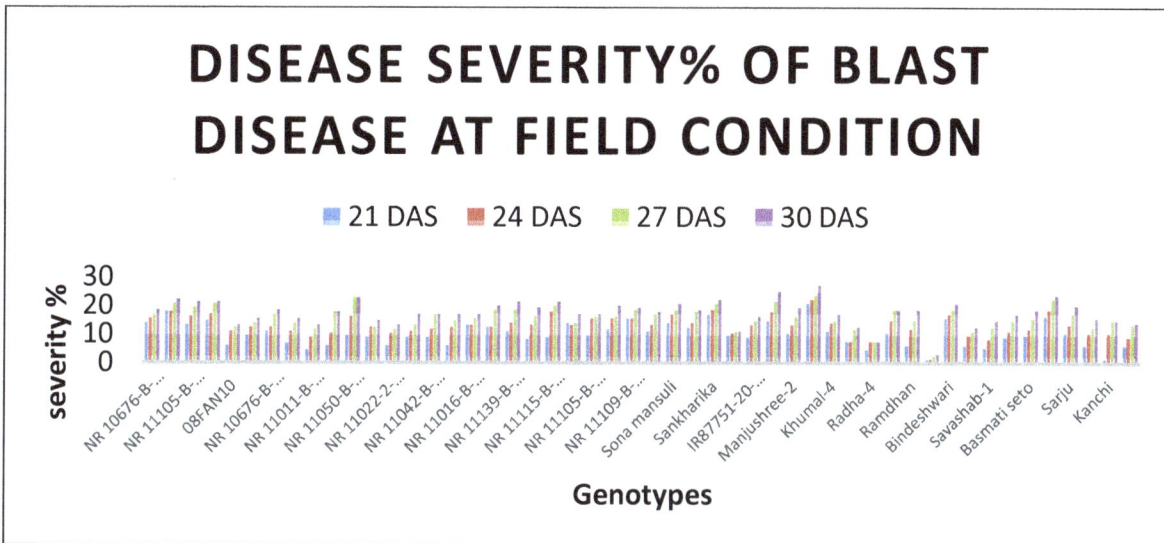

Figure 2: Based on the standard scoring scale of 0-9 developed by IRRI, disease were scored at 21, 24, 27 and 30 DAS and disease severity % was calculated as above mentioned formula(materials and methods). During 21 DAS lowest disease severity was observed on kanchi masuli followed by sabitri whereas on all other days of observation lowest disease severity % was observed on sabitri. However, during all days of observation Taichung-176 showed highest disease severity percentage along with varieties sankharika, pusa basmati, NR 11050-B-B-B-1 and NR 11111-B-B-23-2.

B-B-27 were found highest [18-20]. Sabitri was reported to be most resistant by Chaudary et al. [5]. Genotypes starting with NR initial were breeding lines developed by Nepal Agriculture Research Council (NARC) as they were developed for high hills and this experiment being conducted in the terai region might had induced blast incidence on these lines due to unsuitable temperature to the genotypes.

During 21 DAS lowest disease severity was observed on kanchi masuli followed by sabitri whereas on all other days of observation lowest disease severity % was observed on sabitri. However, during all days of

observation Taichung-176 showed highest disease severity percentage along with varieties sankharika, pusa basmati, NR 11050-B-B-B-1 and NR 11111-B-B-23-2. Experiment by Manandher et al. [11] presented sankharika to be most susceptible variety and established that it is adversely affected by blast pathogen whereas Taichung-176 were found to be highly susceptible variety by Manandhar et al. [9]. Kumar et al. [21] reported pusa basmati as most susceptible to blast disease.

Similarly sabitri showed lowest level of AUDPC value and was categorized as the resistant genotype along with Radha-4 which is

Genotypes	AUDPC1	AUDPC2	AUDPC 3	TOTAL AUDPC
NR 10676-B-1-3-3-3	44.44^{bcdefg}	$48.49^{bcdefghi}$	$53.33^{bcdefghij}$	$146.67^{bcdefghij}$
NR 10490-89-3-2-1	53.33^{ab}	57.78^{abcd}	64.44^{abc}	175.56^{abcd}
NR 11105-B-B-27	44.44^{bcdefg}	53.33^{abcdef}	61.11^{abcde}	$158.89^{bcdefgh}$
NR 11052-B-B-B-B_6	47.78^{abcde}	56.67^{abcde}	63.33^{abc}	167.78^{abcde}
08FAN10	25.56^{hijkl}	$35.56^{ghijklmn}$	38.89^{fghijk}	100.00^{jklm}
NR10769-4-2-2	$33.33^{defghijkl}$	$40.00^{fghijkl}$	$44.44^{cdefghij}$	$117.78^{fghijkl}$
NR 10676-B-5-3	$35.56^{cdefghijk}$	$44.44^{cdefghijkl}$	$53.33^{bcdefghij}$	$133.33^{bcdefghijkl}$
NR 11011-B-B-B-B-3	26.67^{hijkl}	$37.78^{fghijklm}$	$44.44^{cdefghij}$	108.89^{ijklm}
NR 11011-B-B-B-B-2	20.00^{klm}	31.11^{klm}	37.78^{ghijk}	88.89^{lm}
Sugandha-2	24.44^{hijkl}	$42.22^{defghijkl}$	$53.33^{bcdefghij}$	$120.00^{efghijkl}$
NR 11050-B-B-B-1	$38.89^{bcdefghij}$	58.89^{abc}	68.89^{ab}	166.66^{abcdef}
NR 11037-B-B-B-B-5	$32.22^{defghijkl}$	$37.78^{fghijklmn}$	$41.11^{efghijk}$	111.11^{hijklm}
NR 11022-2-2-3-3-1	24.44^{hijkl}	33.33^{ijklm}	37.78^{ghijk}	95.56^{klm}
NR 11092-B-B-B-12	$30.00^{fghijkl}$	36.67^{hijklm}	$45.56^{cdefghij}$	112.22^{ghijkl}
NR 11042-B-B-B-1-1	$31.11^{efghijkl}$	$43.33^{cdefghijkl}$	$51.11^{bcdefghij}$	$125.56^{efghijkl}$
NR 11082-B-B-B-5-3	27.78^{ghijkl}	$41.11^{efghijkl}$	$47.78^{cdefghij}$	116.67^{ghijkl}
NR 11016-B-5-2-3-3-2	$40.00^{bcdefghi}$	$43.33^{cdefghijkl}$	$47.78^{cdefghij}$	$131.11^{cdefghijkl}$
NR 11011-B-B-B-B-6	$37.78^{bcdefghij}$	$46.67^{abcdefghij}$	$57.78^{abcdefghij}$	$142.22^{bcdefghijk}$
NR 11139-B-B-B-21	$37.77^{bcdefghij}$	$48.89^{bcdefghi}$	60.00^{abcde}	$146.67^{bcdefghij}$
NR 11050-B-B-B-B-2	$32.22^{defghijkl}$	$44.44^{cdefghijkl}$	$53.33^{bcdefhij}$	$130.00^{cdefghijkl}$
NR 11115-B-B-31-3	$40.00^{bcdefghi}$	56.67^{abcde}	62.22^{abcd}	$158.89^{bcdefgh}$
NR 11130-B-B-B-19	$41.11^{bcdefgh}$	$41.11^{efghijkl}$	$45.56^{cdefghij}$	$127.78^{defghijkl}$
NR 11105-B-B-16-2	$37.78^{bcdefghij}$	$47.78^{bcdefghij}$	$50.00^{bcdefghij}$	$135.56^{bcdefghij}$
NR 11111-B-B-23-2	$41.11^{bcdefgh}$	$47.78^{bcdefghij}$	$54.44^{bcdefghij}$	$143.33^{bcdefghijk}$
NR 11109-B-B-12-3-2	46.67^{bcdef}	51.11^{bcdefg}	$56.67^{abcdefgh}$	$154.44^{bcdefghi}$
Madhya dhan -845	$36.67^{bcdefghij}$	$45.56^{bcdefghijk}$	$52.22^{bcdefghij}$	$134.44^{bcdefghijkl}$
Sona mansuli	46.67^{bcdef}	53.33^{abcdef}	58.89^{abcdef}	$158.89^{bcdefgh}$
Radha-22	$40.00^{bcdefghi}$	$47.78^{bcdefghij}$	$54.44^{bcdefghij}$	$142.22^{bcdefghijk}$
Sankharika	53.33^{ab}	58.89^{abc}	64.44^{abc}	176.67^{abcd}
Radha-11	$30.00^{fghijkl}$	32.22^{ijklm}	33.33^{jk}	95.56^{klm}
IR87751-20-4-4-2	$33.33^{defghijkl}$	$42.22^{defghijkl}$	$46.67^{cdefghij}$	$122.22^{efghijkl}$
NR 11111-B-B-23	48.89^{abcde}	58.89^{abc}	70.00^{ab}	177.78^{abc}
Manjushree-2	$35.56^{cdefghijk}$	$44.44^{cdefghijkl}$	$53.33^{bcdefghij}$	$133.33^{bcdefghijkl}$
Taichung-176	64.44^{a}	68.89^{a}	76.67^{a}	210.00^{a}
Khumal-4	$37.78^{bcdefghij}$	$43.33^{cdefghijkl}$	$47.78^{cdefghij}$	$128.89^{cdefghijkl}$
Kalo masino	22.22^{jkl}	28.89^{lm}	36.67^{hijk}	87.78^{lm}
Radha-4	17.78^{lm}	22.22^{m}	22.22^{kl}	62.22^{mn}
Sawa mansuli	$37.78^{bcdefghij}$	$50.00^{bcdefgh}$	$55.56^{bcdefghi}$	$143.33^{bcdefghijk}$
Ramdhan	26.67^{hijkl}	$40.00^{fghijkl}$	$50.00^{bcdefghij}$	116.67^{ghijkl}
Sabitri	4.44^{m}	5.56^{n}	7.77^{l}	17.78^{n}
Bindeshwari	48.89^{abcd}	53.33^{abcdef}	58.88^{abcdef}	$161.11^{abcdefg}$
Sukkah-3	23.33^{ijkl}	31.11^{klm}	35.55^{ijk}	90.00^{lm}
Savashab-1	20.00^{klm}	31.11^{klm}	41.11^{efghij}	92.22^{lm}
Jethi mansuli	$30.00^{fghijkl}$	$38.89^{fghijkl}$	$47.78^{cdefghij}$	116.67^{ghijkl}
Basmati seto	$32.22^{defghijkl}$	$41.11^{efghijkl}$	$51.11^{bcdefghij}$	$124.44^{efghijkl}$
Pusa basmati	52.22^{abc}	61.11^{ab}	68.89^{ab}	182.22^{ab}
Sarju	$35.56^{cdefghijk}$	$45.56^{bcdefghijk}$	$55.56^{bcdefghi}$	$136.67^{bcdefghijkl}$
Hardinath	24.44^{hijkl}	34.44^{hijkl}	$42.22^{defghijk}$	101.11^{ijklm}
Kanchi	17.78^{lm}	$37.78^{fghijklm}$	$44.44^{cdefghij}$	100.00^{jklm}
Makwanpure	22.22^{jkl}	33.33^{ijklm}	$41.11^{efghijk}$	96.67^{klm}

Table 1: AUDPC values of rice genotypes.

Category	Range	Genotypes
Resistant	0-70	Sabitri Radha-4
Moderately resistant	71-120	Kalo masino Sukkah-3 NR 11011-B-B-B-B-2 Savashab-1 NR 11022-2-2-3-3-1 Kanchi Mansuli masuli Hardinath Radha-11 08FAN10 Makwanpure NR 11092-B-B-B-12 NR 11037-B-B-B-B-5 NR 11011-B-B-B-B-3 NR 11082-B-B-B-5-3 Ramdhan Jethi mansuli NR 10769-4-2-2 Sugandha-2

Moderately susceptible	121-140	IR87751-20-4-4-2 Basmati seto NR 11042-B-B-B-1-1 NR 11130-B-B-B-19 Khumal-4 NR 11050-B-B-B-B-2 NR 11016-B-5-2-3-3-2 NR 10676-B-5-3 Manjushree-2 Madhya dhan-845 NR 11105-B-B-16-2 Sarju
Susceptible	141-170	Radha-22 NR 11111-B-B-23-2 NR 11011-B-B-B-6 NR 10676-B-1-3-3-3 NR 11139-B-B-B-21 Sawa Mansuli NR 11109-B-B-12-3-2 Bindeshwari NR 11105-B-B-27 NR 11115-B-B-31-3 NR 11050-B-B-B-1 Sona Mansuli NR 11052-B-B-B-B-6
Highly susceptible	171-225	NR 11111-B-B-23-2 Sankharika Pusa Basmati Taichung-176 NR 10490-89-3-2-1

Table 2: Based on the AUDPC value rice genotypes are listed on the five categories from resistant to highly susceptible.

supported by Chaudary et al. [5] suggesting that Sabitri and Radha varieties to be resistant to blast pathogen; whereas Taichung-176 pusa basmati and Sankharika were categorized as the most susceptible varieties which coincides with the result presented by (Manandhar et al. and Manandhar et al.) and Kumar et al. [9,11,21]. Similary, NR 11111-B-B-23-2 and NR 10490-89-3-2-1 were also categorized as the most susceptible and more conclusive result are yet to be drawn of these genotypes.

Conclusion

50 rice genotypes were sown in 6th July in Randomized complete block design at chitwan. The experiment was only limited to seedling stage and its purpose was to identify the resistant and susceptible variety among the different rice genotypes collected all over the country along with some of the breedling lines provided by the NARC khumaltar. Sankharika, Taichung-176, pusa basmati, NR 11111-B-B-23 and NR 10490-89-3-2-1 were found most susceptible and sabitri and Radha-4 were found to be resistant.

As Taichung-176, sankharika was found to be most susceptible to blast on both field and lab condition as NARC has described Taichung-176 as susceptible variety to mid-hills and sankharika to Terai region. Sabitri was found to be most resistant among all genotypes. Further research is recommended on the varieties mentioned above for further certainty; in addition, further research work such as comparison of plant yield with disease can be done and also molecular study of the plant varieties is further recommended.

Acknowledgements

I would like to express my sincere gratitude to prof. Dr. Sundarman Shrestha and Mr. Shankhar Gaire, Faculty of Plant pathology at Agriculture and Forestry University for their guidance and support during all the activity of this research. I would also like to acknowledge Mr. Bikash adhikari for his help during all the field activities; Mr. Ramesh Acharya for his support; Mrs. Shawantana Ghimire Khanal for her encouragement to write this paper.

References

1. Kato H (1974) Epidemiology of rice blast disease. REV plant prot Res 7: 1-20.

2. Rossman AY (1990) Pyricularia grisea the correct name for the Rice Blast disease fungus. Mycologia 82: 509-512.

3. Swaminathan MS (1982) Biotechnology factors in the epidemiology of rice blast. Annual review of phytopathology 13: 139-256.

4. Barman RS, Chattoo BB (2005) Rice blast fungus sequenced. Current Science 89: 930-931.

5. Chaudary B, Shrestha SM, Sharma RC (2001) Resistance in Rice breeding lines to blast fungus in Nepal. Journal of institute of agriculture and animal sciences (Nepal) 2001. Nepal agriculture research journal 6: 49-56.

6. Adhikari TB, Shrestha SM (1986) Blast epidemic in chitwan valley, Nepal. International Rice Research Newsletter 11: 22.

7. Zeigler RS, Leong SA, Teng PS (1994) Rice Blast Disease. Commonwealth Agricultural Bureau International, Wallingford, England.

8. Swelgard JA, Carrol AM, Kang S, Farral L, Chumley FG, et al. (1995) Identification, cloning and characterization of PWL2 of gene for host species specificity in Rice Blast fungus. The plant cell 7: 1221-1233.

9. Manandhar HK, Shrestha K, Amatya P (1992) Seed-borne diseases. In: Plant diseases, seed production and seed health testing in Nepal. Danish Government, Institute of Seed Pathology for Developing Countries, Copenhagen, Denmark. pp: 59-74.

10. Chaudary B (1999) Effect of blast disease on rice yield. Nepal Agricultural Research Journal 3: 8-13.

11. Manandhar HK, Thapa BJ, Amatya P (1985) Efficacy of various fungicides on the control of rice blast disease. Journal of Institute of agriculture and animal sciences 6: 21-29.

12. Pradhanang PM (1988) Outbreak of blast disease at Lumle Agricultural Centre (LAC) and its extension command area (ECA). In: Proceedings of the first national rice blast workshop. National Agricultural Research and Services Center, National Rice Research Program, pp: 61-69.

13. Reissig W, Heinrichs E, Litsinger J, Moody K, Fiedler L (1986) Illustrated guide to integrated pest management in rice in tropical Asia. IRRI, Los Banos, Laguna, Philippines.

14. Chaudhary B, Karki PB, Lal KK (1994) Neck blast resistant lines of Radha-17 isolated. International Rice Research Notes 19: 11.

15. Chaudhary B, Sah DN (1997) Effect of promising rice genotypes on leaf blast disease progression. Nepal Agriculture Research Journal 1: 27-31.

16. Chaudhary B, Sah DN (1998) Efficacy of Beam 75 WP in controlling leaf blast disease at the seedling stage of rice. Nepal Agriculture Research Journal 2: 42-47.

17. Teng PS, Klein-Gebbink HW, Pinnschmidt H (1991) An analysis of the blast pathosystem to guide modelling and forecasting. In: Rice blast modelling and forecasting. International Rice Research institute, Manila, Philippines. pp: 1-30.

18. Kingslover CH, Mackenzie DR, Rush MC (1988) Field testing a computerized forecasting systems for rice blast disease. Phytopathology 78: 931- 934.

19. Sah DN (1989) Effects of flooding and leaf wetness duration on resistance of rice lines to Pyricularia oryzae. Journal of Institute of Agriculture and Animal Sciences 10: 41-48.

20. Suryanarayan S (1966) Environment and the blast disease. Indian phytopathological society 3: 100-114.

21. Kumar N, Singh D, Gupta S, Sirohi A, Ramesh B, et al. (2013) Determination and expression of genes for resistant to blast(M. oryzae) in Basmati and Non-basmati Indica rices (Oryzae sativa L.). African journal of Biotechnology 12: 4098-4104.

26

Molecular Tools for Nursery Plant Production

Peng Jiang*

Horticulture Department, University of Georgia, Athens, Georgia, USA

Abstract

Breeding strategies in nursery plants is lagging behind most of the agricultural crops while molecular methods have been adopted last decade. Identification and verification of varieties for nursery plants were applied by molecular tools. Marker assisted breeding utilizes the DNA markers linked to genes of interest to achieve efficient selection strategies. Marker assisted selection (MAS) is a process whereby a marker is used for indirect selection of genetic determinants of a trait of interest. There are different kinds of molecular markers, such as restriction fragment length polymorphisms (RFLPs), random amplified polymorphic DNA (RAPDs), amplified fragment length polymorphisms (AFLPs), microsatellites and single nucleotide polymorphisms (SNPs). These molecular markers allow high density DNA marker maps. In this review, all of these molecular markers have been applied widely among crops and ornamentals and the advantages and disadvantages have been listed. The best molecular markers are those that distinguish multiple alleles per locus (highly polymorphic) and are co-dominant.

Keywords: Nursery; Production; Ornamentals; Molecular markers; SSR; AFLP; RFLP; RAPD; SRAP

Introduction

Most of the traits of interest for plant breeding programs are quantitative traits. These traits are controlled by many genes and environmental factors. Phenotypic selection is the most common used form of selection in traditional genetic improvement programs. However, by using this method, you will not know which genes are actually being selected. With the development of molecular markers, marker assisted selection (MAS) become increasingly important in the coming years. MAS involve the selection of plants carrying genomic regions that are involved in the expression of traits of interest through molecular markers [1].

Molecular markers can be thought as constant landmarks in the plant genome. There are different kinds of molecular markers, such as restriction fragment length polymorphisms (RFLPs), random amplified polymorphic DNA (RAPDs), amplified fragment length polymorphisms (AFLPs), microsatellites and single nucleotide polymorphisms (SNPs). These molecular markers allow high density DNA marker maps.

There are three types of relationships between the markers and the genes of interest. First, the molecular marker is located within the gene of interest. Second, the marker is in linkage disequilibrium (LD) with gene of interest throughout the whole population. Third, the marker is not in linkage disequilibrium with gene of interest throughout the whole population [2]. This study will give a general review about molecular markers used in nursery plants.

Molecular markers

In genetics, a molecular marker is a fragment of DNA that associated with a certain location within the genome. Molecular markers are usually phenotypically neutral and could identify by techniques such as southern hybridization or PCR. Several different kinds of molecular marker could be applied on plant selection: such as restriction fragment length polymorphisms (RFLPs), is detected by southern hybridization. The principle of RFLPs is detecting a site in a genome where the distance between two restriction sites varies among different individuals. These sites are identified by restriction enzyme digests of chromosomal DNA. It requires a radioactive probe when do southern blotting.

Other methods involve using PCR, such as amplified fragment length polymorphisms (AFLPs) uses restriction enzymes to digest genomic DNA [3]. Usually this technique has three steps: first, digestion of total plant DNA with one or more restriction enzymes and ligation of restriction half-site specific adaptors to all restriction fragments. Second, selective amplification of a subset of these fragments with two PCR primers that have corresponding adaptor and restriction site specific sequences. Third, run the amplicons on a gel matrix, followed by visualization of the band pattern. Random amplified polymorphic DNA (RAPDs) markers are about 10 nucleotide length DNA fragments from PCR amplification of random segments of genomic DNA. RAPDs are able to differentiate between genetically distinct individuals. In recent years, RAPD has been used to characterize the phylogeny of diverse plant and animal species [4]. Single nucleotide polymorphisms (SNPs) refer to a single nucleotide difference in the sequence of a gene or segment of the genome [5]. There are a variety of methods for analyzing SNPs; detection of SNPs can be done without gels, such as high resolution melting method. All of the above molecular markers have been applied widely among crops and ornamentals and the advantages and disadvantages have been listed in Table 1 [6]. The best molecular markers are those that distinguish multiple alleles per locus (highly polymorphic) and are co-dominant (each allele can be observed).

Sequence-related amplified polymorphism (SRAP) is a simple marker technique aimed for the amplification of open reading frames. Based on two-primer amplification, SRAP combines simplicity, reliability, moderate throughput ratio and facile sequencing of selected bands [7].

Current status of applications of molecular markers in nursery plants production

Molecular marker technologies have been widely used in

*Corresponding author: Peng Jiang, Horticulture Department, University of Georgia, Athens, Georgia, USA, E-mail: pjiang@uga.edu

Molecular marker	Advantages	Disadvantages	Codominant (C) or Dominant (D)
Amplified fragment length Polymorphism (AFLP)	Multiple loci High levels of polymorphism generated	Large amounts of DNA required Complicated methodology	D
Simple sequence repeats (SSRs) or microsatellites	Technically simple Robust and reliable Transferable between population	Large amounts of time and labor required for production of primers Usually require polyacrylamide electrophoresis	C
Restriction fragment length polymorphism (RFLP)	Robust Reliable Transferable across populations	Time-consuming, laborious and expensive Large amount of DNA required Limited polymorphism	C
Random amplified polymorphic DNA (RAPD)	Quick and simple Inexpensive Multiple loci from a single primer possible Small amounts of DNA required	Problems with reproducibility Generally not transferable	D
Sequence-related amplified polymorphism (SRAP)	Simple Reliable Moderate throughput ratio Facile sequence of related bands	Time and labor required for production of primers	C

Table 1: Advantages and disadvantages of most commonly-used DNA markers.

Ornamental	Trait	Samples	Methods	Primers	Gene/QTL	Linked marker	Year	Reference
Capsicum annuum L	Erect versus pendant orientated fruit	108 F2:3 individuals	Bulked segregant analysis (BSA) and amplified fragment length polymorphism (AFLP)		Saengryeog 211 (pendant), Saengryeog 213 (erect)	A2C79	2008	[9]
Oil Palm	Genetic diversity	6 Cultivars	Simple sequence repeat (SSR)	20 SSR markers			2012	[10]
Mei (Prunus mume Sieb. Et Zucc.)	Genome-wide characterization and linkage mapping	mei genome	Genome-wide characterization of simple sequence repeats (ssrs)	188,149 ssrs occurring at a frequency of 794 SSR/Mb.			2013	[11]
Ornamental kale (Brassica oleracea L. Var. Acephala)	Artistic diversiform leaf color	500 F2 individuals	Sequence related amplified polymorphism (SRAP)		Re (red leaf)	Me8Em4 Me8Em17Me9Em11	2013	[12]
Cherry plum (myrobalan plum)	Resistance to root-knot nematodes (RKN)				Ma1 and Ma3	SCAL19690 and SCAFLP2202	2004	[13]
Paeonia	Genetic diversity	29 cultivars	Sequence related amplified polymorphism (SRAP)	24 primers		Me8/Em8 Me8/Em1	2008	[14]
Dendrobium (Orchidaceae)	Genetic diversity	31 Chinese Dendrobium species	Sequence-related amplified polymorphism (SRAP)	14 primers	727 loci		2013	[15]
Aechmea gomosepala	Genetic divergence of bromeliad hybrids		Sequence related amplified polymorphism (SRAP)	16 primers	265 loci		2012	[16]

Table 2: Selected examples of gene-marker associated for important traits in ornamentals.

Ornamentals	Trait	Samples	Primers	Year	Reference
Heather (Calluna vulgaris)	Genetic mapping of the "bud-flowering"	Single mapping population	535 AFLP markers	2013	[17]
Evergreen azalea	Genetic diversity	130 genotypes	3 primers (408 polymorphic fragments)	2013	[18]
Mei (Prunus mume Sieb.et Zucc.)	Genetic diversity	65 accessions	64 -primer combination	2012	[11]
Sinningia speciosa	Genetic diversity	24 accessions of S. Speciosa	16 primers	2012	[19]
Viburnum	Interspecific cross		5 primers	2012	[20]
Sacred lotus	Genetic diversity	58 accessions	20 primers	2012	[21]
Spring orchid (Cymbidium goeringii)	Genetic diversity	Two wild populations	15 primer sets	2011	[22]
Aquilegia (Ranunculaceae)	Genetic diversity	64 accessions	16 primers	2011	[23]
Viola suavis	Parallel evolution of white-flowered morphotypes	36 populations	3 primers	2008	[24]
Berberis thunbergii	Influence of invasive populations	85 plants representing five invasive populations.	6 primers	2008	[25]
Ginkgo biloba	Genetic diversity	21 cultivars	64 primers	2006	[26]
Yellow camellia (Camellia nitidissima)	Genetic diversity	6 populations	8 primers	2006	[27]
Aglaonema	Genetic diversity	54 culivars	53 primers	2004	[28]

Table 3: Selected examples of amplified fragment length polymorphisms (AFLPs) marker assisted selection in ornamentals.

ornamental plants. Most of the traits of ornamental importance are quantitative traits with complex inheritance and regulated by several genes, the environment and their interactions. Moreover, improving polygenic traits through MAS is a complex process [8]. Because more than one gene is involved in a quantitative trait, these genes have smaller individual effects on the phenotype. So the effect of the individual genes cannot be easily identified. In the following tables, the reader can find a brief summary of the current status regarding application of MAS in the different ornamentals. Gene-markers associated for important traits in ornamentals are listed in Table 2. Furthermore, marker selections in ornamentals by using amplified fragment length polymorphisms (AFLPs) method are showed in Table 3.

Conclusions

In nursery plants production, the majority of application of molecular marker is used for genetic diversity studies. However, MAS for quantitative traits is a difficult task in ornamentals, as with many other crops. Further advances in molecular technology and genome programs will soon create a wealth of information that can be exploited for the genetic improvement of ornamental crops. High-throughput genotyping, for example, will allow direct selection on marker information based on population-wide LD. Methods to effectively analyze and use this information in selection are still to be developed. The eventual application of these technologies in practical breeding programs will be on the basis of economic grounds, which, along with cost-effective technology, will require further evidence of predictable and sustainable genetic advances using MAS. Until complex traits can be fully dissected, the application of MAS will be limited to genes of moderate-to large effect and to applications that do not endanger the response to conventional selection. Until then, observable phenotype will remain an important component of genetic improvement programs, because it takes account of the collective effect of all genes.

References

1. Young ND (1999) A cautiously optimistic vision for marker-assisted breeding. Mol Breeding, 5: 505-510.

2. Babu R, Nair SK, Prasanna, BM, Gupta HS (2004) Integrating marker-assisted selection in crop breeding - Prospects and challenges. Current Science. 87: 607-619.

3. Meudt HM, Clarke AC (2007) Almost forgotten or latest practice? AFLP applications, analyses and advances. Trends Plant Sci 12: 106-117.

4. Agarwal M, Shrivastava N, Padh H (2008) Advances in molecular marker techniques and their applications in plant sciences. Plant Cell Rep 27: 617-631.

5. Liu CG, Zhang GQ (2006) Single nucleotide polymorphism (SNP) and its application in rice. Yi Chuan 28: 737-744.

6. Collard BCY, Jahufer MZZ, Brouwer JB, Pang ECK (2005) An introduction to markers, quantitative trait loci (QTL) mapping and marker-assisted selection for crop improvement: The basic concepts. Euphytica, 142: 169-196.

7. Li G, Quiros CF (2001) Sequence-related amplified polymorphism (SRAP), a new marker system based on a simple PCR reaction: its application to mapping and gene tagging in Brassica. Theor Appl Genet, 103: 455-461.

8. Nakaya A, Isobe SN (2012) Will genomic selection be a practical method for plant breeding? Ann Bot 110: 1303-1316.

9. Lee HR, Cho MC, Kim HJ, Park SW, Kim BD (2008) Marker development for erect versus pendant-orientated fruit in Capsicum annuum L. Mol Cells, 26: 548-553.

10. Zaki NM, Singh R, Rosli R, Ismail I (2012) Elaeis oleifera Genomic-SSR Markers: Exploitation in Oil Palm Germplasm Diversity and Cross-Amplification in Arecaceae. Int J Mol Sci. 13: 4069-4088.

11. Sun L, Yang W, Zhang Q, Cheng T, Pan H, et al. (2013) Genome-wide characterization and linkage mapping of simple sequence repeats in mei (Prunus mume Sieb. et Zucc.). PLoS One 8: e59562.

12. Wang YS, Liu ZY, Li YF, Zhang Y, Yang XF, Feng H (2013) Identification of sequence-related amplified polymorphism markers linked to the red leaf trait in ornamental kale (Brassica oleracea L. var. acephala). Genet Mol Res, 12: 870-877.

13. Dirlewanger E, Cosson P, Howad W, Capdeville G, Bosselut N et al. (2004) Microsatellite genetic linkage maps of myrobalan plum and an almond-peach hybrid--location of root-knot nematode resistance genes. 109: 827-38.

14. Hao Q, Liu ZA, Shu QY, Zhang R, De Rick J, et al. (2008) Studies on Paeonia cultivars and hybrids identification based on SRAP analysis. Hereditas 145: 38-47.

15. Feng SG, Lu JJ, Gao L, Liu JJ, Wang HZ (2014) Molecular phylogeny analysis and species identification of Dendrobium (Orchidaceae) in China. Biochem Genet 52: 127-136.

16. Zhang F, Ge YY, Wang WY, Shen XL, Yu XY (2012) Assessing genetic divergence in interspecific hybrids of Aechmea gomosepala and A. recurvata var. recurvata using inflorescence characteristics and sequence-related amplified polymorphism markers. Genet Mol Res, 11: 4169-4178.

17. Behrend A, Borchert T, Spiller M, Hohe A (2013) AFLP-based genetic mapping of the "bud-flowering" trait in heather (Calluna vulgaris). BMC Genet 14: 64.

18. Zhou H, Liao J, Xia YP, Teng YW (2013) Determination of genetic relationships between evergreen azalea cultivars in China using AFLP markers. J Zhejiang Univ Sci B, 14: 299-308.

19. Zaitlin D (2012) Intraspecific diversity in Sinningia speciosa (Gesneriaceae: Sinningieae), and possible origins of the cultivated florist's gloxinia. AoB Plants 2012: pls039.

20. Al-Niemi T, Weeden NF, McCown BH, Hoch WA (2012) Genetic analysis of an interspecific cross in ornamental Viburnum (Viburnum). J Hered 103: 2-12.

21. Hu J, Pan L, Liu H, Wang S, Wu Z, et al. (2012) Comparative analysis of genetic diversity in sacred lotus (Nelumbo nucifera Gaertn.) using AFLP and SSR markers. Mol Biol Rep 39: 3637-3647.

22. Huang JL, Zeng CX, Li HT, Yang JB (2011) Isolation and characterization of 15 microsatellite markers from the spring orchid (Cymbidium goeringii) (Orchidaceae). Am J Bot 98: e76-77.

23. Zhu RR, Gao YK, Xu LJ, Zhang QX (2011) Genetic diversity of Aquilegia (Ranunculaceae) species and cultivars assessed by AFLPs. Genet Mol Res 10: 817-827.

24. Mereda P Jr, Hodálová I, Mártonfi P, Kucera J, Lihová J (2008) Intraspecific variation in Viola suavis in Europe: parallel evolution of white-flowered morphotypes. Ann Bot 102: 443-462.

25. Lubell JD, Brand MH, Lehrer JM, Holsinger KE (2008) Detecting the influence of ornamental Berberis thunbergii var. atropurpurea in invasive populations of Berberis thunbergii (Berberidaceae) using AFLP1. Am J Bot 95: 700-705.

26. Wang L, Xing SY, Yang KQ, Wang ZH, Guo YY, et al. (2006) Genetic relationships of ornamental cultivars of Ginkgo biloba analyzed by AFLP techniques. Yi Chuan Xue Bao 33: 1020-1026.

27. Tang S, Bin X, Wang L, Zhong Y (2006) Genetic diversity and population structure of yellow camellia (Camellia nitidissima) in China as revealed by RAPD and AFLP markers. Biochem Genet, 44: 449-461.

28. Chen J, Devanand PS, Norman DJ, Henny R, Chao CC (2004) Genetic relationships of Aglaonema species and cultivars inferred from AFLP markers. Ann Bot 93: 157-166.

27

Impact of Applying Calcium on Yield and Visual Quality of Groundnut (*Arachis hypogaea* L.)

Kirthisinghe JP[1]*, Thilakarathna SMCR[2], Gunathilaka BL[2] and Dissanayaka DMPV[3]

[1]*Department of Crop Science, Faculty of Agriculture, University of Peradeniya, Sri Lanka*
[2]*Department of Agriculture, Kurunegala, Sri Lanka*
[3]*Export Agriculture Department, Narammala, Sri Lanka*

Abstract

Recently, Department of Agriculture (DOA), Sri Lanka had developed four new varieties namely, Thissa, Indi, Walawa and Tikiri. Groundnut is cultivated in the dry and intermediate zones of Sri Lanka mainly under rain fed in October season and, in paddy lands under irrigation in April season. Low yields and the poor quality kernels are the major constraints to cultivation of Groundnut (*Arachis hypogaea L.*) in Sri Lanka. Groundnut is usually grown in well-drained soils with a pH of 6.5. Inadequate and unbalanced supply of nutrients may be one of the reasons for low yields in acidic and sandy soils. Farmers in Dambulla and Maspotha, apply 60-350 kg ha^{-1} of Gypsum for their cultivations and obtained a yield of 2000-2500 kg ha^{-1}. Therefore, this study was conducted to find out the effect of Calcium using Gypsum on the yield and quality of Groundnut in Maspotha divisional secretariat area of Kurunegala district, in the intermediate zone of Sri Lanka. Soil pH and EC were measured to determine the acidity and salinity levels. The field experiment was conducted during the October seasons 2011/2012 and 2012/2013. Randomized Complete Block Design was used with four treatments (0, 125, 175, 250 kg ha^{-1} of Gypsum) and three replicates. The crop management practices were done according to the standard procedures. The nut yield, number of pegs per plant, kernel and shell weight of fifty pegs per plot, dry weight of fifty seeds were measured. The seed quality and filling of seeds in each treatment were also evaluated. The results revealed that with application of 250 kg ha^{-1} of Gypsum increased the soil pH from 4.1 to 5.0 and increase the mean pod dry weight per plot of 40 plants, from 578 to 835 g with better quality kernels.

Keywords: Groundnut; Gypsum; Kernel weight; Shell weight

Introduction

The groundnut, *Arachis hypogaea* L. was originated from South America and presently grown in tropical countries [1]. Recently, Department of Agriculture (DOA), Sri Lanka had developed four new varieties namely, Thissa, Indi, Walawa and Tikiri. Groundnut is cultivated in the dry and intermediate zones of Sri Lanka mainly under rain fed in October season and, in paddy lands under irrigation in April season. It is mainly grown in Moneragala, Hambantota, Kurunagala, Anuradhapura, Mullative, Rathnapura and Puttalum districts [2]. In Sri Lanka, Groundnut is used as an oil crop, as a snack and in confectionaries [2]. Groundnut is grown in well-drained sandy loam and clay loam soils. Deep well-drained soils having pH of 6.5 - 7.0 and high fertility are ideal for groundnut.

According to the DOA [2], the application of Calcium (C_aCO_3) is important for proper kernel development in groundnut. Calcium carbonate can be used as a calcium source, but, compared to Gypsum; it is slow releasing due to less solubility. Therefore, Gypsum ($CaSO_4$-$2H_2O$) can be used at flowering to ensure the adequate availability of Ca in the fruiting zone to enhance the pod development.

Chapman et al. [3] reported that the less amount of soluble calcium in the pegging zone cause low peg formation. The researchers found that the groundnut pegs and pods treated with gypsum had a significantly less pod rot, than the untreated [3]. According to the DOA statistics (2011), the cultivated groundnut extent was 9251 ha and the production was 16800 Mt. with an average yield of 1.8 Mt ha^{-1}.

Farmers in Dambulla area apply 60-350 kg ha^{-1} of gypsum for their cultivations and obtained a yield of 2500 kg ha^{-1}. In Maspotha area, the yield of groundnut is around 750 - 1000 kg ha^{-1}. Therefore, this study was conducted to find out the effect of gypsum for higher yield in Maspotha area.

Methodology

This study was conducted at the Maspotha divisional secretariat area in Kurunegala district, which belongs to IL1a agro ecological region [4] in October season under rainfed conditions for two consecutive years 2011/2012 and 2012/2013. The average annual rainfall was 1100-1400 mm. Soil type was reddish brown earth. The Groundnut variety Thissa was used for this study. The recommended time for planting is in October and April. The recommended fertilizer rate is 30 kg ha^{-1} N, 45 kg ha^{-1} P_2O_5, 45 kg ha^{-1} K_2O [2] and no recommendation for Gypsum.

Raised beds were used in high lands with the spacing of 45 cm×15 cm. The plot size was 1 m×3 m. Two rows of plants were planted as guard rows for each plot. The total number of plants per plot was 40. After 25 days after planting at pegging earthing up was done to improve pod filling. When the crop was matured at 110 days after planting the whole plants were uprooted and pods were collected and dried. The experiment was laid according to Randomized Complete Block Design (RCBD) with four treatments. Each Treatment had three replicates. The treatments were as follows,

T_1 - 0 kg ha^{-1} of gypsum (control)

T_2 - 125 kg ha^{-1} of gypsum

T_3 - 175 kg ha^{-1} of gypsum

T_4 - 250 kg ha^{-1} of gypsum

Soil pH, CEC and Electrical Conductivity (EC) were measured at

---

*Corresponding author: Prof. Kirthisinghe JP, Department of Crop Science, Faculty of Agriculture, University of Peradeniya, Sri Lanka
E-mail: jpkrmk@gmail.com

the planting of seeds, 5 days after application (5 DAA) of gypsum at the pegging stage and at harvest. Yield data were collected at harvesting. Randomly 5 plants were selected from each plot and counted the number of pegs per plant. Pods fresh weights (g), Pods dry weight (g), Kernel weight of fifty pods (g), Shell weight of fifty pods (g) were also measured. Dry weights of the seeds were taken after drying the seeds for three days using solar drying system. The quality of the kernel was assessed visually by sorting and grouping the seeds according to the size of the kernel of 50 pods into large, medium and small.

Data were analysed using the analysis of variance (ANOVA) procedure by statistical analyze system (SAS) and mean separation was done using Duncan's Multiple Range Test (DMRT) at p=0.05.

Results and Discussion

Soil Properties

Initial pH, CEC (cmol kg^{-1}) and EC (micro Siemens cm^{-1}) values showed that the soil was in the acidic range and it was below the recommended pH range (Table 1). The pH and EC values were slightly increased after five days of applying gypsum and slightly decreased at harvest. CEC also increased with applying gypsum and slightly decreased at harvest.

Warren [5] observed that the gypsum will improve the pod filling without changing the soil pH. The researcher also explained that a good soil EC level will be somewhere above 200 µS cm^{-1} and 1200 µS cm^{-1} (1.2 mS cm^{-1}). Any soils <200 does not have enough available nutrients to the plant and may be a sterile soil with minimum microbial activity. An EC above 1200 µS cm-1 may indicate that of high salt fertilizer or perhaps a salinity problem due to lack of drainage.

Plant performances

The seeds were germinated 10 to 14 days of planting. Flowers were formed about 30-40 days after the germination. The days to 50% flowering ranged between 35 to 50 days with a mean value of 41. The pegs developed to a depth of 4 to 5 cm to form pods. The number of pegs per plant significantly increased in T_3 and T_4 when compared to the control and T2 (Table 2). The treatment T_4 showed the highest peg formation.

Stage	Property	T_1	T_2	T_3	T_4
Initial	pH	4.2	4.1	4.2	4.1
Initial	CEC (cmol kg^{-1})	6.0	6.5	6.4	6.3
Initial	EC (µS cm^{-1})	205	220	200	210
5 DAA*	pH	4.3	4.4	4.3	5.0
5 DAA*	CEC (cmolkg^{-1})	7.2	7.5	7.7	7.8
5 DAA*	EC (µS cm^{-1})	230	250	260	270
at harvest	pH	4.3	4.3	4.2	4.8
at harvest	CEC (cmol kg^{-1})	7.1	7.4	7.5	7.7
at harvest	EC (µS cm^{-1})	210	235	245	260

* 5 days after application
Table 1: Chemical properties of soil.

Treatment	Number of pegs per plant	Mean pod fresh weight (g)/plot	Mean pod fresh weight (g)/plant	Mean Pod dry weight (g)/plot	Mean Pod dry weight (g)/plant
T_1	28 $_c$	937 $_d$	27.1 $_c$	578 $_d$	13.8 $_d$
T_2	30 $_c$	1045 $_c$	27.3 $_c$	679 $_c$	14.3 $_c$
T_3	33 $_b$	1235 $_b$	27.8 $_{ab}$	754 $_{ab}$	15.1 $_b$
T_4	35 $_a$	1357 $_a$	28.8 $_a$	835 $_a$	16.7 $_a$
CV%	10	12	11	10	11

Values within a column followed by a common letter are not significantly different at P=0.05, according to DMRT
Table 2: Plant performances.

The results showed that there was a significant difference between T_1, T_2 and T_3 compared to the control. It appears that gypsum requirement for increased peg formation lies more than 125 kg /ha.

The mean pod fresh and dry weight per plots is given in Table 2. The T_4 treatment showed a significantly higher yield compared to other treatments. The T_2 and T_3 treatment showed a significant difference, and T1 treatment without gypsum showed a significantly lower yield than other treatments. Therefore, T_4 treatment with 250 kg ha^{-1} of gypsum could be identified as the best performer.

The treatment T_4 showed significantly higher pods dry weight yield when compared to other treatments (Table 2). The treatments T_3 and T_4 were not significant. The treatment T_1 gave the lowest yield. Therefore, T_4 with 250 kg ha^{-1} of gypsum could be identified as best treatment to obtain higher yields.

The mean kernel weight and mean shell weight

The mean kernel weight showed a significant difference (p<0.05) among treatments (Table 3). The treatment T_4 showed a significantly higher kernel yield and a good quality appearance compared to other treatments. The treatment T_1 gave the lowest yield with half-filled nuts. Therefore, according to the results, the treatment T4 with 250 kg ha^{-1} of gypsum can be recommended as the best treatment to obtain higher kernel yield.

The results showed that the mean shell weight was also significantly different (p<0.05) among treatments (Table 3).

The quality of the kernel

The results showed that the quality of the kernel of 50 pods was significantly different (p<0.05) among treatments (Table 4). With the application of 250 kg ha^{-1} of gypsum the T4 treatment gave better kernel size compared to other treatments. However, all the treatments with application of Ca showed an improvement in kernel size.

Conclusion

The application of 250 kg ha^{-1} of gypsum increased the mean pod dry weight from 578 to 835 g with better quality kernels in pH 4.1 soils at Maspotha divisional secretariat area in Kurunegala district in the intermediate zone of Sri Lanka in October season under rainfed conditions. Thus, with application of 250 kg ha^{-1} of gypsum, groundnuts produce the higher number of pegs per plant and increased kernel weight.

Treatment	mean kernel weight (g)/plant	mean kernel weight (g)	mean shell weight (g)/plant	mean shell weight (g)
T_1	11.4 $_d$	0.41 $_c$	12.0 $_d$	0.43 $_c$
T_2	12.4 $_c$	0.41 $_c$	13.7 $_c$	0.46 $_b$
T_3	14.2 $_b$	0.43 $_b$	16.7 $_b$	0.51 $_a$
T_4	16.2 $_a$	0.46 $_a$	17.7 $_a$	0.51 $_a$
CV%	10	10	11	11

Values within a column followed by a common letter are not significantly different at P=0.05, according to DMRT
Table 3: The mean kernel weight and mean shell weight.

Treatment	Size of the kernel		
	Large (%)	Medium (%)	Small (%)
T_1	20 $_d$	30 $_a$	50 $_a$
T_2	30 $_c$	30 $_a$	40 $_b$
T_3	50 $_b$	20 $_b$	30 $_c$
T_4	60 $_a$	20 $_b$	20 $_c$

Values within a column followed by a common letter are not significantly different at P=0.05, according to DMRT
Table 4: The size of the kernel.

References

1. Reddy PS (1998) Groundnut. Indian Council of Agricultural Research. Krishi, New Delhi. 583.

2. DOA (2006) Annual report. Socio Economic & Planning Centre, Department of Agriculture.

3. Chapman SC, Ludlow MM, Blamy FPC, Fischer KS (1993) Effect of drought during pod filling on utilization of water and on growth of cultivars of Groundnut (ArachishypogaeaL.). Field Crop Research 32: 243-255.

4. Punyawardana BVR (2008) Agro ecological regions and rainfall of Sri Lanka.

5. Warren AD (2011) Gypsum as an Agricultural amendment. General guidelines. Chapter 3, School of Environment and Natural Resources, The Ohio State University, USA.

28

Seasonal Variations on Quality Parameters of Pak Choi (*Brassica rapa* L. subsp. *chinensis* L.)

Funda Eryilmaz Acikgoz*

Department of Plant and Animal Production, Vocational College of Technical Sciences, Namik Kemal University, Tekirdag, Turkey

Abstract

The aim of this study is to determine the changes in quality of pak choi in successive growing seasons. Experiments were carried out in a PE covered cold greenhouse in late autumn-early winter and late winter-early spring growing period in Turkey (41°11' N, 27°49' E). Results showed that morphological features, the leaf area (73.82 mm^2), leaf length (31.21 mm), average leaf width (2.70 mm), maximum leaf width (3.92 mm), mass (95.27 g) and moisture content values of leaves (0.90% w.b.) were higher in late autumn-early winter period but leaf thickness (0.33 mm) contents were higher in late winter-early spring period. Yield characteristics investigated were affected by season and were found relatively higher in late autumn-early winter than late winter-early spring growing period. And it was 5713 kg ha^{-1} in late autumn-early winter growing period and 5034 kg ha^{-1} in late winter-early spring growing period. In point of the dry matter content investigated were affected by season, and were higher in late autumn-early winter than late winter-early spring growing period. It was 12.4% in late autumn-early winter growing period and 10.25% in late winter-early spring growing period. Account of total protein of plants grown in late autumn-early winter (25.16%) was higher than in late winter-early spring (22.45%) growing period. Sowing time did affect ascorbic acid content significantly. It was 44.21 mg.100^{-1} g in late autumn-early winter growing period and was recorded as 38.02 mg.100^{-1} g in late winter-early spring growing period. Color parametres L, a, b values were measured as 27.92; -6.98 and 8.53 in late autumn-early winter and as 21.64; -5.48; 7.49 late winter-early spring. Sowing time did affect color parameters content significantly. Late autumn-early winter growing time result was higher than in late winter-early spring growing time. With regard to mineral content N (4.67%), P (0.65%), Cu (9.32 ppm), Fe (879.6 ppm) and Zn (49.13 ppm) values were higher in plants grown in late autumn-early winter in comparison with plants grown in late winter-early spring as K (4.45%), Ca (2.36%), Mg (0.41%) and Mn (62.25 ppm) values of plants were higher in late winter-early spring growing period. Results of the study indicated that different weather conditions influenced the pak choi yield and the chemical composition of the leaf as well. Pak choi with its relatively short cultivating period, easy growing, and dietary value can be a good alternative crop for late autumn-early winter growing period in cold greenhouses.

Keywords: Pak Choi (*Brassica rapa* L. subsp. *chinensis* L.); Morphological features; Ascorbic acid; Total protein; Color parameters; Yield; Mineral contents

Introduction

As a highly rated leafy variety of vegetables and a marvelous food alternative, brassicas is grown for its enlarged, edible, terminal buds; and is preferably eaten almost everywhere in the world as well [1]. This green vegetable was made known around the world by the efforts of the travelers and immigrants [2-4]. Pak Choi: syn. *Brassica Chinensis* L. (1759), *Brassica Campestris* L. subsp. *Chinensis* (L.) Makino [5], *Brassica rapa* L. subsp. *Chinensis* (L.) Hanelt [6] evolved in China, and its cultivation was recorded since the 5th century AD. It is widely grown in southern and central China, and also Taiwan. This group is a relatively new comer vegetable in Japan where it is still referred to as 'Chinese vegetable' [7]. As a leafy vegetable, Chinese cabbage presents short storing life and therefore it should be produced near the markets. This species can be possibly cultivated in climatic zone of Central Europe from spring to winter because of not having high thermal needs and possessing a rather short vegetation period. Leaves of the crop can be consumed from a stage of transplant, but it is recommended to harvest rosettes after 50 to 60 days from sowing or 30 to 40 days from transplanting [8]. The crisp leaves and thick petioles of bitter taste are excellent for cooking as a boiled vegetable [9]. Food nutrition is becoming one of the most important features in the choice of products in modern conditions. *Brassica* vegetables are characterized by high water content, low caloric value, containing high quality of protein, carbohydrates, fibre, vitamins, minerals, and also secondary plant metabolites. In humans, the last given issues have anti-carcinogenic,

antioxidant, antibacterial and antiviral effects, and they encourage the immune system and reduce inflammation. Nevertheless, *Brassicas* prevent the development of cardiovascular diseases and illnesses associated with ageing as well [10,11]. The objectives of this study have focused on the evaluation of nutritive aspects by determining the changes in quality of *Brassica rapa* L. subsp. *Chinensis* L. in successive growing seasons.

Materials and Methods

Using high tunnel cold greenhouse covered by polyetilen (PE) with UV additive which belongs to Namik Kemal University, Vocational College of Technical Sciences, Plant and Animal Production Department, the experiments were carried out during successive crop seasons: late autumn-early winter and late winter-early spring in Tekirdag city (40°98' N, 27°48' E) Turkey. Research was designed as 3 replications according to randomized block experimental design.

---

*Corresponding author: Funda Eryilmaz Acikgoz, Department of Plant and Animal Production, Vocational College of Technical Sciences, Namik Kemal University, Tekirdag 59030, Turkey
E-mail: feryilmaz@nku.edu.tr (or) fundaea@yahoo.com

The 1340C variety of Pak Choi (Chilternseeds Firm) was used for the research (Figure 1). Seeds were sown in multi-celled trays filled with peat (Klasmann-Deilmann, Potground H, Germany) in October. Some specifications of the used peat are: 160-260 mg/L N, 180-280 mg/L P_2O_5, 200-150 mg/L K_2O, 80-150 mg/L Mg, pH: 6, 0.8% N, 70% organic matter, and 35% C. When the seedlings became 2 to 3 true leaves (21 days for pak choi after seed sowing) they were planted to pre-prepared places in high tunnel cold greenhouse with 10×10 cm intervals and 10 plants in each parcel. Some chemical contents of the soil used in the experimental field can be seen in Table 1. The climate data measured inside the tunnel during the growing of the plants can be seen in Table 2. Plants were harvested 40 days after seed sowing. Since there were no diseases and pests, no pesticides was used during the growing period.

Morphological features

The LI-COR brand LI-3000C model's portable area measurement device was used to measure of the leaf length, width and surface area of Pak Choi plants. A mechanical type micrometer, which has measurement range between 0-25 mm, was used in order to determine leaf thickness, and the digital sliding caliper, which has 0.01 accuracy, was used in order to define stalk thickness. The AND GF-610 brand precision balance with 0.001 accuracy was used for measuring the mass of plants. The measurements were performed using 10 plants with 3 replications by randomly choosing the leaves of the experiment from these plants.

Dry matter

The dry matter content in leaves was determined by drying the sample at 68°C up until constant weight is obtained [12].

Yield

The marketable yield was determined. Harvest was performed only once in particular growing periods, specifying the mass of leaves. The leaves which were fully developed and without damages were treated as they are marketable [13].

Determination of color parameters

Color measurements were performed using Hunter Lab D25LT Color Measurement device, which has big measurement range, and especially which is suitable for color measurements of non-homogeneous materials. The color parameters that are brightness (L) and color coordinates of 'a' and 'b'. L value changes between 0 and 100.0 shows black color and 100 shows white color. Color coordinates of 'a' and 'b' don't have any specific measurement interval and they

Soil Properties	Results
pH	8.01
Salinity (%)	0.07
$CaCO_3$ (%)	2.74
Organic matter (%)	1.35
Ca (%)	0.54
P (ppm)	36.40
K (ppm)	253.80
Mg (ppm)	473.10
Mn (ppm)	5.68
Cu (ppm)	0.81
Fe (ppm)	7.43
Zn (ppm)	0.97

Table 1: Some Chemical Properties of Soil of Growing Soil.

Month	Average temperature (°C)	Maximum temperature (°C)	Minimum temperature (°C)	Average humidity (%)
October	16.01	19.02	13.05	89
November	12.4	14.8	10.0	87
December	9.4	12.5	6.4	88
January	7.1	10.1	4.1	89
February	7.6	10.9	4.3	87
March	9.5	13.0	6.1	90
April	14	17.8	10.2	84

Table 2: Average climate data in unheated greenhouse during the months of the experiment.

can have positive and negative values. The 'a' value represents red-green axis, where positive values are for red color, negative values are for green color, and 0 is neutral. If color coordinate of 'b' is positive it shows yellow color and the negative values show the blue color [14]. The measurements were made using 10 plants and 3 replications were performed on the every plant by randomly choosing the leaves. The 3 replication measurements on the randomly chosen leaves from the same plant were made by 3 replications.

Ascorbic acid

Ascorbic acid was determined in direct plant extracts with 2, 6-dichloroindophenol by visual titration.

Total protein

The content of total proteins was estimated in dry material by the Kjeldahl method [15].

Mineral contents

Leaf samples were analyzed by ICP optical emission spectrometry (ICP-OES) total nitrogen, phosphorus, potassium, calcium, magnesium and some trace elements Fe, Cu, Zn, and Mn in each plant leaf sample [15].

Statistical analysis

All data were analyzed statistically with SPSS software program (v.16.0 for Windows OS) and the differences between practices were compared by using least significant difference (LSD) test at ($p<0.05$) probability [16].

Results and Discussion

Morphological features

The leaf area (73.82 mm^2), leaf length (31.21 mm), average leaf width (2.70 mm), maximum leaf width (3.92 mm), mass (95.27 g)

Figure 1: Pak Choi plants (Original).

and moisture content values of leaves (0.90% w.b.) were higher in late autumn-early winter period but leaf thickness (0.33 mm) contents were higher in late winter-early spring period (Table 3). According to (Maynard et al. and Barillari et al.) and Cengel et al. [17-19] growth and composition of leafy vegetables varies with season or time of year and some environmental and agronomic factors might significantly change the quality of the product. According to (Siomos and Kalisz et al.) and (Kalisz and Kalisz et al.) [20-23] effect of different sowing dates on transplant characteristics were found both for non-heading Chinese cabbage and for heading group of Chinese cabbage.

Yield

Yield characteristics investigated were affected by season and were found relatively higher in late autumn-early winter than late winter-early spring growing period. It was 5713 kg ha^{-1} in late autumn-early winter growing period and 5034 kg ha^{-1} in late winter-early spring growing period (Table 4).

In this experiment, pak choi responded positively to plantings in late autumn-early winter. Plants were grown in lower temperatures and less intensive sunlight therefore they resulted in higher yields. A significant effect of growing period on the yield and quality were stated for many Brassicaceae species: leafy cultivars of Brassica rapa [24]; broccoli [25]; Acikgoz, [26]; red cabbage [27]; cauliflower [28]; Brussels sprouts [29] and Brassica rapa var. narinosa [30].

Dry matter

The dry matter content investigated were affected by season, and were higher in late autumn-early winter than late winter-early spring growing period. It was 12.4% in late autumn-early winter growing period and 10.25% in late winter-early spring growing period. The dry matter content changed significantly amongst Brassica rapa cultivars [31]. Hara and Sonoda, [32] found that grown cabbage had lower dry matter content at higher temperatures.

Physical properties	Growing Date		Mean	LSD$_{0.05}$
	Late autumn-early winter	Late winter-early spring		
Leaf area (mm^2)	73.82a	71.62b	72.72	2.15
Leaf length (mm)	31.21a	29.11b	30.16	2.98
Average leaf width (mm)	2.70a	1.90b	2.33	0.37
Maximum leaf width (mm)	3.92a	2.91b	3.41	0.51
Leaf thickness (mm)	0.33b	0.34a	0.33	0.02
Mass (g)	95.27a	90.36b	92.81	10.1
Moisture content values leaf (% w.b.)	0.90a	0.90a	0.90	0.06

Table 3: Some Morphological Properties of Plants.

	Growing Date		Mean	LSD$_{0.05}$
	Late autumn-early winter	Late winter-early spring		
Yield (kg ha^{-1})	5713a	5034b	5373.5	54.09
Dry matter %	12.4a	10.25b	11.32	6.02
Protein %	25.16a	22.45b	23.80	1.98
Ascorbic acid mg.100^{-1} g	44.21a	38.02b	41.11	2.71
a	-6.98b	-5.48a	-6.23	4.21
b	8.53a	7.49b	8.01	5.71
L	27.92a	21.64b	24.78	2.07
YI	42.54a	39.95b	41.24	2.76

Table 4: Influence of sowing time on the yielding, content of nutritive and color parameters of plants.

Total protein

Total protein of plants grown in late autumn-early winter (25.16%) was higher than in late winter-early spring (22.45%) growing period (Table 4). This result can be related to the N content of plants grown in late autumn-early winter. Rosa and Heaney, [33] detected that, without considering the cultivars, total protein contents of Brassica crops were 94.6 g kg^{-1} DM in spring- summer period and 409.7 g kg^{-1} DM in summer-spring period.

Ascorbic acid

Sowing time did affect ascorbic acid content significantly. It was 44.21 mg.100^{-1} g in late autumn-early winter growing period and was recorded as 38.02 mg.100^{-1} g in late winter-early spring growing period (Table 4). Early findings showed that limited light, clouding and low light intensity have reducing effects on ascorbic acid content of the plant tissues (Lee and Kader and Shinıhara and Suzuki) (Weston, Barth and Tamura) [34-37]. Despite this light is not essential for the ascorbic acid synthesis, the amount and intensity of light during the growing season have a definite influence on the amount of ascorbic acid formed [38].

Color parameters

The measured whiteness or brightness/darkness value (L), the value for greenness (-a), the value for yellowness (b) and the yellowness index (YI) values for leaves of Pak Choi plant were given in Table 4. When the results of 3 replications measurement were analyzed, it was determined that the differences between L, a, b and yellowness index were significant ($p < 0.05$). L, a, b values were measured as 27.92; -6.98 and 8.53 in late autumn-early winter and as 21.64; -5.48; 7.49 late winter-early spring. Sowing time did affect color parameters content significantly. Late autumn-early winter growing time result was higher than in late winter-early spring growing time. According to Fallovo et al. [39], a value (greenness) of leafy lettuce increased owing to lower N levels in the leaves, and it is consistent with this study. Similarly according to Ali et al. the 'b' value resulted in a higher level in non-shaded conditions, and showed deeper color with lower lightness.

Mineral contents

The N (4.67%), P (0.65%), Cu (9.32 ppm), Fe (879.6 ppm) and Zn (49.13 ppm) contents of plants were higher in late autumn-early winter period, and the K (4.45%), Ca (2.36%), Mg (0.41%) and Mn (62.25 ppm) contents were higher in late winter-early spring period (Table 5). Nutrient contents of plants may be changed by environment [40-42]. Light affects the concentration of the elements in the plant by its effect on the amount of photosynthate produced, and alters the ratio of element to dry matter concentration. According to Jones et al. [43] the dilution effect due to production of carbohydrates in full light is rather characterized by reduced concentration for most nutrients, they also reported that total N in spinach leaves was reduced as light increased with no N applied. In some cabbage species, total nitrogen varies from 1.36 to 4.60%, phosphorus from 0.39 to 0.81% and potassium from 2.18 to 3.77% [44]. Kale have more N, P, K and Mg in fall than in spring, and air temperature also affects N uptaken by salad greens [41]. Rosa and Heaney [45] found that, in some Brassica species, with exception of Ca and Mg, the contents of all minerals investigated were higher in fall sowing time than in spring sowing time and mineral contents of plants responded to environmental changes. Caruso et al. [46] states that while Cu accumulated in plant tissues mainly in fall on the other hand Ca accumulated in spring growing period.

	Growing Date			
	Late autumn-early winter	Late winter-early spring	Mean	$LSD_{0.05}$
N (%)	4.67a	4.59b	4.63	1.34
P (%)	0.65a	0.57b	0.61	0.41
K (%)	4.39b	4.45a	4.42	1.23
Ca (%)	2.27b	2.36a	2.31	0.45
Mg (%)	0.38b	0.41a	0.39	0.25
Mn (ppm)	42.13b	62.25a	31.19	2.01
Cu (ppm)	9.32a	7.01b	8.16	0.85
Fe (ppm)	879.6a	139.2b	509.4	12.21
Zn (ppm)	49.13a	37.98b	43.55	2.23

Table 5: Influence of sowing time Mineral contents of plants.

Conclusions

In this study, some morphological properties of plants except leaf thickness were found to be favorable in late autumn-early winter growing period. And content of nutritive elements and color parameters of plants except a (greenness) were rather favorable in late autumn-early winter growing period. In terms of the mineral contents as K, Ca, Mg, and Mn higher results held in late winter-early spring, whereas N, P, Cu, Fe and Zn contents were higher in late autumn-early winter growing period. Pak choi with its relatively short cultivating period, easy growing, and dietary value can be a good alternative crop for late autumn-early winter growing period in cold greenhouses.

References

1. Tirasoglu E, Cevik U, Ertugrul B, Apaydın G, Baltas H, et al. (2005) Determination of trace elements in cole (Brassica oleraceae var. acephale) at Trabzon region in Turkey. J Quantitative Spectroscopy Radiative Transfer 94: 181-187.

2. Nieuwhof M (1969) Cole Crops. Leonard Hill, London, p: 102-104.

3. Balkaya A, Yanmaz R (2005) Promising kale (Brassica oleracea var acephala) populations from Black Sea region Turkey. N Zealand J Crop Horticult Sci 33: 1-7.

4. Adiloglu S, Eryilmaz AF, Adiloglu A (2015) Artan dozlarda azot uygulamasinin mibuna (Brassica rapa var Nipposinica) ve mizuna (Brassica rapa var Japonica) bitkilerinin bazi agronomik ozellikleri, C vitamini, protein ve mineral madde miktari uzerine etkisi. Journal of Agricultural Faculty of Uludag University 29: 1-11.

5. Makino T (1912) Observations on the flora of Japan. Botanical Magazine of Tokyo. 23: 93-102.

6. Hanelt P (1986) Formal and informal classifications of the infraspecific variability of cultivated plants-advantages and limitations. In: Styles BT (ed.) Infraspecific Classification of Wild and Cultivated Plants, pp: 139-156.

7. Dixon GR (2007) Vegetable Brassicas and Related Crucifers. CAB International North American Office, 875 Massachusetts Avenue, 7th Floor, Cambridge, USA, p: 327.

8. Larkcom J (2007) Oriental vegetables. Frances Lincoln Ltd. London, UK.

9. Opena RT, Kuo CG, Yoon JY (1988) Breeding and seed production of Chinese cabbage in the tropics and subtropics. AVRDC, Shanhua, Tairan, Tech Bull. 17: 92.

10. Artemyeva AM, Solovyeva AE (2006) Quality Evaluation Of Some Cultivar Types Of Leafy Brassica Rapa. Acta Horticulturae 706: 121-128.

11. King G, Barker G (2003) Pak choi en son hali.

12. Kacar B, Inal A (2008) Plant analysis. Ankara, Turkey, Nobel Press.

13. Kalisz A, Sekara A, Gil J, Grabowska A, Cebula A (2013) Effect of Growing Period and Cultivar on the Yield and Biological Value of Brassica rapa var Narinosa. Not Bot Horti Agrobo 41: 546-552

14. Anonymous (1996) CIE Lab Color Scale. Application Note-Insight on Color, Hunter Lab 8: 1-4.

15. Jones BJ, Wolf B, Mills HA (1991) Plant Analysis Handbook. Micro-Macro Publishing, USA.

16. Duzgunes O, Kesici T, Kavuncu O, Gurbuz F (1987) The methods of research and experiment methods (Statistic Methods II). Ankara University Agric Fac Publ, p: 1021.

17. Maynard DN, Barker AV, Minotti PL, Peck NH (1976) Nitrate accumulation in vegetables. Advances in Agronomy. 28: 71-118.

18. Barillari J, Cervellati R, Costa S, Guerra MC, Speroni E, (2006) Antioxidant and choleretic properties of Raphanus sativus L sprout (Kaiware Daikon) extract. Journal of Agricultural and Food Chemistry 54: 9773-9778.

19. Cengel M, Okur N, Irmak YF (2009) Organik bag topraklarinda yesil gubre bitkileri ve ciftlik gübresi uygulamalarinin topraktaki mikrobiyal aktiviteye etkileri. Ege Univ Ziraat Fak Derg 46: 25-31.

20. Siomos AS (1999) Planting date and within-row plant sparing effects on pak choi yield and quality characteristics. J Veg Crop Prod 4: 65-73.

21. Kalisz A, Siwek P, Cebula S (2006) Ocena wzrostu i składu chemicznego rozsady kapusty pekińskiej (Brassica pekinensis Rupr.) w zależności od metody hartowania i terminu produkcji. Folia Hort Supl 1: 200-206.

22. Kalisz A (2010) Optymalizacja jakości rozsady a plonowanie kapusty pekińskiej (Brassica pekinensis Rupr.) oraz wybrane elementy modelowania rozwoju roślin. Zesz Nauk UR w Krakowie, 465: 1-121.

23. Kalisz A, Cebula S, Siwek P, Sekara A, Grabowska A, (2014) Effects of Row Covers Using Non-woven Fleece on the Yields Rate of Bolting, and Quality of Heading Chinese Cabbage in Early Spring Cultivation. J Japan Soc Hort Sci 83: 133-141.

24. Acikgoz FE (2012) Determination of yield and some plant characteristics with vitamin C, protein and mineral material content in mibuna (Brassica rapa var nipposinica) and mizuna (Brassica rapa var japonica) grown in fall and spring sowing times. Journal of Tekirdag Agr Fac 9: 64-70.

25. Kaluzewicz A, Krzesinski W, Knaflewski M (2009) Effect of temperature on the yield and quality of broccoli heads. Veg Crops Res Bull 71: 51-58.

26. Acikgoz FE (2011) Influence of different sowing times on mineral composition and vitamin C of some broccoli (Brassica oleracea var italica) cultivars. Sci Res Essays 6: 760-765.

27. Tendaj M, Sawicki K (2012) The effect of the method and time of seedling production on red cabbage (Brassica oleracea L ssp oleracea convar capitata (L) Alef var capitata L f rubra DC) yield. Acta Agrobot 65: 115-122.

28. Cebula S, Kalisz A (1997) Value of different cauliflower cultivars for autumn production in a submontane region as depending upon the planting time. Yields and pattern of cropping. Folia Hort. 9: 3-12.

29. Mirecki N (2006) Influence of the planting dates on chemical composition and yield of Brussels sprouts. Acta Agric Serb 21: 53-61.

30. Kalisz A, Kostrzewa J, Sekara A, Grabowska A, Cebula S (2013) Yield and nutritional quality of several non-heading Chinese cabbage (Brassica rapa var Chinensis) cultivars with different growing period and its modeling. Korean Journal of Horticultural Science and Technology 30: 650-656.

31. Artemyeva AM, Solovyeva AE (2006) Quality Evaluation Of Some Cultivar Types Of Leafy Brassica Rapa. Acta Horticulturae 706: 121-128.

32. Hara T, Sonoda Y (1982) Cabbage-head development as affected by nitrogen and temperature. Soil Sci Plant Nutr 28: 109-117.

33. Rosa E, Heaney R (1996) Seasonal variation in protein, mineral and glucosinolate composition of Portuguese cabbages and kale. Animal Feed Science Tecnology 57: 111-127.

34. Lee SK, Kader AA (2000) Preharvest and postharvest factors influencing vitamin C content of horticultural crops. Postharvest Biology and Technology 20: 207-220.

35. Shinohara Y, Suzuki Y (1981) Effects of light and nutritional conditions on the ascorbic acid content of lettuce. J Japan Soc Hort Sci 50: 239-246.

36. Weston LA, Barth MM (1997) Preharvest factors affecting postharvest quality of vegetables. HortScience 32: 812-816.

37. Tamura A (2004) Effect of air temperature on the content of sugar and vitamin C of spinach and komatsuna. Hort Res 3: 187-190.

38. Gruda N (2005) Impact of environmental factors on product quality of greenhouse vegetables for fresh consumption. Critical Reviews in Plant Sciences 24: 227-247.

39. Fallovo C, Rouphael Y, Rea E, Battistelli A, Colla G (2009) Nutrient solution concentration and season affect yield and quality of lactuca sativa L. var. acephala in floating raft culture. Journal of the science of food and agriculture 89: 1682-1689.

40. Gent MPN (1991) High tunnels extend tomato and pepper production in Connecticut. Connecticut Agricultural Exp Station Bull 893: 1-16.

41. Ali MB, Khandaker L, Oba S (2010) Changes in pigments total polyphenol antioxidant activity and color parameters of red and green edible amaranth leaves under different shade levels. Journal of Food Agriculture & Environment 8: 217-222.

42. Gent MPN (2002) Growth and composition of salad greens as affected by organic compared to nitrate fertilizer and by environment in high tunnels. Journal of Plant Nutrition 25: 981-998.

43. Jones BJ, Wolf B, Mills HA (1991) Plant Analysis Handbook. Micro Macro Publishing, USA.

44. Wells OS (1996) Row cover and high tunnel growing system in The United States. HortTechnology 6: 172-176.

45. Singh J, Upadhyay AK, Bahadur A, Singh KP (2004) Dietary antioxidant and minerals in crucifers. Journal of Vegetable Crop Production 10: 33-38.

46. Caruso G, Villari A, Villari G (2004) Quality characteristics of Fragaria vesca L. fruits influenced by NFT solution EC and shading. Acta Horticulturae 648: 167-174.

29

Morphological Characterization and Yield Traits Analysis in Some Selected Varieties of Okra (*Abelmoschus Esculentus* L. Moench)

Anwanobong Jonathan Eshiet[1*] and Ebiamadon Andi Brisibe[1,2,3]

[1]Plant Genetic Resources and Cell and Tissue Culture Research Laboratory, Department of Genetics and Biotechnology, University of Calabar, Calabar, Nigeria

[2]Department of Biological Sciences, Niger Delta University, Wilberforce Island, P.M.B. 71 Yenagoa, Nigeria

[3]Department of Pharmaceutical Microbiology and Biotechnology, Niger Delta University, Wilberforce Island, P.M.B. 71 Yenagoa, Nigeria

Abstract

Okra (*Abelmoschus esculentus* L. Moench) is an important all-season food crop widely grown throughout the tropical and semi-temperate regions of the world, for its tender green pods and leaves. Unfortunately, the crop rarely reaches its yield potential in most of these areas, primarily due to the use of unimproved cultivars and limited utilization of fertilizers and irrigation inputs. In addition, investments in breeding programmes that are aimed at enhancing its yield in the field are very limited. In the current study, therefore, some agronomic traits and yield components of four cultivars of okra (NHAe-47-4, V35, LD88 and a local variety), were compared in a field plot to aid in the development of selection strategies that could be used for okra improvement. The experiment was laid out in a completely randomized design with each accession, which served as the main factor, replicated five times and grown to maturity. Data collected from the four varieties included number of days to seedling emergence, number of days to flowering, plant height at flowering, number of leaves at flowering, mean number of pods per plant, pod length, mean number of seeds per pod, and mean weight of one hundred seeds, which were individually subjected to analysis of variance (ANOVA) test. The results demonstrated that the okra varieties differed significantly ($p<0.05$) in number of days to flowering (71.75-112 days), plant height at flowering (49.75-128 cm), number of leaves at flowering (7.50-19.33), pod length (3.23-6.83 cm) and one hundred seed weight (3.87-4.42g). There were, however, no significant ($p>0.05$) differences between the cultivars in terms of certain yield characters including average number of pods and average number of seeds per pod per plant. Taken together, the findings from this study will certainly be useful to okra breeders for appropriate selection strategies in cultivar improvement programmes.

Keywords: Accessions; Morphological traits; Okra yield; Selection; Variability

Introduction

Okra (*Abelmoschus esculentus* L. Moench) is one of the most important and widely grown crops found throughout the tropical and sub-tropical regions of the world. It is an annual, erect growing, high yielding crop with numerous cultivars varying in plant height, degree of branching and pigmentation of the various parts, period of maturity, and pod shape and size. It is mainly grown for its tender green pods and leaves, which are cooked and commonly consumed as boiled vegetables [1]. The total commercial production of okra in the world was estimated at 4.8 million tons, with India and Nigeria being the predominant producers [2]. Other minor producers include Pakistan, Ghana, Egypt, Ethiopia, Iran, Iraq, Turkey, Brazil, Guyana, Japan and USA.

Okra constitutes a major economic crop in the West African sub-region owing to its vital importance as a component of various recipes in many local cuisines and delicacies. It has a considerable area under cultivation in Nigeria, for example, where it is traditionally cultivated as a rainy season crop by women often on a wide range of habitats including marginal land along roadsides, backyard gardens, and wastelands [3].

Nutritionally, tender green pods of okra are important sources of vitamins A, B_1, B_3, B_6, C and K, folic acid, potassium, magnesium, calcium and trace elements such as copper, manganese, iron, zinc, nickel, and iodine [4], which are often lacking in the diet of people in most developing countries. The young green pod is a nutritious vegetable containing 86.1% moisture, 9.7% carbohydrate, 2.2% protein, 0.2% fat, 1.0% fiber and 0.8% ash [5]. The tender green pods are also popular in most tropical countries due to their medicinal values as

they contain very high levels of antioxidants including β-carotene, xanthin and lutein [6]. The mucilage substance derived from the wall of tender okra pods has been found to have a good alkaline pH, which contributes to its relieving effect in gastrointestinal ulcer by neutralizing digestive acids [7]. Okra also stabilizes blood sugar levels and helps to control the rate at which sugar is absorbed in the body. In addition, okra mucilage has also been reported as an effective tablet binder [8], plasma replacement and blood volume expander with enormous potential to improve renal function, alleviates renal diseases, and reduces proteinuria [9]. Mature seeds of okra on the other hand contain about 20% protein, with an amino acid profile similar to their composition in soybean. The seeds also contain about 20% oil that is similar in fatty acid composition to cotton seed oil [10] such that a very high potential presently exists for wide cultivation of okra for edible oil production [11].

In spite of its enormous economic benefits, okra rarely reaches its maximum yield potential due to several constraints. Some of the major factors limiting okra production, amongst several others, include the use of locally unimproved varieties, high incidence of pest and disease

*Corresponding author: Anwanobong Jonathan Eshiet, Plant Genetic Resources and Cell and Tissue Culture Research Laboratory, Department of Genetics and Biotechnology, University of Calabar, Calabar, Nigeria, E-mail: andibrisibe24@gmail.com

burden, a narrow genetic base of existing varieties and sub-optimal planting densities [12-14]. As okra production plays a significant role in the rural economies of most tropical countries, for example, in Africa, where it is cultivated, more attention needs to be directed to the selection of high yielding cultivars for edible fruits and seeds as no serious effort has been paid to its improvement in international research programmes in the past. Meanwhile, a high degree of wide morphological variation is known to exist among various accessions of the crop, especially in the West African sub-region [15] where several landraces have been identified.

Improvement of okra requires a broad spectrum of genetic variability from which useful characters can be selected for developing broad-based populations to be used in hybridization programmes towards improvement. Therefore, to harness and utilize useful traits in okra genetic resources, it is essential to assemble, characterize and evaluate many useful varieties in order to maximize their utilization in any crop improvement programme [16], which is highly dependent on the amount of genetic variability that is available in the gene pool. Expectedly, any successful selection programme can only be achieved when there is valid information about the genetic variability of the traits of interest in the crop population. In other words, the availability of genetically based variations for morphological traits and yield-attributing characters is a prerequisite for the development of new cultivars. Driven by the desire to identify genetically good parents to utilize for hybridization to expand genetic variation for selection of superior genotypes in okra, the current study was conducted to investigate variability in some morphological traits and yield characteristics as a precursor to exploring the gene pool in order to formulate a suitable breeding strategy for the development of new cultivars.

Materials and Methods

Plant materials

Three cultivars of okra (NHAe-47-4, V_{35}, and LD88) were obtained courtesy of the National Institute of Horticultural Research (NIHORT), Ibadan, Nigeria. In addition, a local variety was procured from a subsistence farmer in Calabar, Nigeria (4.95° N, 8.32° E, 99 m.a.s.l.) and used in the experiment.

Experimental site, study design and agronomic operations

Seeds of the four okra accessions were soaked in water for 24 h and thereafter planted out in field plots at the Teaching and Research Farm, University of Calabar, Calabar, Nigeria, during the rainy season between June and August in 2012 and 2013, respectively, in a completely randomized design and their performance evaluated at various stages of growth. The different okra varieties served as the main factor with each treatment having 30 plants at a spacing of 100 cm × 30 cm in a plot that was replicated five times. Plots were detached from one another by path distance of 1 m within and 2 m between replications.

The area where the experimental plots were situated features lowland humid tropical conditions that is affected by two opposing air masses, with a bimodal pattern of rainfall with a rainy, wet season between April and October and a dry season from November to March that is characterized by a cold dry dusty wind (harmattan), which blows from the Sahara towards the western coast of Africa, especially in January and February. Temperatures within the experimental site were relatively constant throughout the year, with average highs usually ranging between 22 and 28°C (Table 1). There was also little variance in temperature between daytime and nighttime, as temperatures at night

were generally only a few degrees lower than the high temperature that was typically observed during the daytime. In addition, the experimental site in Calabar averages about 3,900 mm of precipitation annually with a relative humidity (60-82%) that varies directly with the peaks of the bimodal rainfall. Radiation is fairly high and varies according to different periods of the year with values above 1,600 h per year being common [16].

Soil samples collected from different locations of the experimental site were oven-dried, ground and sieved through 2 mm sieve. The particle size distribution of the soil into sand, clay and silt contents was evaluated by a simplified soil size determination method [17]. The soil pH was determined using a potable pH metre in a 1:2.5 soil/water ratio while total nitrogen content was determined using the micro-kjeldahl method [18], and total phosphorus according to Bray 1 method [19]. Calcium and magnesium were determined using an atomic absorption spectrophotometer (model 372, Perkin Elmer) while potassium and sodium were evaluated by flame emission photometry. Organic carbon was determined according to Walkey and Black [20] and the present organic matter was estimated by multiplying the percent organic carbon with a factor 1.724.

Standard agronomic practices such as weeding, using a simple hoe, during the experimental period were adopted. Plants were tended carefully at the different stages of growth until maturity after which twenty stands were chosen at random from the central rows of each plot for the purpose of data collection on the following agronomic parameters:

(i)   Number of days to seedling emergence

(ii)   Number of days to flowering

(iii)   Plant height at flowering (cm)

(iv)   Number of leaves at flowering

(v)   Number of pods per plant

(vi)   Pod length (cm)

(vii)   Average number of seeds per pod

(viii) Mean weight of one hundred (100) seeds (g)

Statistical analysis

Data on parameter means for each plot were pooled and subjected to analysis of variance (ANOVA) test for determination of significant differences among the four okra varieties evaluated. The least significant difference (LSD) test was used to separate significantly different means.

Results

Climatic and edaphic conditions

The intensity of rainfall during the study period (between June and August in 2012 and 2013, respectively) was quite heavy. The highest amount of rainfall and highest number of rainy days were usually recorded in July of either of the two years (Table 1). The average monthly temperature during the study period ranged between 22.1 and 28.7°C while the average relative humidity ranged from 76.3 to 77.2% for the two years (Table 1).

The physic-chemical properties of the soil at the experimental site are given in Table 2. While total nitrogen level was low (0.13 and 0.18%), however, the soil had a moderate level of phosphorus (5.9ppm and 5.8ppm) and a correspondingly low level of potassium (0.20 and

	Average monthly precipitation (mm)	Average monthly temperature (°C)		Average relative humidity (%)
2012		**Maximum**	**Minimum**	
June	3945 (18)*	27.0	22.1	78.4
July	4450 (20)	26.8	24.3	78.9
August	3940 (14)	26.0	22.4	76.3
2013				
June	3940 (17)	27.2	23.0	77.2
July	4390 (21)	28.1	22.4	77.5
August	3925 (12)	28.1	23.2	79.2

* Values in parentheses indicate number of rainy days.
Source: Teaching and Research Farm, University of Calabar, Calabar, Nigeria Meteorological Station

Table 1: Meteorological data for the experimental site in Calabar, Nigeria between June and August, 2013 and 2013.

0.29%) during the study period in 2012 and 2013, respectively. In addition, relatively moderate amounts of calcium and magnesium were also present in all the soil units examined. During the two years, organic matter was low (2.8 and 3.1%), while the pH in water was near neutral (Table 2).

Number of days to seedling emergence

Having noted the prevailing weather and soil conditions at the experimental site, attention was thereafter shifted to the growth parameters of the four okra varieties evaluated. First, there were no significant differences ($p>0.05$) in the mean number of days taken for seedlings of the four varieties of okra to emerge. Cultivars NHAe-47-4 and V_{35} took a mean number of 4.33 days to emerge while LD88 and the local variety took a mean of 4.20 and 3.25 days, respectively. These results are presented in Table 3.

Number of days to flowering

Number of days to flowering differed significantly ($p<0.01$) among the four varieties of okra evaluated. While NHAe-47-4 took the highest mean number of days (112) to flowering, LD88 and the local variety flowered at statistically the same mean number of days (91.8 and 71.75 days, respectively). As shown in Table 3, the local variety may be considered to have matured first and therefore stands a better chance of earlier returns than NHAe-47-4, V_{35} or LD88.

Plant height at flowering

The growth parameters were significantly affected by the okra variety. For example, plant growth measured as plant height (in cm) was different among the four okra varieties (Table 3) evaluated. The tallest plants at flowering were recorded for LD88, with a mean value of 128 cm. This was closely followed by those of NHAe-47-4 and V_{35} with a mean height of 119.67 cm and 118.15 cm, respectively, which were greater than that produced by plants from the local variety with a mean height of 49.75 cm. Mean square estimates for plant height at flowering showed significant ($p<0.05$) differences amongst the four okra varieties evaluated in the current study.

Number of leaves at flowering

Number of leaves also differed significantly ($p<0.001$) among the okra varieties at flowering. NHAe-47-4 had the highest mean number of leaves (19.33) at flowering, followed by V_{35} and LD88 with a mean value of 17.09 and 8.2, respectively, while the local variety had the least mean (7.5) number of leaves at flowering. NHAe-47-4 and V_{35} also produced significantly larger sizes of leaves.

Average number of pods per plant

The four varieties of okra evaluated in the current study had statistically the same ($p>0.05$) average number of pods per plant. NHAe-47-4, V_{35}, LD88 and the local variety had a mean number of 5.67, 4.45, 3.20 and 4.0 pods, respectively, per plant. Though there was no statistical difference in fresh fruit yield/plant between NHAe47-4 and V_{35}, however, NHAe47-4 and V_{35} in 2012 and 2013 produced more fruits than LD88 and the local variety. Comparatively, yields of NHAe47-4 and V_{35} did not differ significantly.

Pod length

Pod length differed significantly ($p<0.01$) among the four okra varieties. Pods of NHAe-47-4, V_{35} and LD88 were longer (6.38, 6.23 and 6.83 cm, respectively) than those of the local variety with a mean length of 3.23 cm.

Average number of seeds per pod

There were no significant ($p>0.05$) differences in the average number of seeds per pod among the four varieties of okra evaluated. NHAe-47-4, V_{35}, LD88 and the local variety had mean values of 34, 36.20, 33.4 and 43.50 seeds, respectively, per pod.

Parameter	Year (2012)	Year (2013)	Analytical method used
Organic matter (%)	2.8	3.1	Walkley-Black
Nitrogen (%)	0.13	0.18	Micro-Kjeldahl
P_2O_5 (ppm)	5.9	5.9	Flame photometer
K (%)	0.20	0.29	Oxidation
Ca (meq/100g)	4.84	5.20	AAS
Mg (meq/100g)	3.36	2.84	AAS
pH (H_2O)	6.7	6.6	pH meter
pH ($CaCl_2$)	5.8	5.8	pH meter
Soil particle size distribution (%)	sand 81, clay 8 and silt 11	sand 80, clay 8 and silt 12	Kettler et al.

ppm = parts per million; AAS = Atomic Absorption Spectrophotometer

Table 2: Physico-chemical properties of the soil at the experimental site in 2012 and 2013.

Parameter	Variety			
	NHAe-47-4	**V_{35}**	**LD88**	**Local variety**
Number of days to seedling emergence	4.33^a ± 0.67	4.33^a ± 0.12	4.20^a ± 0.48	3.25^a ± 0.25
Number of days to flowering	116.45^a ± 7.21	112.00^a ± 8.33	91.80^b ± 5.80	71.75^b ± 4.71
Plant height at flowering (cm)	119.67^a ± 8.21	118.15^a ± 2.18	128.00^a ± 23.84	49.75^b ± 5.76
Number of leaves at flowering	19.33^a ± 2.60	17.09^a ± 1.12	8.20^b ± 0.37	7.50^b ± 0.65
Average number of pods per plant	5.67^a ± 1.20	4.45^a ± 1.20	3.20^a ± 0.49	4.00^a ± 0.41
Pod length (cm)	6.38^a ± 1.13	6.23^a ± 0.19	6.83^a ± 0.60	3.23^b ± 0.13
Average number of seeds per pod	34.00^a ± 3.06	36.20^a ± 0.05	33.40^a ± 2.62	43.50^a ± 7.17
One hundred seed weight (g)	4.25^b ± 0.10	4.22^b ± 0.82	4.42^a ± 0.02	3.87^c ± 0.02

*Means with same superscript letters in each horizontal array are not significantly (P>0.05) different from one another

Table 3: Growth components and yield attributes in four varieties of okra (*Abelmoscus esculentus*).

Mean weight of one hundred seeds

The four varieties differed significantly (p <0.05) in 100 seed weight. While LD88 had the highest weight of 100 seeds (4.42 g), followed by NHAe-47-4 (4.25 g), V_{35} (4.22 g) the local variety had the least weight of a 100 seeds at 3.87 g. These results are presented in Table 3.

Discussion

The results of this study are indicative of the fact that the four varieties of okra evaluated were significantly different in some of the morphological traits but definitely not in most of the yield attributes except for one hundred (100) seed weight. The observation of significant differences in some of the traits is an indication that genetic diversity do exist among the varieties, thereby providing a basis for selection. This is in consonance with what has been reported earlier [21-23], where it was demonstrated that such genetic variability existed amongst okra varieties evaluated in the respective studies on genetic diversity among some okra germplasm.

In this study, the four varieties differed significantly in the number of days to flowering. It has been demonstrated that on a general basis, early flowering is detrimental for overall productivity in okra as the source to sink ratio will be potentially limited for effective photosynthesis [24]. Differences in flowering periods among the varieties in the current study imply that their maturity periods vary. Depending on the desire of the breeder or farmer, appropriate selection can thus be made for either early or late maturing plants.

Plant height of the varieties evaluated was also significantly different. The height of the plant can potentially affect yield as those that are taller are usually more prone to windstorms in the event of heavy seasonal or monsoonal rains. Plant height at flowering and fruiting are of particular interest for breeding programmes because the presence of plants with tall and thin stems will increase the rate of lodging near harvesting and this could lead to loss of dry matter and subsequent decrease in fruit yield. Number of days to bud emergence and plant height at maturity, among other agronomic characters, are some of the most variable traits that are necessary for selection programmes aimed at improving desirable traits in okra [25]. It is suggestive from this that number of days to and plant height at flowering are controlled by the same genetic variables [26,27]. Consequently, selection for dwarf stature may thus be made on the local variety as it was shorter than the other three varieties evaluated in the current study (Table 3).

Variation was also observed in the number of leaves, pod length and one hundred seed weight of the cultivars. Since leaves serve as the sites for photosynthetic activities in any plant, an increase or a decrease in their number may have very serious implications for production of assimilates in the crop. Consequently, a greater number of them in any particular variety would be assumed to produce a better crop yield due to the higher photosynthetic capacity that is brought to bear by an increased leaf area index and a resultant higher fraction of intercepted radiation and its utilization efficiency [21]. The better performance of NHAe47-4 and V_{35} in the current study, amongst other reasons, can thus be attributed to the higher number of leaves as well as their larger leaf sizes (Table 3), which may have enabled these cultivars to produce greater assimilates during their photosynthetic activities. This position is reinforced by the suggestion that NHAe47-4 is a variety bred by NIHORT in Nigeria, which is characterized by early flowering with thick fresh pods, short to medium plant height, deeply lobed leaves and profuse branching [28]. This suggestion is further supported when the published data is examined for the reasons behind the yield increases

that have been reported for this variety. Strictly speaking, NHAe47-4 is very often known to produce higher yield than most local and exotic varieties including V_{35} (an exotic variety that has the same morphological features as those of NHAe47-4), as it is well suited to the local environment in Nigeria. Also one hundred seed weight is usually associated with pod yield and other yield attributes in a seed bearing plant. An increased number of leaves and a higher seed weight would thus be important selection criteria in the improvement of okra accessions.

The tender green pod is often considered as the most important and economical part of okra production since it is utilized as vegetable throughout the world. Consequently, fruit length in consonance with pod number and pod weight are the most important determinants of production or yield. Thus, selection based on these characters will be quite beneficial in okra breeding programmes, especially since three amongst the four cultivars evaluated in this study showed identical results for fruit length, which were statistically similar (Table 3). A similar pattern has been identified earlier, where it was reported that the Sabz pari variety of okra had the most promising result for maximum pod length in that study [29]. These results are also in conformity with others reported earlier[25,30-33], where statistically different results for pod length amongst the different okra cultivars evaluated were observed.

Overall, there were no significant variations in the number of days to seedling emergence, average number of pods per plant and average number of seeds per pod amongst the four okra varieties evaluated. This might be on account of the genetic characteristics of the cultivars. It is most probable that the genes controlling these character traits are dominant in all the varieties, which could have accounted for the absence of variation expressed amongst them. Consequently, these three traits cannot be considered as effective strategies while formulating selection indices for the improvement of okra.

Conclusion

It is apparent on the basis of the findings from this study that NHAe-47-4 and V_{35} okra varieties will be useful in breeding programmes where a higher crop yield is the dominant desire. Conversely, the local okra variety would be preferable where shorter and early maturing plants are desired since it had a comparatively shorter stature and an earlier flowering date than NHAe-47-4, V_{35} or LD88. However, further studies need to be conducted with a wider range of local and exotic okra varieties to enable a more robust selection of useful accessions for breeding programmes, especially on yield improvement.

References

1. Chattopadhyay A, Dutta S, Chatterjee S (2011) Seed yield and quality of okra as influenced by sowing dates. African Journal of Biotechnology 10: 5461-5467.

2. Gulsen OS, Karagul S, Abak K (2007) Diversity and relationships among Turkish germplasm by SRAP and phenotypic marker polymorphism. Biologia 62: 41-45.

3. Adeniji OT, Kehinde OB, Ajala MO, Adebisi MA (2007) Genetic studies on seed yield of West African Okra [Abelmoschus caillei (A. Chev.) Stevels]. Journal of Tropical Agriculture 45: 36-41.

4. Lee KH, Cho CY, Yoon ST, Park SK (2000) The effect of nitrogen fertilizer, plant density and sowing date on the yield of okra. Korean Journal of Crop Science 35: 179-183.

5. Saifullah M, Rabbani MG (2009) Evaluation and characterization of okra (Abelmoschus esculentus L. Moench.) genotypes. SAARC J Agric 7: 91-98.

6. Rahman K, Waseem M, Kashif MS, Jilani M, Kiran G, et al. (2012) Performance

of different okra (Abelmoschus esculentus L.) cultivars under the agro-climatic conditions of Defra Ismail Khan. Pakistan Journal of Science 64: 316-319.

7. Wamanda DT (2007) Inheritance studies in collected local okra (Abelmoschus esculentus L. Moench) cultivars. In: Combining ability analysis and heterosis on diallel cross of okra. African Journal of Agricultural Research 5: 2108-2155.

8. Ofoefule SI, Chukwu AN, Anayakoha A, Ebebe IM (2001) Application of Abelmoschus esculentus in solid dosage forms 1: use as binder for poorly water soluble drug. Indian J Pharm Sci 63: 234-238.

9. Siesmonsma JS, Kouame C (2004) Vegetables. In: Plant Resources of Tropical Africa 2 (Grubben, G.J.H. & Denton, O.A., Eds.) PROTA Foundation, Wageningen, Netherlands/Backhuys Publishers, Leinden, Netherlands/CTA, Wageningen, Netherlands 20-29.

10. Siemonsma JS, Hamon S (2002) Abelmoschus caillei (A. Chev.) Stevels. In Oyen, L.P.A. and Lemmens R.H.M. (eds) Plant Resources of Tropical Africa. Precursor PROTA Programs Wageningen, The Netherlands 27-30.

11. Rao PU (1985) Chemical composition and biological evaluation of okra (Hibiscus esculentus) seeds and their kernels. Qual. Plant Food Human Nutr 35: 389-396.

12. Adejonwo KO, Ahmed MK, Lagoke STO, Karikari SK (1989) Effects of variety, nitrogen and period of weed interference on growth and yield of okra (Abelmoshus esculentus). Nigeria Journal of Weed Science 2: 21-27.

13. Dikwahal HD, Haggai PT, Aliyu L (2006) Effects of sowing date and plant population density on growth and yield of two okra (Abelmoschus esculentus L.) varieties in the northern guinea savanna of Nigeria. Nigerian Journal of Horticultural Science 11: 56-62.

14. Das S, Chattopadhyay A, Chattopadhyay SB, Dutta S, Hazra P (2012) Genetic parameters and path analysis of yield and its components in okra at different sowing dates in the Gangetic plains of eastern India. African Journal of Biotechnology 11: 16132-16141.

15. Omonhinmin CA, Osawaru ME (2005) Morphological characterization of two species of Abelmoschus: Abelmoschus esculentus and Abelmoschus caillei. Genetic Resources Newsletter 144: 51-55

16. Onwueme IC, Sinha TD (1991) Field Crop Production in Tropical Africa. CTA, Ede, The Netherlands 480.

17. Kettler TA, Doran JW, Gilbert T L (2001) Simplified method for soil particle-size determination to accompany soil-quality analyses. Soil Science Society Amer J 65: 849-852.

18. Ma T, Zuazaga G (1942) Micro-Kjeldahl determination of nitrogen. A new indicator and an improved rapid method. Ind Eng Chem Anal Ed 14: 280-282.

19. Bray RH, Kurtz LT (1945) Determination of total, organic, and available forms of phosphorus in soils. Soil Science 59: 39-45 .

20. Walkley A, Black IA (1934) An examination of Degtjareff method for determining soil organic matter and a proposed modification of the chromic acid titration method. Soil Science 37: 29-37.

21. Ahiakpa JK, Kaledzi PD, Adi EB, Peprah S, Dapaah HK (2013) Genetic diversity, correlation and path analyses of okra (Abelmoschus spp. (L.) Moench) germplasm collected in Ghana. International Journal of Development and Sustainability 2: 1396-1415.

22. Adeoluwa OO, Kehinde OB (2011) Genetic variability studies in West African okra (Abelmoschus caillei). Agriculture and Biology Journal of North America 2: 1326-1335.

23. Aladele SE, Ariyo OJ, de Lapena R (2008) Genetic relationships among West African okra (Abelmoschus caillei) and Asian genotypes (Abelmoschus esculentus) using RAPD. African Journal of Biotechnology 7: 1426-1431.

24. Aboagye LM, Isoda A, Nyima H, Takasaki Y, Yoshimura T et al. (1994) Plant type and dry matter production in peanut (Arachis hypogea L.) cultivars: varietal differences in dry matter production. Japanese Journal of Crop Science 63: 289-297.

25. Akinyele BO, Oseikita OS (2006) Correlation and path coefficient analyses of seed yield attributes in okra (Abelmoschus esculentus (L.) Moench). African Journal of Biotechnology 5: 1330-1336.

26. Choudhary UN, Khanvilkar MH, Desai SD, Prabhudesai SS, Choudhary PR et al. (2006) Performance of different okra hybrids under North Konkan coastal zone of Maharashtra. Journal of Soils and Crops 16: 375-378.

27. Hussain S, Muhammad Noor S, Shah A, Iqbal Z (2006) Response of okra (Abelmoschus esculentus) cultivars to different sowing times. Journal of Agricultural and Biological Science 01: 55-59.

28. Iyagba AG, Onuegbu BA, Ibe AE (2012) Growth and yield response of okra (Abelmoschus esculentus (L.) Moench) varieties to weed interference in South-eastern Nigeria. Global Journal of Science Frontier Research Agriculture and Veterinary Sciences 12: 23-31.

29. Akram AM, Shah H (2002) Performance of okra (Abelmoschus esculentus, L.) varieties in the up lands of Balochistan [Pakistan]. Balochistan Journal of Agricultural Sciences (Pakistan) 3: 1-3.

30. Khan H, Khan MU, Khan AU, Anwar KB, Mehmood K (2000) Response of different cultivars of okra (Abelmoschus esculentus L.) to three different sowing dates in the mid hill of Swat Valley. Pakistan Journal of Biological Science 3: 2010-2012.

31. Ashraful AKM, Hossain MD (2006) Variability of different yield contributing parameters and yield of some okra (Abelmoschus esculentus L) accessions. J Agric Rural Dev 4: 119-127.

32. Sachan VK (2006) Performance of okra (Abelmoschus esculentus L.) varieties in mid-hills of Sikkim Orissa. Journal of Horticulture 34: 131-132.

33. Bello D, Sajo AA, Chubado D, Jellason JJ (2006) Variability and correlation studies in okra (Abelmoschus esculentus L.). Journal of Sustainable development in Agriculture and Environment 2: 120-125.

30

Integrated Weed Management in *Rabi* Sweet Corn (*Zea mays* L. var. *Saccharata*)

Mathukia RK[2]*, Dobariya VK[1], Gohil BS[1] and Chhodavadia SK[1]

[1]*Ph.D.Scholars Department of Agronomy, College of Agriculture, Junagadh Agricultural University, Junagadh-362001 Gujarat, India*

[2]*Associate Research Scientist, Department of Agronomy, College of Agriculture, Junagadh Agricultural University, Junagadh-362001, Gujarat, India*

Abstract

A field experiment was conducted during *rabi* 2010-11 at Junagadh (Gujarat, India) to find out most efficient and economical method of weed control in *rabi* sweet corn (*Zea mays* L. var. *saccharata* Sturt). The pre-emergence (PRE) herbicides viz., atrazine, pendimethalin and oxadiargyl were combined either with hand weeding (HW) and interculturing (IC) or with post-emergence (POST) herbicide 2, 4-D (SS) to evolve integrated weed management. The weed flora of the experimental site constituted *Digera arvensis, Cyperus rotundus, Brachiaria* spp., *Asphodelus tenuifolius, Indigofera glandulosa, Amaranthus viridis, Acanthospermum hispidum, Panicum colonum, Launaea nudicaulis, Euphorbia hirta, Chenopodium album, Portulaca oleracea, Dactyloctenium aegyptium* and *Celosia argentea*. The results revealed that physical methods viz., weed free, HW and IC twice at 15 and 30 days after sowing (DAS) as well as integrated methods viz., atrazine @ 0.5 kg a.i. /ha as PRE+HW and IC at 30 DAS and pendimethalin @ 0.9 kg a.i. /ha as PRE+HW and IC at 30 DAS significantly enhanced growth and yield attributes ultimately higher cob and fodder yields over unweeded check. The treatments viz., weed free, HW and IC twice at 15 and 30 DAS, atrazine @ 0.5 kg a.i. /ha as PRE+HW and IC at 30 DAS, and pendimethalin @ 0.9 kg a.i. /ha as PRE+HW and IC at 30 DAS also recorded the lower weed population at 30, 60 DAS and at harvest, dry weight of weed at harvest with lower weed index and higher weed control efficiency and herbicidal efficiency index. These treatments were found economical by recording higher net returns and B: C ratio compared to unweeded check.

Keywords: Sweet corn; *Zea mays* L. var. *saccharata* Sturt; Atrazine; Pendimethalin; Oxadiargyl, 2, 4-D.

Introduction

Maize is considered as the "Queen of Cereals". Being a C_4 plant, it is capable to utilize solar radiation more efficiently even at higher radiation intensity. In Indian agriculture, maize assumes a special significance on account of its utilization as food, feed and fodder besides several industrial uses. Sweet corn (*Zea mays* L. var. *saccharata* Sturt), also called Indian corn, sugar corn and pole corn, is a variety of maize with a high sugar content. Nature of weed problem in *rabi* maize is quite different from that of the rainy season maize. In the rainy season emergence of maize and weed start simultaneously and first 20-30 days are most critical looking to crop-weed competition. Contrarily in the winter maize, weed emerges most often after the first irrigation. However, wider row spacing and liberal use of irrigation and fertilizers lead to more growth of weeds [1]. Yield loss due to weed in maize varies from 28 to 93%, depending on the type of weed flora and intensity and duration of crop-weed competition [2]. Pre-emergence application of herbicides may lead to cost effective control of the weeds right from the start which otherwise may not be possible by manual weeding. The study was carried out to find economically effective method of weed control for realising higher productivity and profitability of *rabi* sweet corn.

Materials and Methods

The experiment was carried out at Instructional Farm, Department of Agronomy, Junagadh Agricultural University, Junagadh (Gujarat, India) during *rabi*-2010-11. The experimental soil was clayey in texture and low in available N and P, and moderate in available potash. Sweet corn variety 'Sugar-75' was used in the experiment. The temperature ranged from 9.7 to 20.6°C during *rabi* season. The crop was sown on 11th December with the seed rate of 15 kg/ha at spacing of 60 cm x 20 cm. Standard package of practices was followed throughout the cropping season. The crop was harvested on 27th March.

To evolve integrated weed management, pre-emergence (PRE) herbicides viz., atrazine, pendimethalin and oxadiargyl were combined either with hand weeding (HW) and interculturing (IC) or with post-emergence (POST) herbicide 2, 4-D (SS) to evolve integrated weed management. The experiment comprised nine treatments, namely, (1) Atrazine @ 0.5 kg a.i./ha as PRE+HW & IC at 30 days after sowing (DAS), (2) Pendimethalin @ 0.9 kg a.i./ha as PRE+HW & IC at 30 DAS, (3) Oxadiargyl @ 90 g a.i./ha as PRE+HW & IC at 30 DAS, (4) Atrazine @ 0.5 kg a.i./ha as PRE+2,4-D (SS) @ 0.5 kg a.i./ha as POST at 30 DAS, (5) Pendimethalin @ 0.9 kg a.i./ha as PRE+2,4-D (SS) @ 0.5 kg a.i./ha as POST at 30 DAS, (6) Oxadiargyl @ 90 g a.i./ha as PRE+2,4-D (SS) @ 0.5 kg a.i./ha as POST at 30 DAS, (7) HW & IC twice at 15 & 30 DAS, (8) weed free and (9) weedy check, were replicated thrice in randomized block design.

Pre-emergence herbicides were applied next day of sowing and post-emergence herbicide was sprayed at 30 DAS. The spraying was done using knapsack sprayer with flat fan nozzle keeping spray volume of 500 L/ha. Weeding was done by labours and interculturing was done by bullock drawn harrow in between two rows of the crop. In manual weed control treatments, weeds were uprooted and removed at 30 DAS as per treatment. In weed free plots, the weeds were removed manually after every seven days for ensuring complete weed free condition.

*Corresponding author: Mathukia RK, Department of Agronomy, College of Agriculture, Junagadh Agricultural University, Junagadh-362001, Gujarat, India
E-mail: rkmathukia@jau.in

Growth and yield attributes as well as cob and fodder yields were recorded at harvest of the crop. Number of weeds (monocots, dicots and sedge) was counted at 30, 60 DAS and harvest using 1 m x 1 m quadrat from each plot. At harvest time, after uprooting of weeds, the weeds were sun-dried completely till reached to constant weight and finally the dry weight was recorded for each treatment and expressed as kg/ha. Weed control efficiency (WCE), weed index (WI) and herbicide efficiency index (HEI) were calculated by the formulae suggested by Kondap and Upadhayay (1985), Gill and Kumar (1969) and Krishnamurthy et al., respectively [3-5].

$$WCE (\%) = \frac{DW_C - DW_T}{DW_C} \times 100$$

Where, DW_C and DW_T are dry matter accumulation of weeds in unweeded control and treated plot, respectively.

$$WI = \frac{Y_{WF} - Y_T}{DW_C} \times 100$$

Where; Y_{WF} and Y_T are the yield from weed-free plot and yield from treated plot, respectively.

$$HEI = \frac{Y_T - Y_C}{Y_C} \times 100$$

Where, Y_t and Y_c are yield from treated and unweeded control plot, respectively.

The data were subjected to statistical analysis by adopting appropriate analysis of variance as described by Gomez and Gomez [6]. Wherever the F values were found significant at 5 per cent level of probability, the critical difference (C.D. at 5%) values were computed for making comparison among the treatment means. The data on weed count were subjected to square root transformation [6]. Gross returns (monetary income from cob and fodder yields), net returns (monetary income obtained after deducting cost of cultivation from gross returns) and

B:C ratio (gross returns divided by cost of cultivation) were calculated using prevailing market price of inputs (including treatments), labour and produce for assessing the economic viability of treatments.

Results and Discussion

Weed flora

The weed flora in the experimental site constituted by monocot weeds viz., *Brachiaria* spp. (19.0%), *Asphodelus tenuifolius* Cav. (9.5%), *Indigofera glandulosa* L. (8.8%), *Panicum coloratum* L. (2.4%) and *Dactyloctenium aegyptium* (L.) Willd (1.3%) and dicot weeds viz., *Digera arvensis* Forsk (21.0%), *Amaranthus viridis* L. (6.0%), *Acanthospermum hispidum* DC. (3.7%), *Launaea nudicaulis* L. (2.3%), *Euphorbia hirta* L. (2.0%), *Chenopodium album* L. (1.6%), *Portulaca oleracea* L. (1.4%), and *Celosia argentea* L. (1.0%) and sedge weed *Cyperus rotundus* L. (20.0%).

Crop growth and yield

Growth and yield attributes as well as cob and fodder yield were significantly influenced by different weed control practices (Table 1). Results showed that significantly the highest cob length (22.95 cm), cob girth (16.25 cm), number of cobs per plant (1.40), number of kernels per cob (275.67), fresh weight of cob (136.06 g), highest dry weight of cob (41.97 g), cob yield (7674 kg/ha) and fodder yield (37659 kg/ha) were recorded under weed free. However, the weed free treatment remained statistically equivalent to HW & IC at 15 & 30 DAS atrazine @ 0.5 kg a.i./ha as PRE+HW and IC at 30 DAS and pendimethalin @ 0.9 kg a.i./ha as PRE+HW & IC at 30 DAS. The improved growth and yield attributes under these treatments might be due to periodical removal of weeds by hand weeding or pre-emergence herbicide supplemented with manual weeding and interculturing as evidenced by less number of weeds and dry weight of weeds (Table 2), which might have maintained high soil fertility status and moisture content by means of less removal

Treatments	Cob length (cm)	Cob girth (cm)	Cobs per plant	Kernels per cob	Fresh weight of cob (g)	Dry weight of cob (g)	Green cob yield (kg/ha)	Green fodder yield (kg/ha)
Atrazine+HW & IC	20.85	14.91	1.28	267.33	130.43	36.40	6271	34572
Pendimethalin+HW & IC	20.81	13.87	1.27	261.00	126.81	34.35	6292	33329
Oxadiargyl+HW & IC	20.71	13.40	1.25	243.03	120.60	28.95	5861	28097
Atrazine+2,4-D (SS)	20.62	13.98	1.26	252.03	126.48	35.28	5997	30333
Pendimethalin+2,4-D (SS)	20.69	13.78	1.25	246.04	126.00	33.53	5986	30309
Oxadiargyl+2,4-D (SS)	20.56	13.61	1.20	235.33	122.10	28.71	5799	27704
HW & IC twice	22.32	15.71	1.35	270.33	132.10	40.36	6642	35769
Weed free	22.95	16.25	1.40	275.67	136.06	41.97	7674	37659
Weedy check	17.47	12.75	1.15	225.33	117.83	25.19	5382	25590
C.D. (P=0.05)	2.79	2.10	0.14	31.66	10.71	7.41	1186	5702

Table 1: Effect of different treatments on growth and yield.

Treatments	Monocot weeds per m²			Dicot weeds per m²			Sedge weeds per m²			Dry weight of weed (kg/ha)
	30 DAS	60 DAS	Harvest	30 DAS	60 DAS	Harvest	30 DAS	60 DAS	Harvest	
Atrazine+HW & IC	2.24 (4.53)	1.83 (2.87)	1.42 (1.53)	1.74 (2.53)	1.42 (1.53)	1.37 (1.41)	2.52 (5.87)	1.83 (2.87)	1.79 (2.73)	322.92
Pendimethalin+HW & IC	2.29 (4.77)	1.89 (3.10)	1.50 (1.77)	1.80 (2.77)	1.60 (2.10)	1.57 (2.00)	2.62 (6.43)	2.05 (3.77)	2.00 (3.60)	431.60
Oxadiargyl+HW & IC	3.14 (9.54)	2.87 (7.87)	2.65 (6.64)	2.74 (7.20)	2.70 (6.87)	2.67 (6.79)	3.50 (11.9)	2.87 (7.87)	2.85 (7.75)	477.08
Atrazine+2,4-D (SS)	2.96 (8.46)	2.78 (7.46)	2.61 (6.46)	2.64 (6.79)	2.55 (6.46)	2.51 (6.28)	3.27 (10.3)	2.96 (8.26)	2.93 (8.13)	412.50
Pendimethalin+2,4-D (SS)	3.04 (8.92)	2.81 (7.58)	2.62 (6.58)	2.76 (7.25)	2.53 (6.25)	2.54 (6.23)	3.26 (10.3)	3.00 (8.52)	2.98 (8.38)	437.85
Oxadiargyl+2,4-D (SS)	3.81 (14.1)	3.64 (12.8)	3.44 (11.3)	3.53 (12.1)	3.36 (10.8)	3.35 (10.7)	4.05 (16.1)	3.89 (14.8)	3.85 (14.6)	525.00
HW & IC twice	2.20 (4.38)	1.58 (2.04)	1.38 (1.42)	1.95 (3.38)	1.34 (1.30)	1.24 (1.05)	1.96 (3.38)	1.73 (2.58)	1.70 (2.46)	183.33
Weed free	0.71 (0)	0.71 (0)	0.71 (0)	0.71 (0)	0.71 (0)	0.71 (0)	0.71 (0)	0.71 (0)	0.71 (0)	0.00
Weedy check	5.52 (30.1)	6.72 (44.8)	6.91 (47.5)	5.05 (25.1)	6.25 (38.8)	6.49 (41.8)	6.29 (39.1)	6.80 (45.8)	6.83 (46.3)	882.64
C.D. (P=0.05)	0.71	0.61	0.62	0.76	0.66	0.68	0.61	0.56	0.54	94.59

Table 2: Effect of different treatments on weed parameters.
Note: $\sqrt{x + 0.5}$ transformation and figures in parenthesis are original values.

Treatment	WI	WCE	HEI	Gross returns (`/ha)	Cost of cultivation (`/ha)	Net returns (`/ha)	B:C
Atrazine+HW & IC	18.28	63.41	16.52	97281	34520	62761	2.82
Pendimethalin+HW & IC	18.01	51.10	16.91	96247	35370	60877	2.72
Oxadiargyl+HW & IC	23.63	45.95	8.90	86708	35170	51538	2.47
Atrazine+2,4-D (SS)	21.85	53.27	11.43	90299	33670	56629	2.68
Pendimethalin+2,4-D (SS)	22.00	50.39	11.22	90170	34370	55800	2.62
Oxadiargyl+2,4-D (SS)	24.43	40.52	7.75	85691	34170	51521	2.51
HW & IC twice	13.45	79.23	23.41	102193	35270	66923	2.90
Weed free	0.00	100.0	42.59	114396	36470	77926	3.14
Weedy check	29.87	0.00	0.00	79410	32870	46540	2.42

Table 3: Effect of different treatments on weed index, weed control efficiency, herbicidal efficiency index and economics

Market Price:

Commodity	`/kg		Herbicide	:	`/kg or L
Urea	:	5.87	Atrazine	:	350
DAP	:	19.50	Pendimethalin	:	400
Green cob	:	10.00	Oxadiargyl	:	1080
Green fodder	:	1.00	2,4-D (SS)	:	400

of plant nutrients and moisture by weeds. These findings are in close conformity with those reported by Sinha et al., Kolage et al., Mandal et al., Kamble et al. and Deshmukh et al. [7-11].

Weed parameters

The weed management treatments significantly influenced the weed population (Table 2). The weed free check recorded the lowest weed population. HW & IC at 15 & 30 DAS also recorded significantly lower weed population, which remained statistically at par with atrazine @ 0.5 kg a.i./ha as PRE+HW & IC at 30 DAS and pendimethalin @ 0.9 kg a.i./ha as PRE+HW & IC at 30 DAS. Except weed free, the lowest dry weight of weed was observed under HW & IC at 15 & 30 DAS, though it was found statistically at par with atrazine @ 0.5 kg a.i./ha as PRE+HW & IC at 30 DAS. A perusal of data presented in Table 2 indicated that besides weed free, HW & IC at 15 & 30 DAS contained minimum WI, while maximum WCE and HEI, closely followed by pendimethalin @ 0.9 kg a.i./ha as PRE+HW & IC at 40 DAS and atrazine @ 0.5 kg a.i./ha as PRE+HW & IC at 30 DAS. This might be attributed to the effective control of weeds under these treatments, which reflected in less number of weeds and ultimately lower weed biomass. In addition to this, dense crop canopy might have suppressed weed growth and ultimately less biomass. The weedy check recorded significantly the highest number and dry weight of weeds owing to uncontrolled condition favoured luxurious weed growth leading to increased density and dry matter of weeds (Table 3). These findings are in close conformity with those reported by Sinha et al., Kolage et al. and Verma et al. [8,12,13].

Economics

The investigated data revealed that maximum net returns of `77926/ha and B:C of 3.14 were realized with weed free treatment, followed by HW & IC at 15 & 30 DAS and atrazine @ 0.5 kg/ha as PRE+HW & IC at 30 DAS. The lowest net returns of ` 46540/ha was accrued under treatment weedy check with B: C value of 2.42. The higher benefit under these treatments might be due to higher production of cob as well as fodder leading to increased monetary returns with comparatively lower cost. These findings are in close vicinity with those reported by Malviya and Singh, Rao et al., and Sunitha et al. [14-16].

Conclusion

On the basis of the results obtained from present field study, it can be concluded that effective management of weeds with profitable green cob and fodder yield of sweet corn in *rabi* season can be obtained by keeping the crop weed free throughout crop period or adopting two hand weeding and interculturing at 15 and 30 DAS. However under paucity of labours, pre-emergence application of atrazine @ 0.5 kg a.i./ha+HW & IC at 30 DAS or pendimethalin @ 0.9 kg a.i. /ha as pre-emergence+HW & IC at 30 DAS would be the better option under south Saurashtra Agro-climatic conditions.

References

1. Porwal MK (2000) Economics of weed-control measures in winter maize (Zea mays). Indian Journal of Agronomy, 45: 433-347.

2. Sharma V, Thakur DR (1998) Integrated weed management in maize (Zea mays) under mid-hill condition of north-western Himalayas. Indian Journal of Weed Science, 30: 158-162.

3. Kondap SM, Upadhyay UC (1985) A Practical Manual of Weed Control. Oxford and IBH Publ. Co., New Delhi.

4. Gill GS, Kumar V (1969) Weed index a new method for reporting weed control trials. Indian Journal of Agronomy, 16: 96-98.

5. Krishnamurthy K, Rajshekara BG, Raghunatha G, Jagannath MK, Prasad TVR (1995) Herbicide efficiency index in sorghum. Indian Journal of Weed Science, 7: 75-79.

6. Gomez K, Gomez A (1984) Statistical Procedures for Agricultural Research. John Willey and Sons, New York.

7. Sinha SP, Prasad SM, Singh SJ (2000) Effect of integrated weed management on growth, yield attributes and yield of winter maize. Journal of Applied Biology, 10: 158-162.

8. Kolage AK, Shinde SH, Bhilare RL (2004) Weed management in kharif maize. Journal of Maharashtra Agricultural Universities, 29: 110-111.

9. Mandal, Subhendu, Mondal, Subimal and Nath, Subhadeep (2004) Effect of integrated weed management on yield components, yield and economics of baby corn (Zea mays). Annals of Agricultural Research, 25: 242-244.

10. Kamble TC, Kakade SU, Nemade SU, Pawar RV, Apotikar VA (2005) Integrated weed management in hybrid maize. Crop Research Hisar, 29: 396-400.

11. Deshmukh LS, Jadhav AS, Jathure RS, Raskar SK (2009) Effect of nutrient and weed management on weed growth and productivity of kharif maize under rainfed condition. Karnataka J.Agri.Sci, 22: 889-891.

12. Sinha SP, Prasad SM, Singh SJ, Sinha KK (2003) Integrated weed management in winter maize (Zea mays) in North Bihar. Indian Journal of Weed Science, 35: 273-274.

13. Verma VK, Tewari AN, Dhemri S (2009) Effect of atrazine on weed management in winter maize-green gram cropping system in central plain zone of Uttar Pradesh. Indian Journal of Weed Science, 41: 41-45.

14. Malviya Alok, Singh Bhagwan (2007) Weed dynamics, productivity and economics of maize (Zea mays) as affected by integrated weed management under rainfed condition. Indian Journal of Agronomy, 52: 321-324.

15. Rao AS, Ratnam M, Reddy TY (2009) Weed management in zero-till sown maize. Indian Journal of Weed Science, 41: 46-49.

16. Sunitha N, Reddy PM, Reddy DS (2011) Influence of planting pattern and weed control practices on weed growth, nutrient uptake and productivity of sweet corn (Zea mays L.). Crop Research Hisar, 41: 13-20.

Phosphorus Sorption Characteristics and External Phosphorus Requirement of Bulle and Wonago Woreda, Southern Ethiopia

Zinabu Wolde[1]*, Wassie Haile[2] and Dhyna Singh[2]

[1]*Graduate student of Soil Science, Hawassa University, P. O. Box 05, Hawassa, Ethiopia*
[2]*Gedeo Zone Agricultural Office, Gedeo Zone, P. O. Box 128, Dilla, Ethiopia*

Abstract

Determination of the P-sorption characteristics of soils is important for economic fertilizer application and to recommend appropriate management strategies for high P-fixing soils. Thus, the objectives of this study were to evaluate P-sorption characteristics of soils occurring in some areas of southern Ethiopia and identify factors contributing to P sorption. Composite surface (0-20 cm) soil samples from 6 sites of two locations were collected. The results revealed that the P-sorption data were fitted well with both Langmuir and Freundlich models with r^2 values of 0.99. But later model was found to better in describing P-sorption data than the former model. The adsorption maxima (Xm) and distribution coefficient (Kf) values of soils ranged from 909-2000 mg PKg^{-1} and 245-487 mg PKg^{-1} based on Langmuir and Freundlich models respectively. The corresponding SPR values ranged between 57-196 mg P kg^{-1} and 71.8-211 mg P kg^{-1} based on Langmuir and Freundlich models respectively. Bonding energy constant (K) of Langmuir model and Freundlich constant (b) ranged from 0.15-1.4 and 0.54-0.66 Lmg^{-1} respectively. The bonding energy constant (K) of Langmuir model was found to be more valuable than Xm in discriminating the study soils as high and low P sorbing soils. Based on both models soils of Bulle had SPR values >150 mg PKg^{-1} and then were classified as high P-fixing soils while soils of Wonago had SPR values <150 mgPKg^{-1} were classified as low P-fixing. The result of path analysis revealed that exchangeable Al and Clay had direct effect on P-sorption parameters of both models. It is concluded that P-sorption models can effectively be used to discriminate soils based on P-fixation ability. However, validation of both models through real time experiments in greenhouse and field is recommended before the models used for large scale.

Keywords: P-Sorption/Fixation; Standard P requirement (EPR); Exchangeable Al; Fe; Langmuir model; Freunlich model

Introduction

In tropical and subtropical acidic soils low Phosphorus (P) availability is a major factor that limits plant growth. When soil P levels are too low, phosphorus deficiency in plants represents a major constraint to world agricultural production [1]. According to [2], soil P deficiency may be due to low P containing parent material from which soil was formed or low inherent P content, high weathering incidence and soil reaction, long term anthropogenic mismanagement through imbalance between nutrient inputs, and P losses by erosion and surface runoff.

Furthermore, the limited availability of P in soils may be attributed to severe P fixation or retention. In which Phosphorus retention by soils often involves complex combination of sorption and precipitation reactions. Initially, the phosphate ions undergo sorption onto organic and inorganic particle surface and then slowly precipitate into less available forms [3]. Depending on this the fate and efficiency of native and applied P therefore remains one of the biggest problems in arable crop production in the tropics. One problem is that fertilizer P can largely be fixed by oxides, hydroxides and oxy-hydroxides of Fe and Al and clay minerals in acidic soil, which makes it less available or effectively unavailable to plants [4].

According to Abayneh [5] Ethiopia has diverse soil resources largely because of diverse topography, climatic conditions and geology. Therefore, successful agriculture to meet the increasing demands of food, fiber, fuel and others requires the sustainable use of soil by managing the nutrient supplying capacity of the soil.

However, the morphological, physical and chemical characteristics of soils of Southern Ethiopia in relation to nutrient retention and management alternatives are not well documented with this accurate

prediction of nutrient requirements is highly desirable, because the cost of fertilizer is a constraint that limiting fertilizer use by small farmers.

Isotherm equations still are a common approach to study P sorption in soils, since their use allows summarizing results, reducing the sorption data to comparable values. Among isotherm adsorption models, Langmuir and Freundlich equations are the one most used for soil characterization and research purposes [6].

Recommending fertilizer requirement of crops based on biological tests alone is unlikely to address the problem of plant nutrition unless used in combination with soil test results. Moreover, [7] stressed that soil testing is essential for accurate and profitable fertilizer recommendations provided that the soil test result correlate to the crop response.

Langmuir and Frendlich models are the most widely used models to describe P-sorption charactatersitics of soils and draw external P-requirements [8]. These models are able to discriminate soils based on their ability to sorb/adsorb P from soil solutions. They also give an insight into strength of P-adsorption on the surface of a particular soil.

However, so far only limited research results [9] are available on the

*Corresponding author: Zinabu Wolde, Graduate student of Soil Science, Hawassa University, P. O. Box 05, Hawassa, Ethiopia
E-mail: sos.zine04@gmail.com

P sorption characteristics and soil factors contributing to P-sorption for Ethiopian soils. Thus, it is essential to determine the P-sorptin isotherm of some soils of southern Ethiopia and model the resulting data for generating valuable recommendation for future use. Therefore, the objectives of this study were: to identify soil with high and low p sorption characteristics, to estimate the soil's standard P requirement based on P- sorption isotherm.

Materials and Methods

Description of study sites

The sites considered in the present study are among the agriculturally important soils in the Gedeo Zone, Southern regions of the country. Only two sites from the Zone (Bulle and Wonago woreda) were selected by stratifying the entire study area based on differences in altitude, crop type, slope gradient and mainly as classified by government for the ease of political management by identifying the potential area and grouping them in to sub kebelles (which called locally "Goxi"). The sampling depth was restricted to the plough layer (0-20 cm) where most of the plant nutrient and the roots of crops are concentrated and actively interact.

Soil sampling and sample preparation

Visual observations of the area were first taken to have a general view of the variations in the study area and representative fields were selected. Following the site selection, 40 subsamples were taken from each site to make one composite sample for each from the depth of 0-20 cm in a Zigzag sampling scheme using an auger. For labeling conventions the soil samples from the Wonago site were (Wo1, Wo2 and Wo3) while sample from, Bulle (Be1, Be2 and Be3). The soil samples were mixed well, air-dried and passed through a 2 mm sieve for the analysis of selected soil physical and chemical properties.

Analysis of selected soil physico-chemical properties

Soil analysis was done following standard procedures adopted by National Soil Testing Center for determining selected soil chemical and physical properties at Hawssa University, College of Agriculture Soil Laboratory and Southern Region Agricultural Bureau soil laboratory.

Soil particle size distribution was determined by the Bouyoucos hydrometer method [10] after destroying OM using hydrogen peroxide and dispersing the soils with sodium hexametaphosphate.

The pH of the soils was measured in water suspension in a 1:2.5 (soil: liquid ratio) potentiometrically using a glass-calomel combined electrode [10]. The Walkley and Black wet digestion method was used to determine soil carbon content and percent soil OM was obtained by multiplying percent soil OC by a factor of 1.724 following the assumption that OM contains 58% carbon [11]. Even though the Olsen method is the most widely used for P extraction under wide range of pH both in Ethiopia and elsewhere in the world [12,13] available soil P was analyzed according to the standard procedure of Bray-II.

Cation Exchange Capacity (CEC) of soil was determined by neutral sodium-acetate saturation and neutral NH_4^+-acetate displacement. Exchangeable basic cations (Ca, Mg, K, and Na) of the soil were determined in the leachate using 1N neutral ammonium acetate adjusted to a pH of 7. Finally, exchangeable Ca^{++} and Mg^{++} were determined in the extract using AAS, whereas K^+ and Na^+ were determined by flame photometer. Exchangeable Al was determined by saturating the soil samples with potassium chloride solution and titrating with sodium hydroxide as described by McLean. Available micronutrients (Fe, Cu,

Zn and Mn) were extracted by DTPA and all these micronutrients were measured on AAS, as described by Lindsay and Norvell [14].

P-Sorption study

Phosphorus sorption characteristics were determined by batch equilibrium methods in which soil samples were agitated with P solutions of known concentrations [15]. Subsamples of soils collected from seven sites (Hagereselam, Damot Gale, and Hawassa zuria, Wonago, Chencha, Bulle and Halaba) that were previously used for soil physico -chemical analysis were used in P-sorption study. All samples were air dried at ambient temperature (Preferably between 20-25°C) crushed and sieved through a 2 mm sieve.

Phosphorus as (KH_2PO_4) was dissolved in a 0.01M solution of Calcium chloride in distilled water. The $CaCl_2$ solution is used as the aqueous solvent phase to improve centrifugation and minimize Cation exchange [16].

According to the methods of Fernandes MLV and Coutinho J to study the sorption of P by soils, 2 g air-dried samples of each soil were placed in 100 ml plastic bottle in order to leave free space for with 25 ml of 0.01M $CaCl_2$ in which the final volume was adjusted to 30 ml [17]. The Continuous mixing was provided during the experimental Period with a constant agitation speed of 350 rpm for better mass transfer with high interfacial area of contact. Afterwards, calculated amount of stock solution of P for each rate was added. The concentrations of the stock solution were 0, 10, 20, 30, 40 and 50 mgl^{-1} P. Each sorption set for P was replicated twice. The mixture was shaken for 30 minutes with maximum speed of 380 rpm and equilibrated for 24 hr. After equilibration time, the suspension was filtered through Whatman paper No. 42 filter paper and the concentration of P in the clear extract was determined by ascorbic acid method. Phosphorus disappeared from the solution was considered as sorbed P which was plotted against P concentration in the solution to obtain a P sorption isotherm. A blank was run for each soil with the same amount of soil and total volume of 30 ml 0.01 $CaCl_2$ solution (without P) was added and it was subjected to the same procedure. This served as a background control during the analysis to detect interfering compounds or contaminated soils. All the tests, including blanks were performed in duplicate.

The P sorption data for the soils were fitted into the following forms of Langmuir equation and Frendlich equation, because linear regression is convenient and best of data-fitting process.

Langmuir equations: $C/X = 1/K.Xm + C/Xm$ (Equation1) [18-20].

Where $C (mgl^{-1})$ is the equilibrium concentration, X (mg kg^{-1}) is the amount of P adsorbed per unit mass of adsorbent, K (Lmg^{-1}) is a constant related to the energy of sorption, and Xm (mg kg^{-1}) is P sorption maximum.

X is calculated as $C_oV_o - C_fV_f$/mass of soil (kg)   (Equation 2)

Where C_o is initial concentration, C_f is final concentration, V is volume of solution. The linear form of equation (1) was obtained by plotting the equilibrium concentration of phosphate(C) against the amount of phosphate adsorbed (X) and the slope of the graph equals to $1/Xm$ and intercept of the graph is equals to $1/K Xm$. But K was easily determined by dividing the slope by intercept.

Freundlich equation: $X = KC^b$ or $logX = logkf + blogC$ (Equation 3)

Where, K and b (b<1) are constants, X (mgkg^{-1}) is the amount of P adsorbed per unit mass of adsorbent, and C (mgl⁻¹) is the equilibrium concentration. The linear form of equation 3 was obtained by plotting LogC against LogX. The slope b (Lkg⁻¹) and intercept LogKf representing respectively. Phosphorus sorption curves were drawn by plotting the quantity of sorbed P against the P concentration in the equilibrated soil solution [21].

The external P-requirement of each soil or the amount of P required for each soil at 0.2 mg/L equilibrium solution of P also known as the standard P requirement (SPR) was calculated based on Langmuir and Freundlich models/Equations developed for each soil. The soil solution P of 0.2 mg/L is the amount of P that should be available in the soil for optimum plant growth [22].

Statistical analysis

Path coefficient analysis to differentiate between correlation and causation and to describe P-sorption as influenced by the various soil properties was carried out using SAS',(2002) software. This analysis allows direct comparison of causal relationships between soil properties and P-sorption by soil. Moreover, it permits partitioning of simple correlation coefficient between dependent (isotherm parameters) and independent (soil properties) variables into direct and indirect effects [23].

Result and Discussion

Soil physico-chemical properties

Selected physico-chemical properties of soils are summarized in Table 1. The soil tests have demonstrated variations among soils in their soil physical and chemical properties. Accordingly, all soils of Bulle and some soils of Wonago were belonged to Clay loam textural class. The pH of the soils ranged from 5.19 to 5.5, According to USDR, almost all of the soils of study area had pH values between lower than or equal to 5.5 and were categorized as very strongly acidic. The soils had OM ranging from 3.4 to 5.7, according to Sahlemedihn ranked 3-5% organic matter (OM) as high and more than 5% as very high [24]. According to this rating, the organic matter content of soils was generally high, which is related to environmental conditions, particularly to vegetation, climate and to the history of cultivation. In line with this, the OM content of the soils of study area was in the high range (Table 1). The high levels of organic carbon are in agreement with observations made by Asnakew et al. who reported that a large proportion (62-100%) of cultivated field soils in the acidic soil had high organic carbon [25].

The highest values of organic carbon could be because of poor drained condition of the soil and high amount of rainfall that reduces the rate of organic matter decomposition in the study sites [26]. The

available P content of the soils ranged from 14.7 to 24.9 mgkg⁻¹ for all soils ranged in acidic which indicates that the soil available P of the study area is low, as described by Birru et al. [9]. From this it needs further treatment in all the acidic soils which is in agreement with the finding of Mengel and Kirkby which stated that values of lower soluble P was indicative of soil capable of significant yield responses to application of appropriate level of P fertilizer [27].

The lower availability P of the surface soils in the southern region may partially be a result of depletion of P through crop removal and improper management of land. The soil exchangeable aluminum was ranged from 0.72 to 3.21 cmolkg⁻¹ and it was highest for the soil of Bulle. The CEC value between 13 to 32.0 cmolkg⁻¹ are high to medium and satisfactory for agriculture with the use of fertilizer and CEC>40 as high to very high and needs only small amounts of lime and potassium fertilizers, the result of experiment show the CEC value ranged in medium.

The soils were low in exchangeable bases, which due to the higher rainfall and seasonal variation normally observed in the region leading to intense leaching of bases and accumulation of exchangeable Al in these soils.

Concerning the micronutrients, the highest contents of available Fe (24.68 mgkg⁻¹) were observed mainly on the soils under Bulle while the rest were low in the study area (Table 1) and the highest (7.29 mgkg⁻¹) of available Mn was obtained in the soils of Bulle (Table 1). Sims and Johnson (1991) indicated that the critical or threshold levels of available Fe and Mn for crop production are 2.5-4.5 mgkg⁻¹ and 1-50 mgkg⁻¹, respectively. Therefore, the results observed in this study seem to be adequate for the production of most crop plants, especially in the soils of Bulle high Fe observed is the indication of high P sorbing capacity of the soil.

P sorption indices

The P sorption data was adequately plotted according to the Freundlich and Langmuir equations for all soils. The data showed a satisfactory agreement with both Freundlich (R²>0.99) and Langmuir (R²>0.99) equation (Table 2).

Sorption behavior was described by the linearzed Langmuir sorption model with regression coefficient (R²>0.93) observed for almost all soils under study (Table 2). The soils differed considerably in sorption characteristics with different location.

Almost all the soils had Sorption maxima ranging from 909 to 2000 mg Pkg⁻¹ which indicates that the soils have high sorption site, but sorption affinity constant, which is the dominant factor showing

Location	Soil texture	pH (H₂O)	CEC (cmolkg⁻¹)	Av. P (Mg/kg)	OM (%)	Exch.Al (cmolkg⁻¹)	Ca	Mg	K	Na	Fe	Zn	Cu	Mn
							Ex. Bases (meq/100 g)				Micronutrient(mgkg⁻¹)			
Wonago														
Wo1	CL	5.4	32.0	24.9	5.7	0.72	5.67	1.7	0.89	0.47	2.23	0.41	1.67	2.22
Wo2	L	5.5	25.8	23	3.8	0.38	4.64	3.1	1.35	0.32	1.59	1.62	2.81	3.51
Wo3	CL	5.3	31.0	19	3.4	0.72	4.97	2.1	2.33	0.45	2.20	1.09	0.98	1.92
Bulle														
Be1	CL	5.39	13	15	5.2	3.12	2.8	1.9	0.26	0.03	24.68	1.99	4.01	3.91
Be2	CL	5.28	14.3	14.7	4.2	3.09	2.6	1.7	0.33	0.05	14.81	1.32	3.29	4.82
Be3	CL	5.19	13.8	15.9	4.8	3.21	3.6	2.1	0.18	ND	22.47	1.21	2.99	7.29

CL: Clay Loam, L: loam, SCL: Sandy Clay Loam, SL: Sandy Loam.
Av.p: Available Phosphorus (Bray-ɪɪ); CEC: Cation Exchange Capacity, Ex.Al: Exchangeable Almunium, OM: Organic Matter, ND: Not Detected

Table 1: Selected physico-chemical properties of soil from two agriculturally important sites with three fields each at Wonago (Wo) and Bulle (Be) in southern Ethiopia.

Location	P sorption indices[LE]				P sorption indices [FI]			
	X_m	K	$SPR_{(0.2)}$	R^2	b	Kf	$SPR_{(0.2)}$	R^2
	$Mg.kg^{-1}$	$L.mg.kg^{-1}$	$Mg.kg^{-1}$		$L.kg^{-1}$	$Mg.kg^{-1}$	$Mg.kg^{-1}$	
Wonago								
Wo1	2000	0.15	57	0.99	0.81	245	71.8	0.99
Wo2	1428	0.26	70	0.95	0.66	265	94	0.96
Wo3	1111	0.45	92	0.92	0.62	322	119	0.99
Bulle								
Be1	1000	1.1	182	0.92	0.54	479	200	0.99
Be2	909	1.4	196	0.92	0.52	487	211	0.99
Be3	909	1.4	196	0.92	0.52	484	208	0.99

LE: Langmuir Equation, FI: Freundlich Equation, Xm: Langmuir Sorption Maximum, K: Bonding Energy, SPR: Standard P Requirement, Kf: Freundlich Surface Coverage.

Table 2: Slope, intercept and coefficient of determination (R^2) of the isotherm models of selected soils.

the bonding energy of the soil to retain P, ranged from 0.15 to 1.4 (l mg^{-1}) and was smaller for all soils of Wonago compared to the soils of Bulle (Table 2).

The amount of P required for maintaining a soil solution concentration of 0.2 mg P kg^{-1}($P_{0.2}$) ranged from 57 to 196 mg P kg^{-1} soil (Table 2).

Soils within a site were classified under the same group, but they differed substantially in the extent to which they fixed P, which shows soils with the same group will not necessarily fix P to the same extent or have similar P fertilizer requirements. This is because soils in nature have variable characteristics, which make them, very complex. This finding is in agreement with B¨uhmann et al. who observed similar sorption trends for soils of South Africa and suggested that for optimum P recommendations soil P fertilization assessments done at field scale level [28].

In addition, the amounts of added P required maintaining a concentration of 0.2 mg P L^{-1} ($P_{0.2}$) in solution, which is Standard P requirement in this study (Table 2), were generally higher than the range reported in other studies.

Sarafaz et al. for example reported values ranging from 50 to 201 mg P kg^{-1} for surface samples from non-cultivated and non-fertilized areas in Ethiopia in which only one sample having a SPR of 123 mg P kg^{-1} fell within this range while others had very low SPR values indicating possible early P saturation of these soils following repeated applications of P fertilizers [29]. This could lead to elevated P levels in the soil solution, which in time could contribute to the eutrophication of freshwater bodies. However, the present study ranges strongly highly acidic to (Table 1) and had high range of SPR, which indicate that acidic soil, have high need of SPR.

The Freundlich parameter, i.e., sorption capacity (logkf) and P sorption energy (b); and correlation coefficient values computed from the data plotted according to logX against logC in equilibration solution were explained. The goodness of fit of the model was ascertained by looking at the R^2 values. All the plots were highly correlated with R^2 values ≥ 0.96 indicating apparent high conformity of the adsorption data to the Freundlich model.

Generally, the Freundlich model seemed fit at all equilibrium concentrations (Table 2). For all the soils of the study area the values of sorption capacity and P sorption energy ranged from 245 to 487 mgkg^{-1} and 0.52 to 0.81 Lkg^{-1}, respectively (Table 2). Since the Freundlich adsorption equation was derived empirically, its parameters (logkf) and (b) have been considered. Despite this, it was proposed that logkf could be considered as capacity factor Sarafaz et al. implying having a

larger logkf value has a larger adsorbing capacity than the one having a smaller logkf value [29]. For practical purpose, the logkf may be used to differentiate soils having different P adsorption capacities. Thus, in this study, all the soils of Bulle had the higher sorption capacity relative to the others. The Freundlich parameter logkf was found to be practically useful parameter in summarizing the adsorption properties of soil over a wide range of equilibrium concentrations. The sorption energy (logkf, Freundlich model) is the value which an indication of the adsorption capacity of the adsorbent and the slope (b) show the effect of concentration on the adsorption capacity and represents adsorption intensity. Therefore, sorption capacity in Freundlich is determines whether the soil is high sorber or not relative to each other and was found to be strong for the high P-sorbing soils.

Figure 1a and 1b and c show the P sorption curves that were obtained by plotting the quantity of P sorbed on the surfaces of against the solution P concentration. According to Sanchez and Goro, soils that adsorb less than 150 mg Pkg^{-1} soil to meet the SPR value of 0.2 mgl^{-1} in soil solution are considered to be low sorbing soil and those adsorbing greater than this value are high P sorbing ones [30]. Accordingly, all soils of Bulle were high P fixers while the soils of Wonago were low P fixers (Table 2). Since the six composite samples used in the study were from two different locations in the Gedeo Zone, there is need for a broader study involving soils from other agro-ecologies in the region in order to confirm the proportions.

Nevertheless, the results suggest that P availability could be compromised in the soils of Bulle with high P-sorption capacity and that measures to mitigate the adverse effects of P-sorption may be necessary to ensure that P is not a limiting factor to crop production where such soils are found.

Figure 1a and 1b. Phosphorus sorption isotherm curves for the high P sorbing soils; vertical arrows indicate SPR of the soils (on the y-axis) at a standard solution P concentration 0.2 mg Pl^{-1}.

The rest soils were grouped as low P- sorbing soils (Figure 1a and 1b). The curve followed a smooth plateau pattern. It is evident from the curves that the rate of P sorption increased with an increase of P concentration, but at a certain point of higher concentration, the level of P became almost constant having no more sorption capacity, it was evident from the observations that during sample collection where the soil color of low sorbing soils did not have the same appearance as that of other soils grouped under high P-sorbing ones. The relative amount of P sorbed was dramatically higher at a low concentration than at a higher concentration as also reported by Tsado [31]. This suggests that the reaction between phosphate and the soil was rapid on initial contact; this perhaps could have been due to a low available P content

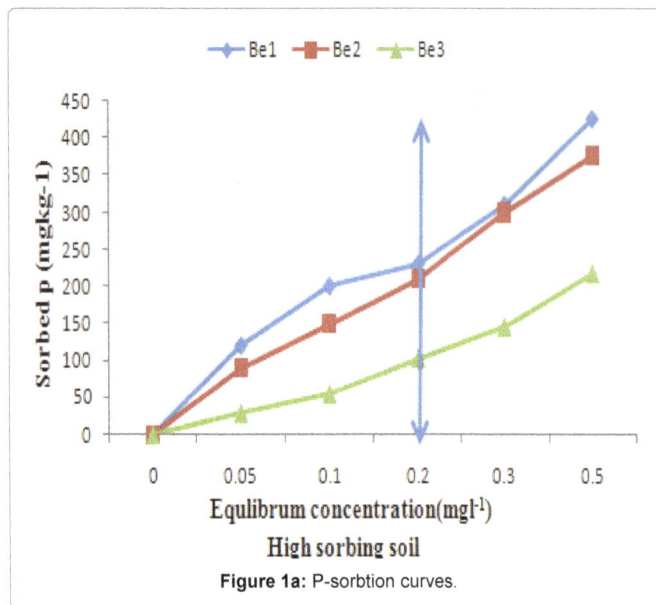

Figure 1a: P-sorbtion curves.

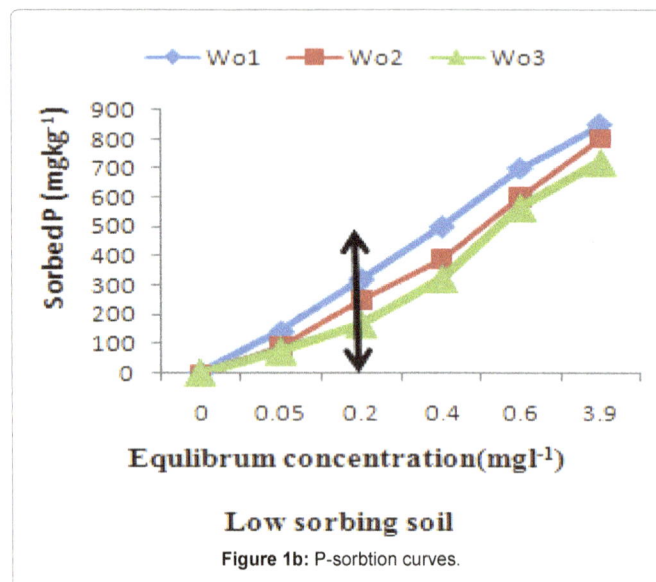

Figure 1b: P-sorbtion curves.

(Table 2) resulting in high adsorption potential at the surface. Similar observations have been reported by Bala [32].

Correlation between P-sorption indices and soil properties

The sorption maximum in the soils of study area was found to correlate negatively with SOM and pH ($p \leq 0.05$). Soil OM may also be able to complex Al, stabilizing the compound and possibly enhancing its ability to sorb P which is in agreement with Darke and Walbridge [33]. Organic matter can also inhibit recrystalization of both Fe and Al oxides, thus indirectly assisting P sorption. In these soils, SOM was positively correlated with both Fe and Al. The effect of SOM is often attributed to complexation with Al or both Al and Fe [33]. Organic matter has the potential to impede or enhance P sorption. If of appropriate size it may occlude sorption sites, either by filling them or by physically blocking them.

Soil texture was also significantly correlated with P sorption. Fine textured soils, such as clays, have higher surface area and thus greater

reactivity. They also tend to have higher concentrations of Fe and Al-oxides than coarser soils. Studies in the southeastern US on acidic soil condition which is similar to the present study have found that fine-texture soils were able to sorb more P than coarser substrates [33]. Soils with higher clay and silt contents had higher Xm.

Soil pH was negatively correlated with Xm, indicating that P sorption increases with decreasing pH. In acidic soils (pH<5.5), more sites on the soil matrix are protonated and Al is liberated, thus enhancing P sorption. According to Birru, pH is a significant predictor of P sorption in soils. As pH is a master variable, controlling many biological and chemical processes, it can also enhance P sorption at high values, the negative correlation with pH was probably observable in the soils of the study area, because the pH ranged from 5.19 to 5.5 (Table 1) [34].

Sorption isotherms indices showed that Xm was negatively correlated with sorption energy of Langmuir (k) (Table 2) which shows that the soils may have high sorption sites. But it may have low sorption energy to hold P on the surface since it is determinant factor showing the soil have high sorption capacity of soil (Table 2). Sorption energy was related with the standard P requirement of soil which is positively correlated ($p \leq 0.01$). Xm is negatively correlated with $SPR_{0.2}$, meaning that the soil having high Xm may have low SPR (Table 2).

Freundlich sorption parameter logkf and b were negatively correlated with SPR of soil which shows that this sorption isotherm didn't determine $SPR_{0.2}$ of the soil, rather show the fitness of the model. It shows the advantage of Langmuir equation over the Freundlich equation in the determination of $SPR_{(0.2)}$. Generally, the Freundlich model seemed fit at all equilibrium concentrations (Table 2) for all the soils. Since the Freundlich equation was derived empirically, its parameters have been considered physically, but for present study it best explains the high and low sorbing soil. Besides, to this, it was proposed that logkf could be considered as capacity factor which is related to sorption capacity of the soil. This is in agreement with the finding Taylor et al. conducted in both acidic and calcareous soil [6].

Path analysis of P sorption

Concerning the correlation and causation between the isotherm parameter (Xm and logkf) and the soil variable (pH, OM, CEC, Exch. Al and Clay), analysis was made and adequate result was found. Additionally an uncorrelated residue (U) was calculated for both models using the equation ($U = \sqrt{1 - R^2}$). Soil pH, Exch.Al and OM were positively correlated ($p \leq 0.01$) with the Xm (Table 3). However, not with the specific isotherm parameters, there are reports that indicate sorption of P increased with decreasing soil pH (acidic soil reaction) and with increasing content of Exchangeable Al [35].

Result of path coefficient analysis, showed that Xm was significantly ($p \leq 0.01$) influenced by the direct effect of soil pH with negligible indirect effect from other soil variable (Table 3). At low pH, the adsorbed P is said to be held tightly because at this pH situation, the physically adsorbed p will slowly convert to the mineral apatite by precipitating and crystallizing into iron and Aluminum P forms [36].

The positive and significant direct effect ($p \leq 0.01$) on the P sorption maximum (Xm) occurred from exchangeable Al (Table 2). The direct effect of Exch.Al and clay on the sorption maximum (0.69, 0.58; $p \leq 0.01$) is an indication of the specific role of Al and clay plays in the processes of P sorption.

Apparently, as concentration of exchangeable Al in the soil

Variables	pH	OM	CEC	Exch.Al	Clay	r	R²	U
pH	0.548	-0.086	0.012	-0.43	0.42	0.937***	0.624**	0.61
OM	-0.291	0.163	-0.006	0.007	-0.87	0.990***		
CEC	-0.143	0.023	-0.045	-0.46	-0.19	0.812**		
Exch.Al	-0.147	-0.004	-0.022	0.69**	-0.33	0.967***		
Clay	0.142	-0.087	0.005	-0.33	0.58**	0.775**		

***, ** values are significant at $p \leq 0.01$ and $p \leq 0.05$; Exch.Al: Exchangeable Aluminum; OM: Organic Matter, CEC: Cation Exchange Capacity; U: Residual of Path Coefficient ($U = \sqrt{1 - R^2}$).

Table 3: Path coefficient analysis for direct effect of soil properties (diagonal and underliend) and indirect effect of other soil variable (off diagonal) on the p sorption maximum (Xm) of langmuhir isotherm.

Variable	pH	OM	Exch.Al	CEC	Clay	r	R²	U
pH	0.14**	-0.064	0.039	0.006	-0.19	-0.61**	0.898**	0.101
OM	-0.073	0.12**	0.003	-0.003	0.39	0.95**		
Exch.Al	-0.037	-0.003	-0.15**	-0.01	0.15	-0.34		
CEC	-0.036	0.016	-0.07	-0.02	0.085	0.65**		
Clay	0.035	-0.064	0.03	0.003	-0.72**	-0.82**		

**Values are significant at $p \leq 0.01$

Table 4: Path coefficient analysis for direct effect of soil properties (diagonal and underliend) and indirect effect of other soil variable (off diagonal) on the p sorption energy (logkf) of Freundlich isotherm.

increases, the soil's surface areas for anion sorption will definitely be larger. As far as the correlation between Xm and both pH and organic matter are concerned, the larger portions were due to the direct effect of exchangeable Al than to the indirect effects (Table 2). However, the modes of action of exchangeable aluminum through pH and SOM might be different. The SOM has strong coating effect on the sorbing surface of the soils and minerals such as Al, Fe and the various silicate clay minerals, and in turn these minerals have immobilizing effect on the functional groups of organic matter as reported by Loganathan et al. [37].

Depending on the output of the path diagram, it is possible to calculate the total effect of soil parameter on Xm. The pH of the soil has direct effect on the Xm (= 0.548, Table 2) and indirect effect on the Xm with organic matter (Table 2). Therefore to calculate the total effect (direct and indirect) of pH on Xm, indirect effect first should be calculated by multiplying the direct effect of pH by indirect effect of OM and then add to the direct effect.

(0.548* - 0.291= -0.159, then -0.159 + 0.548 = 0.389), this value indicate the total effect of pH on Xm, likewise the total effect of soil parameter on Xm could be 0.166, -0.044, 0.462 and 0.385 for OM, CEC, Exch.Al and Clay respectively. The total effect of Exch. Al was higher followed by that of pH and clay content, which was highly related with the soil of Bulle having high Exch. Al than Wonago my finding was in agreement with Henry and Smith who also observed that Al had greater influence on P retention than other [38].

As to that of Frendliuch, energy of P sorption (logkf) was significantly (p ≤ 0.01) influenced by the direct effect of soil pH with negligible indirect effect from the other variable (Table 4). At low pH, the adsorbed p is held tightly because at this pH, the physically adsorbed P will slowly convert to the mineral apatite by precipitating, and crystallizing into Fe and Al forms.

Of the total estimated correlation coefficient between the organic matter and logkf (r=0.95; p ≤ 0.01), the larger portion (0.016) was due to the indirect effect of CEC. Although it was not specifically indicated for which isotherm parameter, Hernandez and Burnham observed the cause of changes in P sorption due to the SOM content as to be related to the changes in clay, the correlation coefficient between OM and logkf

was positive [39]. However, this was also the indirect effect of clay with negligible direct effect from OM. This result might be partly explained by the fact that the rate of decomposition of SOM increases with rising soil pH from acidic to neutral and also clay type were dominant. In this situation, the organic anions released from the decomposing organic matter have the capacity to replace P ions from exchangeable sites. In addition, organic matter has a coating effect upon Fe, Al and other soil particle surface that can form stable complex and thus prevent their subsequent reaction with the P ions.

Conclusion and Recommendation

The results revealed that depending on the SPR value all the soils of Bulle were classified as high P sorbing while the soils of Wonago was low P sorbing. In the study Langmuir model well described P-sorption data, but Xm value is not a good indicator of P-fixing capacity of soil rather the K (bonding energy) that plays important role in determining SPR value and the P-fixation capacity of the soils in the study area. This further implies that it is not necessarily true that soils with high Xm values will have high external P requirements.

It is concluded that in general both Langmuir and Freunlich models found to be effective in describing the P-sorption data. Based on these models the test soils have widely varied in their P-fixation capacity and SPR values. Variations in their Al, Fe, OM, Were found to be the most important factors that accounted significantly for variation in P-sorption by different soils.

Based on the result it is recommended that, there is a need to investigate measures to mitigate against P sorption in the soils identified as high P sorbing soils to ensure that P availability is not compromised in these soils and also Model validation based on greenhouse and field experiments are needed before the results of the current sorption studies are applied on wide scale.

The contrasting differences in the P fixing capacities of the soils suggested that the use of blanket phosphate fertilizer recommendations may not be a good strategy for the study area as it may lead to under-application or over-application of P.

References

1. Palomo L, Claassen N, Jones DL (2006) Differential mobilization of P in

the maize rhizosphere by citric acid and potassium citrate. Soil Biology and Biochemistry 38: 683-692.

2. Fairhurst TR, Lefroy R, Mutert E, Batijes N (1999) The importance, distribution and causes of phosphorus deficiency as a constraint to crop production in the tropics. Agroforestry Forum 9: 2-8.

3. Ravikovitch S (1986) Anion exchange: I. sorption of phosphoric acid ions by soil, In: Robert D. Harter (ed.) Sorption phenomena, Soil Science Series, Van Nostrand Raeinhold Co, New York, pp: 147-166.

4. Shen HH, Shi XC, Wang C, Cao Z (2001) Study on adaptation mechanisms of different crops to low Phosphorus stress. Plant Nutr Fert Sci 7: 172-177.

5. Abayneh E (2001) Application of Geographic Information System (GIS) for soil resource study in Ethiopia. Proceedings of the National Sensitization Workshop on Agro metrology and GIS, Addis, Ababa, Ethiopia, pp: 17-18.

6. Taylor RW, Bleam WF, Tu SI (1996) On the Langmuir phosphate adsorption maximum. Commun Soil Sci Plant Anal 27: 2713-2722.

7. Sharpley AN (1995) Dependence of runoff phosphorus on extractable soil phosphorus. J Environ Qual 24: 920-926.

8. Freundlich PN (1988) Kinetic control of dissolved phosphate in natural rivers and estuaries: A primer on the phosphate buffer mechanism. Limnol Oceanogr 33: 649-668.

9. Birru Yitaferu, Heluf Gebrekidan, Gupta VP (2003) Sorption characteristics of soils of the north-western highlands of Ethiopia. Ethiopian Journal of Natural Resources 5: 1-16.

10. Van Reeuwijk LP, (1992) Procedures for soil analysis (3rd Ed) International Soil Reference and Information Center (ISRIC), Wageningen, the Netherlands, pp: 34.

11. Walkley A, Black IA (1934) An examination of the Degtjareff method for determining soil organic matter and a proposed modification of the chromic acid titration method. Soil Sci 37: 29-38.

12. Landon JR (1991) Booker tropical soil manual: A Handbook for Soil Survey and Agricultural Land Evaluation in the Tropics and Subtropics. Longman Scientific and Technical, Essex, New York. pp: 474.

13. Tekalign Mamo, Haque I (1991) Phosphorus status of some Ethiopian soils, Evaluation of some soil test methods for available phosphorus. Tropial Agriculture 68: 51-56.

14. Lindsay WL, Norvell WA (1978) Development of a DTPA soil test for zinc, iron, manganese and copper. Soil Science Society of American Journal 42: 421-428.

15. Graetz DA, Nair VD (2008) Phosphorus sorption isotherm determination, In: Kovar JL, Pierzynski GM (ed.) Methods of P analysis for soils, sediments, residuals, and waters -So. Coop. Ser. Bull. No. 396, pp: 35-38.

16. Fuhrman JK, Zhang H, Schroder JL, Davis RL, Payton ME (2004) Water soluble phosphorus as affected by soil to extractant ratios, extraction time and electrolyte. Commun. Soil Sci Plant Anal.

17. Fernandes MLV, Coutinho J (1994) Phosphorus sorption and its relationship with soil properties, Trans. 13th World Congress Soil Science. Acapulco, Mexico, 3b: 103-104.

18. Fang F, Brezonik PL, Mulla DJ, Hatch LK (2002) Estimating runoff phosphorus losses from calcareous soils in the Minnesota River basin. J Environ Qual 31: 1918-1929.

19. Kleinman PJA, Sharpley AN (2002) Estimating soil phosphorus sorption saturation data from Mehlich-3 data. Commun Soil Sci Plant Anal 33: 1825-1839.

20. Xu D, Xu J, Wu J, Muhammad A (2006) Studies on the phosphorus sorption capacity of substrates used in constructed wetland systems. Chemosphere 63: 344-352.

21. Fox RL (1981) External phosphorus requirements of crops; in: Chemistry in the Soil Environment. America Society of Agronomy, Madison, Wisconsin 40: 223-240.

22. Bolland K, Otabbong E, Barberis E (2001) Phosphorus sorption in relation to soil properties in some cultivated Swedish soils. Nutrient. Cycling Agroecosyst 59: 39-46.

23. Basta NT, Pantone DJ, Tabatabi MA (1993) Path analysis of heavy metal sorption by soil . Agronomy Journal 85: 1054-1057.

24. Sahlemedhin Sertsu (1999) Draft guideline for regional soil testing laboratories. NFIA, Addis, Ababa, Ethiopia.

25. Asnakew Wldeab, Tekalign Mamo, Mengesha Bekele, Tefera Ajema (1991) Soil fertility management studies on wheat in Ethiopia. Ethiopia, pp: 137-141.

26. Mandiringana OT, Mnkeni PNS, Mkile Z, Van averbeke, Ranst WEV et al. (2005) Mineralogy and Fertility status of selected soils of the Eastern Cape Province, South Africa. Communications in Soil Science and Plant Analysis 36: 2431-2446.

27. Mengel K, Kirkby EA (1996) Principles of Plant Putrition. Panimo publishing yields, in a Mediterranean climate. Agronomy Journal 86: 221-226.

28. Buhmann C, Beukes DJ, Turner DP (2006) Plant nutrient status of soils of the Lusikisiki area, Eastern Cape Province. South African Journal of Plant and Soil 23: 93-98.

29. Sarafaz M, Abid M, Mehdi SM (2009) External and internal phosphorus requirement of wheat in Rasulpur soil series of Pakistan. Soil and Environ 28: 38-44.

30. Sanchez P, Goro U (1980) In: The role of phosphorus in agriculture. Symposium proceeding.ASA, CSSA, SSSA, Madison, Wisconsin, USA, pp: 471-514.

31. Tsado PA, Osunde OA, Igwe CA, Adeboye MKA, Lawal BA (2012) Phosphorus sorption characteristics of some selected soil of the Nigerian Guinea Savanna. University of Nigeria,Nsukka, Enugu State .International Journal of AgriScience Vol. 2: 613-618.

32. Bala A (1992) The effect of phosphate on zinc and copper sorption by a tropical soil. M.Sc. Dessertation. Department of Soil Science, UK. pp: 33.

33. Darke AK, Walbridge MR (2000) Al and Fe biogeochemistry in a floodplain forest: implications for P retention. Biogeochemsitry 51: 1-32.

34. Birru Yitaferu (2000) Phosphorus status and sorption characteristics of soils of the north-western highlands of Ethiopia. Msc thesis, School of Graduate Studies, Alemaya University, Alemaya, Ethiopia. pp: 106.

35. Tekalign Mamo, Haque I (1987) Phosphorous status of some Ethiopia soils. Sorption characteristics, Plant and Soil 102: 261-266.

36. Archer J (1988) Phosphorus In: John Archer(ed.) Crop nutrition and fertilizer use, (2ndedn) Farming Press. Ltd, USA. pp: 57-64.

37. Loganathan P, Isirimah NO, Norachuku DA (1987) Phosphorus sorption by Ultisols and Inceptisols of the Niger delta in Southern Nigeria. Soil Science 144: 330-338.

38. Henry PC, Smith MF (2002) Phosphorus sorption study of selected South African soils. South Africa Journal of Plant and Soil 19: 61-69.

39. Hernandez LD, Burnham CP (1974) The effect of pH on sorption in soils. Journal of Soil Science 25: 207-216.

32

Long-Term Tillage and Crop Rotation Impacts on a Northern Great Plainsmollisol

Ibrahim MA, Alhameid AH, Kumar S*, Chintala R, Sexton P, Malo DD and Schumacher TE

Department of Plant Science, South Dakota State University, Brookings, South Dakota (SD), USA

Abstract

Soil properties can be altered by tillage and rotation, however, these effects cannot be detected in short-term studies. This study was conducted to assess the long-term (14 years) tillage and rotation impacts on selected soil surface properties. A long-term experimental site comprised of two tillage systems [no tillage (NT) and conventional tillage (CT)], and three crop rotations [corn (*Zeamays*)-soybean (*Glycinemax*), corn-soybean-wheat (*Triticumaestivum*), and corn-soybean-wheat-alfalfa (*Medicagosativa*)] were used for the present analysis. Surface (0-15 cm) soil samples were collected every year from 1991 through 2004 and analyzed for soil organic matter (SOM), available P, available K, and nitrate (NO_3^--N). Results indicated that SOM concentration (averaged across all years) under NT (37 g kg^{-1}) was significantly higher compared to that of CT (36 g kg^{-1}). However, overall crop rotation did not impact SOM. Soil P concentration under NT (208 mg kg^{-1}) was significantly higher (8.3%) than that of CT (191 mg kg^{-1}). Available P concentration was the highest in the 2-year-rotation, intermediate in the 3-year-rotation, and lowest in the 4-year-rotation. Tillage system did not significantly impact NO_3^- concentration; nonetheless, its concentration was the highest in the 4-year-rotation followed by the 3-year-rotation, and the lowest was in the 2-year-rotation. The available K concentration under NT was not significantly different from that under CT; however, its concentration under the 2-year-rotation (340 mg kg^{-1}) was significantly lower than those under the other two rotations. This 14-year tillage and rotation study had minimal impact on surface soil properties at this location.

Keywords: No-tillage; Conventional tillage; Soil nutrients; Soil organic matter

Introduction

Tillage and crop rotation systems impact SOM and soil nutrients; however, the impact of these systems cannot be observed in the short term experiments but rather in the long term experiments. Tillage and crop rotation impacts on soils depend on multiple factors such as soil type, cropping system, residue management, and environmental conditions [1,2]. No-tillage (NT) management can increase soil organic carbon (SOC) compared with conventional tillage [3]. Applying different crop rotations is important in recycling nutrients because plants differ in their ability of taking up nutrients from different depths, and releasing these nutrients after death and decay [4]. Increased SOM in a tillage system can be achieved via adding more organic materials to the soil, decreasing the rate of SOM decomposition, or a combination of both [5] Pulverizing soil surface via conventional tillage (CT) mixes the organic materials for the depth of plowing (25 cm depth), and increases the aeration and microbial activity, which in turn may lead to decreased SOM in the upper 15 cm of the soil [6].

During the first few years in long-term experiments, NT and CT may have the same content of SOM; nonetheless, CT may have more SOM compared with NT [7]. However, after several years of using NT system, SOM concentrations may significantly increase for the surface depth (0-15 cm) compared with that under CT [8,9]. observed a significant increase in SOM in the upper layer of soil under NT compared with that under CT after 20 years of continuous corn. In another long-term (30 years) study, Godsey et al. [10] found a significant increase in SOM concentrations in the surface layer of soil under NT compared with that under CT.

Similarly, tillage and crop rotation may impact soil nutrients contents and their availability to crops. These nutrients can be added to soils in the form of fertilizers, manure, leguminous crops, or crop residues. Soil nutrients usually include N, P, and K. Nitrogen is one of the most important nutrients for the crop growth. However, when its availability exceeds plant requirements, the excess reactive N moves freely to the ground water (in the form of NO_3^-), or volatiles to the atmosphere (in the form of N_2O), and negatively impacts the environment. For example, N is of a major environmental concern because of NO_3^-, which could be leached to the ground water and decrease its quality [11]. The CT system enhances SOM Therefore, innovative and holistic management practices are needed for minimizing the off-site movement of applied N. These practices include: (i) multi-purpose cover crops facilitating biological N fixation, (ii) organic farming improving soil biology diversity and soil health, and (iii) use of conservation tillage and diverse cropping systems. These alternative practices help in minimizing N and other nutrient losses from agro ecosystems, and greatly benefit the environment. Therefore, introducing these conservation practices to existing cropping systems is helpful in reducing N losses. The leached NO_3^- in the tile drainage was found to be smaller under NT compared with that under CT in a corn-soybean rotation [12].

Soil management systems greatly affect the distribution of K vertically and horizontally in the soil profile [13,14]. Previous studies indicated that K content in the surface layer of 3-5 cm was often higher than other depths under NT [13]. In contrast, under CT system, the distribution of K in the soil profile is relatively uniform [15]. The source of P in soils could be inorganic (e.g., fertilizers or parent materials) or organic (e.g., released P from decomposed plant residue)

*Corresponding author: Kumar S, Department of Plant Science, South Dakota State University, Brookings, South Dakota (SD), USA
E-mail: sandeep.kumar@sdstate.edu

[16]. Immobile nutrients such as P could be concentrated at the soil surface as a result of applying NT [17]. Mineralization of SOM may significantly affect the concentration of some nutrients like P and N, but does not affect K concentration Calonego et al. [18] and Karlen et al. [19] found that there was no difference between K concentration under NT and CT system. At the soil surface, nutrients concentrations are higher but they may decrease with depth under NT; however, CT may result in a homogenous distribution of these nutrients with depth [7,13] in their work in Spain found that P and K concentrations at the soil surface (0-15 cm) were higher under NT compared with CT, which could be due to the high SOM concentration under NT as well as maintaining the applied P fertilizers. Further, crop rotation may also impact the distribution of nutrients such as P and K in the soil profile [13] The present study was conducted with the specific objective to assess the long-term (14 years) tillage and crop rotation impacts on soil organic matter and soil nutrients.

Materials and Methods

Experiment location and design

The experimental site is located at the Southeast Research Farm of the South Dakota State University South Dakota located at Clay County (43° 02' 58" N, 96° 53' 30" W), South Dakota (Figure 1). The experiment was initiated in 1990 to assess the impact of different tillage systems and crop rotations on the long term production and economics of cropping systems. The first growing season started in the spring of 1991. The experiment was conducted on Egan soil series (Fine-silty, mixed, superactive, mesic Udic Haplustolls) [20]. The study soil series occupies 3,931,019 acres (or 99.6% of the total area occupied by this soil series in USA), and it is nearly level with a slope of 0-1%. The daily average temperature ranges from -14.1°C in January to 31.8°C in July, and the mean annual precipitation is 627.4 mm [21].

The experimental site has 80 plots distributed randomly in a complete block design. Each plot has a width of 20 m and a length of 100 m. The experimental plots were designed to be large so that field operations could be carried out using commercial sized farm equipment. The experiment had three different tillage systems which were no till (NT), conventional till (CT), and ridge till (RT). Ridge till system had only a two year crop rotation of corn (*Zea mays* L.) – soybean (*Glycine max*. L.). In the fall of every year after harvest, residues of corn, soybean, and wheat were disked and chiseled in all of the conventionally tilled plots. The RT plots were excluded from this study because it had only one rotation system. Both NT and CT had three rotation systems, which were a two year rotation of corn-soybean, a three year rotation of corn-soybean-wheat (*Triticum aestivum* L.), and a four year rotation of corn-soybean-wheat-alfalfa (*Medicago sativa* L.) (Figure 1).

Soil sampling and analyses

It is of importance to mention that the data we used here were collected from 1991 to 2004, but they have not been published in scientific journals. Also, not all of the soil properties were analyzed every year; that is why we selected the years that have data of the same soil properties. Soil samples were collected every fall after harvesting the crops from 1991 to 2004 from each plot. Three cores of soil samples from each plot were collected at a depth of 0-15 cm using a 3.5-cmdiameter and 50-cm-tall hand probe(Inc. JMC Soil Samplers) and mixed together to make a composite sample. Composited soil samples were labeled, sealed in plastic zip-lock bags, and transported to the laboratory. Every year, after bringing the soil samples to the

laboratory, all of them were air dried, ground, and sieved to pass a 2-mm sieve. All of the analyses were carried out using the soil fine fraction (<2 mm in diameter).

Soil organic matter (SOM) was measured using the loss on ignition (LOI) method [22]. Briefly, 10 g of each soil sample was weighed in an aluminum crucible, transferred to a muffle at a temperature of 450-500°C for 4 h, and then the loss of weight was determined. P was extracted using a 0.5 M $NaHCO_3$ solution and then the extraction was measured colorimetrically [23]. Nitrate was determined using a nitrate-specific ion electrode [24]. Available K was extracted by 1 M NH_4OAc at pH 7.0, and it was determined using an atomic absorption (AA) [25].

Statistical analysis

Soil organic matter and soil nutrients that measured in all of the tillage and crop rotation systems of each year were analyzed separately and together. An analysis of variance (ANOVA) was conducted using the SAS software (version 9.3, SAS Institute, Cary, NC, USA). An estimate for the least significant difference (Duncan's LSD) between two tillage systems and three crop rotations were obtained using the 'GLIMMIX procedure' in SAS. Statistical differences were declared significant at the α = 0.05 level.

Results and Discussion

Tillage and crop rotation impacts on soil organic matter

The NT system increased SOM (37 g kg⁻¹) for the 0-15 cm depth after 14 years compared to CT system (36 g kg⁻¹) (Figure 2A). This was attributed to the mixing of SOM content within the plowing depth (25 cm) and the increased aeration under CT treatment leading to SOM decomposition [26,27]. Compared with SOM concentration under NT in 1991, its concentration significantly increased across the years of the experiment under all of the crop rotation systems. For example, under the 4-year-rotation, the highest and lowest SOM concentrations of 39.3 and 32.6 g kg⁻¹ were observed in 2004 and 1991, respectively (Table 1). Furthermore, SOM concentrations under CT followed atrend similar to those under NT (Table 1). In 2002, however, SOM concentration under NT and CT drastically declined, which could be attributed to the higher temperature in that year leading to more oxidation of SOM. Concentrations of SOM under CT were lower than those under NT in almost all of the years of the experiment, which could be due to the mixing action of the surface layer to the depth of the plowing (25 cm) and the increased aeration and oxidation under CT [28] (Table 1). However, the differences between SOM concentrations under NT vs

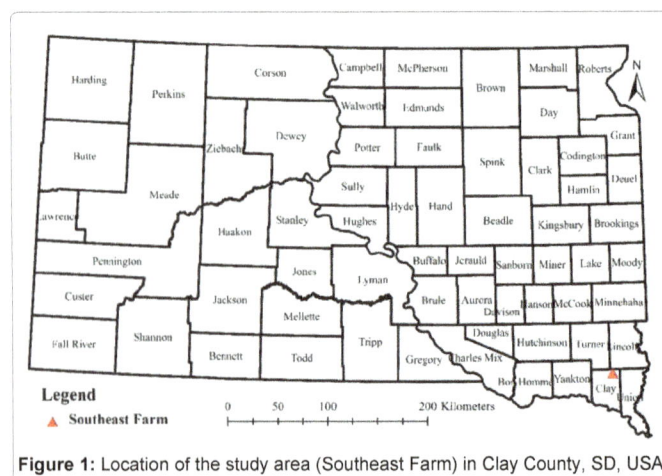

Figure 1: Location of the study area (Southeast Farm) in Clay County, SD, USA.

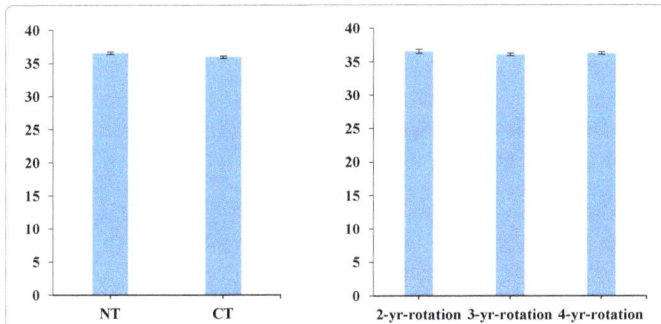

Figure 2: Soil organic matter (SOM) as influenced by long-term tillage (A) and rotation systems (B) for the 0-15 cm depth. Error bars represent standard errors.

Treatments	SOM (g kg⁻¹)					
	1991	1994	1995	1997	2002	2004
4-year-rotation (corn-soybean-wheat-alfalfa)						
NT	32.6dA†	37.1bA	38.5aA	38.6aA	34.6cA	39.2aA
CT	33.5dA	35.1cB	38.6aA	36.8bB	33.3dA	37.6abB
3-year-rotaion (corn-soybean-wheat)						
NT	34.3dA	36.0cA	39.3aA	37.6bA	33.9dA	37.9bA
CT	32.2cB	35.7bA	38.2aA	37.9aA	32.4cB	37.6aA
2-year-rotation (corn-soybean)						
NT	32.0eB	36.3cA	39.1aA	37.1bcA	33.9dA	38.9abA
CT	34.0cA	37.1bA	38.9aA	37.6abA	33.9cA	38.8aA

†Means with different lower case letters within a row are significantly different at P<0.05. Higher case letters show the differences between NT and CT for each year and under each crop rotation separately.

Table 1: Soil organic matter (SOM) concentration as impacted by no-tillage and conversional tillage systems under three different rotation systems for the 0-15 cm depths.

CT for individual years were mostly not significant especially under the 2 and 3-year-rotations (Table 1). In the 4-year-rotation; however, SOM concentration under NT was significantly higher by 5.4, 4.7, and 4.1% than that under CT in 1994, 1997, and 2004, respectively (Table 1).

Generally, SOM concentrations under the three different crop rotations (corn-soybean, corn-soybean-wheat, and corn-soybean-wheat-alfalfa) were not significantly different among each other (Figure 2B), which agreed with what was observed by Karlen et al. [29]. Moreover, data in Table 2 indicated that there was no impact of the crop rotations under NT system on SOM concentration in almost all of the years [30]. In contrast, crop rotations under CT impacted SOM concentration in some of the years. For example, SOM concentrations under the 2-year-rotation were significantly higher than that under the 3 and 4-year-rotation systems for 1994, 2002, and 2004 (Table 2). The majority of SOM under the 2-year-rotation was sourced from corn. Organic materials sourced from corn have wider C/N ratio and decompose slower compared with those sourced from legumes [30]. That may interpret why the 2-year-rotation had some higher values of SOM under CT compared with those under the other two rotations under CT. Overall, NT system enhanced the accumulation of SOM compared with CT. However, crop rotation systems did not significantly impact SOM concentration.

Tillage and crop rotations impacts on available phosphorus

Averaged across all the years, the available P concentrations under NT vs CT systems were significantly different. For example, available P concentration of 208 mg kg⁻¹ under NT was significantly higher (8.4%)

compared with that of 191 mg kg⁻¹ under CT (Figure 3A). Similar findings were also found by Wyngaard et al. [31]. In their work on tillage effects on soil properties of a Mollisol in Argentina. On a yearly basis, available P concentration was significantly higher under NT vs CT under all of the rotation systems (Table 3). For example, in the 4-year-rotation, P concentrations were higher under NT and these were 40, 30, 22, 32.6, and 21.5% higher than those under CT for 1994, 1995, 1997, 2002, and 2004, respectively. This high concentration of available P under NT vs CT could be due to the application of P fertilizer to crops every year which may cause an accumulation of P at the soil surface as a consequence of the low mobility of P and due to the mixing of the plowing depth, which distributed the P content within a 25 cm depth. In contrast, in the yearly basis in the 2 and 3-year-rotations, NT and CT exhibited no significant impact on available P concentration [32] (Table 3).

Available P concentrations differed significantly among all of the three crop rotations. For example, its concentration under the 2-year-rotation of 236 mg kg⁻¹ was significantly higher than that under the 3-year-rotation of 211 mg kg⁻¹, which was higher than that of 172 mg kg⁻¹ under the 4-year-rotation (Figure 3B), similar results were found in Canada [30], Across the years of the experiment under NT and for the 4-year-rotation, P concentration under NT significantly decreased from 1991 to 1997, but significantly increased in both 2002 and 2004 (Table 4). However, in the 2 and 3-year-rotations under NT, available P concentration generally increased from 1991 to 2004 (Table 4). In contrast, available P concentration generally increased from the beginning of the experiment to 2004 under NT and CT in both the 2 and 3-year-rotations (Table 4). Data in Table 4 under NT revealed that there was almost no significant difference in available P concentrations under all of the three crop rotations. However, under CT, available

Treatments	SOM (g kg⁻¹)					
	1991	1994	1995	1997	2002	2004
No tillage						
4-yr-rot	32.6b†	37.1a	38.5a	38.6a	34.6a	39.2a
3-yr-rot	34.3a	36.0a	39.3a	37.6a	33.9a	37.9b
2-yr-rot	32.0b	36.3a	39.1a	37.1a	33.9a	38.9ab
Conventional tillage						
4-yr-rot	33.5a	35.1b	38.6a	36.8a	33.3b	37.6b
3-yr-rot	32.2b	35.7b	38.2a	37.9a	32.4b	37.6b
2-yr-rot	34.0a	37.1a	38.9a	37.6a	33.9a	38.8a

†Means with different letters within a column under each tillage system are significantly different at P<0.05.

Table 2: Soil organic matter (SOM) concentrations as impacted by crop rotations under no-tillage and conversional tillage systems for the 0-15 cm depths.

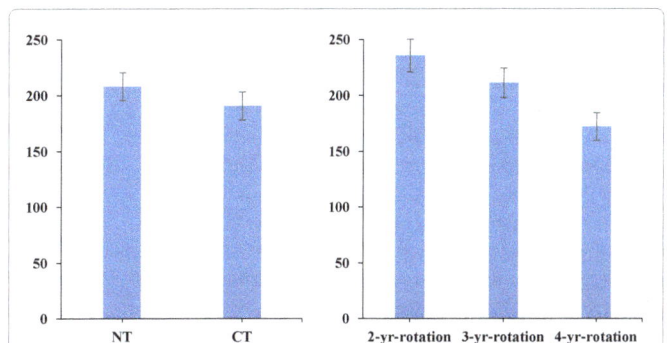

Figure 3: Available phosphorus (P) as influenced by long-term tillage (A) and rotation systems (B) for the 0-15 cm depth. Error bars represent standard errors.

Treatments	P (mg kg⁻¹)					
	1991	1994	1995	1997	2002	2004
4-year-rotation (corn-soybean-wheat-alfalfa)						
NT	196cA†	150dA	149dA	179cdA	243bA	284aA
CT	144bB	90cB	103cB	140bB	164bB	226aB
3-year-rotaion (corn-soybean-wheat)						
NT	148cdB	128dB	157cdA	232bA	172cA	325aA
CT	213bA	195bA	167bA	202bA	195bA	290aA
2-year-rotation (corn-soybean)						
NT	138cA	165bcA	128cA	183bcA	198bA	388aA
CT	168bA	148bA	148bA	190bA	175bA	418aA

†Means with different lower case letters within a row are significantly different at P<0.05. Higher case letters show the differences between NT and CT for each year and under each crop rotation separately

Table 3: Soil available phosphorus (P) concentration as impacted by no tillage vs conventional tillage under three rotation systems at the soil surface (0-15 cm).

P concentrations under the 4-year-rotation were significantly lower compared with those under the other two crop rotations (Table 4), which could be attributed to the higher demand of P for the [33] under the 4-year-rotation than that under the other two crop rotations.

Tillage and crop rotations impacts on soil nitrate (NO₃⁻ -N)

In general, averaged across all of the years, the nitrate (NO₃⁻) concentrations for the soil surface (0-15 cm) under the two tillage systems were not significantly different (Figure 4A). Similarly, from 1991 to 2004, nitrate concentrations under NT for the same crop rotation were not significantly different. Nitrate concentration in 1995, however, was significantly lower than those in the other years, which could be due to the very wet weather in 1995, which caused a delay in planting all of the crops as well as leaching NO₃⁻ from the surface layer of the soil (Table 5). Although SOM concentration under CT significantly increased from 1991 to 1995, nitrate concentration significantly decreased from 1991 to 1995 (Tables 1 and 5). This opposite relation could be attributed to the nitrate downward movement under CT (especially during the very wet weather in 1995). Generally, in each separate year under all of the crop rotations, nitrate concentrations under NT were not significantly different from those under CT (Table 5). Overall, CT exhibited higher but not significantly different NO₃⁻ concentrations than those under NT [34], which could be ascribed to the higher rate of decomposing SOM, which released more nitrate under CT compared with NT.

Generally, nitrate concentrations were significantly different among all of the applied three crop rotations (Figure 4B). For example, NO₃⁻ concentration in the 4-year-rotation of 18 kg ha⁻¹ was significantly higher than that of 16 kg ha⁻¹ in the 3-year-roatation, which was significantly higher than that of 15 kg ha⁻¹ in the 2-year-rotation of (Figure 4B). Such a trend could be attributed to the narrow C/N ratio in the legumes – soybean and alfalfa – in the 4-year-rotation which caused faster decomposition of SOM and released more nitrate [35-37]. Similarly, in each separate year, under the NT system, nitrate concentrations in the 4-year-rotation were significantly higher than those under the other two rotations Table 6 which also could be attributed to the narrow C/N ratio in the organic materials of the leguminous crops in the 4-year-rotation. Nevertheless, nitrate concentrations under CT were almost not significantly different among the three crop rotation systems (Table 6).

Tillage and crop rotations impacts on soil potassium

Averaged across all of the 14 years of the experiment, there was no significant difference in available K concentrations between the NT (359

mg kg⁻¹) and CT system (347 mg kg⁻¹) (Figure 5A). Across the years of the experiment, available K concentration was decreasing under both of the tillage systems (Table 7). For example, its concentration under NT within the 4-year-rotation in 2004 was 34.4% lower than that in 1991. During the experiment, crops were consuming K from the soil while there was no supplemental K applied as a fertilizer, which could interpret the decreased trend in K concentration from 1995 to 2004 under all of the treatments. In each separate year within the 4-year-rotation, K concentration was significantly higher under NT than that under CT in 1994, 1995, and 1997. This could partially attributed to the reason that the content of K in the NT plots were higher at the beginning compared with that of CT plots [30] (Table 7). In the 2 and 3-year-

Treatments	P (mg kg⁻¹)					
	1991	1994	1995	1997	2002	2004
No tillage						
4-yr-rot	196a†	150a	149a	179b	243a	284b
3-yr-rot	148a	128a	157a	232a	172b	325b
2-yr-rot	138a	165a	128a	183ab	198ab	388a
Conventional tillage						
4-yr-rot	144b	90b	103b	140b	164b	226c
3-yr-rot	213a	195a	167a	202a	195a	290b
2-yr-rot	168ab	148ab	148a	190a	175ab	418a

†Means with different letters within a column under each tillage system are significantly different at P<0.05.

Table 4: Soil available phosphorus (P) concentrations as impacted by crop rotations under no tillage and conventional tillage at the soil surface layer (0-15 cm).

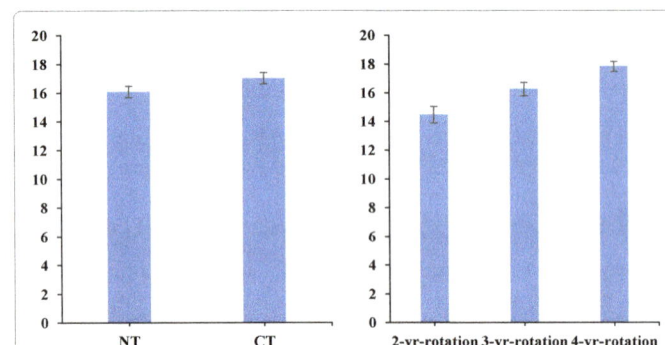

Figure 4: Soil nitrate (NO₃⁻) as influenced by long-term tillage (A) and rotation systems (B) for the 0-15 cm depth.

Treatments	NO₃⁻ (kg ha⁻¹)					
	1991	1994	1995	1997	2002	2004
4-year-rotation (corn-soybean-wheat-alfalfa)						
NT	17.8abA†	19.3aA	15.0bA	20.7aA	20.8aA	17.5abB
CT	19.9aA	15.0bB	11.0cB	18.3aA	17.9abA	20.8aA
3-year-rotaion (corn-soybean-wheat)						
NT	17.2aA	16.4aA	11.1bA	15.7aA	13.6abA	15.7aA
CT	22.1aA	19.4abA	10.5cA	16.8bA	18.0abA	19.0abA
2-year-rotation (corn-soybean)						
NT	13.5abA	15.3aA	11.5bA	12.1bB	12.9bA	12.8bB
CT	12.3dA	16.2bA	12.8cdA	15.6bcA	14.9bcdA	23.3aA

†Means with different lower case letters within a row are significantly different at P<0.05. Higher case letters show the differences between NT and CT for each year and under each crop rotation separately.

Table 5: Soil nitrate (NO₃⁻) Concentration differences between no tillage and conventional tillage under three rotation systems for the soil surface layer (0-15 cm).

Treatments	NO$_3^-$ (kg ha^{-1})					
	1991	1994	1995	1997	2002	2004
No tillage						
4-yr-rot	17.8a†	19.3a	15.0a	20.7a	20.8a	17.5a
3-yr-rot	17.2ab	16.4a	11.1b	15.7b	13.3b	15.7ab
2-yr-rot	13.5b	15.3a	11.5b	12.1b	12.9b	12.8b
Conventional tillage						
4-yr-rot	19.9a	15.0b	11.0a	18.3a	17.9a	20.8a
3-yr-rot	22.1a	19.4a	10.5a	16.8a	18.0a	19.0a
2-yr-rot	12.3Bb	16.2ab	12.8a	15.6a	14.9a	20.3a

†Means with different letters within a column under each tillage system are significantly different at $P<0.05$.

Table 6: Soil nitrate (NO$_3^-$) concentrations as impacted by crop rotations under no tillage and conventional tillage at the soil surface layer (0-15 cm).

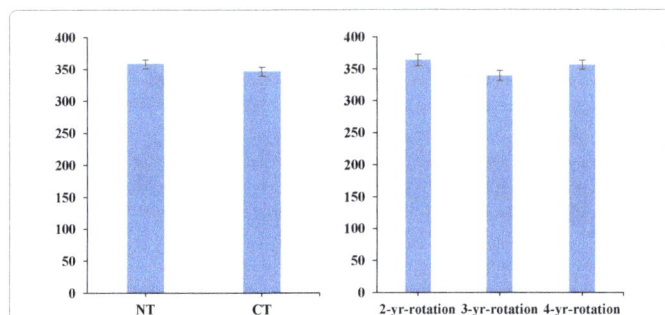

Figure 5: Soil available K as influenced by long-term tillage (A) and rotation systems (B) for the 0-15 cm depth. Error bars represent standard errors.

Treatments	K (mg kg^{-1})					
	1991	1994	1995	1997	2002	2004
4-year-rotation (corn-soybean-wheat-alfalfa)						
NT	419bA†	412bA	472aA	372cA	326dA	275eA
CT	359abB	336bcB	364aB	317cB	311cA	315cA
3-year-rotaion (corn-soybean-wheat)						
NT	351aA	362aA	379aA	313bA	286bcA	277cB
CT	351abA	387aA	380aA	340bcA	308cA	329bcA
2-year-rotation (corn-soybean)						
NT	390bA	451aA	389bA	337cA	275dA	270dA
CT	357bA	433aA	412aA	333bA	322bcA	293cA

†Means with different lower case letters within a row are significantly different at $P<0.05$. Higher case letters show the differences between NT and CT for each year and under each crop rotation separately.

Table 7: Soil potassium (K) concentration differences between no tillage and conventional tillage under three rotation systems for the soil surface layer (0-15 cm).

Treatments	K (mg kg^{-1})					
	1991	1994	1995	1997	2002	2004
No tillage						
4-yr-rot	419a†	412a	472a	372a	326a	275a
3-yr-rot	351b	362b	379b	313b	286b	277a
2-yr-rot	390a	451a	389b	337ab	275b	270a
Conventional tillage						
4-yr-rot	359a	336b	364b	317a	311a	315a
3-yr-rot	351a	387ab	380b	340a	308a	329a
2-yr-rot	357a	433a	412a	333a	322a	293a

†Means with different letters within a column under each tillage system are significantly different at $P<0.05$.

Table 8: Soil available potassium (K) concentrations as impacted by crop rotations under no tillage and conventional tillage at the soil surface layer (0-15 cm).

rotation for each separate year, however, available K concentration under NT and CT was not significantly different (Table 7). Within the crop rotations, the 3-year-rotation had the lowest concentration of available K (340 mg kg^{-1}) which was significantly different from the other two rotations, and there was no significant difference between its concentration in the 2-year-rotation (365 mg kg^{-1}) and 4-year-rotation (357 mg kg^{-1}) (Figure 5B). Data in Table 8 indicated that the available K concentrations under NT in each separate year were significantly lower under the 3-year-rotation system than those under the other two rotations. However, under the CT system, the available K concentrations were not significantly different among the three crop rotation systems.

Conclusions

In the long term experiment (14 years) data showed that the SOM concentration at a depth of 0-15 cm under NT system was significantly higher than that under CT. However, there was no significant difference in SOM concentrations between under crop rotations. Soil P concentration under NT was significantly higher compared with that under CT. The 2-year-rotation had the highest P concentration followed by the 3-year-rotaion, and the lowest concentration was observed in the 4-year-rotation. Nitrate concentration under NT was not significantly different from that under CT; however, the 4-year-rotation had the highest NO$_3^-$ concentration followed by the 3-year-rotation, and the lowest concentration was observed in the 2-year-rotation. The K concentration was not significantly different under NT and CT; however, the 3-year-rotation had the lowest K concentration compared with the other two rotations.

Acknowledgements

The authors acknowledge the financial support from the Agriculture Experiment Station of the South Dakota State University.

References

1. Johnson MG, Levine ER, KernJS (1995) Soil organic matter: Distribution, genesis, and management to reduce greenhouse gas emissions. Water Air Soil Pollut 82: 593-615.

2. Paustian K, Collins HP, Paul EA (1997) Management controls on soil carbon. In: Paul EA (eds.) Soil Organic Matter in Temperate Ecosystems: Long Term Experiments in North America. CRC Press, Boca Rotan, FL 15-49.

3. Logan, TJ, Lal R, Dick WA (1991) Tillage systems and soil properties in North America. Soil and Tillage Research 20: 241-270.

4. Eltz FLF, Norton LD (1997) Surface roughness changes as affected by rainfall erosivity, tillage, and canopy cover. Soil Sci 61: 1746-1754.

5. Follett RF (2001) Soil management concepts and carbon sequestration in cropland soils. Soil Tillage Res 61: 77-92.

6. Sainju UM, Singh BP, Whitehead WF, Wang S (2006) Carbon supply and storage in tilled and non-tilled soils as influenced by cover crops and nitrogen fertilization. J Environ Qual 35: 1507-1517.

7. Martin-Rueda I, Munoz-Guerra LM, Yunta F, Esteban E, Tenorio JL, et al. (2007) Tillage and crop rotation effects on barley yield and soil nutrients on a CalciortidicHaploxeralf. Soil and Tillage Research 92: 1-9.

8. Kumar S, Kadono A, Lal R, Dick W (2012) Long-term no-till impacts on organic carbon and properties of two contrasting soils and corn yields in Ohio. Soil Sci Soc Am J 76: 1798-1809.

9. Ismail L, Blevins RL, Frye WW (1994) Long-term no-tillage effects on soil properties and continuous corn yields. Soil Sci Soc Am J 58: 193-198.

10. Godsey CB, Pierzynski GM, Mengel DB, Lamond RE (2007) Changes in soil pH, organic carbon, and extractable aluminum from crop rotation and tillage. Soil Sci Soc Am J 71: 1038-1044.

11. Campbell CA, Myers RJK, Curtin D (1995) Managing nitrogen for sustainable crop production. Fertilizer research 42: 277-296.

12. Weed DAJ, Kanwar RS (1996) Nitrate and water present into continuous corn and flowing from root-zone soil. J Environ Qual 25: 709-719.

13. Holanda FSR, Mengel DB, Paula MB, Carvalho JG, Berton JC(1998) Influence of crop rotations and tillage systems on phosphorus and potassium stratifications and root distribution in the soil profile. Commun. Soil Sci Plant Anal 29: 2383-2394.

14. Zhou G, Yin X, Verbree DA (2014) Residual effects of potassium to cotton on corn productivity under no-tillage. Agronomy Journal 106: 893-903.

15. Cruse RM, YakleGA, ColvinTC, TimmonsDR, Mussleman AL (1983) Tillage effects on corn and soybean production in farmer-managed, university-monitored field plots. J Soil Water Conserv 38: 512-514.

16. O'Halloran IP, Stewart JWP, De Jong E (1987) Changes in P forms as influenced by management practices. Plant Soil 100: 113-126.

17. Robbins SG, Voss RD (1991) Phosphorus and potassium stratification in conservation tillage systems. J Soil Water Conserv 46: 298-300.

18. Calonego JC, Rosolem CA (2013) Phosphorus and potassium balance in a corn-soybean rotation under no-till and chiseling. Nutrient Cycling in Agro ecosystems 96: 123-131.

19. Karlen DL, Berti WR, Hunt PG, Matheny TA (1989) Soil-Test Values after eight years of tillage research on a Norfolk Loamy Sand. Commun. Soil Sci Plant Anal 20: 1413-1426.

20. Natural Resources and Conservation Service (NRCS) Staff (2015) a Official soil series description.

21. Natural Resources and Conservation Service (NRCS) Staff (2015) South Dakota online soil survey manuscripts.

22. Mikha MM, Vigil MF, Liebig MA, Bowman RA, McConkey B, et al. (2006) Cropping system influences on soil chemical properties and soil quality in the Great Plains. Renew Agric Food Syst 21: 26-35.

23. Olsen SR, Cole CV, Watanabe FS, Dean LA (1954) Estimation of available phosphorus in soils by extraction with sodium bicarbonate. United States Department of Agriculture Circular 939: 18-19.

24. Gelderman RH, Beegle D (1998) Nitrate-nitrogen. In: Brown JR (ed.), recommended chemical soil test procedures for the North Central Region. NCR Publ 221: 17-20.

25. Warncke D, Brown JR (1998) Potassium and other basic cations. In: Brown JR (ed.), recommended chemical soil test procedures for the North Central Region. NCR Publ 221: 31-33.

26. Carter MR (2005) Long-term effects on cool-season soybean in rotation with barley, soil properties, and carbon and nitrogen storage for fine sandy loams in the humid climate of Atlantic Canada. Soil Tillage Res 81: 109-120.

27. Junior CC, Corbeels M, Bernoux M, Piccolo MC, Neto MS, et al. (2013) Assessing soil carbon storage rates under no-tillage: Comparing the synchronic and diachronic approaches. Soil and Tillage Research 134: 207-212.

28. Scopel E, Findeling A, Guerra EC, Corbeels M (2005) Impact of direct sowing mulch-based cropping systems on soil carbon, soil erosion and maize yield. Agronomy for Sustainable Development 25: 425-432.

29. Karlen DL, Berry EC, Colvin TS, Kanwar RS (1991) Twelve-year tillage and crop rotation effects on yields and soil chemical properties in northeast Iowa. Communications in Soil Science & Plant Analysis 22: 1985-2003.

30. Van Eerd LL, Congreves KA, Hayes A, Verhallen A, Hooker DC (2014) Long-term tillage and crop rotation effects on soil quality, organic carbon, and total nitrogen. Canadian Journal of Soil Science 94: 303-315.

31. Wyngaard N, Echeverría HE, Rozas HR, Divito GA (2012) Fertilization and tillage effects on soil properties and maize yield in a Southern Pampas Argiudoll. Soil and Tillage Research 119: 22-30.

32. Deubel A, Hofmann B, Orzessek D (2011) Long-term effects of tillage on stratification and plant availability of phosphate and potassium in a loess chernozem. Soil Tillage Res 117: 85-92.

33. Hejcman M, Kunzová E, Šrek P (2012) Sustainability of winter wheat production over 50 years of crop rotation and N, P and K fertilizer application on illimerizedluvisol in the Czech Republic. Field Crops Research 139: 30-38.

34. Dao TH (1998) Tillage and crop residue effects on carbon dioxide evolution and carbon storage in a paleustoll. Soil Sci 62: 250-256.

35. Carpenter-Boggs L, Pikul JL, VigilMF, Riedell WE (2000) Soil nitrogen mineralization influenced by crop rotation and nitrogen fertilization. Soil Sci Soc Am J 64: 2038-2045.

36. Power JF, Doran JW, Wilhelm WW (1986) Uptake of nitrogen from soil fertilizer and crop residues by no-till corn and soybean. Soil Sci Soc Am J 50: 137-142.

37. Riedell WE, Osborne SL, Pikul JL (2013) Soil Attributes, Soybean Mineral Nutrition, and Yield in Diverse Crop Rotations under No-Till Conditions. Agronomy Journal 105: 1231-1236.

33

Sorghum (*Sorghum bicolor* L.) Growth, Productivity, Nitrogen Removal, N- Use Efficiencies and Economics in Relation to Genotypes and Nitrogen Nutrition in Kellem- Wollega Zone of Ethiopia, East Africa

Sheleme Kaba Shamme, Cherukuri V Raghavaiah*, Tesfaye Balemi and Ibrahim Hamza

Department of Plant Sciences, College of Agriculture and Veterinary Sciences, PO Box 19, Ambo University, Ambo, West Shoa Zone, Ethiopia, East Africa

Abstract

Sorghum is an important cereal crop, which requires high dose of nitrogen for optimum growth and productivity, specially under rainfed farming situation in tropical regions. Field experiment was conducted at Haro Sabu Agricultural Research Center during main cropping season of 2014 with an objective to investigate Nitrogen Use Efficiency (NUE) of two improved and a local sorghum cultivar in relation to graded rates of N levels and to investigate their effect on yield, N uptake and economics. The treatments comprised factorial combination of four nitrogen rates (0, 46, 92 and 138 kg N ha^{-1}) and three sorghum genotypes (Lalo, Chemada and Local varieties) tested in a Factorial Randomized Block Design with three replications. The results revealed that there was significant effect of N rates on days to 50% flowering, days to 50% physiological maturity, Lodging percentage, leaf area at 90 and 120 DAS, leaf area index, number of green leaves plant^{-1}, biological yield, grain yield, harvest index and nitrogen use efficiency. There was significant interaction effect of N rates and sorghum genotypes on most of parameters studied. Significantly higher grain productivity was obtained in response to the application of 92 kg N ha with Lalo variety in comparison with the rest of the genotype × N-rate combinations. Genotypic variations in N uptake, partitioning and NUE in plant parts like leaves, stems and grain were noted. Increase in the rate of applied N enhanced N uptake, nitrogen utilization efficiency, and N harvest index; while higher rates decreased N use efficiency, N uptake efficiency, N recovery efficiency, and Agronomic efficiency. Economic analysis indicated higher net return with the application of 92 kg N ha^{-1} and Lalo genotype accrued the highest net return and benefit: cost ratio than Local variety.

Keywords: Productivity; Nitrogen rates; Nitrogen uptake; Partitioning; Nitrogen use efficiency; Sorghum genotypes

Introduction

Sorghum (*Sorghum bicolor* L.) is a drought tolerant and nutritious cereal crop usually cultivated for food, feed and fodder by subsistence farmers in Ethiopia. Elsewhere in the world, especially in semi- arid tropical (SAT) regions, where the production is constrained by low and erratic rainfall and low soil fertility, it is grown and consumed as staple food and is also used in the production of a variety of by-products like alcohol, edible oil, confectionary items and sugar. Cereals are the major food crops in Ethiopia, both in terms of the coverage and volume of production [1]. Sorghum is the fifth most important cereal crop worldwide. In the year 2005, sorghum was grown worldwide on 43,727,353 ha with an output of 58,884,425 metric tons [2]. Out of the total grain crop production area, 79.46% (8.1 million hectares) was under cereals. In Ethiopia Tef, maize, sorghum and wheat were raised on 22.08% (2.2 million hectares), 15.00% (1.5 million hectares), 14.43% (nearly 1.5 million hectares) and 14.35% (nearly 1.5 million hectares) of the grain crop area, respectively [1]. Cereals contributed 86.86% (more than 116.2 million quintals) of the grain production. Maize, wheat, Tef and sorghum made up 24.93% (33.4 million quintals), 16.58% (22.2 million quintals), 16.26% (21.8 million quintals) and 16.24% (21.7 million quintals) of the grain production in the same order and the Ethiopian national average yield was 14.81 q/ha [1].

Among the macro nutrients essential for crop growth, nitrogen (N) is a very mobile element in the soil, due to its susceptibility to leaching, de nitrification, and volatilization losses. Excessive use of N fertilizer can lead to pollution of water bodies and may lead to soil acidification. Balanced and efficient use of applied N is of paramount importance in the overall nutrient management system than any other plant nutrient in order to reduce its negative impact on the environment. Besides, even under the best management practices, 30%-50% of the applied

nitrogen is lost through different routes [3], and hence more fertilizer needs to be applied than actually needed by the crop to compensate for the loss. The transitory loss of N not only causes loss to the farmer but also causes irreversible damage to the environment [4,5]. High rates of chemical fertilizer cause environmental pollution [6]. Increased rate of N application was reported to delay flowering, physiological maturity, and increased crop lodging across genotypes in sorghum. Thus, it warrants for a need to optimize the level of N fertilizer to be applied, with special reference to cereals like sorghum. The substantial increase in fertilizer prices many fold over the recent decade dictates that farmers need to use high nitrogen use efficient (NUE) sorghum varieties to reduce the cost incurred on N fertilizer as a major nutrient input under rain fed situations. The use of N efficient sorghum cultivars increased the production of sorghum -based cropping system in West Africa, and similar trend may be witnessed once N efficient sorghum cultivars are identified for Ethiopian conditions.

Differences exist both between and within species in terms of efficient use of mineral nutrients for growth and development. Hybrids that can exhibit superior yields at low N rates are necessary. While some genotypes are capable of performing well under nutrient stress

*Corresponding author: Cherukuri V Raghavaiah, Department of Plant Sciences, College of Agriculture and Veterinary Sciences, Ambo University, Ethiopia
E-mail: cheruraghav@yahoo.in

conditions, others performed poorly. Genotypic differences for nutrient use, especially NUE have been recognized in many crops including sorghum. However, the mechanisms that explain how these genotypes are able to satisfy all their biosynthetic and maintenance needs, with smaller amount of nutrient than required by other genotypes is still not well studied [7]. The objectives of this study were to find out the effect of varied N rates on the productivity of sorghum genotypes; to determine the N removal of genotypes; to delineate the N use efficiency of genotypes and to compute the relative economics of genotypes and N fertilizer application in sorghum production.

Materials and Methods

Description of experimental site

The current field investigation was conducted at Kellem-Wollega zone of Ethiopia to determine the effect of graded nitrogen rates and genotypes on growth, productivity, N removal, nitrogen use efficiency and their relative economics in sorghum during the main cropping season of 2014 at Haro Sabu Agricultural Research Center (HSARC). The center is located in western Ethiopia, Oromia regional state at a distance of 550 km west of Addis Ababa. It lies at a latitude of $8°52'51''$ N, longitude $35°13'18''$ E and altitude of 1515 m above sea level. The center has a warm humid climate with average minimum and maximum temperature of 14°C and 30°C, respectively. The area received an average annual rainfall of 1000 mm with a uni- modal distribution pattern, most of the rain being received from April to October. The soil type of the experimental site is reddish brown, with a pH of 5.53. The area is mainly characterized by coffee- based and crop+livestock mixed farming system comprising non legumes like Coffee, maize, sorghum, finger millet; legumes such as haricot bean, soybean, besides sesame, banana, mango and sweet potato.

Description of the experimental materials

Plant materials: In the present study, Sorghum varieties Lalo and Chemada and a Local variety, which is adapted to the agro-ecology of the area, were used. Varieties Lalo and Chemada are the most promising hybrids released by Bako Agricultural Research Centre in 2006 and 2013, respectively. Both the genotypes have wider adaptability and grow well at altitudes ranging from 1500 to 1900 meters above sea level with annual rain fall of 1100 to 1200 mm. The cultivars mature in about 199 and 180 days, the seed colors are brownish-red and creamy for Lalo and Chemada, respectively. The potential yield of Lalo was 48 q/ ha^{-1} at research farm and 35 q ha^{-1} at farmer's field, and the yield of Chemada was 32 q ha^{-1} at research farm and 25 q ha^{-1} at farmer's field.

Experimental design and plot management: The experimental field was ploughed and harrowed by a tractor to get a fine seedbed and leveled manually before the field layout was made. Sowing was done on June 7, 2014 and the seeds were sown at a spacing of 75 cm between rows and 15 cm between plants. The nitrogen fertilizer source used was urea (46% N) which was applied by drilling in two splits, half at 14 days after emergence and the remaining half at knee high stage along the rows of each plot to ensure that N is uniformly distributed. The treatments comprising factorial combination of three varieties (Lalo, Chemada and Local) and four N rates (0, 46, 92 and 138 kg N ha^{-1}) were arranged in a Randomized Block Design with three replications. The plot size was 3 m long and 4.5 m wide that could accommodate 6 rows. The four rows in the net plot were set aside for data collection to eliminate any border effects. Phosphorus fertilizer in the form of triple super phosphate (TSP) at the recommended rate of 100 kg P$_2$0$_5$ ha^{-1} was uniformly applied to all plots by drilling in the rows at the time of

sowing. Weeds were managed by hand weeding after weed emergence and late-emerging weeds were also removed by hand hoeing to avoid competition with the sorghum plants for the N applied. Sorghum plants in the l net plot area (9 m^2) were harvested at normal physiological maturity. The ear heads in each plot were harvested manually and were separately threshed.

Soil sampling and analysis: Soil samples were collected prior to planting from a depth of 0-30 cm in a zigzag pattern randomly from the experimental field using soil auger. Composite samples were prepared for analysis to determine the soil physico-chemical properties of the experimental site at Nekemte zonal Soil Laboratory. The composited soil samples were air-dried, ground and sieved to pass through a 2 mm sieve. Total nitrogen was determined following Kjeldahl procedure [8], the soil pH (in water suspension) by a digital pH meter [9], organic carbon by wet digestion method [10], available phosphorous by Olsen method [11], Cation exchange capacity (CEC) by ammonium acetate method [8], and soil texture by Bouyoucons Hydrometer method [12].

Crop data collection

Phonological data: The days to 50% flowering was taken as the time from the date of planting until half of the plant populations in the plot started to flower. The days to 90% physiological maturity was taken as the time from date of planting until 90% of the plant leaves turned yellow and the lower most head started drying.

Crop growth parameters: The total leaf area was recorded at 60, 90 and 120 DAS, and determined by multiplying leaf length and maximum width of leaf adjusted by a correction factor of 0.75 as suggested by McKee. Number of leaves was counted from randomly selected five pre-tagged plants after plant emergence and their average was taken as the number of leaves plant^{-1}. The Leaf Area Index (LAI) was calculated from five randomly selected plants by dividing the leaf area by its respective ground area at 120 DAS. The sorghum height of five randomly selected pre-tagged plants (cm) was measured at 60, 90 and 120 DAS from the ground level up to the head/tip of the plant.

Yield and yield components: Five pre-tagged randomly selected plants were considered for determination of above ground dry biomass weight by drying in sunlight for ten days till a constant dry weight was attained. Lodging was recorded at the time of harvest from four net plot rows, and thousand seeds of sorghum was counted and weighed (g) using sensitive balance from the bulk of the seeds of sorghum and adjusted to 13.5% moisture level. Number of panicles/plant was also recorded from five pre-tagged randomly selected plants. Number of effective tillers / plant was counted at physiological maturity. Grain yield (q/ha) was recorded after harvesting from the central four rows of the net plot of 3 m × 3 m=9 m^2. Seed yield was adjusted to 13.5% moisture using moisture tester (Dickey-john) and converted to quintal ha^{-1} for statistical analysis.

Adjusted yield=Actual yield × 100-M/100-D;

where M, is the measured moisture content in grain and D is the designated moisture content (13.5%).

D is the designated moisture

Harvest index was calculated as: HI (%)=Sy/by × 100;

where HI is harvest index, SY is seed yield and BY is above ground dry biological yield.

Nitrogen use efficiency and components of N use computations

Above ground portion of plants from each plot was randomly

sampled at physiological maturity as described by Vanderlip [13]. At each sampling, the leaves were separated from the stems. In addition, heads and the above ground vegetative parts were dried at 60°C in a forced air oven for 72 h. to a constant weight. The oven dry samples were ground using rotor mill and allowed to pass through a 0.5 mm sieve to prepare a sample of 10 g. The straw and grain samples were analyzed for N concentrations from each plot separately using Kjeldahl method as described by Jackson [14]. Nitrogen uptake in grain and straw were calculated by multiplying N content with the respective straw and grain yield ha⁻¹. Total N uptake, by whole biomass was obtained by summing up the N uptake by grain and straw and was expressed as kg ha⁻¹. Total N content in straw and grain samples was determined using Micro Kjeldahl's apparatus and was used to calculate N use efficiency according to Moll et al. [15] and Ortiz-Monasterio et al. [16]. The different efficiencies of nitrogen were arrived at employing the following formulae:

Nitrogen Use Efficient (NUE) (Kg Kg⁻¹)=Grain weight/N supplied

Nitrogen Utilization Efficient (NUTE) (Kg Kg⁻¹)=Grain weight/N total in plant

Nitrogen Uptake Efficient (NUPE) (Kg Kg⁻¹)=N total in plant / N supplied

Percent N Recovery=[N uptake (fertilized plot)-N uptake (un-fertilized plot)] / [N applied] × 100

Nitrogen Harvest index (%)=Grain N content/N total in plant

N uptake of grain=N concentration of grain × grain yield (kg ha⁻¹)/100

N uptake of Straw=N concentration of Straw × Straw yield (kg ha⁻¹)/100

AE=[Grain yield of fertilized treatment (kg ha⁻¹) - Grain yield of unfertilized plot (kg ha⁻¹)] / [Fertilizer applied (Kg/ha)

Where N supplied is the fertilizer N applied+N supplied in the soil, AE=Agronomic efficiency.

Economic analysis

Economic analysis was made following CIMMYT methodology [17]. The cost of 100 kg urea (1087 birr), 100 kg TSP (1415 birr), dry biomass in ton ha⁻¹ (200 birr) and sorghum grain price of 450 birr per 100 kg were used for the benefit: cost analysis. The analysis of data in relation to different factors of production under test viz; genotypes and N fertilizer rates were computed in terms of: 1. Gross return (Birr ha⁻¹) from total economic produce and by products obtained from the crops included in the cropping system are calculated based on the local market prices at harvest, 2. Net return (Birr ha⁻¹)=(Gross return – Cost of Production), 3. Cost of Production (Birr ha⁻¹), 4. Benefit: Cost ratio (Gross return/Cost of Production), 5. Per day Productivity (kg ha⁻¹), (Grain Yield/Crop duration), 6. Return/Birr Investment (Net return/ Production Cost) were determined.

Statistical analysis

All data collected were subjected to the analysis of variance (ANOVA) using SAS [18] version 9.1. Where treatment means are significant, the Tukey test at alpha=5% was adopted for mean comparison.

The model for randomized complete block design is:-

$$y_{ij}=\mu+\tau_i+\beta_j+\varepsilon_{ij}$$

Where:

y_{ij}=An observation in treatment i, and block j

μ=the overall mean

τ_i=the effect of treatment i

β_j=the fixed effect of block j

ε_{ij}=random error

Results and Discussion

Physicochemical properties of the soil

The results of soil analysis showed that the soil had a moderate total nitrogen content of 0.24 (%). According to Tekaligne et al. [19] soil total N availability of less than 0.05% was considered as very low, 0.05-0.12% as poor, 0.12-0.25% as moderate and more than 0.25% as high. The soil has organic matter of 5.0% which was considered as medium. According to Berhanu [20], soil organic matter content of less than 0.8% is considered as very low, 0.8-2.6 as low, 2.6-5.2% as medium and more than 5.2% as high. Thus, the soil available P content of experimental sites was 6.28 ppm, and according to Tekalign et al. [19] soils with available P <10, 11-31, 32-56, >56 ppm as low, medium, high and very high, respectively. The average CEC of the soil was 23 Cmol (+)/kg, which is rated as medium. The soil cation exchange capacity describes the potential fertility of soils and is an indicator of the soil texture, organic matter content and the dominant types of clay minerals present. In general, soils high in CEC contents are considered as agriculturally fertile. According to Landon, top soils having CEC greater than 40 Cmol (+)/kg are rated as very high and 25-40 Cmol (+)/kg as high, 15-25, 5-15 and <5 Cmol (+)/kg of soil are classified as medium, low and very low, respectively. The analytical results also indicated that the textural class of the experimental site was clay loam with a proportion of 37% sand, 25% clay and 38% silt. Thus, the textural class of the experimental soil is ideal for sorghum production and the soil reaction (pH) of the experimental site was 5.53 showing moderate acidity, but it is within the optimum range for sorghum production, i.e., 5.5 -7.0 [21].

Effect of nitrogen fertilizer rates on growth of sorghum genotypes

Plant height: Nitrogen application rates significantly influenced plant height at 90 DAS, but the effect was not significant at 60 DAS and 120 DAS. Sorghum genotypes differed significantly in plant height (Table 1) at various growth stages. Plant height increased linearly and significantly with the increase in the rate of nitrogen application from 0 through 138 kg N ha⁻¹. The increase in plant height following increased

Treatment	Plant height (cm)		
Nitrogen (N) rates	60DAS	90DAS	120DAS
0 kg N ha⁻¹	34.54b	146.82b	263.33b
46 kg N ha⁻¹	39.39ba	165.77b	275.30ba
92 kg N ha⁻¹	40.72ba	196.30a	284.88a
138 kg N ha⁻¹	41.60a	204.99a	290.90a
Sorghum varieties			
Lalo	43.86a	227.42a	299.73a
Local	38.74ba	158.01b	276.54b
Chemada	34.59b	149.98b	259.53b
CV%	18.06	11.92	7.88

DAS=Days after sowing

Table 1: Effect of nitrogen rates on plant height at different growth stages of sorghum genotypes.

N application rate indicates maximum vegetative growth of the plants under higher N availability. These results are in agreement with the results obtained by Akbar et al. [22] found that plant height in maize increased with increase in N rate. However, in contrast to the results of this study, Sadeghi and Bahrani [23] reported that increase in N rate had no significant effect on plant height. The variance in the results obtained in this study and that of Sadeghi and Bahrani [23] might be due to the difference in the range of population stand, soil fertility status, and the crop varieties used.

Number of green leaves plant[-1]: The graded nitrogen rates did not significantly influence green leaves /plant, whereas sorghum genotypes significantly differed in the number of green leaves per plant. There was a significant interaction effect of nitrogen rates and sorghum genotypes on the number of green leaves per plant (Table 2). The highest number of green leaves (14.0) was recorded with the application of 46 kgN ha[-1] with Local variety, while the lowest (10.67) was recorded with 92 kg N/ha in Lalo variety.

Number of effective tillers: The number of effective tillers significantly varied with sorghum genotypes, but not with the nitrogen fertilizer rates. The number of effective tillers was significantly greater (2.18) with Local variety followed by Lalo variety (1.2) (Table 3). The current result is in agreement with that of Botella et al. [24] reported that stimulation of tillers with high application rate of nitrogen might be due to its positive effect on cytokinin synthesis. In the present study the tillers tended to increase with enhanced N rates. Genene [25] also reported higher tillers and maximum survival percentage of tillers with increasing N application in bread wheat.

Table 3 indicates effect of nitrogen rates on number of effective tillers plant[-1], leaf area at 60 DAS, thousand seed weight (g) and panicle number per plant of sorghum genotypes.

Total leaf area: There was significant effect of nitrogen rates on leaf area at 90 DAS, but not at 60 and 120 DAS. Sorghum genotypes also significantly differed in leaf area at 60 DAS (Table 3) and 90 DAS, but the variations were nullified at 120 DAS. However, there was significant interaction effect of sorghum genotypes and nitrogen fertilizer rates on leaf area at 90 and 120 DAS. At 90DAS, the highest total leaf area per plant of 3193.3 cm² was obtained in Lalo variety with 46 kg N ha[-1] while the lowest leaf area per plant of 1955.9 cm² was obtained in variety Chemada with 138 kgN/ha (Table 4). At 120 DAS, the highest total leaf area per plant of 3694 cm² was obtained from variety Lalo with 132 kgN/ha, and the lowest leaf area per plant of 2631.9 cm² was obtained in variety Lalo with 92 kgN/ha (Table 5). The increasing nitrogen fertilizer rates did not result in increment of leaf area at all stages of growth as observed in this study. The result of the current study is in variance with the findings of Demir et al. [26] reported that leaf area increased with increasing N levels.

Leaf Area Index (LAI): There was no main factor effect of nitrogen fertilizer rates and sorghum genotypes on leaf area index (Table 6). However, nitrogen rates and sorghum genotypes significantly interacted to influence this parameter. The highest LAI for Lalo (3.28 cm²) was recorded with the application of 138 kg N ha[-1], while the lowest (2.15 cm²) was recorded from variety Chemada with no N application. Generally, an increasing trend in LAI was observed with increased N application rates. The increase in LAI was possibly due to the improved leaf expansion in plants through application of optimum nitrogenous fertilizers. Similar to this finding, Haghighi et al. [27] reported an increasing trend in LAI in maize due to an increase in N fertilizer application rates.

Nitrogen rates	Sorghum varieties			Mean
	Lalo	Local	Chemada	
0 kg N ha[-1]	2869.4ba	2485.1bc	2376.1dc	2576.89
46 kg N ha[-1]	3193.8a	2715.3bac	2361.5dc	2756.87
92 kg N ha[-1]	2521.0bc	2728.3bac	2805.0bac	2684.77
138 kg N ha[-1]	2799.0bac	2338.9dc	1955.9d	2364.6
Mean	2845.8	2566.9	2374.63	2595.78
CV%	11.27			

Table 4: Interaction effect of nitrogen rates on Leaf area (cm²) at 90 DAS of sorghum genotypes.

Nitrogen rates	Sorghum varieties			Mean
	Lalo	Local	Chemada	
0 kg N ha[-1]	3229.8ba	3169.2bac	3204.5ba	3201.17
46 kg N ha[-1]	3092.7bc	2813.6bc	2814.2bc	2906.83
92 kg N ha[-1]	2631.9c	2859.3bc	3229.0ba	2906.73
138 kg N ha[-1]	3694.9a	2950.8bc	2638.2c	3094.63
Mean	3162.33	2948.23	2971.48	3027.34
CV%	10.89			

Table 5: Interaction effect of nitrogen rates on Leaf area (cm²) at 120 DAS of sorghum genotypes.

Nitrogen rates	Sorghum varieties			Mean
	Lalo	Local	Chemada	
0 kg N ha[-1]	2.87ba	2.82bac	2.15ba	2.85
46 kg N ha[-1]	2.75bc	2.50bc	2.50bc	2.58
92 kg N ha[-1]	2.34c	2.54bc	2.87ba	2.58
138 kg N ha[-1]	3.28a	2.62bc	2.38bc	2.76
Mean	2.81	2.62	2.65	2.69
CV%	10.88			

Nitrogen rates	Sorghum varieties			Mean
	Lalo	Local	Chemada	
0 kg N ha[-1]	11.67bdc	14.00a	13.67a	13.11
46 kg N ha[-1]	11.33dc	14.00a	12.67bac	12.67
92 kg N ha[-1]	10.67d	13.67a	14.00a	12.78
138 kg N ha[-1]	12.00bdc	13.00ba	11.33dc	12.11
Mean	11.42	13.67	12.92	12.67
CV%	7.33			

Table 2: Interaction of nitrogen rates and sorghum genotypes[-1] on number of green leaves/ plant.

Treatment Nitrogen rates	Number of effective tillers per plant	Leaf area(cm²) 60DAS	Thousand seed weight(g)	Panicle number per plant
0 kg N ha[-1]	1.36a	884.8a	26.07ba	59.67a
46 kg N ha[-1]	1.47a	1028.5a	24.17c	61.11a
92 kg N ha[-1]	1.54a	1092.5a	26.57a	62.89a
138 kg N ha[-1]	1.51a	1054.5a	24.53bc	59.00a
Sorghum varieties				
Lalo	1.20b	1192.8a	32.25a	72.67a
Local	2.18a	1005.4ba	22.52b	59.25b
Chemada	1.03b	847.0b	21.24b	50.08c
CV%	22.46	25.31	3.41	9.4

Table 3: Effect of nitrogen rates on number of effective tillers plant[-1], leaf area at 60 DAS, thousand seed weight (g) and panicle number per plant of sorghum genotypes.

Table 6: Interaction effect nitrogen rates and sorghum genotypes on Leaf area index.

Effects of nitrogen on yield and yield components

Biological yield: The results of the analysis of variance showed that biological yield (BY) of sorghum genotypes was not significantly influenced by the effect of nitrogen rates. However, there was significant interaction effect of nitrogen rates and sorghum genotypes on this parameter (Table 7). Biological yield (BY) is a function of photosynthetic rate and proportion of the assimilatory surface area. The increase in biological yield with increase in rate of N might be due to better crop growth rate, LAI and accumulation of photo assimilate due to maximum days to maturity by the crop, which ultimately produced more biological yield. Biomass yield generally increased significantly with the increase in the rate of nitrogen across the increasing frequency of application. The variety Lalo recorded the highest biomass yield (23443 g plot^{-1}) with the application of 92 kg N ha^{-1} and the lowest biomass yield was obtained from variety Chemada (1793 g plot^{-1}) with the application of 138 kg N ha^{-1}. The application of the highest level of N resulted in less biomass yield compared to 92 kg N ha^{-1}. This result however, is not in agreement with that of Haftom et al. [28].

Thousand seed weight: It is an important yield determining component and reported to be a genetic trait that is influenced least by environmental factors [29]. The analysis of variance showed that the main effect of both nitrogen rates on thousand seed weight and that of sorghum genotypes was significant. The highest 1000 seed weight (32.25 g) was obtained from variety Lalo followed by variety Local (22.52 g). However, the lowest 1000 seed weight (21.24 g) was obtained from the Chemada. On the other hand, the highest 1000 seed weight was observed with 92 kg N ha^{-1} and the lowest 1000 seed weight with 46 kg N ha^{-1} and 138 kg N ha^{-1} (Table 3). Contrary to the finding of this study; Melesse [30] reported no significant effect of the application of different rates of nitrogen fertilizer on 1000 kernel weight of bread wheat.

Panicle number/plant: Panicle number per plant is an important yield attributes of sorghum that contributes to grain yield. Crops with higher panicle number could have higher grain yield. Panicle number was significantly influenced by the main effect of the sorghum genotypes but not by the main effect of nitrogen rates as well as the interaction of the two factors (Table 3). Among the genotypes, Lalo had significantly greater number of panicles /plant than Local variety and Chemada.

Grain yield: The results showed that the sorghum grain yield was significantly influenced by the nitrogen rates, sorghum varieties and their interaction (Table 8). The highest grain yield was recorded in the variety Lalo with the application of 92 kg N ha^{-1} (47.72 q ha^{-1}) and the lowest grain yield was recorded by variety Chemada with no nitrogen application (24.27 q ha^{-1}). Thus Lalo and Local varieties showed response up to 92 kgN/ha, whereas Chemada responded up to 42 kgN/ha. The results of this study are consistent with result of Sage and Pearcy who reported that a well-balanced supply of N results in

Nitrogen rates	Sorghum varieties			Mean
	Lalo	Local	Chemada	
0 kg N ha^{-1}	41.63	40.16	24.27	35.35
46 kg N ha^{-1}	45.97	41.10	31.54	39.54
92 kg N ha^{-1}	47.72	45.27	30.83	41.27
138 kg N ha^{-1}	41.87	41.08	28.93	37.29
Mean	44.3	41.9	28.89	38.36
LSD (0.05)	2.82			
CV%	4.13			

Table 8: Interaction effect of nitrogen rates on grain yield hectare^{-1} (quintal) of sorghum genotypes.

Nitrogen rates	Sorghum varieties			Mean
	Lalo	Local	Chemada	
0 kg N ha^{-1}	20.64a	17.11 cb	10.24f	15.99
46 kg N ha^{-1}	21.02a	18.12 cd	12.29ef	17.14
92 kg N ha^{-1}	22.0a	19.63ab	13.38f	18.34
138 kg N ha^{-1}	21.3a	18.97 ed	12.85e	17.71
Mean	21.24	18.46	12.19	17.3
CV%	8.54			

Table 9: Interaction effect of nitrogen rates on harvest index of sorghum genotypes.

higher net assimilation rate and increased grain yield as also found by Al-Abdulsalam [31]. Corroborating the results of this study, Blankenau et al. [32] stated that proper rate and time of N application are critical for meeting crop needs, and considerable opportunities exist for yield improvement.

Harvest index (HI): The physiological efficiency or translocation of assimilates from source into economic sinks is known as harvest index (HI). The effect of nitrogen rates on harvest index was not significant, but the sorghum varieties differed significantly in harvest index. The interaction effect of nitrogen rates and sorghum cultivar on harvest index was highly significant (Table 9). Lawrence [33] reported that harvest index in maize increases when nitrogen rates increased. In the present experiment, with the increase in the rate of nitrogen application, harvest index increased significantly upto 92 kgN/ha. This indicates significantly lower biomass partitioning to grain production when N was increased beyond certain level. The lower mean HI values in this experiment with the higher N application might indicate the need for the enhancement of biomass partitioning through genetic improvement. In tandem with the results of this study, Abdo [34] reported highest harvest index from treatments with the lowest rate of nitrogen application. Among the test genotypes, Lalo had greater HI followed by Local and Chemada indicating genotypic variations in partitioning efficiency.

Nitrogen uptake and nitrogen use efficiency partitioning of sorghum genotypes

The current study revealed differential quantities of nitrogen partitioned into different plant parts (grain, stem and leaves) (Figure 1). The removal of N in both grain and straw showed a distinct increasing trend with increase in N supply up to 92 kgN/ha, beyond which it showed a marginal increase (Table 10). In terms of Nitrogen use efficiency (NUE) Sorghum genotypes differed distinctly. Many studies have reported variation for NUE and components of NUE at high and low N inputs [35] as well as significant effect of genotype and N fertilization [36]. NUE and components of NUE were influenced by environments where soil test results show low residual N. Reduction in NUE with increasing N supply could result from reduction in N uptake

Nitrogen rates	Sorghum varieties			Mean
	Lalo	Local	Chemada	
0 kg N ha^{-1}	20459a	20131ba	22123a	20904.33
46 kg N ha^{-1}	21046a	20529a	20100ba	20558.33
92 kg N ha^{-1}	23443a	21784ba	21087a	22104.67
138 kg N ha^{-1}	21099ba	20237a	17934b	19756.67
Mean	21511.75	20670.25	20311	20831
CV%	10.01			

Table 7: Interaction effect of nitrogen rates and sorghum genotypes on biological yield plot^{-1} (g).

Parameter	Lalo variety				Local variety				Chemada variety			
	0 kg N ha^{-1}	46 kg N ha^{-1}	92 kg N ha^{-1}	138 kg N ha^{-1}	0 kg N ha^{-1}	46 kg N ha^{-1}	92 kg N ha^{-1}	138 kg N ha^{-1}	0 kg N ha^{-1}	46 kg N ha^{-1}	92 kg N ha^{-1}	138 kg N ha^{-1}
Nitrogen use efficiency (NUE) kg kg^{-1}	-	90.74	49.07	25.94	-	134.67	68.97	40.66	-	58.73	29.55	18.5
Nitrogen utilization efficiency (NUTE) in kg kg^{-1}	-	869.22	1010.82	695.69	-	1400.25	1691.18	1087.75	-	510.45	629.39	524.08
Nitrogen uptake efficiency (NUPE) kg kg^{-1}	-	0.104	0.05	0.037	-	0.096	0.041	0.037	-	0.12	0.047	0.035
Nitrogen uptake of grain kg ha^{-1}	23.61	77.72	94.48	93.46	67.78	98.43	105.71	107.63	15.45	52.57	41.87	58.27
Nitrogen uptake of straw kg ha^{-1}	179.84	735.95	521.64	1254.24	195.28	692.98	1061.31	759.79	198.62	633.15	713.13	516.1
Nitrogen harvest index (NHI)in %	0.29	0.39	0.47	0.51	0.25	0.36	0.44	0.37	0.29	0.37	0.36	0.49
Nitrogen recovery efficiency(NRE) %	-	117.63	79.21	50.62	-	66.63	41.23	28.88	-	80.43	28.72	31.03
Agronomic efficiency kg ha^{-1}	-	0.12	0.075	-0.26	-	0.05	0.024	-0.13	-	0.082	0.0067	-0.065

Table 10: Nitrogen up take and use efficiency and, N harvest index of sorghum genotypes as influenced by different N levels.

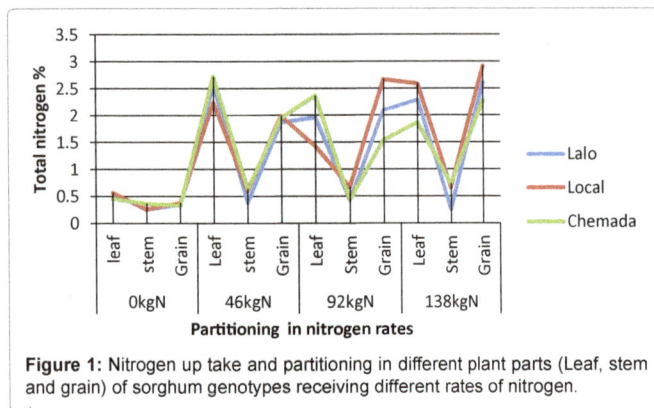

Figure 1: Nitrogen up take and partitioning in different plant parts (Leaf, stem and grain) of sorghum genotypes receiving different rates of nitrogen.

efficiency, N utilization efficiency and N retention efficiency. Nitrogen use efficiency is reported to be higher at lower N rates and decrease at higher N rates. This may indicate that plants are unable to absorb N when applied in excess quantities, because their absorption mechanism might have been saturated [37]. Under this condition, there exists the probability that more N will be subject to loss through ammonia gas, leaching or de-nitrification.

Among the genotypes, the Local variety tended to remove more N in both grain and straw than the improved genotypes Lalo and Chemada. This is in agreement with the findings of Nakamura et al. [38] reported that N absorption was regulated by root activities, higher and low–N conditions among grain sorghum genotypes. Greater N accumulation in the grain was associated with higher grain yields and NUE. Nutrient uptake by sorghum is influenced by several factors including nutrient availability, soil water availability, soil organic matter, soil chemical and physical properties, type of previous crop, plant population and the genotype. By and large, in the present study, with increase in the rate of N application there was an increase in the N removal in grain and straw, more N utilization efficiency, and greater N harvest index of sorghum genotypes, whereas, concomitantly there was a decrease in N use efficiency, N uptake efficiency, N recovery efficiency, and Agronomic efficiency (Table 10).

Nitrogen Use Efficiency (NUE): Irrespective of genotype, the NUE was maximum with 46 kgN/ha, and with every increase in N rate it decreased. Among the test genotypes, Local variety exhibited

maximum NUE (134.6) followed by Lalo (90.7) and the least was with Chemada (58.7) at 46 kg N ha^{-1}.

Nitrogen utilization efficiency (NUTE): It increased with increased N rate up to 92 kg/ha, and there after it dwindled at high rate of N application. With respect to genotypes, the Local variety showed maximum NUTE at 9 kg N ha^{-1}(1691), followed by Lalo (1010) and Chemada (629).

Nitrogen uptake efficiency (NUPE): It decreased with every increment in the rate of applied N, and the maximum was obtained with 46 kg N ha^{-1} in all the genotypes. Among the genotypes, Chemada showed maximum efficiency (0.12) followed by Lalo (0.104) and Local variety (0.096).

Nitrogen harvest index (NHI): This parameter distinctly improved with every increment in N rate from 0 through 138 kgN/ha in all the test genotypes. The improved genotype Lalo possessed higher NHI at varied rates of applied N followed closely by Chemada, while the local variety had the least NHI.

Nitrogen recovery efficiency (NRE): It declined substantially with every increase in the rate of N applied from 46 kg N ha^{-1} to 138 kg/ha. Lalo revealed maximu NRE (117.6) followed by Chemada (80.40) and Lovcal variety (66.6) receiving 46 kg N ha^{-1}.

Agronomic efficiency (AE): With every increase in N application rate, the Agronomic efficiency showed diminishing trend similar to that of N uptake efficiency, and N recovery efficiency. With respect to genotypes, the improved genotype Lalo offered maximum agronomic efficiency (0.12 kg/ha) than Chemada (0.082 kg/ha) and Local variety (0.05 kg/ha) with application of 46 kg N ha^{-1}. Beyond 138 kg N ha^{-1}, the agronomic efficiency tended to be negative.

Economics of sorghum genotypes and N fertilizer

To assess the cost and benefit associated with different treatments, the partial budget technique of CIMMT was applied on yield and biomass results. Based on this technique, as clearly shown it was found that the Lalo variety with a net benefit of 13979.39 Birr ha^{-1} was found to be the most profitable (Table 11), particularly when grown with a rate of 92 kg N ha^{-1} (Table 12). The partial budget analysis also indicated that 92 kg N ha^{-1} resulted in maximum relative net return of 12434.54 Birr ha^{-1} followed by 46 kg N ha^{-1} with (12237.04) Birr ha^{-1} relative net return. The least relative net return was recorded with 138 kg N ha^{-1}.

Sorghum Varieties	Grain yield (Q ha^{-1})	Dry biomass yield of sorghum (t ha^{-1})	Gross return (Birr ha^{-1})	Cost of Production (Birr ha^{-1})	Net return (GR – PC) (Birr ha^{-1})	Benefit: cost ratio (GR/PC Eth. Birr	Per day Productivity (GY/ CD kg ha^{-1})	Return/Birr Investment (NR/PC) ETB	Total variable cost	Net return (Eth. Birr ha^{-1})
Lalo	44.3	23.74	22730	16750.61	5979.39	1.36	24.98	0.36	16750.61	13979.39
Local	41.6	24.12	31860	16750.61	15109.39	1.9	35.69	0.9	16750.61	12109.39
Chemada	29.14	23.98	16896.5	16750.61	145.89	1.01	15.73	0.01	16750.61	545.89

Table 11: Economic analysis of sorghum genotypes.

Nitrogen rates applied	Grain yield (Q ha^{-1})	Dry biomass yield of sorghum (t ha^{-1})	Gross return (Birr ha^{1})	Cost of Production (Birr ha^{-1})	Net return (GR - PC Birr ha^{-1})	Benefit: cost ratio (GR/PC Birr)	Per day Productivity (GY/ CD kg ha^{-1})	Return/Birr Investment (NR/PC) ETB	Total variable cost	Net return (Birr ha^{-1})
0 kg N ha^{-1}	40.26	23.56	22829	10932.46	11896.54	2.08	25.4	1.09	10932.46	11896.54
46 kg N ha^{-1}	43.57	23.25	24256.5	12019.46	12237.04	2.02	26.42	1.02	12019.46	12237.04
92 kg N ha^{-1}	45.26	25.87	25541	13106.46	12434.54	1.95	26.57	0.95	13106.46	12434.54
138kg N ha^{-1}	39.15	22.99	22215.5	14193.46	8022.04	1.56	23.65	0.57	14193.46	8022.04

GR=Gross return; PC=Production cost; GY=Grain yield; CD=Crop duration; NR=Net return

Table 12: Economic analysis of N Fertilizer rates.

Conclusion

Sorghum genotypes investigated in the current study differed significantly in various growth, yield, and yield parameters, N use efficiency, N removal and economics. The maximum number of effective tillers (2.18) was recorded in response to nitrogen applied at the rate of 92 kg N ha^{-1} with the local variety followed by Lalo (1.2). The highest LAI was recorded with maximum application of N (138 kg N ha^{-1}) while Chemada variety scored the lowest LAI without fertilizer application. By and large, an increasing trend in LAI was observed with increased N application rates. Biological yield (BY) is a function of photosynthetic rate and proportion of the assimilatory surface area. The increase in biological yield with increase in rate of N might be due to better crop growth rate, LAI and accumulation of photo assimilate due to maximum days to maturity by the crop, which ultimately produced more biological yield. Biomass yield generally increased significantly with the increase in the rate of nitrogen. The Lalo variety recorded highest biomass yield when plants were supplied with 92 kg N ha^{-1}, and the lowest biomass yield was scored by Chemada variety fertilized with 138 kg N ha^{-1}. The application of highest level of N resulted in less biomass yield compared to 92 kg N ha. This might be due to the effect of lodging resulted from too high amount of N fertilizer that encourage vegetative growth and height leading to lodging before the translocation of dry matter to economic yield. Thousand seed weight was significantly increased with increase in the rate of nitrogen application. Panicle number was significantly influenced by the sorghum genotypes but not by nitrogen rates and the interaction of the two factors as well.

Nitrogen rates and sorghum genotypes interacted on grain yield and harvest index, the highest grain yield was with variety Lalo fertilized with 92 kg N ha^{-1} and the lowest being Chemada variety with no N application. With increase in the rate of nitrogen application, harvest index decreased significantly. In the current study there is differential effect of nitrogen on partitioning of assimilates into grain, stem and leaves. NUE and components of NUE influenced environments where soil test results show low residual N. The results indicated that there was genotypic difference in N uptake in plant parts (leaves, stems and grain). By and large, in the present study, with increase in the rate of N application there was an increase in the N removal in grain and straw, more N utilization efficiency, and greater N harvest index of sorghum genotypes. Whereas, concomitantly there was a decrease in N use efficiency, N uptake efficiency, N recovery efficiency, and Agronomic

efficiency. The economic analysis showed that variety Lalo with more net return was found to be the most profitable compared to the local variety. The partial budget analysis indicated that 92 kg N ha^{-1} resulted in maximum relative net return followed by 46 kg N ha^{-1}. Based on the genotypic difference in N uptake, partitioning and NUE in plant parts (leaves, stems and grain), and general performance of sorghum, the application of N fertilizer at the rate of 92 kg N ha^{-1} is suggested. Though these findings are zonal –specific, they can be applicable to similar agro-eco zones of Ethiopia [39]. The variables that can potentially affect these findings can be biotic and abiotic factors like rain fall, temperature, and soil which determines the moisture holding capacity and nutrient supply under rain fed farming system. In the years to come, research on sorghum should focus on relative drought tolerance of genotypes, sorghum–based cropping systems, and response to K, Ca, S and micro nutrients to alleviate soil health problems and sustainable production.

References

1. Central Statistical Authority, Agricultural Sample Survey (2006) Report on the area and major production for major crops of Private peasant holdings. Addis Ababa, Ethiopia.

2. FAO (2005) Scaling soil nutrient balance. Enabling same level application for African realities: FAO Fertilizer and Plant Nutrition Bulletin 15, FAO, Rome.

3. Stevenson FJ (1985) The nitrogen cycle in soil: Global and ecological aspects. In Stevenson, F J ed. Cycles of Soils. A Wiley Inter Science Publication. Wiley, New York, USA, pp: 106-153.

4. Kessel CV, Pennock DJ, Farrel RE (1993) Seasonal variation in de nitrification and nitrous oxide evolution at the landscape scale. Soil Sci Soc Am J 57: 988-995.

5. Gosh BC, Bhat R (1998) Environmental hazards of nitrogen loading in wetland rice fields. Environment Pollution 102: 123-126.

6. William PH (1992) The role of fertilizers in environmental pollution. In: Proc. Int Symp on nutrient management for sustainable productivity. Dept Soil, Punjab Agric Univ, Ludhiana, India, pp: 195-215.

7. Maranville JW, Pandey RK, Sirifi S (2002) Comparison of nitrogen use efficiency of a newly developed sorghum hybrid and two improved cultivars in the Sahel of West Africa. Comm in Soil Sci Plant Anal 33: 1519-1536.

8. Cottenie A (1980) Soil and plant testing as a basis for fertilizer recommendations. FAO Soil Bulletin, Rome 38: 61-100.

9. Page AL (1982) Methods of soil analysis- part two. Chemical and microbiological properties. American Society of Agronomy 3: 128-133.

10. Walkley A, Black CA (1934) An examination of Digestion of Degrjareff Method for Determining Soil Organic Matter and Proposed Modification of the Chromic Acid Titration Method. Soil Science 37: 29-38.

11. Olsen SR, Cole CV, Watanabe FS, Dean LA (1954) Estimation of available phosphorus in soils by extraction with sodium bicarbonate. US Dept Agric, p: 18.

12. Bouyoucos GJ (1962) Hydrometer method improved for making particle size analysis. Agronomy journal 54: 464-465.

13. Vanderlip RL (1993) How a sorghum plant develops. Cooperative extension service. Kansas Agricultural Experiment Station, Manhattan, Kansas, USA.

14. Jackson ML (1973) Soil Chemical Analysis. Prentice Hall Grice. Englewood Cliffs, USA, p: 284.

15. Moll RH, Kamprath EJ, Jackson WA (1982) Analysis and interpretation of factors which contribute to efficiency of nitrogen utilization. Agronomy Journal 74: 562-564.

16. Ortiz-Monasterio I, Sayre KD, McMahon MA (1997) Genetic progress in wheat yield and nitrogen use efficiency under four nitrogen rates. Crop Science 37: 898- 904.

17. CIMMYT (International Maize and Wheat Improvement Center) (1988) From Agronomic Data to Farmer Recommendations: An Economics Training Manual. Completely Revised Edition, Mexico, DF, p: 79.

18. SAS (Statistical Analysis System) (2002) SAS user guide, statistics SAS Inc. Cary. North Carolina, USA.

19. Tekalign M, Haque L, Aduayi EA (1991) Working Document: Soil, Plant, Fertilizer, Animal Manure and Compost Analysis Manual, International Livestock center for Africa, No. B13, Addis Ababa, Ethiopia.

20. Berhanu D (1980) The physical criteria and their rating proposed for land evaluation in the highland region of Ethiopia. Land use planning and regulatory department, Ministry of Agriculture. Addis Ababa, Ethiopia.

21. Onwueme IC, Sinha TD (1991) Field Crop Production in Tropical Africa. CTA, Wageningen, Netherlands.

22. Akbar F, Wahid Akhtar A, Ahmad S (1999) Optimization of method and time of nitrogen application for increased nitrogen use efficiency and yield in maize. Pakistan Journal of Botany 31: 337-334.

23. Sadeghi H, Bahrani MJ (2002) Effects of plant density and N rates on morphological characteristics and protein contents of corn. Iranian Journal of Agriculture Science 33: 403-412.

24. Botella MA, Cerda AC, Lips SH (1993) Dry matter production, yield and allocation of carbon-14 assimilate by wheat as affected by nitrogen source and salinity. Agronomy Journal 35: 1044-1049.

25. Genene G (2003) Yield and quality response of bread wheat varieties to rate and time of nitrogen fertilizer application at Kulumsa, southern Ethiopia. MSc Thesis, Alemaya University, Alemaya, Ethiopia.

26. Demir K, Yanmaz R, Ozcoban M, Kutuk AC (1996) Effects of different organic fertilizers on yield and nitrate accumulation in spinach. GAP 1. Vegetable Symposium, Sanliurfa, Turkey, pp: 256-257.

27. Haghighi BJ, Yarmahmodi Z, Alizadeh O (2010) Evaluation of the effects of biological fertilizer on physiological characteristics and yield and its components of corn (Zea mays L.) under drought stress. American Journal of Agricultural Biological Science 5: 189-193.

28. Haftom G, Mitiku H, Yamoah CH (2009) Tillage frequency, soil compaction and N-fertilizer rate effects on yield of tef (Eragrostis tef (Zucc.) Trotter). Ethiopian Journal of Science 1: 82-94.

29. Ashraf A, Khaid A, Ali K (1999) Effects of seeding rate and density on growth and yield of rice in saline soil. Pak Biol Sci 2: 860-862.

30. Melesse H (2007) Response of bread wheat (Triticum aestivum L.) varieties to N and P fertilizer rates. MSc Thesis, Haramaya University, Haramaya, Ethiopia.

31. Al-Abdul Salam MA (1997) Influence of nitrogen fertilization rates and residual effect of organic manure rates on the growth and yield of wheat. Arab Gulf Journal of Science Research 15: 647-660.

32. Blankenau K, Olfs HW, Kuhlmann H (2002) Strategies to improve the use efficiency of mineral fertilizer nitrogen applied to winter wheat. Journal of Agronomy and Crop Sci 188: 146-154.

33. Lawrence JR, Ketterings QM, Cherney JH (2008) Effect of nitrogen application on yield and quality of corn. Agronomy Journal 100: 73-79.

34. Abdo W (2009) Effect of different rates of nitrogen fertilizer on yield, yield related traits and quality of durum wheat (Triticum turgidum L. var. durum). Plant Soil 257: 205-217.

35. Gitelson AA, Buschmann C, Litchtender HK (1998) Leaf chlorophyll fluorescence corrected for re–absorption by means of absorption and reflectance measurement. J Plant Physiol 152: 283-296.

36. Le Gouis J, Béghin D, Heumez E, Pluchard P (2000) Genetic differences for nitrogen uptake and nitrogen utilization efficiencies in winter wheat. Eur Agron J 12: 163-173.

37. Fageria NK, Slaton NA, Baligar VC (2003) Nutrient management for improving low land rice productivity and sustainability. Adv Agron 80: 63-152.

38. Nakamura T, Adu–Gyamfi JJ, Yamamoto A, Ishikawa S, Nakano H, et al. (2002) Varietal differences in root growth as related to nitrogen uptake by sorghum plants in low–nitrogen environment. Plant Soil 245: 17-24.

39. Sheleme KS, Raghavaiah CV, Tesfaye B, Hamza I (2015) Differential productivity response of rain fed sorghum(Sorghum bicolor L.) genotypes in relation to graded levels of nitrogen in Kellam Wollega zone of Ethiopia, East Africa. Int J Life Sciences 3: 306-316.

34

Using Spatial Analysis of ANR1 Gene Transcription Rates for Detecting Nitrate Irregularities in Cherry Tomatoes (*Solanum lycopersicum* var. *cerasiforme*) Organic Greenhouses

Amir Mor-Mussery[1,2*], Orit Edelbaum[2], Arie Budovsky[3] and Jiftah Ben Asher[4]

[1] The Department of Geography and Environmental Development, Ben Gurion University of the Negev, Bee'r Sheva, Israel
[2] TheDepartment of Soil and Water Sciences, Hebrew University, Rehovot, Israel
[3] Technological Center, Biotechnology Unit, Beer Sheva, Israel
[4] Katif R&D Center for Coastal Deserts Development Ministry of Science, Israel

Abstract

Organic fertilizers differ from the chemical ones by high inconsistency in their mineralization rates in the soil. This intrinsic property occasionally results in formation of distinct 'soil patches' different from one another with regard to the concentrations of soluble nutrients. This feature together with the different nutritional requirements during the crop's physiological development complicates the designing of appropriate organic fertilization scheme. To overcome these difficulties a methodology was designed and tested based on analysis of molecular indicators. As a case study we tested cherry tomato (*Solanum lycopersicum* var. *cerasiforme*) organically grown in greenhouse located in Netzer Hazani village, Northern Negev, Israel at 2001. As a model nutrient we chose the soluble nitrate, due to its importance for the crop growth and development. We found that the best indicator of nitrate content were the ANR1 gene transcription rates which values were best correlated to the measured soluble nitrate in soil and the crop's needs during development. Practically, the spatial analysis helped identifying the surpluses and deficits of soluble nitrate in soil patches which were subsequently treated in a quick and precise manner. Implementing this methodology on other crops and nutrients will allow constructing accurate and economical fertilization scheme which decrease the damages to ecosystems.

Keywords: ANR1 gene; Gene transcription rate; Soil soluble nutrients; Cherry tomatoes (*Solanum lycopersicum* var. *cerasiforme*); Spatial analyses in organic farming

Introduction

The organic farming requires use of fertilizers from organic sources such as compost, animal manure, animal urine, etc. [1]. Due to the complex structure of most of these fertilizers and the differences in their types, origin and composition, it is difficult to assess their mineralization rates. As a result, the organic fields are characterized by 'areal soil patches' with wide range in the levels of the soluble nutrients, as opposed to the homogeneity observed after chemical fertilization [2]. Therefore, there is a high chance that parts of the fields will have mineral content irregularities (deficits and surpluses) which could damage the crop growth and reproduction [3]. In the case of extreme surpluses, the minerals could even infiltrate to the ground water [4]. Due to this pattern of patches distribution, random sampling will not be useful for locating the irregular plots [5-7]. Additionally, the nutritional needs of the crop change during its physiological growth [8]. A summary for this state was given by Diacono and Montemurro [1] with relation to Nitrogen in their work entitled "Organic farming: challenge of timing Nitrogen availability to crop Nitrogen requirements" (for other nutrients see the following references: Potassium- Stockdale et al. [9]; Sulphur-Scherer [10]).

An alternative diagnosis to the soil nutrients analysis is based on the visual analysis of crop leaf state [11]. The biggest advantage of Wolf (1982) diagnosis is reflection of the plant nutrients needs in the tested physiological state. Still, this diagnosis suffers from inaccuracy and in many cases the results are lately achieved and the farmer cannot use them for planning his following practices. In the spirit of Wolf (1982) approach we measured the transcription rates of the nutrient dependent genes in the plant young leaves. As opposed to the plant's DNA which remains un-changed during the plant life cycle (although claims have been made for possible changes in the chromosomes order, and even in the DNA sequences itself in higher plants in some cases; Walbot and Cullis [12], the mRNA amounts may change abruptly in response to the plant's surrounding ecosystem (for changes in gene expression due to climate changes see- Lawlor and Mitchell [13], soil nutrients-Lambers et al. [14], and atmospheric gases composition and radiation-Willekens et al. [15], and even due to different soil practices- Park et al. [16]. In the recent years measuring of the mRNA levels has become a routine and rapid practice which takes only several minutes based upon use of commercially available low cost kits [17]. The analysis is based on sampling flag leaves which are characterized by rapid changes in molecular expression [18]. In the framework of the analysis, the cells walls of the flag leaves are broken [19], the mRNAs of the carrier proteins are processed [20], and the mRNA amounts of the examined genes is calculated (in our case, those related to the plant nutrition).

Thus, the aim of this study was to collect and analyze the transcription data from the field in order to locate fast the heterogeneous patches in organic sandy soil greenhouses based on the crop's physiological stage.

Materials and Methods

Site of study

The study (an episode of long term study on growing organic crops in sandy soil and arid climate) was carried out in Netzer Hazani village,

---

*Corresponding author: Amir Mor-Mussery, Ben Gurion University of the Negev, Bee'r Sheva, Israel, E-mail: Amir.mussery@gmail.com

Northern Negev with rainfall of 400-500 mm year^{-1} (Longitude-15°34', Latitude-20°31'). The upper soil horizon is based on dunes' sand 10-15 m thick [21], of uniform pattern (Particles content: Sand: 93%, Silt: 0.5% and Clay: 0%, Organic matter: 6.5%, pH: 7.9 and EC: 1.7, 'Water holding capacity': 30-40% and 'Field capacity': 10-35%). The site of study was greenhouse of 30 × 118 m size, planted with trellised Cherry tomatoes Solanum lycopersicum var. cerasiforme, type '495' developed by Zeraim Gedera (distance between lines: 1 m and between the plants: 0.5 m, planting depth: 0.2 m). The seeds were sprouted and grown on organic bed for three weeks separately, and planted in the greenhouses at 15.02.2001. Prior to planting, 2.5 ton of 'compost' (the compost contains plant residues and animal manure, with relative ratio of 3:1, respectively), was mixed with the greenhouse soil at rate of 2.5 ton ha^{-1}. The elements concentrations of the compost determined by Elemental analyzer EA1108 Eager [22] were as follows: N-5.53%, C-13.197%, H-6.24%, S-2.06%, C/N ratio of 1:3), Gaskell and Smith [23].

For maximizing the mRNA production in the crop leaves, despite the problematic aspects of the soil bed Tsoar [24], additional two organic fertilizers were added: During the growth season 0.15 Kg m^{-2} of Guano from Sunleaves© co. were added (Elements concentration from dry matter, based on supplier data is as follows: N-12%, P-11%, K-2%, www.sunleaves.com/Home/About).

Additionally, mount of 2 L was added to each plant (also called 'cow urine', composed of 95% water and 2.5% Urea, collected from cows breeder, Silva et al. [25] at planting and during the growth season, in order to assure that the essential nutritional elements were adequately supplied. The remaining agro technology implementation was based on organic farming protocols Baker [26].

As the key analyzed nutritional molecule, we chose the Soil Soluble Nitrate (SSNt) concentration due to the crops' high Nitrate consumption rates from the soil by most of the crops as compared to the other nutritional compounds [27]. This choice was also evident in view of the tight correlation between the crops' physiological development stage and their consumption rates of soil Nitrate [28]. Finally, the studied crop Cherry tomatoes (Solanum lycopersicum var. cerasiforme) was chosen due to its fast-growing rates, and rapid changes in different physiological stages [29]. For additional description of the agrotechnology see Mor-Mussery et al. [30].

Soil nitrate sampling and analysis

Soil samples, each weighing approximately 0.5 Kg, were taken in between the surface and 0.2 m depth at 07.04.2001. Overall, 27 samples were taken from the whole greenhouse in three sampling lines (distance of 10 m between the lines). The average distance between the samplings was 5 m. The extreme samples were taken 5 m from the edges to diminish the side effects. The soil samples were dried and mixed with distilled water (70 ml water per 100 gr of soil) for 24 hours. Afterwards, the soil solution was analyzed using spectrophotometer and compared to calibrated samples of known concentrations [22].

Plant tissue sampling and molecular analyses

From the whole greenhouse a representative area of 154 m^2 (14 × 11 m with ~150 plants) was chosen (due to the research limitations) for the molecular analysis. From this area nine plants with the same physiological state were randomly marked based on gridding scheme, Miesch [31]. Immediately after the flower beds appearance (at 10.04.2001, eight weeks after planting, early in the morning when the genes' transcription rates are minimally affected by sunlight radiation; Jansen et al. [32]) two flag leaves from each chosen plant were sampled, as suggested by Gregersen

and Holm [33]. The leaves were frozen in liquid Nitrogen tank. The frozen tissues of the sampled plants were used for extracting the total RNA (RNA$_{Total}$) using Quiagen® RNeasy mini kit (www.qiagen.com/shop/sample-technologies/rna-sample-technologies/total rna/rneasy-mini-kit). Afterwards, analyses were performed in order to assess the RNA$_{Total}$ solutions purity. Primers known from scientific documentation to belong to eight genes associated with Nitrogen acquisition and assimilation were scanned on hybridization gel for identifying their binding specificity [34]. The primers themselves were synthesized by the authors using data from the Gene bank (www.ncbi.nlm.nih.gov). The genes were: ANR1 (gene related to roots elongation; Mathews et al. [35]); NRT1-2 (genes encoding for Nitrate transporter protein; Hildebrandt et al. [36]); Nia (gene which encodes the Nitrate reductase enzyme; Hildebrandt et al. [36]), and AMT1-3 (genes encoding ammonium transporter proteins; Von Wirén et al. [37]). The results of the gene specification analysis demonstrated that the synthesized primers of Nia (related to nitrate assimilation) and ANR1 (related to nitrate acquisition) genes demonstrated high specificity of hybridization, while the other primers bound to many unrelated genes. After this preliminary step, two analyses were performed - a PCR analysis for detecting the total amount of the Nia and ANR1 genes in each leave tissue, and Reverse Transcription PCRs (RT-PCR) for calculating the mRNA amounts related to these genes. The ratio between the mRNA and its DNA found in each examined tissue represents the Gene Transcription Rate (GTR), for further description see Spitzer-Rimon et al. [38].

Spatial analysis principles

The spatial analysis is based upon the assumption that plots in a given area have reciprocal influence on one another. As stressed earlier, in the field this influence manifests itself by the presence of areal soil patches. Specifically, for this study, the SSNt analysis, was performed as follows:

1. Measuring the soil's Nitrate soluble concentrations and integrating it with the spatial locations.

2. Calculating the semi-variances of the Nitrate and graphing it against the separation distances of the measured plots - called 'semi-variogram'.

3. The obtained semi-variances were used for calculating the Nitrate values all over the tested area based on 'Simple Kriging' interpolation [39]. Finally, the values were presented as distribution map divided into seven equal values groups, which allows locating the soil's soluble nitrate irregularities (see also, Kravchenko et al. [40].

The described methodology was performed similarly for the Nia and ANR1 GTRs.

4. Crossing over the values maps of the SSNt and mRNAs concentrations belonging to different genes by using 'Multi-layer analysis' allowed assessing the spatial interactions between them, and subsequently choosing the most reliable genes for the SSNt state [41].

These analyses were performed by the ArcGIS Geostatistical Analyst®^{1} ver 10. and GS+® ver. 9.

*Any tool, equipment, fertilizer, software etc. mentioned in the paper do not reflect the authors' preferences.

Results

Figure 1 represents the SSNt concentrations distribution at the site of study after the spatial analysis. The patchiness pattern of the different

Using Spatial Analysis of ANR1 Gene Transcription Rates for Detecting Nitrate Irregularities in Cherry Tomatoes...

183

SSNt concentrations is easily demonstrated with strips shape. Visual comparison of the calculated data from this study to the findings of Frias-Moreno et al. and Flores et al. [3,42] demonstrate that the SSNt concentrations of the 180-223 mg Kg^{-1} range are in the norm, which reflects more than 50% of variation from the studied area. The low values are mainly observed at the southern part, while the surpluses are found in the northern part.

The Nia and ANR1 GTRs are demonstrated in Figure 2A and 2B, respectively. The GTRs highest values for ANR1 are in the west side of the tested area, while the highest ones for the Nia were in the eastern part. Inorder to get more precise correlation, a 'pixel to pixel' analysis (termed as 'Multi-layer analysis', Barbosa et al. [41]) was performed, by defining 7 SSNt value groups for both genes analyses.

While the correlation between SSNt' concentrations and the ANR1's GTR could be defined by polynomial shape, with regard to the Nia we got linear shape (Figure 3B), both with regression coefficient of r^2=0.97 (Figure 3A).

Discussion

The significant correlations between the Nia and ANR1 transcription rates and SSNt concentrations demonstrate the usability of the described methodology in the case of Nitrate with relation to cherry tomatoes. This is true even in the case of organic management which is characterized by high heterogeneity in the soluble nutrient concentrations of the soil [2]. Crews and People [28] stressed that one of the biggest challenges of the present agriculture is "synchronizing the nitrogen supply and the crop demands", so the next step should be identifying the most indicative gene (in this case either the Nia or the ANR1) of the plant's SSNt needs. This data is valuable mostly before the ripening for planning immediate practices in order to get yield maximization (such as addition of cow urine or fish powder to correct SSNt deficits, CMG). In order to obtain description of the cherry tomatoes SSNt needs at ripening, we used Flores et al. data which define the norm for tomato crop between 180-220 ppm SSNt. In addition, Flores et al. [42] stressed that small deviations from this range could harm the plant's development. Comparisons of the correlation patterns between the Nia and ANR1 GTRs, with the SSNt needs (Figure 3) as described by Flores et al. [42] revealed that the most indicative gene is the ANR1. Our observations could be explained by the tight correlation of the ANR1 expression levels to the fruit ripening Jaakola et al. [43] which makes the ANR1 gene more sensitive to irregularities. This is in contrast to the Nia GTR which has increased

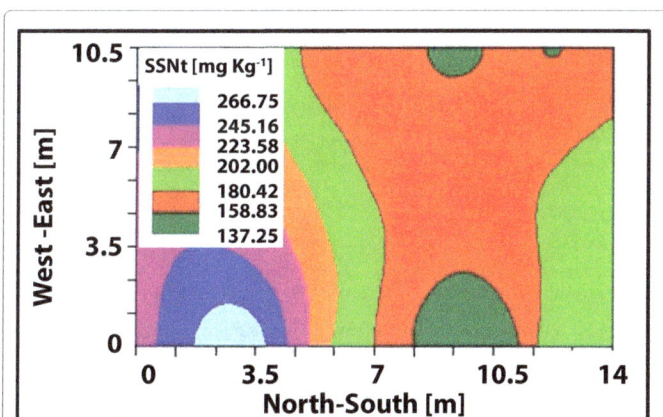

Figure 2: The spatial distribution of the cherry tomatoes Gene Transcription Rates (GTR) at ripening based on the spatial analysis of the flag leaves from the analyzed cherry tomato plants at the site of study.

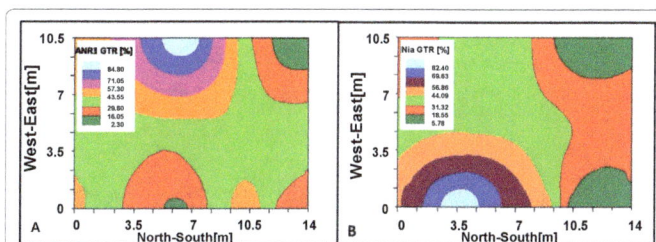

Figure 3: The correlations between the cherry tomatoes' Nia and ANR1 Gene Transcription Rate (GTR) and the Soil Soluble Nitrate (SSNt) content (A- correlation to ANR1 GTRs and B- correlation to Nia GTRs). The lines on bars' heads represent the S. Errs for the different genes transcription rates in the defined SSNt groups.

in the linear pattern even in the area of tomato surpluses which could be explained by its correlations to the vegetative growth of the tomatoes making it less sensitive to the SSNt surpluses (Frias-Moreno et al. [3].

For the farmer, the most important consideration of his work is the total costs and analysis duration. In Table 1 we compare the costs and durations of three methods for detecting the crop needs for nitrate in organic greenhouses. The first method is based on analyzing the total Nitrogen in the crop vegetative tissue. The second one is based on the soil soluble nitrate content, and the third one is based on "related nitrate" genes transcription rates (such as that of the ANR1 gene). It is easily demonstrable that mRNA analysis is not only the cheapest, but also the fastest method so the farmer can 'repair' different extremes in a fast manner.

Still, several points have to be taken into account while discussing the findings of this study. The cherry tomato variety used by us ('495' from Zeraim Gadera') is a newly introduced variety, and so the accurate data on the plant Nitrate needs were missing. In order to fill this gap, the authors used data from Frias-Moreno et al. [3] and Flores et al. [42] studies which were implemented on different varieties, so there is a reasonable chance that the actual SSNt needs of variety '495' are different. Secondly, as Flores et al. stressed, there is an impact of the ratio between the Nitrate and Ammonia concentrations on the plant SSNt needs. Thirdly, the processing methodologies, soil characteristics, etc. have high effect on the SSNt needs (Frias-Moreno et al. [3]; Beaudoin et al. [44]).

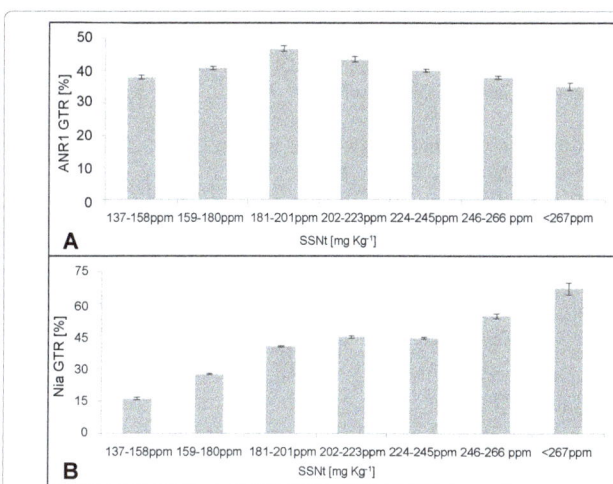

Figure 1: The distribution of the Soil Soluble Nitrate (SSNt) patches over the studied area, based on the spatial analysis of the soil samples collected at the site of study.

	Total Nitrogen content analysis in vegetative tissues	Soil soluble nitrate content analysis	ANR1 transcription rate analysis with kit
Field working			
Duration	1 hour	1 hour	1 hour
Manpower*	35$	35$	35$ (Iced CO_2:15$)
Lab working			
Duration	Tissues drying-2 Dy, Extraction- 1 Dy, Analysis- 1 Dy	Soil drying-1 Dy, Extraction- 1 Dy, Analysis-1 Dy	2 hours[2]
Costs	800$ (40$ per sample)	600$ (40$ per sample)[3]	Kit-110$ (5.5$ per sample)[1]; Manpower-40$ (20$ hour[-1])
Spatial An.	Several minutes	Several minutes	Several minutes
Total (for 20 samples)			
Total duration, costs	4 Days, 835$	3 Days, 635$	3 hours, 200$

Dy- days, An. Analysis; *The costs for the 'Total Nitrogen content analysis in vegetative tissue' and 'Soil soluble nitrate content' are based on data supplied by Gilat Field Services Lab® (http://www.bns.org.il), while the mRNA analysis costs are based on separation of the manpower and kit costs for Quiagen® RNeasy mini kit 250 samples

Table 1: Comparisons between different methods for spatially assessing the crop's nitrate requirements based on common set for field study of 20 samples).

Despite the above-mentioned reservations, future studies based on the principles described in this work will allow constructing comprehensive databases, containing for each crop the most indicative gene which mode of expression will be correlative with certain physiological stage of the crop such as vegetative growth, flourishing, ripening, and the reproductive stage [45]. The databases will also provide information on other factors having impact on the yield such as soil types, soil and irrigation, water salinity, and pH rates, so that the local farmers could adjust the relevant parameters for maximal production from their fields [46,47].

As stressed earlier, the costs of the mRNA extraction kits and subsequent analyses continue to decrease. Thus, the gene expression profiles could be obtained in a fast and easy manner [17].

Conclusions

We believe that by applying data regarding the crop soluble nutrients consumption amounts in a given physiological state, one can benefit from the gene transcriptional analysis presented here. This methodology is crucial mainly for nutrients characterized by difference between their fixed content and the soluble forms in the common fertilizers, as demonstrated here on Nitrate from organic source. The presented spatial analyses could be easily adjusted for many different fields such as greenhouses, or open areas under any processing scheme.

Acknowledgments

The paper is dedicated to Hilberg family from Nezer Hazani village for their extended help for implementing this research.

References

1. Diacono M, Montemurro F (2010) Long-term effects of organic amendments on soil fertility. A review. Agron Sust Dev 30: 401-422.

2. Tandon HLS (2005) Methods of analysis of soils, plants, waters, fertilizers and organic manures. Fertilizer Development and Consultation Organization.

3. Frias-Moreno N, Nuñez-Barrios A, Perez-Leal R, Gonzalez-Franco AC, Hernandez-Rodriguez A, et al. (2014) Effect of Nitrogen deficiency and toxicity in two varieties of tomatoes (Lycopersicum esculentum L.). Agric Sci 5: 1361-1368.

4. Addiscott TM, Whitmore AP, Powlson DS (1991) Farming, fertilizers and the nitrate problem. CAB International (CABI), p: 170.

5. Heinze S, Raupp J, Joergensen RG (2010) Effects of fertilizer and spatial heterogeneity in soil pH on microbial biomass indices in a long-term field trial of organic agriculture. Plant Soil 328: 203-215.

6. Watson CA, Atkinson D, Gosling P, Jackson LR, Rayns FW (2002) Managing soil fertility in organic farming systems. Soil Use and Management 18: 239-247.

7. Pang XP, Letey J (2000) Organic farming challenge of timing nitrogen availability to crop nitrogen requirements. Soil Sci Soc America J 64: 247-286.

8. Le Bot J, Adamowicz S, Robin P, Andriolo JL, Gary C (1997) Modelling nitrate uptake by greenhouse tomato crops at the short and long time scales. II. Modelling Plant Growth, Environmental Control and Farm Management in Protected Cultivation. ISHS Acta Hortic 456: 237-246.

9. Stockdale EA, Shepherd MA, Fortune S, Cuttle SP (2002) Soil fertility in organic farming systems-fundamentally different?. Soil Use and Management 18: 301-308.

10. Scherer HW (2001) Sulphur in crop production-invited paper. Europ J Agron 14: 81-111.

11. Wolf B (1982) A comprehensive system of leaf analyses and its use for diagnosing crop nutrient status. Comm Soil Sci Plant Anal 13: 1035-1059.

12. Walbot V, Cullis CA (1985) Rapid genomic change in higher plants. Ann Rev Plant Physiol 36: 367-396.

13. Lawlor DW, Mitchell RA (2000) Crop ecosystem responses to climatic change: wheat. Climate change and global crop productivity, CAB International Press, pp: 57-80.

14. Lambers H, Raven JA, Shaver GR, Smith SE (2008) Plant nutrient-acquisition strategies change with soil age. Trends Ecol Evol 23: 95-103.

15. Willekens H, Van Camp W, Van Montagu M, Inze D, Langebartels C, et al. (1994) Ozone, sulfur dioxide, and ultraviolet B have similar effects on mRNA accumulation of antioxidant genes in Nicotiana plumbaginifolia L. Plant Physiol 106: 1007-1014.

16. Park W, Scheffler BE, Bauer PJ, Campbell BT (2012) Genome-wide identification of differentially expressed genes under water deficit stress in upland cotton (Gossypium hirsutum L.). BMC Plant Biology 12: 90.

17. Rump LV, Asamoah B, Gonzalez-Escalona N (2010) Comparison of commercial RNA extraction kits for preparation of DNA-free total RNA from Salmonella cells. BMC Res Notes 3: 211.

18. Zhou J, Wang X, Jiao Y, Qin Y, Liu X, et al. (2007) Global genome expression analysis of rice in response to drought and high-salinity stresses in shoot, flag leaf, and panicle. Plant Molecular Biology 63: 591-608.

19. Lin Y, Fuerst O, Granell M, Leblanc G, Lórenz-Fonfría V, et al. (2013) The substitution of Arg149 with Cys fixes the melibiose transporter in an inward-open conformation. Biochimica et Biophysica Acta (BBA)-Biomembranes 1828: 1690-1699.

20. Lin Y, Bogdanov M, Tong S, Guan Z, Zheng L (2016) Substrate selectivity of lysophospholipid transporter LplT involved in membrane phospholipid remodeling in Escherichia coli. Journal of Biological Chemistry 291: 2136-2149.

21. Ministry of the Environment Planning Department (1999) Coastal Zones Management in Israel.

22. Eager (2001) Elemental Analyzer, Reference Manual, Eager Press, USA.

23. Gaskell M, Smith R (2007) Nitrogen sources for organic vegetable crops. Hort Technol 17: 431-441.

24. Tsoar H (1990) The ecological background, deterioration and reclamation of desert dune sand. Agric Ecosyst Environ 33: 147-170.

25. Silva RG, Cameron KC, Di HJ, Hendry T (1999) A lysimeter study of the impact of cow urine, dairy shed effluent, and nitrogen fertiliser on nitrate leaching. Aus J Soil Res 37: 357-370.

26. Baker B (2009) Organic practice guide. Organic materials review institute.

27. Ullrich WR (1992) Transport of nitrate and ammonium through plant membranes. In: Mengel K, Pilbeam DJ (eds.), Nitrogen Metabolism of Plants, pp: 121-137.

28. Crews TE, Peoples MB (2005) Can the synchrony of nitrogen supply and crop demand be improved in legume and fertilizer-based agroecosystems? A review. Nutr Cycl Agroecosys 72: 101-120.

29. Hobson G, Grierson D (1993) Tomato. Biochemistry of Fruit Ripening, pp: 405-442.

30. Mor-Mussery A, Budovsky A, Ben-Asher J (2014) Geostatistical methods to evaluate the response of cherry tomato to soil nitrate. Int J Organic Agric Res Dev 9: 8-19.

31. Miesch AT (1976) Methods of sampling, laboratory analysis, and statistical reduction of data. Geological Survey of Missouri.

32. Jansen MA, Gaba V, Greenberg BM (1998) Higher plants and UV-B radiation: balancing damage, repair and acclimation. Trends in plant science 3: 131-135.

33. Gregersen PL, Holm PB (2007) Transcriptome analysis of senescence in the flag leaf of wheat (Triticum aestivum L.). Plant Biotech J 5: 192-206.

34. Velculescu VE, Zhang L, Vogelstein B, Kinzler KW (1995) Serial analysis of gene expression. Science 270: 484-487.

35. Mathews H, Clendennen SK, Caldwell CG, Liu XL, Connors K, et al. (2003) Activation tagging in tomato identifies a transcriptional regulator of anthocyanin biosynthesis, modification, and transport. The Plant Cell 15: 1689-1703.

36. Hildebrandt U, Schmelzer E, Bothe H (2002) Expression of nitrate transporter genes in tomato colonized by an arbuscular mycorrhizal fungus. Physiol Plant 115: 125-136.

37. Von Wirén N, Lauter FR, Ninnemann O, Gillissen B, Walch Liu P, et al. (2000) Differential regulation of three functional ammonium transporter genes by nitrogen in root hairs and by light in leaves of tomato. The Plant J 21: 167-175.

38. Spitzer-Rimon B, Marhevka E, Barkai O, Marton I, Edelbaum O, et al. (2010) EOBII, a gene encoding a flower-specific regulator of phenylpropanoid volatiles› biosynthesis in petunia. Plant Cell 22: 1961-1976.

39. Bohling G (2005) Kriging Kensas Geology Press.

40. Kravchenko AN, Bullock DG, Boast CW (2000) Joint multifractal analysis of crop yield and terrain slope. Agron J 92: 1279-1290.

41. Barbosa JPRAD, Rambal S, Soares AM, Mouillot F, Nogueira JMP, et al. (2012) Plant physiological ecology and the global changes. Ciênciae Agrotecnologia 36: 286-269.

42. Flores P, Carvajal M, Cerdá A, Martínez V (2001) Salinity and ammonium/ nitrate interactions on tomato plant development, nutrition, and metabolites. J Plant Nutr 24: 1561-1573.

43. Jaakola L, Poole M, Jones MO, Kämäräinen-Karppinen T, Koskimäki JJ, et al. (2010) A SQUAMOSA MADS box gene involved in the regulation of anthocyanin accumulation in bilberry fruits. Plant Physiol 153: 1619-1629.

44. Beaudoin N, Saad JK, Van Laethem C, Machet JM, Maucorps J, et al. (2005) Nitrate leaching in intensive agriculture in Northern France: Effect of farming practices, soils and crop rotations. Agric Ecosys Environ 111: 292-310.

45. Chaïb J, Devaux MF, Grotte MG, Robini K, Causse M, et al. (2007) Physiological relationships among physical, sensory, and morphological attributes of texture in tomato fruits. J Exp Bot 58: 1915-1925.

46. Afzal M, Yousaf S, Reichenauer TG, Kuffner M, Sessitsch A (2011) Soil type affects plant colonization, activity and catabolic gene expression of inoculated bacterial strains during phytoremediation of diesel. J Hazard Mater 186: 1568-1575.

47. Liu B, Mørkved PT, Frostegård Å, Bakken LR (2010) Denitrification gene pools, transcription and kinetics of NO, N_2O and N_2 production as affected by soil pH. FEMS Microbiol Ecol 72: 407-417.

35

The Influence of Intercropping Sorghum with Legumes for Management and Control of Striga in Sorghum at Assosa Zone, Benshangul Gumuz Region, Western Ethiopia, East Africa

Getahun Dereje[1]*, Tigist Adisu[1], Megersa Mengesha[1] and Tesfa Bogale[2]

[1]*Assosa Agricultural Research Centre, Assosa, Ethiopia*
[2]*Jimma Agricultural Research Centre, Jimma, Ethiopia*

Abstract

Sorghum is an important cereal crop and occupies third place in production after Maize and Tef in Ethiopia. *Striga hermonthica* reduces sorghum yields by competing for water, nutrients, space, light and photosynthates with the host plants. Information on the influence of intercropping sorghum with legumes for management and control of striga in sorghum in the Assosa Zone is scanty. On farm experiments were they are conducted at three locations, for three years, to investigate the effect of eleven treatments (Intercrop ground nut with Sorghum 1:1 and Simultaneous planting, Intercrop ground nut with sorghum 1:1 and Relay planting, Intercrop ground nut with Sorghum 2:1 and Simultaneous planting, Intercrop ground nut with Sorghum 2:1 and Relay planting, Intercrop soybean with sorghum 1:1 and Simultaneous planting, Intercrop soybean with sorghum 1:1 and Relay planting, Intercrop soybean with sorghum 2:1 and Simultaneous planting, Intercrop soybean with sorghum 2:1 and Relay planting, Sole soybean, Sole Sorghum and Sole Ground nut.) and was laid out in a randomised complete block design (RCBD) with three replication. Significant influence on the grain yield of sorghum due to treatment application was recorded. During the all season, the sorghum/legume intercrop had the highest sorghum yield. The sorghum/Ground nut intercrop out yielded than the sorghum/soybean intercrop at all growing season. The gross income and Land equivalent ratio indicates greater economic benefit with this intercropping groundnut in 1:1 proportion and simultaneous planting than sole planting. As a result, intercropping groundnut in 1:1 proportion and simultaneous planting for the control striga is essential, ideal and useful to small-scale farmers, in order to achieve sustainable crop production.

Keywords: Sorghum; *Striga hermonthica*; Intercropping; Yield; Land equivalent ratio

Introduction

Striga hermonthica reduces yields by competing for water, nutrients, space, light and photosynthates with the host plants El-Halmouch et al. [1]. In Africa, crop yield losses associated with Striga related activities is about 40% and represents an annual loss of cereals worth US$7 to 13 billion Khan et al. [2]. In East Africa, *S. hermonthica* is the most important species causing an estimated 20-100% total loss for maize, sorghum and millet Emechebe and Ahonsi [3]. *Striga hermonthica* (Del.) Benth (*Scrophulariacae*) is one of the major production constraints in the subsistence sorghum producing farmers in Ethiopia. This problem is further aggravated by the inherent low soil fertility, recurrent drought and overall natural resource degradation because of decades of continuous cereal monoculture and deforestation Fasil [4]. Many environmental factors including soil temperature and moisture status, may affect the growth and development of Striga species either acting directly on the weed or by mediation through the host Oryokot [5]. In sub-Sharan Africa, Striga is exasperated by its exquisite adaptation to the climatic conditions of the semi-arid tropics, its high fecundity, and longevity of its seed reserves in tropical soils Gebisa [6], Gebisa et al. [7]. This is, therefore, reducing the seed bank from the infested field must be considered as one strategy for effective control of Striga.

Controlling *Striga* like other root parasites is such a challenge because the weed can do much damage to the host crop before emerging above the ground. Cultural, mechanical, chemical and biological control measures are available to regulate the parasite population. However, few of these techniques can provide complete Striga eradication and it is usually necessary to use a combination of these methods most relevant to the farming system Parker and Riches [8], Bebawi [9].

Similarly, in Benishangule-Gumuze region, for instance, where *striga* resistant varieties and chemical controls are not possible and affordable, the loss is still occurred and becoming devastating. Although sorghum yield loss due to striga has not been well documented in the region, it could reach up to total failure of the crop depend on the severity of infestation in the region (personal observation). Hence, farmers have no option rather to shift to other crops even though sorghum is their staple food and major crops. Sorghum covers about 25.9% of total area cultivated in the region (CSA, 2014/15). Moreover, these area do not have recommended and effective control methods against striga to their site. On the other hand, even if there has been a released resistant sorghum variety, they didn't adapt well in this region. Hence, looking for cost effective and applicable control measure against this serious weed is of a paramount importance for the resource poor farmers in the region. An effective Striga management program should include a reduction in the number of *Striga* seed from the heavily infested soils and prevention of further seed multiplication [10]. Soil active herbicides have been identified which give partial control of *S. hermonthica* before it emerges from the soil. Dicamba has been used to effectively control *S. asiatica* in the USA Eplee and Norris [11]. In Kenya, Odhiambo and Ransom [12] found Dicamba to be effective

*Corresponding author: Getahun Dereje, Assosa Agricultural Research Centre, Assosa, PO Box 265, Ethiopia, E-mail: getahundereje11@yahoo.com

only when applied at the peak of *Striga* germination and attachment. Rotating or intercropping Sorghum with trap crops such as Soy beans (*Glycine max* L.), groundnut (*Arachis hypogea* L.), Bambara nut (*Vigna subterranea* (L.)Walp), sunflower (*Helianthus annus* L.) and cowpea (*Vigna unguiculata* (L.) Walp) may help to reduce the number of *S. hermonthica* seed in the soil Ejeta and Butler [13], Odhiambo and Ransom [12].

Carson [14] showed that intercropping of sorghum (*Sorghum bicolor* (L.) Moench) and groundnut significantly reduced *Striga hermonthica* (Del) Benth emergence. This was associated with a decrease in soil temperature in the intercropped plots. In Gambia, alternating sorghum or millet (*Pennisetum americanum* (L.) K. Schum) with groundnut resulted in a low *S. hermonthica* infestation Lagoke et al. [15]. Soybeans (*Glycine max* L.), cotton (*Gossypium hirsutum* L.) and bambara nut when grown as intercrops with maize or sorghum, are known to induce abortive germination of *S. hermonthica* seeds, with a consequent reduction in infestation Ejeta and Butler [13]. This has advantages of depleting the seed bank and ensuring that no new seed is added to the soil Sauerborn [16].

Therefore, this research activity was initiated with the objective of determine the best intercrop legume with sorghum by its control ability of striga and yield advantage and to determine appropriate spatial and temporal arrangements on legume and sorghum intercropping.

Materials and Methods

Description of the study area

This experiment was conducted at Assosa Zone, on three woreda's s at Bambasi, Assosa, and Homosha in western Ethiopia during the main rainy season of 2012, 2013 and 2014. They are found at an altitude ranging between 1300-1470 masl with the minimum and maximum temperatures of 14.5 and 28.8°C, respectively. The average annual rainfall of 1358 mm of which 1128.5 mm were received between May and October during the cropping season.

Experimental details

The experimental land was well prepared. Initially, seed were planted in drilling methods and latter thinned to one plant per hill. The spacing was 0.75 m and 0.25 m between rows and plants, respectively. A total of eight rows were kept on each plot. Each plot and block was separated by 0.75 m and 1.5 m, respectively. Local Sorghum variety those farmers practices was used for the experiment. Important agronomic practices like hoeing and weeding were uniformly applied to all experimental plots as often as required.

The experiment was conducted on farmer's fields where striga infestations have been relatively higher. The study was involving three factors: two legumes (soybean and ground nut), two spatial arrangements of sorghum to legume rows ratio (1:1 and 2:1), and two temporal arrangements (Simultaneous and relay planting, planting legumes 3-4 weeks after sorghum planting). In addition sole stand of all crops as the legumes and the sorghum were included as a treatment in all replications. All the treatments were laid down in RCBD with three replications. In all intercropping treatments where the legumes' rows are beside sorghum rows half the recommended row spacing of sorghum was maintained i.e., 37.5 cm. Each plot area was 4 m × 6 m space between blocks and plots was 1.5 m and 1 m, respectively.

The treatments are intercropped groundnut in 1:1 and simultaneous, and relay planting; intercropped groundnut in 2:1 and simultaneous, and relay planting; intercropped soybean in 1:1 and simultaneous planting, and relay planting; intercropped soybean in 2:1 and simultaneous and relay planting; Sole soybean, sorghum and Groundnut making a total treatment of eleven treatments.

Data collection and analysis

Prior to the field experimentation, ten to fifteen per sites random samples (0-30 cm depth) were collected and a composite soil sample was made. Similarly, post crop harvest soil samples were collected from each plot receiving different treatments for selected soil physical and chemical analysis (pH, OC, N, P, K, CEC, texture and bulk density). Striga count will be recorded in all physiological stages of the main crop phenology. Yield and yield determinant factors of both crops (sorghum and the intercrop) were also taken from all plots. Land equivalent ration and economic advantage of the cropping system was computed and Farmer's assessment were done by inviting equal number of female and male farmers to evaluate and share their opinions at mid grain filling stages of the crop.

Analysis of variance was carried out for the yield studied following statistical procedures appropriate for the experimental design using SAS computer software. Whenever treatment effects will be significant, the means will be separated using the least significant difference (LSD) procedures test at 5% level of significance.

Results and Discussion

Soil physico-chemical properties

The soil pH of the study site ranges from very strongly acidic (4.78) to neutral (7.4). At Amba 14, where the soil pH is very strongly acidic at a pH of 4.78, there is possibility of Al toxicity and deficiency of certain plant nutrients while at Amba 5 is neutral.

The total nitrogen ranges from 0.15 to 0.18 and characterized by low according to Landon [17]. Similarly, the available P ranges from 0.15 to 11.05 characterized by low and marginally medium, respectively. The organic carbon (OC) of these soils varies from 1.68 to 3.12 and fall in the ranges of very low to low Landon [17]. The exchangeable K of the soil ranges from 0.13 cmol$^{(+)}$ kg^{-1} to 0.42 cmol$^{(+)}$ kg^{-1}, indicating that these soils has had deficient to adequate ranges respectively. The very low OC and associated low N and available P indicated that the soils in the study area are poor in nutrient supplying power to the growing crops. This could be due to continuous cultivation and lack of incorporation of organic materials.

Population density of *Striga hermonthica*, Sorghum yields and Land Equivalent Ratio (LER)

Population density of striga: Generally, the highest number of *S. hermonthica* were recorded in sole sorghum (777/m²) while the sorghum/legume intercrop repress the presence of striga and resulting lower its density at all sites during the growing season. Intercropping sorghum with ground nut able to reduce the emergence of striga by 11.98-70.24% while with soya bean it varies from 2.95 to 54.21% with 2:1 and 1:1 spatial arrangements, respectively. In ground nut, however, increasing the temporal arrangement from simultaneously to relay helps to suppress the emergence of striga while for soya bean it favors. On the contrary, spatial arrangements do have similar effect on the emergence of striga for both crops as intercropped with 1:1 suppress striga significantly than 2:1 (Table 2).

The lower number of *S. hermonthica* plants that emerged in the intercrops during the growing season indicated a reduced potential for overall flower and capsule production and, consequently, a reduced capacity of increasing the *S. hermonthica* seed bank in the soil. It is, however, important to note that the number of *S. hermonthica* plants that emerged represents an unknown and often variable percentage of the total number of *S. hermonthica* plants that actually parasitize the host's roots. The decreased number of *S. hermonthica* plants in the Sorghum/legume intercrop may be attributed to the suicidal germination caused by the germination stimulant produced by the legume roots. In addition to being a trap crop, Ground nut provides shade which smothers the witch weed thereby reducing its vigor. These two factors are detrimental to the growth and development of *S. hermonthica* plants. Striga transpires less when shaded, thereby reducing the amount of nutrients and water drawn from the maize Stewart and Press [18].

Therefore, it can be concluded that in order to control striga emergence if the intercropped is ground nut, it is better to go for relay cropping and if the crop is soya bean simultaneous planting with 1:1 spatial arrangement helps to suppress the emergence of striga.

Sorghum grain yields: During the all season, the sorghum/legume intercrop had the highest sorghum yield. The sorghum/Ground

nut intercrop out yielded than the sorghum/soybean intercrop at all growing season. Sorghum grain yields during the growing season were significantly different (P<0.05) in all treatments at all sites. The highest grain yield was obtained from the intercropped when compare to the sole. Sorghum grain yields were almost similar in the intercrops during the 2013 and 2014 growing season and this may indicate that in a good season, the type of legume intercropped with sorghum does not affect sorghum grain yields. However, there may be differences in the beneficial effects on sorghum yields in subsequent seasons due to residual nitrogen, which would have been fixed by the particular legume.

Partial Land Equivalent Ratios (PLER): Summarizes the PLER of Sorghum from the various intercrops for the all season. Intercropping sorghum and legumes at all sites resulted in greater than unit (above 1) PLER's for Sorghum (Tables 1-4).

Conclusions and Recommendations

The Sorghum/ground nut intercrop exhibited a higher potential for suppressing *S. hermonthica* emergence as compared to the other

No.	Districts	pH	CEC	OC (%)	N (%)	P	K
1	Amba 14	4.78	17.2	1.68	0.15	3.4	0.192
2	Homosha	7.02	35.53	3.12	0.179	11.05	0.2
3	Amba-5	7.4	25.84	2.48	0.168	7.5	0.13

Table 1: Soil physico-Chemical analysis of some parameters of soil prior to cropping.

Treatments	*Striga* Emergence plants /24 m²	Grain Yield (kg/ha)			LER	Gross Income (ETB)
		Sorghum	Soybean	Ground nut		
S+GN (1:1) & SP	295.10	1213	-	828	2.12	17214
S+GN (1:1) & RP	231.20	1433	-	515	2.23	14778
S+GN (2:1) & SP	683.90	934	-	502	1.55	11628
S+GN (2:1) & RP	615.10	863	-	243	1.30	8094
S+SB (1:1) & SP	355.80	423	897	-	1.11	9714
S+SB (1:1) & RP	395.10	1072	175	-	1.54	7832
S+SB (2:1) & SP	659.30	945	418	-	1.52	9014
S+SB (2:1) & RP	754.10	1155	134	-	1.63	8002
SSB	-	-	1661	-	-	13288
SS	777.00	747	-	-	-	4482
SGN	-	-	-	1663	-	19956
LSD	493.45	737	306	235	-	-
Sign (0.05)	**	**	**	**	-	-
CV	121.51	98.52	109.35	73.47	-	-
SE	69.91	97.18	285.36	246.32	-	-

Table 2: The influence of intercropping on *striga* emergence, yield, LER and income in 2012.

No.	Treatments	*Striga* Emergence plants /24 m²	Grain Yield (kg/ha)			LER	Gross Income (ETB)
			Sorghum	Soybean	Ground nut		
1	S+GN (1:1) & SP	69	2890.3	-	695.8	2.1	25691.4
2	S+GN(1:1) & RP	86.7	2629.2	-	332.6	1.67	19766.4
3	S+GN (2:1) & SP	311	1988.9	-	339.9	1.34	16012.2
4	S+GN (2:1) & RP	231	2398.6	-	134.9	1.39	16010.4
5	S+SB (1:1) & SP	202.3	2202.8	855.6	-	1.54	20061.6
6	S+SB (1:1) & RP	317.7	2222	328.5	-	1.33	15960
7	S+SB (2:1) & SP	284.7	2318	425	-	1.42	17308
8	S+SB (2:1) & RP	211	2390	526	-	1.5	18548
9	SSB	-	-	2420	-	-	19360
10	SS	575.3	1863.9	-	-	-	11183.4

No.	Treatments	Striga Emergence plants /24 m²	Sorghum	Soybean	Ground nut	LER	Gross Income (ETB)
11	SGN	-	-	-	1257.7	-	15092.4
12	LSD	372	1230	630.4	241.6	-	-
13	Sign(0.05)	*	NS	**	**	-	-
14	CV	62	29.6	43	23	-	-
15	SE	14.38	103.61	387.55	198.22	-	-

Table 3: The influence of intercropping on striga emergence, yield, LER and income in 2013.

No.	Treatments	Striga Emergence plants /24 m²	Grain Yield (kg/ha)			LER	Gross Income (ETB)
			Sorghum	Soybean	Ground nut		
1	S+GN (1:1) & SP	25.3	2126.7	-	557.5	2.05	19450.2
2	S+GN (1:1) & RP	73.7	2087.9	-	318.7	1.82	16351.8
3	S+GN (2:1) & SP	158.7	1873.6	-	305.4	1.39	14906.4
4	S+GN (2:1) & RP	152	2476.1	-	142.9	1.96	16570.8
5	S+SB (1:1) & SP	58.7	1826.5	669.6	-	1.68	16318.2
6	S+SB (1:1) & RP	120.7	2124.9	205.9	-	1.67	14396.6
7	S+SB (2:1) & SP	167	2120.4	199.7	-	1.66	14320
8	S+SB (2:1) & RP	219.3	1874.7	88.3	-	1.43	11954.6
9	SSB	-	-	2040.5	-	-	16324
10	SS	479.7	1351.8	-	-	-	8110.8
11	SGN	-	-	-	1147.5	-	13770
12	LSD	173	897.7	72	215	-	-
13	Sign(0.05)	*	NS	**	**	-	-
14	CV	77.9	25	14.3	48.7	-	-
15	SE	44.6	348.4	363.96	176.16	-	-

Note:
*=Significant, **=Highly significant
S+GN (1:1): SP=Intercrop ground nut with Sorghum 1:1; Simultaneous planting.
S+GN (1:1): RP=Intercrop ground nut with sorghum 1:1; Relay planting.
S+GN (2:1): SP=Intercrop ground nut with Sorghum 2:1; Simultaneous planting.
S+GN (2:1): RP=Intercrop ground nut with Sorghum 2:1; Relay planting.
S+SB (1:1): SP=Intercrop soybean with sorghum 1:1; Simultaneous planting.
S+SB (1:1): RP=Intercrop soybean with sorghum 1:1; Relay planting,
S+SB (2:1): SP=Intercrop soybean with sorghum 2:1; Simultaneous planting.
S+SB (2:1): RP=Intercrop soybean with sorghum 2:1; Relay planting, SSB=Sole soybean, SS=Sole Sorghum; SGN=Sole Ground nut. For calculated the gross income: price of sorghum=6 birr/Kg, Price of Ground nut=12 birr/kg and Price of soybean 8 birr/kg.

Table 4: The influence of intercropping on striga emergence, yield, LER and income in 2014.

intercrops. Therefore, intercropping groundnut in 1:1 proportion and simultaneous planting for the control striga is essential, ideal and useful to small-scale farmers, in order to achieve sustainable crop production.

References

1. El Halmouch Y, Benharrat H, Thalouarn P (2005) Effect of root exudates from different Tomato genotypes on Broomrape (O. aegyptiaca) seed germination and turbacle development. Crop Protection 1: 1-7.

2. Khan ZR, Picket JA, Wadhams L, Muyekho F (2001) Habitat management strategies for the control of cereal stem borers and Striga in maize in Kenya. Insect Science and its Application 21: 375-380.

3. Emechebe AM, Ahonosi MO (2003) Ability of excised roots and stem pieces of maize, cowpea and soybean to course germination of Striga hermonthica seeds. Crop Protection 22: 347-353.

4. Reda F (2002) Striga hermonthica in Tigray (northern Ethiopia): Prospects for control and improvement of crop productivity through mixed cropping. Free University.

5. Oryokot J (1993) Striga: Strategies of its control - a review. In: Proceedings of the African Crop Science Society 1: 224-226.

6. Gebisa E, Patrick JR (2010) Marker Assisted and Physiology Based Breeding for Resistance to Root Parasitic Orobanchaceae. In: Parasitic Orobanchaceae. Purdue University, West Lafayette, USA, pp: 369-391.

7. Gebisa E, Butler LG, Babiker AGT (1992) New approaches to control of Striga: Striga research at Purdue University, Agricultural Experimental Station, Purdue University, West Lafayette, USA, p: 47907.

8. Parker C, Riches CR (1993) Parasitic Weeds of the World: Biology and Control. CAB International. Wallingford, p: 332.

9. Bebawi FF (1987) Cultural practices in witchweed management. In: Parasitic Weeds in Agriculture Striga. Musselman LJ (ed.). Striga. CRC Press, Florida, USA 1: 159-169.

10. Ramaiah KV, Vasudeva RMJ (1983) Proceedings of the International Workshop on Striga. ICRISAT, Patancheru, India, pp: 53-56.

11. Eplee RE, Norris RS (1987) Chemical Control in Striga in parasitic weeds in Agriculture. CRC Press, Florida, USA 1: 173-182.

12. Odhiambo GD, Ransom JK (1994) Long term strategies for Striga control. Fourth Eastern and Southern Africa Regional Maize Conference, Nairobi, Kenya, pp: 263-266.

13. Ejeta G, Butler LG (1993) Host parasite interactions throughout the Striga life cycle, and their contributions to Striga resistance. African Crop Science Journal 1: 75-80.

14. Carson AG (1989) Effect of intercropping sorghum and groundnut on density of Striga hermonthica in the Gambia. Tropical Pest Management 35: 130-132.

15. Lagoke STO, Parkinson V, Aguinbiade RM (1988) Parasitic weeds and their control methods in Africa: Combating Striga in Africa. Proceedings of the International Workshop Organised by IITA, ICRISAT and IRDC. Ibadan, Nigeria, pp: 3-14.

16. Sauerborn J (1999) Legumes used for weed control in agroecosystems in the tropics. Plant Research and Development 50: 74-82.

17. Landon JR (1991) Booker tropical soil manual. A handbook for soil survey and agricultural land evaluation in the tropics and subtropics. Longman scientific and technical, New York, USA.

18. Stewart GR, Press MC (1990) The physiology and biochemistry of parasitic angiosperms. Annual Review of Plant Physiology 41: 127-151.

36

Trichoderma harzianum as a Growth Promoter and Bio-Control Agent against Fusarium oxysporum f. sp. tuberosi

Walid S Nosir[1,2*]

[1]Department of Plant Pathology, Faculty of Agriculture, Cornell University, Ithaca, NY, USA
[2]Horticulture Department, Zagazig University, Zagazig City, Egypt

Abstract

Trichoderma harzianum and *Bacillus brevis* were tested separately and in combination against infection of Tuberose bulbs by *F. oxysporum* f. sp. *tuberose* in soilless culture using Perlite as the substrate. The efficiency of both of antagonists against bulb rot was evaluated based on vegetative and root growth parameters and on flowering parameters. *T. harzianum* effectively suppressed disease and also enhanced plant growth, leading to increased flower production and quality. *Bacillus brevis* enhanced plant growth when tested alone. The mixture of antagonists reduced the effectiveness of *T. harzianum* in disease control. Numbers of *T. harzianum* colony-forming units (CFU) in the substrate and on bulbs were higher when applied alone compared with treating with both antagonists. No *T. harzianum* was detected in the substrate by 120 day after planting. *Bacillus brevis* CFU recovered from bulbs were lower when inoculated in combination with *T. harzianum* and *F. oxysporum* f. sp. *tuberosi*. However, *Bacillus brevis* CFU were not detected in the substrate of bulbs treated with both *T. harzianum* and *Bacillus brevis* and inoculated with *F. oxysporum* f. sp. *tuberosi*. It was concluded that *T. harzianum* provided an efficient and effective control of *F. oxysporum* f. sp. *tuberosi* bulb rot of Tuberose plants.

Keywords: Tuberose; *Trichoderma harzianum*; *Aneurinibacillus migulanus*; *F. oxysporum*; f. sp. *tuberosi*; Soilless culture; CFU

Introduction

Soil-borne plant pathogens are responsible for severe damage in vegetable and cut flower production. Gladiolus corm rot caused by the fungal pathogen *Fusarium oxysporum* f. sp. *gladioli* is a serious problem in gladiolus production, causing huge financial losses to growers [1]. For example, Raiz et al. [2] reported 100% disease incidence and 20% plant mortality, with reductions in shoot and root biomass of 63% and 100%, respectively, when *Gladiolus grandiflorus* corms grown in a pot culture system were inoculated with *F. oxysporum* f. sp. *gladioli*.

Gladiolus corm rot symptoms usually appear in the generation following the introduction of infected corms. Successful suppression of *F. oxysporum* requires not only the prevention of the pathogen from infecting a particular crop, but also the interruption of the pathogen life cycle, which may begin from spores in the substrate used for cultivation. Chemical fungicides alone do not give satisfactory disease control for *F. oxysporum* f. sp. *gladioli* because the pathogen is protected in the xylem tissues, which is difficult for many chemical fungicides to penetrate [1].

The successful suppression of *F. oxysporum* is not only needed to prevent pathogen from invasion, but to prevent that pathogen new life cycle, which starts from an established spores in plant connectively vessels. Using any normal chemical fungicides will kill only the aerial hyphae, but will not kill the dormant spores in the plant. Tuberose bulb rots symptoms appear normally in the next generation after using infected corms in planting. Biological control agents follow different mode of actions including nutrient compotation, antibiotic production, enzymes that can laze fungal cell wall enzymes, and induced host resistance [3,4].

Introduction of two or more biocontrol agents (BCA) to the rhizosphere, assuming that each has different ecological requirements, may facilitate disease control without affecting the efficacy of a single antagonist under different conditions, and may result in increased efficiency [5]. Most biological control studies deal with one antagonist, although attempts to apply more than one antagonist have been reported.

Elad and Zimand [6] found that increasing the number of antagonists helped in disease control efficiency. For example, the combining of four bacterial antagonists and conidia of *Trichoderma* spp. on strawberry reduced *Botrytis cinerea* spore germination markedly [7].

The efficiency of biological control agents in mixtures was related to complementary modes of action of combined organisms [8]. Using a mixture of fluorescent Pseudomonas spp strains decreased Gaeumannomyces graminis var. tritici infection on wheat by 70% compared to the control Weller [9] and Harman [10] reported that both of *Trichoderma viride* and *T. harzianum* promoted the growth of *Crossandra infundibuliformis* var. Danica as the flower production and leaves length and weight were significantly increased compared with the control and also, reduced the *F. oxysporum* wilt incidence. Also, Khan et al. [11] showed that the treated tomato seeds with *T. harzianum* increased the growth length width and root dry weight. The efficiency of using yeast (Pichia guilermondii) to suppress *Botrytis cinerea* ranged between 38% and 98% for both strains when they are applied separately but using the mixture suppressed the pathogen by 80 to 90%. Thus, application of more than one biocontrol agent is suggested as a reliable means of increasing the disease suppression [5].

Similarly, the success of the pair of antagonists used in the previous studies may be attributable to their complimentary mode of action. Using the mixture of plant growth-promoting rhizopacteria resulted in more stable rhizosphere community. Using mixture of certain beneficial strains with antagonism as the main mechanism of action have provided

*Corresponding author: Walid S Nosir, Department of Plant Pathology, Faculty of Agriculture, Cornell University, Ithaca, NY, USA
E-mail: Waleedsabry2000@yahoo.com

good disease suppression compared to single strain treatments that happen as different antagonists use several mechanisms and may suppress several pathogens [12]. A mixture of *fluorescent pseudomonads* and non-pathogenic isolates of *F. oxysporum* were effective in reducing the density of pathogenic F. oxysporum populations in soils, despite the beneficial isolates being ineffective when used separately [12].

In order to make successful use of BCAs for disease control, understanding of the biology and ecology of the organisms is required, to include survival time, fluctuations in population size, and distribution in or on the crop.

The effectiveness of biological disease control depends not only on the availability of suitable BCAs, but also on methods for introducing and strategies for maintaining population levels and activities of these organisms in associations with the target crop(s) [13]. Using an appropriate technique for introduction of the antagonist into the environment, therefore, plays a key role in the success of biological control on crops; moreover that the efficiency of the BCA is also related to the costs involved [14]. It has long been recognized that successful disease control may be affected by changing the behavior of fungal plant pathogens in the rhizo- or phyllosphere.

The aim of the work reported here was to investigate the efficacy of *T. harzianum* and *Bacillus brevis*, separately or together, in controlling *F. oxysporum* f. sp. *tuberose*.

Experiment Procedure

Plant materials

Tuberose bulbs (variety, spring size 10; Tylore Bulb, Co., UK) were surface sterilized by removing the husk and immersing in 70% aqueous ethanol for 1 min followed by 20% NaOCl for 20 min before rinsing under running tap water for 6 hours. Bulbs were subsequently rinsed in 3 changes of distilled water. Bulbs were cultivated in an open soil-less system using 1.5 liter plastic pots (17 cm) placed 25 cm × 25 cm apart on the glasshouse bench and filled with Perlite, previously autoclaved at 105 kPa for 1 hour at 121°C. Nutrient solution (Heavy harvest Bloom, Hydro empire, UK) was pumped with a submerged pump at a rate of 1700 l/ h. and passed through a UV sterilization unit (12 l/m, 30 jm/hour, Filpumps, Aberdeen, UK). The drip irrigation nozzles and all piping were sterilized in 20% NaOCl for 20 min prior to use. Irrigation was timed to operate 5 times during daylight hours, running for 1 min every 2 hours. Each experiment comprised 20 replicate plants per treatment organized as a complete randomised block design. The plants were grown in controlled glasshouse conditions at 22°C.

Preparation of antagonist and pathogen inoculation

Trichoderma harzianum isolate T22 and *Bacillus brevis*was was prepared as described by Nosir et al. [15]. *F. oxysporum* f. sp. *tuberosi* was isolated from the husks of purchased *Tuberose* plants bulbs and maintained on PDA at 22°C, with routine sub–culturing at 15 day intervals. Subcultures were prepared by inoculating fresh PDA in 9 cm diam. Petri dishes with 1 cm diam. disks of colonized PDA plus mycelium, cut from the edge of an actively growing, 7 day old colony.

Antagonist inoculation

Bulbs were inoculated as described previously by Nosir et al. [15].

Pathogen inoculation

Bulbs inoculated with antagonists were subsequently inoculated with *F. oxysporum* f. sp. *tuberosi* by removing a 10 mm diam, 5 mm deep piece of tissue from the exterior of the bulb and replacing it with a plug of PDA plus fungal mycelium of the same dimensions.

Data collection

Leaf numbers and areas were recorded at 10 day intervals. Bulb and rooting characteristics recorded included root length, bulb dry weight, and lesion area on the bulb. The lesion area was measured by calculated the length and the width of the necrotic tissues; infected areas were defined by the softness and rotting of the tissue. Flower spike length, dry weight of the inflorescence, and days from inoculation to flowering were also measured. After obtaining the fresh weights, samples were dried at 70°C to constant weight to obtain dry weights.

Re-isolation of antagonists

T. harzianum selective medium was prepared according to the method of Williams et al. [16]. *Bacillus brevis*s elective medium was prepared following Edwards and Seddon [17].

Fungal and bacterial colonization in soil and bulbs: The extent of colonization of bulbs by BCAs or *F. oxysporum* f. sp. *tuberosi* was examined at intervals of 30 days after treatment and planting. Inoculated plants were uprooted and the Perlite attached to the roots carefully removed. One g Rockwall was placed into a 30 ml plastic sterile universal tube containing 10 ml 50 mM phosphate buffer, pH 7.0. The tubes were vortexed for 1 min at maximum speed, placed on a rotary shaker at 150 rpm for 30 min, and the suspensions subsequently diluted to 10^6. Aliquots of suspension (0.1 ml) were plated onto both *Trichoderma* and *Bacillus brevis* selective media. Colonies isolated from the suspensions were considered as the external rhizosphere population. To determine bulb colonization, 1 g bulb tissue from the inoculated sites was rinsed in running tap water, dried briefly on paper towels, weighed and homogenized using a mortar and pestle in 10 ml 50 mM phosphate buffer, pH 7.0. The homogenate was serially diluted to 10^4. Aliquots (0.1 ml) of the diluted suspension were plated on the selective media for each microorganism. *T. harzianum* was counted on Trichoderma selective medium and *Bacillus brevis* on Bacillus selective medium at 30 days intervals to monitor the efficiency of colonisation by the antagonists. Data from fungal and bacterial population density counts were $\log_{10}$ transformed before analysis.

Chemical analysis

Samples of vegetative leaves at the beginning of flowering were dried at 70°C and used for chemical analysis. Samples of the bulbs were dried at 70°C and the leaves were subjected to chemical analysis as follows: total nitrogen percentages in the leaves and bulbs were determined according to Naguib [18]. Total phosphorus percentages in the leaves and corms were determined according to Troug and Mayer [19]. Total potassium percentages in the leaves and bulbs were determined according to Jackson [20]. Chlorophyll A, B and carotenoids were measured according to Wettestein [21].

Statistical analysis

Glasshouse experiments and biological assays were organized in complete block designs. The results represent the mean of experiments run in two seasons. Statistical analyses were conducted using the general linear model procedures of SPSS version 16. Experiments were analyzed using analysis of variance (ANOVA). Significance was evaluated at P<0.05 for all tests. Mean separation was tested using the Tukey HSD test.

Results

Vegetative growth

At the end of the vegetative growth stage, Tuberose bulbs inoculated with *T. harzianum* had significantly greater numbers of larger leaves (Figure 1a and 1b); (P<0.001), compared to control bulbs. Treatment with *Bacillus brevis* suspensions also significantly (P< 0.001) increased plant growth; leaf area, and number increased by 49.4% and 28.31%, respectively.

Leaf area and number were also significantly greater in plants from bulbs treated with *T. harzianum* spore suspension and subsequently inoculated with *F. oxysporum* f. sp. *Tuberosi* (P<0.001). In other hand, treatment with *Bacillus brevis* followed by inoculation with *F. oxysporum* f. sp. *tuberosi* resulted in the production of significantly fewer leaves, compared with those treated with *T. harzianum* (P<0.001). Mixing *T. harzianum* and *Bacillus brevis* in the *F. oxysporum* f. sp. *tuberosi* inoculated bulbs lead to a significant decrease (P<0.001) in leaf area and number, compared with bulbs treated with *T. harzianum* or with *Bacillus brevis* alone.

Rooting characteristics

Root length and bulb dry weights were significantly increased in plants treated with *T. harzianum* alone (Figure 2a and 2b) (P<0.001),

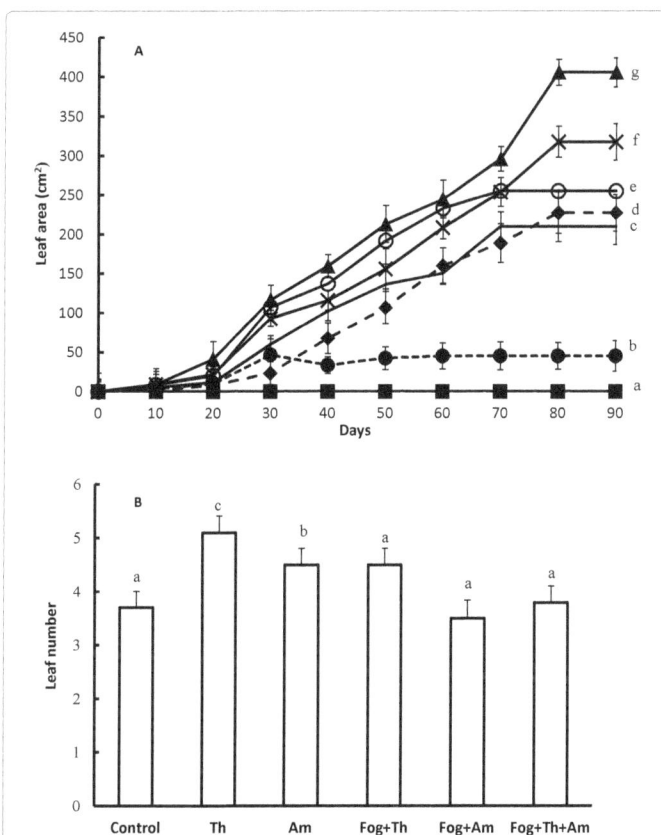

Figure 1: Effects of biological control agents on Tuberose foliage. (A) Effects on leaf area [-♦-, Control; -■-, *F. oxysporum* f. sp. *tuberosi*; -▲-, *T. harzianum*; -X-, *A. migulanus*; -O-, *F. oxysporum* f. sp. *tuberosi* + *T. harzianum*; -●-, *F. oxysporum* f. sp. *tuberosi* + *A. migulanus*; + *F. oxysporum* f. sp. *tuberosi* + *T. harzianum*+ *A. migulanus*]. (B) Effects on leaf numbers [Th: *T. harzianum*; Am: *A. migulanus*; Fog: *F. oxysporum* f.sp. *tuberosi*]. Data represents the means of 20 replicate plants ± SD. Lines or bars with different lower case letters are significantly different (P<0.05).

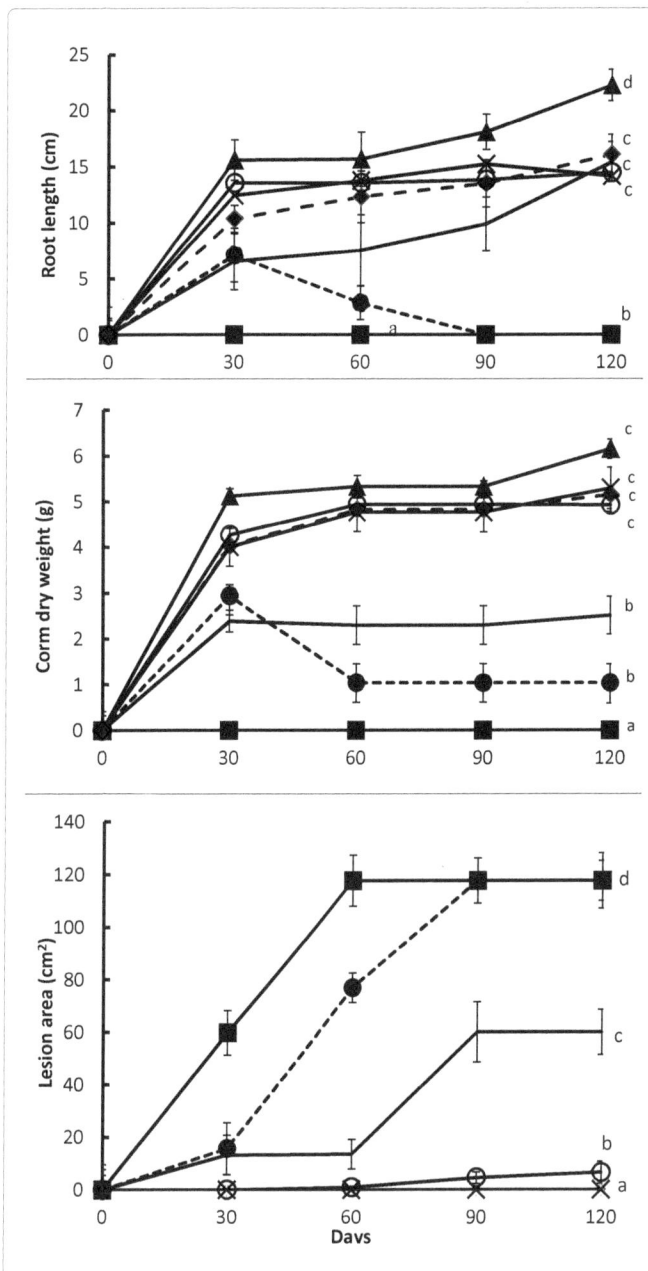

Figure 2: Impacts of biological control agents on Tuberose root and corm development. (A) Root lengths (B) Dry weight of corms; (C) lesion area on corms. area [-♦-, Control; -■-, *F. oxysporum* f. sp. *tuberosi*; -▲-, *T. harzianum*; -X-, *A. migulanus*; -O-, *F. oxysporum* f. sp. *tuberosi* + *T. harzianum*; -●-, *F. oxysporum* f. sp. *tuberosi* + *A. migulanus*; + *F. oxysporum* f. sp. *tuberosi* + *T. harzianum*+ *A. migulanus*]. Different lower case letters indicate significant differences between treatments (P<0.05).

compared with other treatments. This effect was noticeable by 30 days after treatment. Moreover, *T. harzianum* significantly reduced lesion areas caused by *F. oxysporum* f. sp. *tuberosi* inoculation (P<0.001), compared with bulbs inoculated with *F. oxysporum* f. sp. *tuberosi* alone (Figure 2c). No lesion expansion was detected in the inoculated bulbs treated with *T. harzianum* at the first sampling time (30 days), although small lesions were present at 60, 90 and 120 days after inoculation.

In contrast, no differences were found in plants treated with *Bacillus brevis* alone, compared with the controls. Treatment with *Bacillus brevis* alone also had little biological control effect against the pathogen: bulbs treated with *Bacillus brevis* and inoculated with *F. oxysporum* f. sp. *tuberosi* began to form roots normally, but by 60 days after inoculation, all rooting parameters were decreased. After that time, no roots were recorded and the plants died. At the 30 and 60 days sampling times, lesion areas on *F. oxysporum* f. sp. *tuberosi* inoculated bulbs treated with *Bacillus brevis* were significantly reduced compared with bulbs inoculated with *F. oxysporum* f. sp. tuberosi alone (P<0.001); at 90 and 120 days after treatment, however, lesion areas were the same in these two treatments.

When bulbs were treated with the mixture of antagonists and inoculated with the pathogen, the biological control potential of *T. harzianum* was reduced compared with the *T. harzianum-F. oxysporum* f. sp. *tuberosi* treatment (P<0.001). A combination of *T. harzianum* and *Bacillus brevis*was significantly less effective in reducing lesion size compared to the treatment with *T. harzianum* alone (Figure 2c) (P<0.001).

Flower production

Flower quality, assessed on length and dry weight of the flower spike, was significantly increased in plants treated with *T. harzianum* (Figure 3a and b) (P<0.001). Treatment with *Bacillus brevis* also led to increased flower spike height and dry weight (P<0.001) compared to the control treatments. Flowers arising from bulbs treated with *T. harzianum* and *F. oxysporum* f. sp. *tuberosi* were also of significantly greater weight and height (P<0.001) compared with those from the control bulbs.

On contrary, bulbs inoculated with *F. oxysporum* f. sp. *tuberosi* alone, or treated with *Bacillus brevis* followed by inoculation with *F. oxysporum* f. sp. *tuberosi*, died before reaching the flowering stage. Using the mixture of *T. harzianum* and *Bacillus brevis* and inoculating with *F. oxysporum* f. sp. *tuberosi* lead to a significant decrease (P<0.001) in the flower height and dry weight compared with bulbs treated with *T. harzianum* or *Bacillus brevis* alone.

Changes in population of the biological control agents with time after inoculation

Populations of *T. harzianum* were similar in the rhizosphere of the bulbs treated with the fungal antagonist alone, or with both *T. harzianum* and *F. oxysporum* f. sp. *tuberosi* throughout the time course of the experiment (Figure 4a), with CFU of *T. harzianum* in the substrate maintained at approximatly 5-7 $\log_{10}$ per g Perlite (Figure 4a). When both *T. harzianum* and *Bacillus brevis* were co-inoculted, following by *F. oxysporum* f. sp. *tuberosi*, however, the *T. harzianum* population declined in the rhizosphere from day 60 after treatment. CFU of *T. harzianum* were significantly lower (P<0.001) on the surfaces of bulbs which were also treated with *Bacillus brevis* than on those treated with *T. harzianum* alone, or also inoculated with *F. oxysporum* f. sp. *tuberosi* (Figure 4b).

No *Bacillus brevis* was found in the substrate from around bulbs treated with the combination of the bacterium, *T. harzianum* and *F. oxysporum* f. sp. *tuberosi* (Figure 4c) by 30 days after treatment. Differences in CFU recovered from the *Bacillus brevis*alone or *Bacillus*

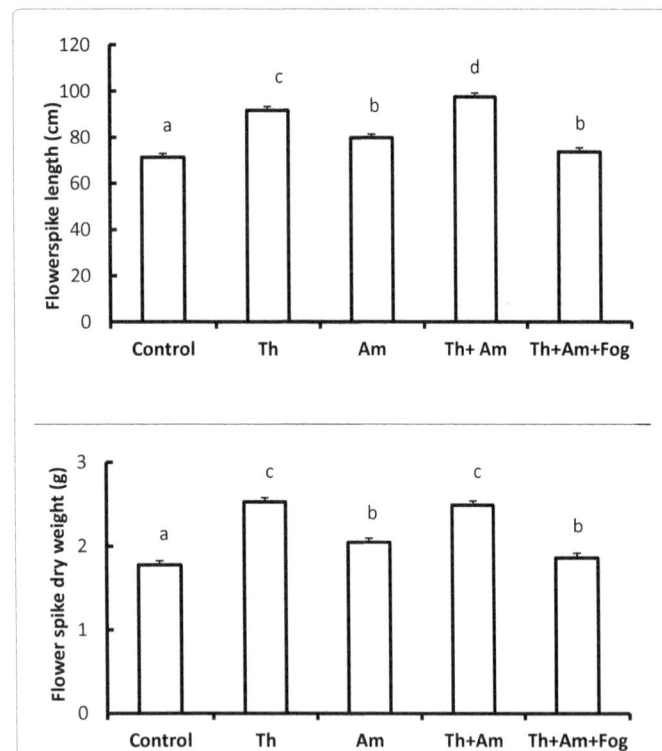

Figure 3: Effects of treatment of Tuberose plants bulbs with biological control agents on (a) Flower dry weight and (b) Length of flower spike. Co, Control; T, *T. harzianum*; B, *Bacillus brevis*; F+T, *F. oxysporum* f. sp. tuberosi+ *T. harzianum*; F+A, *F. oxysporum* f. sp. tuberosi + *Bacillus brevis*; F+T+A, *F. oxysporum* f. sp. *tuberosi* + *T. harzianum*+ *Bacillus brevis*; Data represents the mean ± SD. Different lower case letters indicate significant differences between treatments (P<0.05).

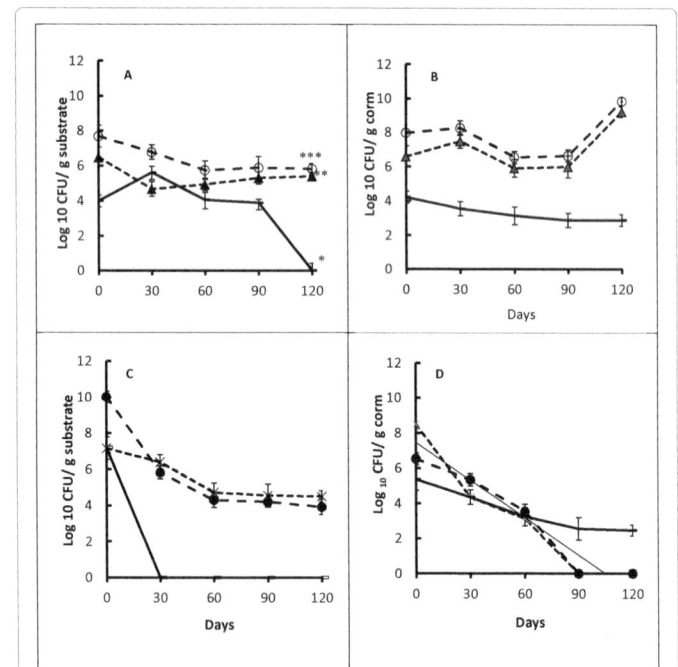

Figure 4: Changes in population of the biological control agents with time after inoculation. (A, B) Changes in *Trichoderma harzianum* populations (Log₁₀ CFU/ g fresh weight) when inoculated alone (– ▲--), with *Fusarium oxysporum* f. sp. *tuberosi* (– –o– –), or with *F. oxysporum* **f. sp.** *tuberosi* and *Aneurinobacillus migulanus* (—×—) in the Perlite substrate comprising the rhizosphere (A), and on the corm surface (B). (C,D) Changes in *A. migulanus* populations (Log₁₀ CFU/ g fresh weight) when inoculated alone (--O--), with *Fusarium oxysporum* f.sp. *tuberosi* (– –•– –), or with *F. oxysporum* **f.sp.** *tuberosi* and *T. harzianum* (———)in the Perlite substrate comprising the rhizosphere (A) and on the corm surface (B) Data represent the means of 10 replicates per treatment ± SE; different letters indicate significant differences between treatment effects (P<0.05).

| Treatment | Nitrogen | Phosphorus | Potassium | Chlorophyll pigments mg/100 mg fresh weight | | |
	bulbs	leaves	bulbs	A	B	Carotenoids
CO	23.45 a	2.17 a	3.54 a	0.45 a	0.64 a	0.25 a
Th	38.45 d	3.87 d	5.34 d	0.654 a	0.65 a	0.32 a
Am	31.32 b	3.21 b	4.34 b	0.567 a	0.54 a	0.43 a
Fog+Th	34.54 c	3.32 b	5.34 c	0.654 a	0.65 a	0.34 a
Fog+Am	33 c	3.65 c	4.98 b	0.587 a	0.43 a	0.36 a
Fof+Th+Am	28.56 b	3.16 b	3.98 a	0.487 a	0.45 a	0.32 a

*Different letters indicate significant differences between treatment effects (P< 0.05).

Table 1: Impacts of biological control agents on Tuberose in nutrient film technique on chemical composition of Tuberose bulbs.

brevis and *F. oxysporum* f. sp. *tuberosi* treatments, however, were not significant (P>0.05).

Bacillus brevis CFU were recovered from the bulb surface in all treatments, up to 60 days after inoculations (Figure 4d). By day 90, however, no *Bacillus brevis* CFU were present on bulbs treated with *Bacillus brevis* alone, or those treated with *Bacillus brevis* and inoculated with *F. oxysporum* f. sp. *tuberosi*. In the treatment combining *Bacillus brevis* and *T. harzianum*, with *F. oxysporum* f. sp. *tuberosi* inoculation, *Bacillus brevis* CFU numbers declined with time, but were still detectable 120 days after treatment.

Data of the Impacts of biological control agents on Tuberose in nutrient film technique on chemical composition of Tuberose bulbs are shown in Table 1. The results indicate that using Trichoderma, Bacillus separately or in combination increased the plant content of Nitrogen, Phosphorus, and Potassium in presence or absence of the *F. oxysporum* as a pathogen significantly (P>0.05). Moreover, the total Chlorophyll, Carotinoids, had had higher values in the bulbs treated with the beneficial microorganisms in presence or absence of *F. oxysporum* significantly (P>0.05).

Discussion

The work reported here demonstrated that the application of *Trichoderma harzianum* T22 to bulbs of Tuberose prior to planting in soil-free culture for flower production could provide an effective method for the control of *Fusarium oxysporum* f. sp. *tuberosi* bulb rot. Not only was *T. harzianum* able to suppress bulb rot, it also improved various production parameters of Tuberose, including increasing leaf area and number and root lengths, and improving flowering spike size. Treatment with the bacterial antagonist *Bacillus brevis* also lead to production of higher leaf areas and numbers, and improved flowering, but the bacterium did not give any significant control of bulb rot. However, combining *Bacillus brevis* with *T. harzianum* showed no improvement in biological control of *F. oxysporum* f. sp. *tuberosi*; in fact the effectiveness of *T. harzianum* control was decreased in the presence of the inoculated bacteria. It is possible that gramicidin S, the major cyclic peptide antibiotic produced by *Bacillus brevis* [17] inhibited *T. harzianum* activity, through an immediate effect due to exposure at the time of inoculation, or that the *Bacillus brevis* occupied locations on the further work is required to substantiate this proposal.

It is evident from the results obtained here, that these antagonists controlled the pathogen by a combination of direct and indirect effects. A combination of direct and indirect effects in antagonism of plant pathogens by biological control agents, including competition with pathogen conidia for nutrients at early stages [22] and a later direct impact on pathogen enzyme activity was suggested by Harman [10]. Induction of host resistance may also occur at the second stage of activity Chang et al. [23] also indicated that *T. harzianum* inhibited elongation of *Botrytis cinerea* germ tubes. Zhang and Xue [24] found

the culture filtrates of *B. subtilis* strain SB24 significantly reduced Sclerotinia sclerotiorum, a major soybean *Sclerotinia* stem rot in Canada by 50- to 75%. They added that the cell suspension and broth culture preparations significantly reduced the disease severity by 45 to 90% at concentrations ranging from 5×10^6 to 10^9 CFU ml^{-1}, which supports this work finding about the role of culture filtrates in the antagonistic mechanism.

Migulanus also increased the various growth parameters assessed here, compared with controls, although the overall enhancement was less than found with *T. harzianum*. This effect is also well-known with certain combinations of bacteria [10]. The modes of action of these plant-growth promoting bacteria are thought to be similar to those discussed above for antagonistic and beneficial fungi, although disease control or host resistance induction effects were not evident with *Bacillus brevis* on tuberose inoculated with *F. oxysporum* f. sp. *tuberosi*. Several other *Bacillus* spp. and strains, however, are thought to promote plant growth through the release of signaling compounds that increase the efficiency of photosynthesis [24].

Biological control efficiency depends upon the establishment and maintenance of a threshold population of antagonists on planting material or in soil [25]. This work has shown clearly that *T. harzianum* will control Fusarium bulb rot of Tuberose, even when the pathogen is introduced into large wounds on the bulbs, although how the BCA prevented expansion of *F. oxysporum* f. sp. *tuberosi* from the inoculation point was not clear. Mixing *T. harzianum* and *B. brevius* results agree with the findings of Nosir et al. [15] regarding mixing two antagonists could be an efficient strategy in increasing the efficiency the antagonistic action. In this study mixing *T. harzianum* and *A. migulanus* as antagonistic reduced the effectiveness of *T. harzianum*, however, the co-inoculation of seeds with a mixture of *R. leguminosarum* and *T. viride* reduced chocolate spot disease and enhanced nodulation, nitrogenase activity and nitrogen fixing bacterial population in the rhizosphere. In addition to, the improvements in the physiological activities (photosynthetic pigments, total phenol and polyphenol oxidase), improved plant growth and yield. On average, this treatment recorded about 57% reduction in chocolate spot disease and 67% increase in Gladiolus corms, compared to control plants Nosir et al. [15].

The results of the interaction tests confirm the ability of *T. harzianum* to produce both volatile and non-volatile compounds that show antifungal activity towards *F. oxysporum* f. sp. *gladioli*. The nature of these antifungal compounds, and their role in pathogen suppression requires further investigation. *T. harzianum* proved a more effective inhibitor of *F. oxysporum* f. sp. *tuberosi* growth than *B. brevis*. These results agree with previous reports on the mechanisms of suppression of *T. harzianum* follows at different stages of infection. The inhibition of germination and germ tube elongation of *F. oxysporum* f. sp. *gladioli* were limited in the presence of *T. harzianum*. The higher efficacy of *T. harzianum* compared with *B. brevis* appeared to be due to the different employed mechanisms in *F. oxysporum* f. sp. *gladioli* suppression; however, *B. brevis* depends only on antibiotic mechanism. *T. harzianum* spores number in the substrate is a crucial factor in its ability to suppress *F. oxysporum* f. sp. *gladioli* growth and development by space or nutrient competition, also more active spores means more metabolites secretion.

In contrast to previous work on Fusarium wilt of tomato caused by *F. oxysporum* f. sp. *lycopersici* [26], *Bacillus brevis* did not reduce Fusarium bulb rot of *Tuberose*. Compared with *T. harzianum*, *Bacillus brevis* requires relatively high temperatures 37°C for optimal growth and the production of gramicidins, the cyclic peptide antifungal metabolites active against fungal pathogens [17]. In previous work, *Bacillus brevis* was shown to give protection against grey mould on various host species [27]. Despite the higher average temperatures in the glasshouse used for the present work, however, no such control occurred in this host-pathogen system. Moreover, Seddon et al. [27] examining the ability of *Bacillus brevis* to control pathogens of foliage and fruit, whereas the host-*F. oxysporum* f. sp. *Tuberosi* occurs in the rooting substrate, where conditions differ greatly from those on aerial plant parts.

The results obtained in this study were comparable with those from previous work on Tuberose [28,29] demonstrated control of *F. oxysporum* f. sp. *tuberosi* by *T. harzianum* and *T. viride* under field conditions. The present work, however, gave much greater control of the disease than reported previously. The more stable environmental conditions in the glasshouse may have been responsible for the increased efficacy of the BCA. Improvements in plant growth and flower production in Tuberose plants following treatment with *T. harzianum* and *T. viride*, amongst other soil amendments, were also reported in previous work [30]. The results obtained from chemical analysis confirm and support the research hypothesis, which demonstrate the efficacy of BCAs in controlling soil pathogens by enhancing the plant growth and increasing the available nutrients to the plants.

Conclusion

The use of BCAs of soil borne plant pathogens in the field has given variable results [31], it appears that many isolates of *Trichoderma* provide good control of disease in the field [32], along with the added benefits of improved plant growth discussed above. The strain used in this work, *T. harzianum* T22, is well-known to provide excellent protection against root pathogens in many crop plants [7] and is tolerant of fungicides [33]. Enhanced flower production by *T. harzianum* T22 could clearly be of great benefit to growers and should be optimized as far as possible for deployment under field conditions.

References

1. Ram R, Manuja S, Dhyani D, Mukherjee D (2004) Evaluations of fortified fungicide solutions in managing bulb rot disease of Tuberose plants caused by *Fusarium oxysporum*. Crop protection 23: 783-788.

2. Raiz T, Khan N, Javaid A (2007) Effect of incorporation of allelopathic plants leaf residues on mycorrhizal colonization and Tuberose plants disease. Allelopathy Journal 20: 61-70.

3. Whipps JM (2001) Microbial interactions and biocontrol in the rhizosphere. J Exp Bot 52: 487-511.

4. Fravel D, Olivain C, Alabouvette C (2003) Fusarium oxysporum and its biocontrol. New Phytologist 157: 493-502.

5. Guetsky R, Shtienberg D, Elad Y, Dinoor A (2001) Combining biocontrol agents to reduce the variability of biological control. Phytopathology 91: 621-627.

6. Elad Y, Zimand G (1993) Use of Trichoderma harzianum in combination or alternation with fungicides to control cucumber grey mould (Botrytis cinerea) under commercial greenhouse conditions. Plant Pathology 42: 324-332.

7. Benítez T, Rincón AM, Limón MC, Codón AC (2004) Biocontrol mechanisms of Trichoderma strains. Int Microbiol 7: 249-260.

8. Pierson EA, Weller DM (1994) Use of mixtures of Fluorescent pseudomonads to suppress take-all and improve the growth of wheat. Phytopathology 84: 940-947.

9. Weller DM, Cook RJ (1983) Suppression of take-all of wheat by seed treatments with fluorescent pseudomonads. Phytopathology 73: 463-469.

10. Harman GE (2000) Myths and dogmas of biocontrol: changes in perceptions derived from research on Trichoderma harzianum T-22. Plant Disease 84: 377-93.

11. Khan T, Mujeebur C, Rahman N, Mustafa L, Uzma H (2005) Bulb rot and yellows of gladiolus and its biomanagement. Phytopathologia Mediterranea 44: 208-215.

12. Lemanceau P, Alabouvette C (1993) Suppression of fusarium wilts by fluorescent pseudomonads: mechanisms and applications. Biocontrol Science and Technology 3: 219-234.

13. Ramani B, Reeck T, Debez A, Stelzer R, Huchzermeyer B. et al. (2006) Aster tripolium L. and Sesuvium portulacastrum L.: two halophytes, two strategies to survive in saline habitats. Plant Physiol Biochem 44: 395-408.

14. Mishra K, Mukhopadhyay N, Fox RTV (2000) Integrated and biological control of gladiolus corms rot and wilt caused by Fusarium oxysporum. Annals of Applied Biology 137: 361-364.

15. Nosir W, McDonald J, Woodward S (2011) An interaction effect between Trichoderma harzianum, and Aneurinobacillus migulanus against Fusarium oxysporum f. sp. gladioli on Trichoderma secondary metabolites. American Journal of Applied Sciences 5: 436-445.

16. Williams J, Clarkson JM, Mills PR, Cooper RM (2003) A selective medium for quantitative reisolation of Trichoderma harzianum from Agaricus bisporus compost. Appl Environ Microbiol 69: 4190-4191.

17. Edwards SG, Seddon B (2001) Mode of antagonism of Brevibacillus brevis against Botrytis cinerea in vitro. J Appl Microbiol 91: 652-659.

18. Naguib MI (1969) Colorimetric determination of nitrogen components of plant tissues. Journal of the Faculty of Science. Cairo University 34: 16–24.

19. Troug E, Mayer AH (1939) Improvement in the denies colorimetric method for phosphoric and arsenic. Industrial and Engineering Chemistry Analytical Edition 1: 136-139.

20. Jackson ML (1970) Soil Chemical Analysis. Englewood Cliffs, NJ: Prentice Hall.

21. Wettestein D (1957) Chlorophyll-letale und der submikroskopis-che form Wechsel der plastiden Exptl (Chlorophyl analysis of plant plastides). Cell Research 12: 427-433.

22. Mandeel Q, Baker R (1991) Mechanisms involved in biological control of Fusarium wilt of cucumber with strains of nonpathogenic Fusarium oxysporum. Phytopathology 81: 462-469.

23. Chang YC, Baker R, Kleifeld O, Chet I (1986) Increased growth of plants in the presence of the biological control agent Trichoderma harzianum. Plant Disease 70: 145-148.

24. Zhang H, Xie X, Kim MS, Kornyeyev DA, Holaday S, et al. (2008) Soil bacteria augment Arabidopsis photosynthesis by decreasing glucose sensing and abscisic acid levels in planta. Plant J 56: 264-273.

25. Whipps JM (2001) Roots and their environment. Microbial interactions and biocontrol in the rhizosphere. Journal of experimental Botany 52: 497-511.

26. Chandel S, Allan EJ, Woodward S (2010) Rhizosphere competence of Brevibacillus brevis in the control of Fusarium oxysporum f. sp. lycopersici on tomato roots. Journal of Phytopathology 158: 470-478.

27. Seddon B, McHugh RC, Schmitt A (2000) Brevibacillus brevis–A novel candidate biocontrol agent with broad spectrum antifungal activity. In: Brighton Crop Protection Conference. Pests & Diseases 2: 563-570.

28. Singh PP, Shin YC, Park CS, Chung YR (1999) Biological control of fusarium wilt of cucumber by chitinolytic bacteria. Phytopathology 89: 92-99.

29. Sharma S, Chandel S, Tomar M (2005) Integrated management of Fusarium yellows of Gladiolus caused by Fusarium oxysporum f. sp gladioli Snyder & Hans. under polyhouse conditions. Integrated Plant Disease Management. pp: 221-229.

30. Raj H, Upmanyu S (2006) Solarization of soil amended with residues of cabbage leaves and bulb treatment with fungicides for management of wilt (Fusarium oxysporum) of Tuberose Plants (Tuberose Plants grandiflorus). Indian Journal of Agricultural Sciences 76: 307-311.

31. De Boer M, Van dersluis I, Van loon LC, Bakker PAHM (1999) Combining fluorescent Pseudomonas spp. strains to enhance suppression of fusarium wilt of radish. European Journal of Plant Pathology 105: 201-210.

32. Howell CR (2003) Mechanism employed by Trichoderma species in the biological control of plant diseases: the history and evaluation of current concepts. Plant Disease 87: 4-10.

33. Lugtenberg B, Kamilova F (2009) Plant-growth-promoting rhizobacteria. Annu Rev Microbiol 63: 541-556.

37

Use of HPPD-inhibiting Herbicides for Control of Troublesome Weeds in the Midsouthern United States

Clay E. Starkey, Jason K. Norsworthy and Lauren M. Schwartz*

Department of Crop, Soils, and Environmental Sciences, University of Arkansas,1366 West Altheimer Drive, Fayetteville, AR 72701, USA

Abstract

Transgenic crops provide cotton and soybean producers additional weed control options for many of the most problematic weeds in midsouthern United States (U.S.). production systems. The expected commercialization of 4-hydroxyphenylpyruvate dioxygenase (HPPD)-resistant soybean in 2017 and cotton in 2020 will provide producers the option to apply HPPD-inhibiting herbicides that will offer an alternative mechanism of action for previously hard-to-control weeds. Experiments were conducted in 2010 and 2011 to determine the efficacy of HPPD-inhibiting herbicides applied preemergence (PRE) or postemergence (POST) for control of problematic weeds of cotton and soybean in the mid southern US. PRE experiments were conducted to understand the length and degree of control of Palmer amaranth and barnyardgrass that could be expected with HPPD-inhibiting herbicides compared with current standards on silt loam and clay soil textures. The HPPD herbicides evaluated included mesotrione, tembotrione, and isoxaflutole compared to several standards currently labeled in soybean. In the POST experiment, applications of isoxaflutole, tembotrione, glyphosate, and two rates of glufosinate applied alone and both HPPD herbicides combined with glyphosate or glufosinate were evaluated for control of Palmer amaranth, barnyardgrass, hemp sesbania, and yellow nutsedge. When herbicides were applied PRE, the HPPD-inhibiting herbicides and the current standard treatments all provided greater than 90% control of Palmer amaranth 4 weeks after treatment (WAT) on both soil textures. Barnyardgrass control with HPPD-inhibitors was generally weaker than the current standards with the exception of mesotrione which proved to be comparable to the standards 4 WAT. In the POST experiment, all treatments, except for glyphosate alone, provided excellent (>85%) control of Palmer amaranth less than 10-cm in height. Barnyardgrass, yellow nutsedge, and hemp sesbania were effectively controlled with HPPD-inhibiting herbicides with and without glufosinate or glyphosate.

Keywords: HPPD-inhibiting herbicides; Preemergence; Postemergence; Tank-mix

Introduction

Options for weed control in midsouthern U.S. crops were broadened with the introduction of transgenic crops, specifically glyphosate-resistant soybean and cotton in 1996 and 1997, respectively. The adoption of glyphosate-resistant crops came with a dramatic shift in herbicide use patterns, most notably the almost sole reliance on glyphosate [1]. Glyphosate is a non-selective herbicide that inhibits the 5-enolpyruvylshikimate-3-photsphate synthase (EPSPS) within a plant. Producers were allowed to apply up to 3.3 kg ae ha^{-1} yr^{-1} over multiple application timings [2]. Due to the fact that glyphosate applications are cheap, effective, and simple [3], applications were being made multiple times per year in cotton and soybean and thus replaced tank mixtures of herbicides, tillage, and residual herbicides in the late 1990s and early 2000s [1,4,5]. Extensive and often exclusive use of glyphosate created an increasing number of glyphosate-resistant weeds [6]. In order to mitigate weed resistance to glyphosate, new mechanisms of action are being sought that can be integrated into current or future cropping systems. In a survey conducted by Norsworthy et al. [7] in Arkansas, cotton consultants overwhelmingly expressed the importance of a need for new tools for resistant weed management.

Another transgenic option for producers to apply an effective broad-spectrum herbicide in crop was the release of glufosinate-resistant crops. Glufosinate-resistant crops allow for over-the-top application of glufosinate, which inhibits glutamine synthetase in sensitive plants [8].

In 2017 and 2020, soybean and cotton are expected to be released that are resistant to a mechanism of action currently used in corn (*Zea mays* L.) and grain sorghum (*Sorghum bicolor* L.) production, 4-hydroxyphenylpyruvate dioxygenase (HPPD)-inhibiting herbicides.

HPPD-inhibiting herbicides prevent the formation of homogentisate in the formation of chloroplasts and carotenoids [9,10]. Enzymatic inhibition results in a bleaching effect in plants due to the absence of carotenoid biosynthesis [11]. HPPD-inhibiting herbicides are known to be broad spectrum, often controlling both grass and broadleaf species. This technology will provide soybean and cotton producers with another option for control of troublesome weeds. These HPPD-resistant crops will eventually possess resistance to glyphosate and glufosinate [12]. The combination of these traits will provide producers additional options to combat the resistant weeds currently infesting cotton and soybean fields.

In a survey of midsouthern U.S. cotton consultants in 2011, of the most problematic weeds in cotton, Palmer amaranth, hemp sesbania, yellow nutsedge, and barnyardgrass were ranked among the top 10 [13]. Palmer amaranth has evolved wide-spread resistance to glyphosate and ALS-inhibiting herbicides making POST over-the-top control impossible in glyphosate-resistant cotton [14]. Applications of glyphosate to control troublesome weeds, such as hemp sesbania and yellow nutsedge, have been marginal depending on rate and size of the

*Corresponding author: Lauren M. Schwartz, Department of Crop, Soils, and Environmental Sciences, University of Arkansas, 1366 West Altheimer Drive, Fayetteville, AR72701, USA, E-mail: lmschwar@uark.edu

plant at application [15,16]. Applications of glufosinate on both hemp sesbania and yellow nutsedge have proven very effective [16,17].

Barnyardgrass is a problematic weed due to its ability to germinate and grow under a wide variety of conditions [18]. It has been predicted that barnyardgrass will eventually evolve resistance to glyphosate [19]. The addition of HPPD-resistant cotton and soybean could be an additional tool that can be used to combat weed resistance. The weed spectrum shift caused by glyphosate-resistant crops has affected the entire southern US. where cotton and soybean are two of the principle crops [20]. The objectives of this research were to evaluate alternative options in the use of HPPD-inhibiting herbicides for crops likely to be labeled in the near future. This research also aims to explore the most efficient method of application to control four of the most troublesome weeds in Arkansas: Palmer amaranth, barnyardgrass, hemp sesbania, and yellow nutsedge.

Materials and Methods

Length and degree of control with pre-applied HPPD-inhibiting herbicides compared to current herbicide standards

Experiments were conducted during the summers of 2010 and 2011 to determine the length of residual control with HPPD-inhibiting herbicides compared to the current PRE-applied herbicides commonly used in midsouthern US. soybean production systems. Experiments were conducted at the University of Arkansas Northeast Research and Extension Center (NEREC) in Keiser, AR in 2010 on a Sharkey (very fine, smectitic, thermic Chromic Epiaquerts, pH 6.5, OM 3.8%) and 2011 on a Sharkey-Steele (very fine, smectitic, thermic Chromic Epiaquerts, pH 6.7, OM 3.3%). Experiments were also conducted at the

University of Arkansas Agricultural Research and Extension Center (AAREC) in Fayetteville, AR in 2010 on a Captina silt loam (fine-silty, siliceous, active, mesic, Typic Fragiudults, pH 6.4, OM 1.8%), in 2011 on a Johnsburg silt loam (fine-silty, mixed active, mesic, Aquic Fragiudults, pH 6.5, OM 1.4%), and in 2011 at the University of Arkansas Pine Tree Branch Experiment Station (PTBES) near Colt, AR on a Calloway silt loam (fine-silty, mixed active thermic Aquic Fraglossudalfs, pH 6.5, OM 2.2%). Soil samples from the top 10 cm were analyzed from all locations to determine soil properties on all five experimental sites (Table 1). Soil organic matter (OM) was determined using loss on ignition [21].

Experiments conducted in 2010 and 2011 at the AAREC and in 2010 at the NEREC where plots were overhead irrigated. The trials were conducted during the spring and early summer at times that would be typical for crop production in the region. In 2011 at NEREC and PTBES, the experiment was surface irrigated. Surface irrigation involved building a levee around the field and applying enough water inside the levee to saturate the soil in the experimental site to activate treatments and germinate weed seeds. The experimental design was a randomized complete block with four replications with the herbicide treatments evaluated within each soil texture. The experimental plots were 1 m wide by 2 m long separated by 2 m alleys between the plots and four replications at all locations. The front 1 by 1 m of each plot was sown with 3,000 barnyardgrass seeds and the remaining 1 by 1 m² was sown with approximately 5,000 Palmer amaranth seeds prior to applying the herbicides. All seeds were lightly incorporated with a rake to approximately a 1.5-cm depth. Barnyardgrass seed was obtained from Azlin Seed Service (Leland, MS 38756), and Palmer amaranth seed was collected from an infested field at AAREC the previous fall. Herbicide treatments for the clay and silt loam soils are shown in Tables 2 and 3, respectively. Phytotoxicity was visually rated on a scale of 0 to 100%,

Location	Year	Sand	Silt	Clay	Soil organic matter	Soil texture	Soil pH
		------------ g g⁻¹ --------			--%--		
Fayetteville	2010	0.23	0.49	0.28	1.8	Silt loam	6.4
	2011	0.27	0.50	0.23	1.4	Silt loam	6.5
Keiser	2010	0.09	0.22	0.69	3.8	Clay	6.5
	2011	0.18	0.20	0.62	3.3	Clay	6.7
Pine Tree	2011	0.05	0.67	0.28	2.2	Silt loam	6.5

Table 1: Soil properties from a 0- to 10-cm depth at Fayetteville, Keiser, and Pine Tree, Arkansas in 2010 and 2011.

Herbicide treatment	Rate	Palmer amaranth Control[a]							
		2010				2011			
		4 WAT		8 WAT		4 WAT		8 WAT	
	g ai ha⁻¹	—————————%—————————							
Isoxaflutole	105	93	a	75	cd	98	ab	69	ab
Tembotrione	92	94	a	82	abc	90	c	55	abc
Thiencarbazone + isoxaflutole	37 + 92	96	a	92	abc	100	a	89	ab
Mesotrione	210	96	a	80	bc	100	a	99	a
S-metolachlor	1784	99	a	89	abc	100	a	70	ab
Pendimethalin	1704	98	a	55	d	93	bc	23	c
Fomesafen	280	95	a	98	ab	93	bc	52	bc
Sulfentrazone + metribuzin[b]	202 + 303	99	a	100	a	100	a	99	a
S-metolachlor + metribuzin	1987 + 473	99	a	100	a	100	a	97	a
S-metolachlor + fomesafen	1217 + 266	99	a	100	a	97	abc	66	ab
Flumioxazin	71	97	a	90	abc	100	a	73	ab
S-metolachlor + mesotrione	1873 + 185	95	a	99	a	100	a	99	a
Chlorimuron + flumioxazin + thifensulfuron	23 + 72 + 7	95	a	88	abc	100	a	93	a

[a]Means within a column followed by the same letter are not significantly different based on Fisher's protected LSD (P ≤ 0.05).

[b]Industry standards for soybean that were included in this trial were sulfentrazone + metribuzin, S-metolachlor + metribuzin, S-metolachlor + fomesafen, flumioxazin, and chlorimuron + flumioxazin + thifensulfuron.

Table 2: Palmer amaranth control with residual herbicides at 4 and 8 weeks after treatment (WAT) on a clay soil at Keiser, AR in 2010 and 2011.

Herbicide treatment	Rate	Palmer amaranth Control[a]							
		2 WAT		4 WAT		6 WAT		10 WAT	
	g ai ha[-1]	------------------------------ % ------------------------							
Isoxaflutole	88	91	a	98	a	66	cd	74	abc
Tembotrione	92	90	ab	93	ab	55	d	55	c
Thiencarbazone + isoxaflutole	37 + 92	100	a	100	a	69	bcd	50	c
Mesotrione	210	100	a	100	a	82	abc	87	ab
S-metolachlor	1335	100	a	99	a	85	abc	85	ab
Pendimethalin	1119	79	b	86	b	77	abcd	56	c
Fomesafen	280	99	a	99	a	98	a	91	a
Sulfentrazone + metribuzin[b]	151 + 227	96	a	99	a	91	ab	87	ab
S-metolachlor + metribuzin	1545 + 368	100	a	99	a	91	ab	88	ab
S-metolachlor + fomesafen	1217 + 266	100	a	100	a	99	a	92	a
Flumioxazin	71	99	a	99	a	93	ab	65	bc
S-metolachlor + mesotrione	1873 + 185	100	a	100	a	95	a	91	a
Chlorimuron + flumioxazin + thifensulfuron	23 + 72 + 7	99	a	99	a	94	ab	89	a

[a]Means within a column followed by the same letter are not significantly different based on Fisher's protected LSD (P ≤ 0.05).

[b]Industry standards for soybean that were included in this trial were sulfentrazone + metribuzin, S-metolachlor + metribuzin, S-metolachlor + fomesafen, flumioxazin, and chlorimuron + flumioxazin + thifensulfuron.

Table 3: Palmer amaranth control with residual herbicides at 2, 4, 6, and 10 weeks after treatment (WAT) on a silt loam soil at Fayetteville, AR averaged over 2010 and 2011.

with 0 being no plant injury and 100 complete control. Weed control in plots was rated weekly for 8 to 10 weeks after application, which is the length of time generally needed for soybean and cotton to achieve a dense crop canopy [22-24]. Barnyardgrass and Palmer amaranth seedlings m[-2] were counted in 2010 and 2011. At Pine Tree, adequate Palmer amaranth failed to emerge in 2011. All Palmer amaranth and barnyardgrass counts were reported as a percent of the total relative to the non treated control to compensate for variation differences in germination from seed sources between years. Data were analyzed across years within a soil texture or locations within a soil texture for both weed species using JMP V. 9.0.0. Means were then separated using Fisher's protected LSD.

POST HPPD-inhibiting herbicides applied alone and in combinations with glufosinate or glyphosate

Field studies were conducted in 2010 and 2011 at the AAREC during the spring and early summer at times that would be typical for crop production in the region. For both years, the experimental area was tilled, bedded, and then the beds were knocked down to a 30-cm wide surface using a bed conditioner. The row width of the implements used at the AAREC was changed in the winter of 2010; therefore, the summer of 2010 row centers were 1 m apart and in 2011 row centers were 0.9 m. These trails were conducted in fields that had a history of small-plot weed control research evaluations. After conducting a trial, the following year the field is fallowed before conducting additional evaluations. The experiment was conducted as a randomized complete block with factorial treatment structure arrangement of 4 POST herbicide timings and 11 herbicide treatments with four replications both years. Plot dimensions were 30 cm by 3.5 m with a non-planted row separating the plots and a 1 m alley between replications. In 2010, the beds were hand-sown to glyphosate-resistant (GR) Palmer amaranth, hemp sesbania, and barnyardgrass. Each plant species were sown in two 1 m length rows on the left and right side of the bed separated by 15 cm to minimize competition among weeds. Glyphosate-susceptible (GS) Palmer amaranth, hemp sesbania, and barnyardgrass were planted in the same manner in 2011 as in 2010. The GR Johnsongrass did not germinate in 2010 and therefore was not included in the 2011 planting. GS Palmer amaranth was used in 2011 due to lack of sufficient GR seed for this experiment. The hemp

sesbania and barnyardgrass seed sown both years was purchased from Azlin Seed Service and was not resistant to any herbicide used in this experiment based on a previous resistance screen. The GR Palmer amaranth used in 2010 was collected from a known GR accession at the AAREC in Washington County, AR. A natural population of yellow nutsedge was present both years. Plots were planted in fields with access to overhead irrigation to provide adequate moisture for weed seed germination both years.

All herbicides were applied with a CO_2-pressurized backpack sprayer calibrated to deliver 140 L ha[-1] with Teejet 110015XR flat-fan nozzles (TeeJet XR110015 flat-fan nozzle, Spraying Systems Co., Wheaton, IL 60189) spaced 48 cm apart at a pressure of 276 kPa. Herbicide rates were chosen based on recommendations in the Arkansas 2010 Weed and Brush Control MP-44 [25]. Application timings were based on size of the fastest growing weed in the plot, which was Palmer amaranth. Both years the applications were applied between the hours of 10:00 AM and 4:00 PM based on work done by Sellers et al. [26] determined that between 4 hours following sunrise to 4 hours prior to sunset is optimum time for application of glufosinate. In 2010, Palmer amaranth sizes were 2.5- to 7.5-, 25- to 38-, and 38- to 50-cm tall at application. In 2011, Palmer amaranth size at application was 2.5 to 10-, 30- to 45-, and 45 to 65-cm. Yellow nutsedge, hemp sesbania, and barnyardgrass were all 2.5 to 7.5 cm for both years at the first application timing.

Treatments applied for both years were isoxaflutole plus a methylated seed oil (MSO) at 105 g ai ha[-1] + 1% v/v, respectively, tembotrione plus a MSO at 92 g ai ha[-1] + 1% v/v, respectively, two rates of glufosinate (450 and 595 g ai ha[-1]), and glyphosate at 860 g ae ha[-1]. Isoxaflutole and tembotrione were also applied with both rates of glufosinate and the single rate of glyphosate for a total of 11 herbicide treatments. Additionally, a non treated control was included to allow weed control to be visually assessed on a 0 to 100% scale, with 0 representing no control and 100 being plant death. Weed control was evaluated 3 weeks after each application. The timing of application across years differed slightly; therefore, data were analyzed separately by year. Fisher's protected LSD was used to separate means across herbicide treatments and timings.

Results and Discussion

Length and degree of control with pre-applied HPPD-inhibiting herbicides compared to current herbicide standards

The effect of year and location and their interaction with herbicide was non significant for Palmer amaranth and barnyardgrass control for the silt loam soil; thus, the control data were pooled over years and locations. Control for both Palmer amaranth and barnyardgrass on the clay soil differed by year; therefore, means were separated by year.

Under overhead irrigation, thiencarbazone + isoxaflutole, a standard in corn, and S-metolachlor + mesotrione controlled Palmer amaranth equal to all non-HPPD-containing treatments at 8 WAT (Table 2). In 2010, tembotrione, mesotrione, and isoxaflutole provided 82, 80, and 75% control, respectively; however, all were well below the industry standards, which provided ≥ 90% control on the clay soil 8 WAT (0.62 g g⁻¹ clay). When surface irrigation was used to activate the herbicides in 2011 at Keiser, control for all treatments 4 WAT were greater than 90%. At 8 WAT, control differed considerably by treatment; mesotrione, S-metolachlor+mesotrione, thiencarbazone + isoxaflutole, and isoxaflutole were all comparable to the industry standards. Tembotrione alone was the only HPPD-inhibiting herbicide that did not provide control of Palmer amaranth comparable to the industry standards. Tembotrione is currently recommended as a POST product in corn; hence, the lack of extensive residual control was not surprising. The combination of S-metolachlor + mesotrione provided 91% control or above for both years. The high control is likely from the S-metolachlor portion of the combination since when applied alone S-metolachlor provided at least 90% control both years.

All treatments were able to provide at least 4 weeks of >90% control of Palmer amaranth on the silt loam soil at Fayetteville (Table 3). Palmer amaranth control with the HPPD-inhibiting herbicides isoxaflutole and mesotrione were comparable to the non-HPPD-inhibiting herbicides at 10 WAT on the silt loam soil. When mesotrione was applied with S-metolachlor, effective Palmer amaranth control (>90%) was obtained

through 10 WAT. Tembotrione alone did not provide comparable Palmer amaranth control to the industry standards at 10 WAT. The addition of thiencarbazone to isoxaflutole did not increase control or length of control of Palmer amaranth likely because the population of Palmer amaranth evaluated in this experiment is resistant to ALS-inhibiting herbicides.

When end-of-season counts were conducted, the Palmer amaranth densities differed tremendously among treatments (Table 4). This is to be expected as there was no crop competition to provide a canopy to assist the herbicides in preventing late-season emergence. The fact that some treatments provided a high level of control through 10 WAT is evidence that season-long control may occur in some instances when some of the herbicides evaluated here are used in HPPD-resistant soybean or cotton.

Isoxaflutole and tembotrione did not provide adequate residual control of barnyardgrass through 4 WAT when applied alone (Table 5). Barnyardgrass control with mesotrione, isoxaflutole, and tembotrione on the clay soil ranged from 53 to 75% in 2010 at 4 WAT. Mesotrione was among the herbicide treatments supplying the highest level of barnyardgrass control at 4 WAT in 2010 and at 4 and 8 WAT in 2011.

Barnyardgrass on a silt loam soil treated with thiencarbazone+isoxaflutole and S-metolachlor+mesotrione resulted in greater than 90% control 2 WAT and residual control continued to remain high through 10 WAT (Table 6). The extended control may have been partially a result of control provided by the ALS-inhibitor thiencarbazone and the chloroacetamide S metolachlor that are marketed as a premix with these HPPD herbicides. Barnyardgrass control with the HPPD-inhibiting herbicides alone ranged from 13 to 53% at 10 WAT, which was markedly less than the level of control obtained with many of the industry standards.

There was a tremendous amount of variability in the barnyardgrass counts among plots on both soil textures, resulting in less detectable differences among herbicide treatments than observed with control data (Table 7). Late season barnyardgrass densities in plots treated with HPPD-inhibiting herbicides alone did not differ from the non treated control, and barnyardgrass densities in HPPD-treated plots

Herbicide treatment	Rate[b]	Keiser (clay) 2010		2011		Fayetteville (silt loam) 2010		2011	
	g ai ha⁻¹	------------------------- % of nontreated-------------------							
Isoxaflutole	105/88*	50	cde	13	cd	38	a	28	a
Tembotrione	92	100	a	40	ab	35	a	14	bc
Thiencarbazone + isoxaflutole	37 + 92	23	bcd	12	d	18	bc	8	d
Mesotrione	210	44	abc	7	d	7	d	32	a
S-metolachlor	1784/1335*	54	bcd	10	d	13	cd	7	d
Pendimethalin	1704/1119*	8	ef	44	a	24	b	8	d
Fomesafen	280	50	def	5	d	17	bcd	1	d
Sulfentrazone + metribuzin[c]	202/151 + 303/227*	0	f	0	d	10	cde	11	c
S-metolachlor + metribuzin	1987/1545 + 473/368*	0	f	3	d	8	de	5	d
S-metolachlor + fomesafen	1217+266	4	ef	20	c	1	e	2	d
Flumioxazin	75	9	ef	0	d	4	e	17	bc
S-metolachlor + mesotrione	1873 + 185	8	def	1	d	6	e	11	c
Chlorimuron + flumioxazin + thifensulfuron	23 + 72 + 7	67	def	2	d	3	e	10	cd

(header: Palmer amaranth Density[a])

[a] Means within a column followed by the same letter are not significantly different based on Fisher's protected LSD (P ≤ 0.05).

[b] '*' Represents different rate for clay or silt loam soil texture where the higher rate is for the clay soil texture.

[c] Industry standards for soybean that were included in this trial were sulfentrazone + metribuzin, S-metolachlor + metribuzin, S-metolachlor + fomesafen, flumioxazin, and chlorimuron + flumioxazin + thifensulfuron.

Table 4: Late season Palmer amaranth density relative to the nontreated control as influenced by choice of residual herbicide in 2010 and 2011 at Keiser and Fayetteville, AR.[a]

Herbicide treatment	Rate	Barnyardgrass Control[a]							
		2010				2011			
		4 WAT		8 WAT		4 WAT		8 WAT	
	g ai ha^{-1}	————%————							
Isoxaflutole	105	55	bc	34	e	73	d	80	abc
Tembotrione	92	53	c	39	e	19	f	30	d
Thiencarbazone + isoxaflutole	37 + 92	72	abc	59	e	90	abc	97	ab
Mesotrione	210	75	abc	65	cde	86	abcd	99	a
S-metolachlor	1784	97	a	93	abcd	89	abcd	89	abc
Pendimethalin	1704	96	a	93	abcd	91	abc	40	d
Fomesafen	280	93	a	96	ab	40	e	60	bcd
Sulfentrazone + metribuzin[b]	202 + 303	99	a	96	a	79	cd	98	a
S-metolachlor + metribuzin	1987 + 473	97	a	95	abc	95	a	99	a
S-metolachlor + fomesafen	1217 + 266	96	a	95	ab	83	bcd	60	dc
Flumioxazin	71	83	ab	68	bcde	80	bcd	81	abc
S-metolachlor + mesotrione	1873 + 185	94	a	93	abcd	91	ab	99	a
Chlorimuron + flumioxazin + thifensulfuron	23 + 72 + 7	61	bc	60	de	83	bcd	94	abc

[a]Means within a column followed by the same letter are not significantly different based on Fisher's protected LSD (P ≤ 0.05).

[b]Industry standards for soybean that were included in this trial were sulfentrazone + metribuzin, S-metolachlor + metribuzin, S-metolachlor + fomesafen, flumioxazin, and chlorimuron + flumioxazin + thifensulfuron.

Table 5: Barnyardgrass control with residual herbicides at 4 and 8 weeks after treatment (WAT) on a clay soil at Keiser, AR in 2010 and 2011.

Herbicide treatment	Rate	Barnyardgrass Control[a]					
		2 WAT		6 WAT		10 WAT	
	g ai ha^{-1}	————%————					
Isoxaflutole	88	51	d	34	c	55	cd
Tembotrione	92	70	c	0	d	13	f
Thiencarbazone + isoxaflutole	37 + 92	98	a	94	a	91	a
Mesotrione	210	92	ab	29	c	30	ef
S-metolachlor	1335	99	a	90	a	83	a
Pendimethalin	1119	93	a	74	ab	59	bcd
Fomesafen	280	84	b	20	cd	16	f
Sulfentrazone + metribuzin[b]	151 + 227	97	ab	73	ab	76	abc
S-metolachlor + metribuzin	1545 + 368	99	a	89	a	90	a
S-metolachlor + fomesafen	1217 + 266	100	a	98	a	90	a
Flumioxazin	71	97	ab	48	bc	50	de
S-metolachlor + mesotrione	1873 + 185	93	ab	85	a	79	ab
Chlorimuron + flumioxazin + thifensulfuron	23 + 72 + 7	94	ab	39	c	53	d

[a]Means within a column followed by the same letter are not significantly different based on Fisher's protected LSD (P ≤ 0.05).

[b]Industry standards for soybean that were included in this trial were sulfentrazone + metribuzin, S-metolachlor + metribuzin, S-metolachlor + fomesafen, flumioxazin, and chlorimuron + flumioxazin + thifensulfuron.

Table 6: Barnyardgrass control with residual herbicides at 2, 6, and 10 weeks after treatment (WAT) on a silt loam soil in 2011 averaged over Fayetteville, AR and Pine Tree, AR.

alone were often greater than those in plots treated with the herbicides currently labeled for use in soybean. Therefore, it is likely that some of the herbicides that are currently being used in soybean today will continue to be needed once HPPD-resistant soybean or cotton is commercialized.

POST HPPD-inhibiting herbicides applied alone and in combinations with glufosinate or glyphosate

The accession of Palmer amaranth used in 2010 was different than that used in 2011. While both were expected to have resistance, the 2011 accession was, in fact, susceptible to glyphosate at 860 g ha^{-1}, which was later confirmed in a greenhouse trial (data not shown). When plants began to emerge, Palmer amaranth quickly overtook most of the natural weed population and other planted weeds. Following trial establishment, it was soon apparent that in addition to the Palmer amaranth that was planted in the 1-m rows, both fields had an abundance of a natural Palmer amaranth population. It has been well documented that *Amaranthus* has a very prolific growth habit,

especially Palmer amaranth [27,28]. The excess Palmer amaranth in the field soon outgrew the other planted weed species, eventually shading them. Hence, the first application at the smallest weed size timing was the only application that provided effective spray coverage to all four of the planted weed species.

Palmer amaranth control

Palmer amaranth control differed by weed size each year; therefore, data are presented separately by year. Within each year, there was a herbicide treatment by timing interaction for Palmer amaranth. In 2010, glyphosate at 860 g ae ha^{-1} was the only treatment to provide less than 85% control of Palmer amaranth when the size was 2.5- to 7.5-cm tall (Table 8). The lack of a control with glyphosate was a result of the Palmer amaranth being from a resistant population. Isoxaflutole and tembotrione alone provided ≥ 94% control when applied alone in both 2010 and 2011 (Table 9). In 2010, the addition of glyphosate to either isoxaflutole or tembotrione did not increase glyphosate-resistant Palmer amaranth control over tembotrione or isoxaflutole alone when the plants were 2.5-

Herbicide treatment	Rate	Barnyardgrass Density (m^{-2})			
		clay[a]		silt loam[b]	
		Keiser		Fayetteville and Pine Tree	
	g ai ha^{-1}			———%———	
Isoxaflutole	105	86	ab	85	a-d
Tembotrione	92	100	a	81	a-d
Thiencarbazone + isoxaflutole	37 + 92	62	ab	53	d
Mesotrione	210	91	ab	72	bcd
S-metolachlor	1780	12	c	85	a-d
Pendimethalin	1700	16	c	55	cd
Fomesafen	280	9	c	100	a
Sulfentrazone + metribuzin[c]	25 + 38	8	c	76	a-d
S-metolachlor + metribuzin	1990 + 473	17	c	92	ab
S-metolachlor + fomesafen	1220 + 266	13	c	100	a
Flumioxazin	71	62	b	50	a-d
S-metolachlor + mesotrione	1870 + 185	10	c	71	bcd
Chlorimuron + flumioxazin + thifensulfuron	23 + 72 + 7	73	ab	97	ab

[a]Barnyardgrass density was not assessed at Keiser in 2011.

[b]Barnyardgrass data did not differ within soil textures; thus the silt loam locations data were pooled. Letters of separation were calculated by the counts of total barnyardgrass emergence at the end of the season. Means within a column followed by the same letter are not significantly different.

[c]Industry standards for soybean that were included in this trial were sulfentrazone + metribuzin, S-metolachlor + metribuzin, S-metolachlor + fomesafen, flumioxazin, and chlorimuron + flumioxazin + thifensulfuron.

Table 7: Percent of total barnyardgrass emergence as influenced by choice of residual herbicide at Keiser, AR in 2010 and 2011[b] and at Fayetteville and Pine Tree, AR in 2011.

Herbicide treatment	Rate	Control					
		Plant height (cm)[b]					
		2.5 to 7.5		25 to 38		38 to 50	
	g ai or ae ha^{-1}			----------------------%----------------------			
Isoxaflutole	105	94	a	53	b-f	43	b-g
Tembotrione	92	98	a	62	b	35	d-g
Glufosinate	450	90	a	51	b-f	30	fg
Glufosinate	595	85	a	51	b-f	37	c-g
Glyphosate	860	33	fg	61	b	33	efg
Isoxaflutole + glufosinate	105 + 450	95	a	55	b-e	42	b-g
Isoxaflutole + glufosinate	105 + 595	99	a	48	b-f	25	g
Isoxaflutole + glyphosate	105 + 860	98	a	53	b-f	43	b-g
Tembotrione + glufosinate	92 + 450	96	a	56	bcd	49	b-f
Tembotrione + glufosinate	92 + 595	89	a	59	bc	38	c-g
Tembotrione + glyphosate	92 + 860	86	a	50	b-f	44	b-g

[a]Control was assessed at 3 wk after treatment for each herbicide application timing.

[b]Means across all plant height columns followed by the same letter did not differ significantly when using Fisher's protected LSD (P ≤ 0.05).

Table 8: Palmer amaranth control in 2010 at Fayetteville, AR with POST applications of herbicides at three timings.[a]

Herbicide treatment	Rate	Control					
		Plant height (cm)[b]					
		2.5-10		30-45		45-65	
	g ai or ae ha^{-1}			----------------------%----------------------			
Isoxaflutole	105	96	a	51	def	35	ef
Tembotrione	92	95	a	59	cde	58	def
Glufosinate	450	96	a	49	def	48	def
Glufosinate	595	97	a	51	def	36	ef
Glyphosate	860	100	a	88	ab	33	f
Isoxaflutole + glufosinate	105 + 450	99	a	52	def	60	cde
Isoxaflutole + glufosinate	105 + 595	99	a	38	ef	44	ef
Isoxaflutole + glyphosate	105 + 860	100	a	84	abc	36	ef
Tembotrione + glufosinate	92 + 450	100	a	50	def	48	def
Tembotrione + glufosinate	92 + 595	100	a	47	def	61	cde
Tembotrione + glyphosate	92 + 860	100	a	53	def	70	bcd

[a]Control was assessed at 3 wk after treatment for each herbicide application timing.

[b]Means within columns and across all plant height columns followed by the same letter did not differ significantly when using Fisher's protected LSD (P ≤ 0.05).

Table 9: Palmer amaranth control in 2011 with POST herbicides applied three timings.[a]

to 7.5 cm. Reduced activity of glufosinate on small Palmer amaranth (<7.5 cm) in 2010 can be attributed to reduced absorption due to a low relative humidity (38%) at application as shown by Coetzer et al. [29]. At the larger sizes of Palmer amaranth, neither HPPD herbicides alone or in combination with glyphosate or glufosinate resulted in acceptable control. Since this research was conducted there has been a study that shows there is no antagonism from glufosinate and tembotrione at a 1x field rate when applied to 7-cm tall Palmer amaranth [30]. Applications to Palmer amaranth plants larger than 25 cm, in either 2010 or 2011, resulted in insufficient levels of control. No herbicide or combination of herbicides in either year provided >70% Palmer amaranth control when plants were at least 25 to 30 cm tall at application, except for glyphosate alone and in combination with isoxaflutole in 2011 on the glyphosate-susceptible biotype. Based on the Palmer amaranth control provided by the combination of glyphosate or glufosinate with each of HPPD herbicide it appears that combination may be antagonistic on Palmer amaranth because the levels of control with the combination are similar to the control when each herbicide was applied alone.

Barnyardgrass control

Barnyardgrass control was only rated at the first application timing of 2.5- to 7.5-cm in 2010 and 2.5- to 10-cm in 2011 because of shading by Palmer amaranth at later timings. The year by treatment interaction was significant; therefore, data are presented by year. In 2010, isoxaflutole, tembotrione, isoxaflutole + glufosinate at both rates, isoxaflutole + glyphosate, and tembotrione + glufosinate at both rates provided ≥ 80% barnyardgrass control (Table 10). Glufosinate at either 450 or 595 g ha-1 did not provide more than 70% control. In 2011, all herbicide treatments provided 96 to 99% barnyardgrass control. Based on this research, isoxaflutole and tembotrione appear to be good post emergence options for controlling barnyardgrass if applications are made according to manufacturer's recommendations only.

Yellow nutsedge and hemp sesbania control

The year by treatment interaction for both yellow nutsedge and hemp sesbania was not significant; hence, data were pooled over years. There were no differences among herbicide treatments for yellow nutsedge or hemp sesbania control, with yellow nutsedge control ranging

from 74 to 90% and hemp sesbania control ranging from 91 to 99% (Table 10). Hence, it is does not appear that the addition of tembotrione or isoxaflutole to glyphosate or glyphosate will improve yellow nutsedge or hemp sesbania control. However, it should be noted that mixing two mechanisms of action that provide effective weed control is a strategy that is commonly recommended to reduce the risk of herbicide resistance evolving [31]. While no herbicide-resistant hemp sesbania has ever been documented, ALS-resistant yellow nutsedge was recently confirmed in Arkansas [32]. Although all treatments provided adequate control, the additional HPPD-inhibiting mechanism of action could be integrated into many integrated weed management systems to help delay resistance.

Summary

The objectives of this research were to determine the length and degree of weed control with HPPD-inhibiting herbicides that could eventually be used in HPPD-resistant cotton and soybean as an alternative or additional mechanism of action for control of problematic and resistant weeds. Results showed that there are still multiple options for the effective control of some of the most problematic weeds of mid southern US. row crops. Palmer amaranth, barnyardgrass, and hemp sesbania can be effectively controlled with the correct combination of herbicides and alternating mechanisms of action. Since this was a non-crop study, there was no weed-crop competition and it is likely that the addition of a crop to these experiments would have resulted in even greater weed suppression.

Although the adoption rate of HPPD-resistant crops by producers remains to be seen, it is an effective option for control of both resistant and susceptible weeds if applied at the correct timing. When used in the correct manner and with the right combination of herbicides, HPPD inhibitors will bring an extra effective mechanism of action to crops to combat an ever increasing problem of herbicide resistance. While HPPD-inhibiting herbicide use is limited in the Mid south, the need for expanded use of these herbicides in more crops will help to mitigate current resistance challenges. The commercialization of HPPD-resistant crops will not be the sole answer to the problematic and resistant weeds currently inundating Mid south production fields; however, it will be an option for producers who have been limited in their herbicide options.

Herbicide treatment	Rate	Barnyardgrass[b] 2010		2011		Yellow nutsedge[d]	Hemp sesbania[d]
	g ai/ae ha-1	----%----					
Isoxaflutole	105	97	a	96	c	84	96
Tembotrione	92	88	ab	98	b	74	99
Glufosinate	450	69	bc	99	a	75	97
Glufosinate	595	26	d	99	a	80	91
Glyphosate	860	66	c	99	a	83	96
Isoxaflutole + glufosinate	105 + 450	96	a	99	a	87	99
Isoxaflutole + glufosinate	105 + 595	99	a	99	a	87	99
Isoxaflutole + glyphosate	105 + 860	99	a	99	a	87	99
Tembotrione + glufosinate	92 + 450	84	abc	99	a	90	99
Tembotrione + glufosinate	92 + 595	80	abc	99	a	89	98
Tembotrione + glyphosate	92 + 860	65	c	99	a	90	95

[a]Weed species of plants at application were 2.5 to 7.5 cm and 1 to 2 lf for all three species.
[b]The year by herbicide treatment interaction was significant for barnyardgrass control; hence, data are presented by year.
[c]Means are separated using Fisher's protected LSD (P ≤ 0.05).
[d]Means for yellow nutsedge and hemp sesbania were not significant based on ANOVA (α=0.05).
Table 10: Yellow nutsedge, barnyardgrass, and hemp sesbania control 3 weeks after POST treatment at Fayetteville, AR.[a]

References

1. Young BG (2006) Changes in herbicide use patterns and production practices resulting from glyphosate-resistant crops. Weed Technol 20: 301-307.

2. Anonymous (2011) Roundup POWERMAX® herbicide label for use on Roundup Ready Flex Cotton®. St. Louis, MO: Monsanto Technology LLC 3.

3. Duke SO, Powles SP (2009) Glyphosate-resistant crops and weeds: Now and in the future. AgBioForum 12: 346-357.

4. Beckie HJ (2006) Herbicide-resistant weeds: Management tactics and practices. Weed Technol 20: 793-814.

5. Dill GM, Cajacob CA, Padgette SR (2008) Glyphosate-resistant crops: adoption, use and future considerations. Pest Manag Sci 64: 326-331.

6. Heap I (2015) International survey of herbicide resistant weeds.

7. Norsworthy JK, Smith KL, Scott RC, Gbur EE (2007) Consultant perspectives on weed management needs in Arkansas cotton. Weed Technol 21: 825-831.

8. Mallory-Smith CA, Retzinger EJ Jr (2003) Revised classification of herbicides by site of action for weed resistance management strategies. Weed Technol 17: 605-619.

9. Grossmann K, Ehrhardt T (2007) On the mechanism of action and selectivity of the corn herbicide topramezone: a new inhibitor of 4-hydroxyphenylpyruvate dioxygenase. Pest Manag Sci 63: 429-439.

10. Viviani F, Little JP, Pallet KE (1998) the mode of action of isoxaflutole II. Characterization of the inhibition of carrot 4-hydroxyphenylpyruvate dioxygenase by the diketonitrile derivative of isoxaflutole. Pestic Biochem Physiol 62: 125-134.

11. Pallett KE, Cramp SM, Little JP, Veerasekaran P, Crudace AJ, et al. (2001) Isoxaflutole: the background to its discovery and the basis of its herbicidal properties. Pest Manag Sci 57: 133-142.

12. Stuebler H, Kraehmer H, Hess M, Schulz A, Rosinger C (2008) Global changes in crop production and impact trends in wed management – an industry view. Proc 5th Int Weed Sci Cong 1: 309-319.

13. Riar SR, Norsworthy JK, Steckel LE, Stephenson DO, Bond JA (2013) Consultant perspectives on weed management needs in midsouthern United States cotton. Weed Technol 27: 778-787.

14. Sosnoskie LM, Kichler JM, Wallace R, Culpepper AS (2009) Multiple resistance to glyphosate and an ALS inhibitor in Palmer amaranth in GA. Proc Beltwide Cotton Conf 1351-1352.

15. Jordan NR, Jannink JL (1997) Assessing the practical importance of weed evolution: a research agenda. Weed Res 37: 237-246.

16. Nelson KA, Renner KA (2002) Yellow nutsedge (Cyprus esculentus) control and tuber production with glyphosate and ALS-inhibiting herbicides. Weed Technol 16: 512-519.

17. Corbett JL, Askew SD, Thomas WE, Wilcut JW (2004) Weed efficacy evaluations for bromoxynil, glufosinate, glyphosate, pyrithiobac, and sulfosate. Weed Technol 18: 443-453.

18. Keeley PE, Thullen RJ (1991) Biology and control of black nightshade (Solanum nigrum) in cotton (Gossypium hirsutum). Weed Technol 5: 713-722.

19. Bagavathiannan MV, Norsworthy JK, Smith KL, Neve P (2011) Modeling the evolution of glyphosate resistance in barnyardgrass in Arkansas cotton. Summ of Arkansas Cotton Research 26-29.

20. Webster TM, Nichols RL (2012) Changes in the prevalence of weed species in major agronomic crops in the southern United States: 1994/1995 to 2008/2009. Weed Sci 60: 145-157.

21. Dean WE (1974) Determination of carbonate and organic matter in calcareous sediments and sedimentary rocks by loss on ignition: comparison with other methods. J Seed Res 44: 242-248.

22. Holt JS, Orcutt DR (1991) Functional relationships of growth and competitiveness in perennial weeds and cotton (Gossypium hirsutum). Weed Sci 39: 575-584.

23. Jha P, Norsworthy JK (2009) Soybean canopy and tillage effects on emergence of Palmer amaranth (Amaranthus palmeri) from a natural seed bank. Weed Sci 57: 644-651.

24. Reddy KN, Boykin JC (2010) Weed control and yield comparisons of twin- and single-row glyphosate-resistant cotton production systems. Weed Technol 24: 95-101.

25. Scott RC, Boyd JW, Smith KL, Seldon G, Norsworthy JK (2014) Recommended chemicals for weed and brush control.

26. Sellers BA, Smeda RJ, Johnson GW (2003) Diurnal fluctuations and leaf angle reduce glufosinate efficiacy. Weed Technol 17: 302-306.

27. Keeley PE, Carter CH, Tullen RJ (1987) Influence of planting date on growth of Palmer amaranth (Amaranthus palmeri). Weed Sci 35: 199-204.

28. Horak MJ, Loughin TM (2000) Growth analysis of four Amaranthus species. Weed Sci 48: 347-355.

29. Coetzer E, Al-Khatib K, Loughin TM (2001) Glufosinate efficacy, absorption, and translocation in amaranth as affected by relative humidity and temperature. Weed Sci 49: 8-13.

30. Botha GM, Burgos NR, Gbur EE, Alcober EA, Salas RA, et al. (2014) Interaction of glufosinate with 2,4-D, dicamba, and tembotrione on glyphosate-resistant Amaranthus palmeri. Amer Jour Exper Agri 4: 427-442.

31. Norsworthy JK, Ward SM, Shaw DR, Llewellyn RS, Nichols RL, et al. (2012) Reducing the risks of herbicide resistance: best management practices and recommendations. Weed Sci 60: 31-62.

32. Wilson MJ, Norsworthy JK, Johnson DB, McCallister EK, Devore JD, et al. (2010) Herbicide programs for controlling ALS-resistant barnyardgrass in Arkansas rice. B. R. Wells Rice Research Series 581: 153-157.

38

The Assessment of Genetic Parameters: Yield, Quality Traits and Performance of Single Genotypes, of Tuberose (*Polianthes Tuberosa*)

Ranchana P*, Kannan M and Jawaharlal M

Department of Floriculture and Landscaping, TNAU, Coimbatore, India

Abstract

An experiment was laid out in a Randomized Block Design (RBD) with three replications to determine genetic parameters and performance of single tuberose cultivars such as "Calcutta Single, Hyderabad Single, Kahikuchi Single, Mexican Single, Navs.ari Local, Phule Rajani, Prajwal, Pune Single, Shringar and Variegated Single" under Coimbatore, India, conditions during 2011-2012. The results of the experiment revealed that 'Prajwal' performed best in certain parameters including days to bulb sprouting, weight of bulb, weight of bulblets per clump, number of leaves per plant, days to spike emergence, flowering duration, spike length, rachis length, number of florets per spike, length of the floret, weight of florets per spike, number of spikes/m^2 and yield of florets per plot (2 * 2 m). The parameters including flowering duration, weight of florets per spike and number of florets per spike showed high phenotypic and genotypic coefficients of variation. Further, a high heritability coupled with high genetic advance as per cent of mean were observed for flowering duration, weight of florets per spike, number of florets per spike and rachis length.

Keywords: Tuberose; Single types; Variability; Heritability; Genetic advance

Introduction

Tuberose (*Polianthes tuberosa*) is one of the most important cut flowers in India. It is an ornamental bulbous plant, native to Mexico and belongs to the family Amaryllidaceae. There are only two types of tuberose (Single and Double) cultivated in the world. Ornamental plants have prime importance in maintaining ecological balance and checking pollution in the surrounding environments. About 45% of the world's trade in floriculture products is contributed by cut flower. In India, it occupies a prime position in the floriculture industry. The waxy white flowering spikes of single as well as double types of tuberose impregnate the atmosphere with their sweet fragrance and because of the longer keeping quality of flower spikes of the double types [1,2] tuberose is in great demand for making floral arrangement and bouquets in the major cities of India. It is widely grown as a specimen for exhibition and for cut flower. Single types of tuberose are cultivated on a large scale in Tamil Nadu, Karnataka, West Bengal and Maharashtra. To a lesser extent it is also grown in Andhra Pradesh, Haryana, Delhi, Uttar Pradesh and Punjab. Valuable natural aromatic oil is extracted from the flowers for the high cost perfume industry. Its essential oil is exported at an attractive price to France, Italy and other countries [1], as long as there is no synthetic flavour to replace its fragrance. There are only a few varieties and hybrids of tuberose under cultivation viz., "Calcutta Single, Calcutta Double, Hyderabad Single, Hyderabad Double, Kahikuchi Single, Mexican Single, Navs.ari Local, Pearl Double, Phule Rajani, Prajwal, Pune Single, Shringar, Suvasini, Vaibhav and Variegated Single". As the commercial cultivation of tuberose is gaining importance, introduction and identification of high yielding varieties is necessary. Varieties which perform well in one region may not do well in other regions of varying climatic conditions [3]. Hence, it is important to study morphological variation and performance of genotypes in a new location to enhance the efficiency of a breeding programme. For a sound breeding programme, critical assessments of the nature and extent of genetic variability in the germplasm and assessment of the heritability and genetic advance of the important yield contributing characters in a crop are essential [4]. Hence, the present investigation was undertaken to understand the relative performance of ten genotypes of single tuberose and the variability present within them.

Materials and Methods

The present study was carried out at the Botanical gardens, Tamil Nadu Agricultural University, Coimbatore, India, during the year 2011-2012. The location is situated at 11° 02" N latitude, 76° 57" E longitude and 426.76 m above mean sea level. The experimental material consists of ten genotypes of tuberose including "Calcutta Single, Hyderabad Single, Kahikuchi Single, Mexican Single, Navs.ari Local, Phule Rajani, Prajwal, Pune Single, Shringar and Variegated Single". The experiment was laid out in a randomized block design (RBD) with three replications. Before initiating the experiment, the soil was brought to a fine tilt with four deep ploughings. Weeds, stubbles, roots etc., were removed. At the time of the last ploughing, Farm yard manure was applied at the rate of 25 t ha^{-1}. After levelling, raised beds of 1 m width and 1 m length were formed and medium sized bulbs (3.0-3.5 cm diameter) of about 25 grams were planted at a spacing of 45 m x 20 m which accommodates 11 plants per m^2. Standard cultural practices were followed throughout the experimentation. The data were recorded on ten plants from each genotype collected at random in each replication for 15 characters viz., days to sprouting (days), bulb weight (g), number of bulblets/ clump, weight of bulblets/ clump (g), plant height (cm), number of leaves per clump, days to spike emergence, flowering duration, spike length (cm), rachis length (cm), number of florets /spike, length of the floret, weight of the florets/spike, number of spikes/m2, yield of florets/ plot (2 m*2 m) and the phenotypic coefficient of variation (PCV) and genotypic coefficient of variation (GCV) were calculated as suggested by Burton

*Corresponding author: Ranchana P, Department of Floriculture and Landscaping, TNAU, Coimbatore, India, E-mail: ranchanahorti@gmail.com

[5]. The heritability (h²) in broad sense was calculated according to Lush [6] and expressed as percent and the genetic advance as percent mean were calculated as suggested by Johnson et al. [7].

Results and Discussion

Vegetative characteristics

The mean performance of the cultivars for vegetative growth (Table 1) reflected the variation among the cultivars. Significantly less number of days to bulb sprouting (12.00) was recorded in 'Prajwal', followed by 'Variegated Single' (13.23) while longer days were observed in 'Hyderabad Single' (16.48). Plant height was highest (117.50cm) in 'Variegated Single followed by Prajwal' (113.05 cm). This is in accordance with the results of Gudi [8] and Vijayalaxmi et

variation in spike length in different cultivars might be due to variation in their genetic factor. 'Prajwal' also showed the highest number of florets/ spike (43.00) followed by Shringar (42), while the lowest was observed in "Mexican Single" (17). The largest growth in floret length was also observed in Prajwal (6.40 cm) while the lowest was observed in Pune Single (6.10 cm). This finding is in consonance with Vijayalaxmi et al. [9] in tuberose. The weight of florets/ spike was also largest in Prajwal (74.80 g) followed by Shringar (51.48 g). This might be due to the increased number of florets/ spike. Among single genotypes of tuberose, Prajwal showed the highest in yield of florets/ plot (2* 2 m) (4.40 tonnes, respectively). The highest yield registered by Prajwal might be due to its capacity to produce more number of florets per spike, floret length and weight of florets / spike.

S.NO	Genotypes	Days taken for sprouting of bulb	Bulb weight at planting (g)	Number of bulblets/clump	Weight of bulblets/ clump	Plant height	No. of leaves per plant
1.	Calcutta Single	14.380	345.230	19.29	72.29	78.89	243.00
2.	Hyderabad Single	16.483	350.290	30.28	116.07	80.96	220.00
3.	Kahikuchi Single	13.670	318.460	20.76	73.12	77.30	252.00
4.	Mexican Single	15.780	400.183	13.54	70.37	91.77	238.00
5.	Navs.ari Local	14.590	328.450	24.43	89.32	92.85	242.00
6.	Phule Rajani	13.793	380.160	34.16	121.08	72.50	251.00
7.	Prajwal	12.120	456.200	22.31	144.92	113.05	260.00
8.	Pune Single	15.970	392.650	25.29	110.23	110.07	232.00
9.	Shringar	14.260	332.497	27.80	122.82	91.75	245.00
10.	Variegated Single	13.230	398.250	26.78	115.23	117.50	253.00
	SE(D)	0.42	10.26	1.52	4.72	2.57	6.74
	CD (0.5)	0.88	21.55	3.22	9.84	5.39	14.15

Table 1: Performance of tuberose genotypes (single) for vegetative growth and bulb production (2010- 2011).

al. [9] Prajwal produced the highest number of leaves/ plant (260) in first year after planting followed by Variegated Single (253), while the lowest number of leaves was recorded in Hyderabad Single (220). The differences among the varieties for vegetative characters are attributed to the variation in their genetic makeup [10]. The weight of bulb was the highest in Prajwal (456.20 g) followed by Mexican Single (400.18 g).

The increased weight of bulb might be due to balanced partitioning of dry matter between floral parts and the storage organs. The cultivars differed significantly with respect to number and weight of bulblets produced per plant. 'Phule Rajani' had the highest number of bulblets (34.16) followed by Hyderabad Single (30.28) while Mexican Single had the least number (13.54). The variation in the number of bulblets produced per plant might be due to its genetic character and the results are in consonance with the findings of Ramachandrudu and Thangam and Vijayalaxmi et al. [9,11]. The weight of bulblets was larger in Prajwal (144.92) followed by Shringar (122.82) while the smallest was recorded in 'Mexican Single' (70.37). The higher relative growth potential of Prajwal may be the probable reason for the large weight of bulblets.

Floral characteristics

The mean performance of the cultivars in floral characteristics (Table 2) reflected the variation among the cultivars. The least number of days to spike emergence was observed in Prajwal (78) while the longest was observed in Calcutta Single (94). The duration of flowering was high in Prajwal (17 days) followed by Shringar (15 days). This is in line with the findings of Patil et al. [12] 'Variegated Single' produced spikes with the highest length of 102.50 cm followed by Prajwal (98.05 cm) while the shortest was found in Kahikuchi Single (52.50 cm). The

Variability, heritability and genetic advance

Variability in a population is a prerequisite especially for characters under genetic improvement. The success of plant breeding programmes largely depends on the amount of genetic variability present in a given crop for the character under improvement [13]. Generally, phenotypic coefficient of variation (PCV) was higher than the corresponding genotypic coefficient of variation (GCV) for all the attributes under study, indicating that traits interacted with environment (Table 3). Similar results were reported by Gurav et al. and Vijayalaxmi et al. [9,14] in tuberose. The PCV and GCV were the highest for flowering duration (35.79 vs. 35.62) followed by the weight of florets per spike (32.43 vs. 32.27) and number of florets per spike (32.01 vs. 31.80) suggesting that these characters are under genetic control. Hence, these characters can be relied upon through phenotypic selection for further improvement. The PCV was higher than the GCV for all the characters under study, indicating the role of environment in the expression of genotype. Similar results were reported by Mishra et al. [15] in dahlia and Sheela et al. [16] in *heliconia*. The low values of PCV and GCV were recorded for characters including the length of florets (3.11 vs. 1.45) number of leaves per plant (5.50 vs. 4.34) and days to spike emergence (6.60 vs. 5.60) and days to bulb sprouting (9.74 vs. 9.08). These findings indicate that minimal variation exist among the genotypes for o these characters.

High heritability coupled with high genetic advance was observed for flowering duration (99.07 vs. 73.04), weight of florets per spike (99.02 vs. 66.15), number of florets per spike (98.72 vs. 65.08) and rachis length (98.47 vs. 54.46) (Table 3). This indicates the lesser influence of environment on these characters and the prevalence of additive gene action in their inheritance. Hence, these traits are

S.No.	Genotypes	Days to spike emergence	Flowering duration (days)	Spike length (cm)	Rachis length	Number of florets/spike	Length of the floret	Weight of florets per spike (g)	Number of spikes/m²	Yield of florets/plot (2* 2 m) (kg)
1.	Calcutta Single	94.00	8.26	63.89	16.75	25.00	6.30	29.25	31.00	2.53
2.	Hyderabad Single	90.00	6.37	65.96	15.30	43.00	6.20	32.76	34.50	2.72
3.	Kahikuchi Single	86.00	10.00	62.30	18.38	38.00	6.30	44.46	33.00	4.01
4.	Mexican Single	88.00	7.00	76.77	21.27	17.00	6.20	28.32	32.20	2.51
5.	Navs.ari Local	92.00	8.98	77.85	27.30	45.00	6.30	33.93	25.75	2.79
6.	Phule Rajani	85.00	8.90	52.50	23.48	40.00	6.30	47.20	35.00	4.06
7.	Prajwal	78.00	17.00	98.05	28.52	47.00	6.40	74.80	47.00	4.40
8.	Pune Single	82.00	9.28	95.07	35.75	37.00	6.10	43.66	39.00	3.90
9.	Shringar	83.00	15.00	76.77	22.32	42.00	6.30	51.48	40.00	4.26
10.	Variegated Single	81.00	7.12	102.50	20.86	28.00	6.20	39.78	33.65	3.78
	SE(D)	1.45	0.28	2.15	3.62	1.12	0.18	1.11	8.11	0.09
	CD (0.5)	2.93	0.58	4.51	7.24	2.36	0.37	2.34	16.24	0.19

Table 2: Performance of tuberose genotypes (single) for floral and yield parameters (2010-2011).

S.NO.	Characters	GCV	PCV	HERT	GA (%) OF MEAN
1	Days to bulb sprouting	9.08	9.74	88.83	17.43
2	Bulb weight	17.16	17.50	96.24	34.69
3	Number of bulblets/clump	4.34	5.50	62.16	17.05
4	Weight of bulblets/clump	5.60	6.60	71.99	19.78
5	Plant height	35.62	35.79	99.07	73.04
6	Number of leaves per plant	21.64	21.91	97.58	44.04
7	Days to spike emergence	26.64	26.85	98.47	54.46
8	Flowering duration	31.80	32.01	98.72	65.08
9	Spike length	1.45	3.11	78.23	15.39
10	Rachis length	32.27	32.43	99.02	66.15
11	Number of florets/spike	16.34	16.68	95.00	42.96
12	Length of the floret	21.71	21.96	98.00	44.22
13	Weight of florets per spike	32.27	39.74	88.83	17.43
14	Number of spikes/m2	16.34	17.50	96.24	34.69
15	Yield of florets/plot	21.71	5.50	62.16	17.05

Table 3: Estimates of variability and genetic parameters for flower yield and its components

suitable for selection. High heritability with moderate genetic advance were recorded for yield of florets/ plot (2* 2 m) (98.00 vs. 44.22), spike length (97.58 vs. 44.04), plant height (96.24 vs. 34.69) and number of spikes/m² (95.00 vs. 42.96) indicate the presence of both additive and non-additive gene actions, and simple selection would offer the best possibility for the improvement of these trait. The estimate of heritability was high with low genetic advance as percentage of mean for days to sprouting (88.83 vs. 17.43), length of the floret (78.23 vs. 15.39), days to spike emergence (71.99 vs. 19.78) and number of leaves per plant (62.00 vs. 17.00). The high heritability could be due to non-additive gene effects and a strong influence of the environment. Hence, there is limited scope for selection in these traits Sheikh et al. [17] reported similar results in Iris.

References

1. Sadhu MR, Bose TK (1973) Tuberose for most artistic garlands. Indian Hort 18: 17-20.

2. Benschop M, De Hertogh A, Le Nard M (1993) The physiology of flower bulbs(Edn) Elsevier, Amsterdam, The Netherlands 589-601

3. Kamble BS, Reddy BS, Patil RT, Kulkarni BS (2004) Performance of gladiolus (Gladiolus hybridus Hort.) cultivars for flowering and flower quality. Journal of Ornamental Horticulture 7: 51-60.

4. Nair M, Dwivedi VK (2006) Genetic variability studies in gladiolus. Journal of Asian Horticulture 2: 235-238.

5. Burton GW (1952) Quantitative inheritance in grasses. Proc 6th Intl Grassld. Congr 1: 227-283.

6. Lush JL (1940) Intra-sire correlation and regression of offspring on dams as a method of estimating heritability of characters. Proc. Amer. Soc. Animal Prodn 33: 293-301.

7. Johnson HW, Robinson HF, Comstock RE (1955) Estimates of Genetic and Environmental Variability in Soybeans. Agronomy Journal 47: 314-318.

8. Gudi G (2006) Evaluation of tuberose varieties. Thesis submitted to University of Agricultural Sciences, Dharwad, and Karnataka.

9. Vijayalaxmi M, Manohar Rao A, Padmavatamma AS, Siva Shanker A (2010) Evaluation and variability studies in tuberose (Polianthes tuberosa L.) single cultivars. Journal of Ornamental Horticulture 13: 251- 256.

10. Swaroop K (2010) Morphological variation and evaluation of gladiolus germplasm. Indian Journal of Agricultural Sciences 80: 742-745.

11. Ramachandrudu K, Thangam M (2009) Performance of tuberose (Polianthes tuberosa L.) cultivars in Goa. Journal of Horticultural Sciences 4: 76-77.

12. Patil VS, Munikrishnappa PM, Tirakannanavar S (2009) Performance of growth and yield of different genotypes of tuberose under transitional tract of north Karnataka. Journal of Ecobiology 24: 327-333.

13. Falconer DS (1981) Introduction to Quantitative genetics. Ronalds press Company, New York.

14. Gurav SB, Katwate SM, Singh BR, Kahade DS, Dhane AV et al. (2005) Quantitative genetic studies in tuberose. Journal of Ornamental Horticulture 8: 124-127.

15. Misra RL, Verma TS, Thakur PC, Singh B (1987) Variability and correlation studies in dahlia. Indian Journal of Horticulture 44: 269-273.

16. Sheela VL, Rakhi R, Jayachandran Nair CS, Sabina George T (2005) Genetic variability in heliconia. Journal of Ornamental Horticulture 8: 284-286.

17. Sheikh MQ, John AQ (2005) Genetic variability in Iris (Iris japonica thumb.) Journal of Ornamental Horticulture, 8: 75-76.

Permissions

The contributors of this book come from diverse backgrounds, making this book a truly international effort. This book will bring forth new frontiers with its revolutionizing research information and detailed analysis of the nascent developments around the world.

We would like to thank all the contributing authors for lending their expertise to make the book truly unique. They have played a crucial role in the development of this book. Without their invaluable contributions this book wouldn't have been possible. They have made vital efforts to compile up to date information on the varied aspects of this subject to make this book a valuable addition to the collection of many professionals and students.

This book was conceptualized with the vision of imparting up-to-date information and advanced data in this field. To ensure the same, a matchless editorial board was set up. Every individual on the board went through rigorous rounds of assessment to prove their worth. After which they invested a large part of their time researching and compiling the most relevant data for our readers.

The editorial board has been involved in producing this book since its inception. They have spent rigorous hours researching and exploring the diverse topics which have resulted in the successful publishing of this book. They have passed on their knowledge of decades through this book. To expedite this challenging task, the publisher supported the team at every step. A small team of assistant editors was also appointed to further simplify the editing procedure and attain best results for the readers.

Apart from the editorial board, the designing team has also invested a significant amount of their time in understanding the subject and creating the most relevant covers. They scrutinized every image to scout for the most suitable representation of the subject and create an appropriate cover for the book.

The publishing team has been an ardent support to the editorial, designing and production team. Their endless efforts to recruit the best for this project, has resulted in the accomplishment of this book. They are a veteran in the field of academics and their pool of knowledge is as vast as their experience in printing. Their expertise and guidance has proved useful at every step. Their uncompromising quality standards have made this book an exceptional effort. Their encouragement from time to time has been an inspiration for everyone.

The publisher and the editorial board hope that this book will prove to be a valuable piece of knowledge for researchers, students, practitioners and scholars across the globe.

List of Contributors

Atif sarwar and Shahid Javed Butt
Department of Horticulture, PMAS-Arid Agriculture University, Rawalpindi, Pakistan

Endale Hailu, Tadesse Sefera, Negussie Tadesse, Anteneh Boydom, Daniel Kassa and Tamene Temesgen
Ethiopian Institute of Agricultural Research, P.O.Box 2003, Ethiopia

Gezahegne Getaneh
Addis Ababa University, Salale Campus, P.O.Box 2003, Ethiopia

Beyene Bitew
Debre Birhan Research center, Ethiopia

Marta Giner and Jesús Avilla
Department of Forest Science and Crop Production, University of Lleida, Lleida (Spain) and Department of Crop Protection, IRTA Center, Lleida, Spain

Mercè Balcells
Department of Chemistry, University of Lleida, Lleida, Spain

Benjamin SA and Bradford BJ
Department of Animal Sciences and Industry, Kansas State University, Manhattan, KS 66506, USA
Department of Agronomy, Kansas State University, Manhattan, KS 66506, USA

Yan Chen
LSU Agricultural Center Hammond Research Station, 21549 Old Covington Highway, Hammond, LA 70403, USA

Richard Story
LSU Agricultural Center Department of Entomology, 404 Life Sciences Building, LSU, Baton Rouge, LA 70803, USA

Michelle Samuel-Foo
University of Florida, IR-4 Southern Region Program, Gainesville, FL 3261, USA

Tadeos Shiferaw, Fano Dargo and Abdurhman Osman
College of Dry Land Agriculture, Department of Dryland Crop Sciences, Jigjiga University, P.O. Box, 1020, Jigjiga, Ethiopia

Mohammed Amin, Sileshi Fitsum, Thangavel Selvaraj and Negeri Mulugeta
Ambo University, Ethiopia

Nahida Jelali and Mohamed Gharsalli
Laboratory of Plant Adaptation to Abiotic Stress, CBBC, Borj-Cedria Technopark , BP 901, 2050 Hammam-Lif, Tunisia

Marc El Beyrouthy
Faculty of Agricultural and Food Sciences, Holy Spirit University of Kaslik, P.O. Box 446, Jounieh, Lebanon

Marta Dell'orto
Dipartimento di Scienze Agrarie e Ambientali - Produzione, Territorio, Agroenergia, Università degli Studi di Milano, Via Celoria 2, I-20133 Milano, Italy

Wissem Mnif
LR11-ES31 Biotechnology and Bio-Enhancement of Geo Resources, Higher Institute of Biotechnology of Sidi Thabet Sidi Thabet BiotechPole, 2020, Tunisia Manouba University, Tunisia

Ranchana P, Kannan M and Jawaharlal M
Department of Floriculture and Landscaping, HC& RI, TNAU, Coimbatore-3, India

Jagendra Singh
Research Scholar Senior Research Fellow Directorate of Research Services, RVSKVV Race Course Road, Gwalior 474002, MP, India

Ravi Sharma
Ex-Principal ESS College, Dr B.R. Ambedkar University (formerly Agra University), Dayalbagh, Agra, India

P. Tripathy, Sahoo BB, Das SK, Priyadarshini A, Patel D and Dash DK
All India Network Research Project on Onion and Garlic, College of Horticulture, (OUAT), Chiplima, Sambalpur-768025, Odisha, India

Miheretu Fufa
Adami Tullu Agricultural Research Center; Plant Biotechnology Team; P.O. Box 35, Zeway, Ethiopia

Ashu Singh and Sengar RS
Tissue culture Lab, College of Biotechnology, University of Agriculture & Technology, Meerut, India

Miheretu Fufa
Oromia Agricultural Research Institute, Sinana Agricultural Research Center; Horticulture and Spice Technology Generation Team, Ethiopia

Sonia Plaza-Wüthrich and Regula Blösch
Institute of Plant Sciences, University of Bern, Altenbergrain 21, 3013 Bern, Switzerland

Zerihun Tadele
Institute of Plant Sciences, University of Bern, Altenbergrain 21, 3013 Bern, Switzerland
Institute of Biotechnology, Addis Ababa University, P.O. Box 32853, Addis Ababa, Ethiopia

Janusz Gołaszewski and Dariusz Załuski
Department of Plant Breeding and Seed Production, University of Warmia and Mazury in Olsztyn, Poland

Krystyna Żuk-Gołaszewska
Department of Agrotechnology, University of Warmia and Mazury in Olsztyn, Poland

Getachew Mekonnen
College of Agriculture and Natural Resources, Mizan Tepi University, Mizan Teferi, Ethiopia

Sharma JJ, Lisanework Negatu and Tamado Tana
College of Agriculture and Environmental Sciences, Haramaya University, Dire Dawa, Ethiopia

Seelavarn Ganeshan
Mauritius Sugar Industry Research Institute, Réduit, Moka, Mauritius

Hudaa Neetoo
Faculty of Agriculture, University of Mauritius, Réduit, Moka, Mauritius

Enrique Salazar Sosa
Technology Institute of Torreon, Mexico

Jesus Luna Anguiano and Enrique Salazar Melendez
Agricultural Sciences and Forestry, Mexico

Hector I Trejo Escareno, Miguel A. Gallegos Robles, Jose Dimas Lopez Martinez and Orona Castillo Ignacio
College of Agriculture and Animal Husbandry of Durango University of Durango State (FAZ-UJED), Ejido Venecia Municipal of Gomez Palacio, Durango. Km 28 Gomez Palacio-Tlahualilo, Mexico

Gandhi S
Scientist Department of FRM, India

Mehta M
College of Home Science, CCS Haryana Agricultural University, HISAR, Haryana, India

Dahiya R
DES Home Science Krishi Vigyan Kendras, Fatehabad, Haryana, India

Victor Afari-Sefa
AVRDC - The World Vegetable Center, Eastern and Southern Africa, P. O. Box 10, Duluti Arusha, Tanzania

Elvis Asare-Bediako
University of Cape Coast, School of Agriculture, Department of Crop Science, Private Mail Bag Cape Coast, Ghana

Lawrence Kenyon
AVRDC- The World Vegetable Center, Headquarters P. O. Box 42 Shanhua Tainan 74199, Taiwan

John A. Micah
University of Cape Coast, School of Agriculture, Department of Agricultural Economics and Extension Private Mail Bag Cape Coast, Ghana

Aby N, Traoré S, Kobénan K, Kéhé M, Thiémélé DEF and Gnonhouri G
Laboratory of Entomology and Plant Pathology, National Centre of Agronomic Research, Research Station of Bimbresso; 01 BP 1536 Abidjan 01

Badou J and Koné D
Laboratory of plant physiology, University of Félix Houphouët Boigny Abidjan, 01 BP 852 Abidjan 01

Solomon Abera Gebrie
Ethiopian Institute of Agricultural Research, Holoeta Research Center, Holeta, Ethiopia

Jiang P, Chen Y and Wilde HD
Horticulture Department, University of Georgia, Athens, Georgia, USA

Khanal Sabin, Subedi Bijay, Bhandari Amrit, Giri Dilli Raman and Shrestha Bhuwan
Institute of Agriculture and Animal Science, Tribhuwan Univeristy, Chitwan, Nepal

Neupane Priyanka, Shrestha Sundar Man and Gaire Shankar Prasad
Department of Plant Pathology, Faculty of Agriculture, Agriculture and Forestry University, Nepal

Peng Jiang
Horticulture Department, University of Georgia, Athens, Georgia, USA

Kirthisinghe JP
Department of Crop Science, Faculty of Agriculture, University of Peradeniya, Sri Lanka

Thilakarathna SMCR and Gunathilaka BL
Department of Agriculture, Kurunegala, Sri Lanka

Dissanayaka DMPV
Export Agriculture Department, Narammala, Sri Lanka

Funda Eryilmaz Acikgoz
Department of Plant and Animal Production, Vocational College of Technical Sciences, Namik Kemal University, Tekirdag, Turkey

Anwanobong Jonathan Eshiet
Plant Genetic Resources and Cell and Tissue Culture Research Laboratory, Department of Genetics and Biotechnology, University of Calabar, Calabar, Nigeria

Ebiamadon Andi Brisibe
Plant Genetic Resources and Cell and Tissue Culture Research Laboratory, Department of Genetics and Biotechnology, University of Calabar, Calabar, Nigeria
Department of Biological Sciences, Niger Delta University, Wilberforce Island, P.M.B. 71 Yenagoa, Nigeria
Department of Pharmaceutical Microbiology and Biotechnology, Niger Delta University, Wilberforce Island, P.M.B. 71 Yenagoa, Nigeria

Dobariya VK, Gohil BS and Chhodavadia SK
Ph.D.Scholars Department of Agronomy, College of Agriculture, Junagadh Agricultural University, Junagadh-362001 Gujarat, India

Mathukia RK
Associate Research Scientist, Department of Agronomy, College of Agriculture, Junagadh Agricultural University, Junagadh-362001, Gujarat, India

Zinabu Wolde
Graduate student of Soil Science, Hawassa University, P. O. Box 05, Hawassa, Ethiopia

Wassie Haile and Dhyna Singh
Gedeo Zone Agricultural Office, Gedeo Zone, P. O. Box 128, Dilla, Ethiopia

Ibrahim MA, Alhameid AH, Kumar S, Chintala R, Sexton P, Malo DD and Schumacher TE
Department of Plant Science, South Dakota State University, Brookings, South Dakota (SD), USA

Sheleme Kaba Shamme, Cherukuri V Raghavaiah, Tesfaye Balemi and Ibrahim Hamza
Department of Plant Sciences, College of Agriculture and Veterinary Sciences, PO Box 19, Ambo University, Ambo, West Shoa Zone, Ethiopia, East Africa

Amir Mor-Mussery
The Department of Geography and Environmental Development, Ben Gurion University of the Negev, Bee'r Sheva, Israel
TheDepartment of Soil and Water Sciences, Hebrew University, Rehovot, Israel

Orit Edelbaum
TheDepartment of Soil and Water Sciences, Hebrew University, Rehovot, Israel

Arie Budovsky
Technological Center, Biotechnology Unit, Beer Sheva, Israel

Jiftah Ben Asher
Katif R&D Center for Coastal Deserts Development Ministry of Science, Israel

Getahun Dereje, Tigist Adisu and Megersa Mengesha
Assosa Agricultural Research Centre, Assosa, Ethiopia

Tesfa Bogale
Jimma Agricultural Research Centre, Jimma, Ethiopia

Walid S Nosir
Department of Plant Pathology, Faculty of Agriculture, Cornell University, Ithaca, NY, USA
Horticulture Department, Zagazig University, Zagazig City, Egypt

Clay E. Starkey, Jason K. Norsworthy and Lauren M. Schwartz
Department of Crop, Soils, and Environmental Sciences, University of Arkansas,1366 West Altheimer Drive, Fayetteville, AR 72701, USA

Ranchana P, Kannan M and Jawaharlal M
Department of Floriculture and Landscaping, TNAU, Coimbatore, India

Index

CPSIA information can be obtained
at www.ICGtesting.com
Printed in the USA
BVHW061936270519
549349BV00003B/102/P

9 781641 160667